C.T. Russell
Editor

New Horizons

Reconnaissance of the Pluto-Charon System and the Kuiper Belt

Previously published in *Space Science Reviews* Volume 140, Issues 1–4, 2008

Springer

C.T. Russell
Institute of Geophysics & Planetary Physics
University of California
Los Angeles, CA, USA

Cover illustration: NASA's New Horizons spacecraft was launched on 2006 January 19, received a gravity assist during a close approach to Jupiter on 2007 February 28, and is now headed for a flyby with closest approach 12,500 km from the center of Pluto on 2015 July 14. This artist's depiction shows the spacecraft shortly after passing above Pluto's highly variegated surface, which may have black-streaked surface deposits produced from cryogenic geyser activity, and just before passing into Pluto's shadow when solar and earth occultation experiments will probe Pluto's tenuous, and possibly hazy, atmosphere. Sunlit crescents of Pluto's moons Charon, Nix, and Hydra are visible in the background. After flying through the Pluto system, the New Horizons spacecraft could be re-targeted towards other Kuiper Belt Objects in an extended mission phase.
This image is based on an original painting by Dan Durda.

Back cover illustration: The New Horizons spacecraft was launched aboard an Atlas 551 rocket from the NASA Kennedy Space Center on 2008 January 19 at 19:00 UT.

Library of Congress Control Number: 2008944238

DOI: 10.1007/978-0-387-89518-5

ISBN-978-0-387-89517-8 e-ISBN-978-0-387-89518-5

Printed on acid-free paper.

1

springer.com

Contents

Foreword

New Horizons: Reconnaissance of the Pluto–Charon System and the Kuiper Belt

C.T. Russell

Originally published in the journal Space Science Reviews, Volume 140, Nos 1–4, 1–2.
DOI: 10.1007/s11214-008-9450-0

Exploration is mankind's imperative. Since the beginnings of civilization, men and women have not been content to build a wall around their settlements and stay within its confines. They explored the land around them, climbed the mountains, and scanned the horizons. The boldest among them pushed exploration to the most distant frontiers of the planet. As a result, much of the Earth was inhabited well before the days of the renowned European explorers of the 15th and 16th centuries. Exploration did not cease, after the circumnavigation of the globe; it continued to the present. Today explorers are going in new directions, not just east and west, north and south. They explore backward in time and upward in space. Archeology explores the shorter time scales, and geochemistry the longer time scales of geophysical events: asteroidal and cometary collisions, magnetic reversals, continental formation and more. However, on Earth we cannot go back indefinitely, for much of the evidence of the very earliest days has been lost.

To go further back in time we must first go up, climb out of the Earth's gravitational potential well, and use our spacecraft to scan new horizons on the surfaces of our planetary neighbors. Like the Earth, these planets are dynamic, evolving spheres, large enough that their interiors have undergone significant thermal evolution. The terrestrial planets, Mercury, Venus, Earth, and Mars have aged the most. The outer gas giants have also changed, and because of their enormous size, most evidence of past processes is difficult to obtain. Fortunately, the smaller planetary bodies in the asteroid belt and those in the furthest reaches of the solar system, the ones that have evolved the least, are accessible with current technology.

To enable bold initiatives such as a flight to the outermost classical planet, Pluto, NASA established the New Frontiers program. This volume describes the first successful response to New Frontiers opportunity. The very appropriately named New Horizons mission, by the year 2020, extends planetary exploration to the Pluto–Charon system and one or two Kuiper Belt objects. Selected in 2000 and launched in 2006, New Horizons is well on its

C.T. Russell (✉)
University of California, Los Angeles, CA 90095, USA
e-mail: ctrussel@igpp.ucla.edu

way to arrival at Pluto on July 14, 2015, fifty years to the day after the first successful Mars reconnaissance by Mariner 4 in 1965.

This collection of articles describes the New Horizons mission, its spacecraft, instruments and scientific investigations. It begins with a historical overview of the events leading to the New Horizons mission by the Principal Investigator, Alan Stern. This introduction to the mission is followed by a description of the spacecraft by G.H. Fountain and colleagues. A critical aspect of the mission is the Jupiter gravity assist that was essential in speeding the spacecraft on its way to the Pluto–Charon system. The mission design that enabled the gravity assist and the observations to be obtained at Pluto and Charon is described by Y. Guo and R.W. Farquhar. Next H. Weaver and colleagues describe the science payload and L.A. Young and colleagues describe the anticipated scientific investigations. The volume closes with seven articles describing the instruments and investigations. D.C. Reuter et al. discuss Ralph; the visible and infrared imager. S.A. Stern et al. describe the UV imaging spectrograph. A.F. Cheng et al cover the long-range reconnaissance imager, LORRI. G.L. Tyler et al., present the radio science experiment, REX. D.J. McComas et al. describes the solar wind instrument. R.L. McNutt et al. discusses the energetic particle spectrometer science investigation, PEPSSI. Finally M. Horanyi and colleagues describe the Student Dust Counter.

The success of this volume is due to many people; but first of all, the editor wishes to thank the authors who had the difficult job of distilling the thousands of documents and the millions of facts such missions produce, into highly readable articles. The editor also benefited from an excellent group of referees who acted as a test readership, refining the manuscripts provided by the authors. These referees included M. A'Hearn, J.D. Anderson, S. Auer, K. Baines, M. Belton, D. Byrne, M.C. de Sanctis, A. Driesman, A. Fedorov, J. Goldstein, J. Gosling, L. Iess, B. Klecker, E.A. Miller, S. Mottola, B. Perry, B.R. Sandel, D. Seal, S. Solomon, and J. Woch.

Equally important has been the strong support this project has received at Springer and the extra effort expended by Fiona Routley, Randy Cruz and Harry Blom. At UCLA we were skillfully assisted by Marjorie Sowmendran who acted as the interface between the authors, the referees and the publishers.

The New Horizons Pluto Kuiper Belt Mission: An Overview with Historical Context

S. Alan Stern

Originally published in the journal Space Science Reviews, Volume 140, Nos 1–4, 3–21.
DOI: 10.1007/s11214-007-9295-y

Abstract NASA's New Horizons (NH) Pluto–Kuiper Belt (PKB) mission was selected for development on 29 November 2001 following a competitive selection resulting from a NASA mission Announcement of Opportunity. New Horizons is the first mission to the Pluto system and the Kuiper belt, and will complete the reconnaissance of the classical planets. New Horizons was launched on 19 January 2006 on a Jupiter Gravity Assist (JGA) trajectory toward the Pluto system, for a 14 July 2015 closest approach to Pluto; Jupiter closest approach occurred on 28 February 2007. The ~400 kg spacecraft carries seven scientific instruments, including imagers, spectrometers, radio science, a plasma and particles suite, and a dust counter built by university students. NH will study the Pluto system over an 8-month period beginning in early 2015. Following its exploration of the Pluto system, NH will go on to reconnoiter one or two 30–50 kilometer diameter Kuiper Belt Objects (KBOs) if the spacecraft is in good health and NASA approves an extended mission. New Horizons has already demonstrated the ability of Principal Investigator (PI) led missions to use nuclear power sources and to be launched to the outer solar system. As well, the mission has demonstrated the ability of non-traditional entities, like the Johns Hopkins Applied Physics Laboratory (JHU/APL) and the Southwest Research Institute (SwRI) to explore the outer solar system, giving NASA new programmatic flexibility and enhancing the competitive options when selecting outer planet missions. If successful, NH will represent a watershed development in the scientific exploration of a new class of bodies in the solar system—dwarf planets, of worlds with exotic volatiles on their surfaces, of rapidly (possibly hydrodynamically) escaping atmospheres, and of giant impact derived satellite systems. It will also provide other valuable contributions to planetary science, including: the first dust density measurements beyond 18 AU, cratering records that shed light on both the ancient and present-day KBO impactor population down to tens of meters, and a key comparator to the puzzlingly active, former dwarf planet (now satellite of Neptune) called Triton which is in the same size class as the small planets Eris and Pluto.

Keywords Pluto · New Horizons · Mission · Kuiper belt

S.A. Stern (✉)
NASA Science Mission Directorate, 300 E Street SW, Washington, DC 20546, USA
e-mail: alan.stern@nasa.gov

1 Mission Overview

New Horizons is a flyby reconnaissance mission to provide the first in situ exploration of the Pluto system and KBOs. It is the first mission in NASA's New Frontiers series of medium-class, robotic planetary exploration missions. New Horizons is a Principal Investigator led mission; the author of this article is the mission PI.

The New Horizons flight system consists of a single, ~400 kg spacecraft featuring redundant subsystems and seven scientific instruments. The spacecraft is powered by a Radioisotope Thermoelectric Generator (RTG).

The top level science goals of the New Horizons mission are, in priority order, to:

- Reconnoiter the Pluto system for the first time.
- Sample the diversity of KBOs by making one or more KBO flybys after the Pluto system flyby.
- Obtain Jupiter system science during the JGA; and
- Obtain cruise science along the route to Pluto and through the KB traverse.

The specific scientific measurement objectives of the mission were developed by NASA's Outer Planets Science Working Group (OPSWG; S.A. Stern chair) in 1992, and slightly refined and then re-ratified by the Pluto Kuiper Express (PKE) mission Science Definition Team (SDT) in 1996 (J.I. Lunine, chair; Lunine et al. 1996). These objectives were adopted by NASA for the mission Announcement of Opportunity (AO) that led to the selection of New Horizons (NASA 2001).

The full suite of New Horizons mission science objectives described in that AO were ranked in three categories, called Group 1, Group 2, and Group 3. This categorization was first developed by OPSWG (then denoting the rank categories as Group IA, Group IB, and Group IC). Group 1 objectives represent an irreducible floor for the mission science requirements at the Pluto system. Group 2 goals add depth and breadth to the Group 1 objectives and are termed highly desirable. The Group 3 objectives add further depth and are termed desirable, but are of distinctly lower priority than the Group 2 objectives. These various objectives can be briefly summarized in Table 1.

Each of the Group 1 objectives was defined by the SDT in significantly more detail, giving measurement requirements that included resolutions, SNRs, dynamic ranges, etc., as appropriate. They will be described and discussed in the accompanying article by Young et al.

Since Pluto's small moons Nix and Hydra (Weaver et al. 2006) were not known in the 1990s when these objectives were constructed, they are not included. New Horizons, however, is in the enviable position of knowing about and being able to plan science observations for Nix and Hydra with almost 10 years advance notice from their discovery (itself motivated by New Horizons encounter planning) to arrival at the Pluto system. Therefore, the mission team is treating the compositional and geologic mapping of these bodies, along with other aspects of their study (orbit refinement, temperature measurements, etc.) as additional objectives. In addition to enhancing what we know about the Pluto system itself, the comparative study of Nix and Hydra to KBOs is expected to be particularly illuminating.

This article provides an overview of the mission, describing its history, its goals, its architecture and development, and its first two years of flight in summary form. Other articles in this Space Science Reviews volume provide more in-depth background regarding the mission, the mission science, and the spacecraft and its instrument payload. Other articles in this volume include a description of the mission's scientific objectives (Young et al. 2007), the instrument payload (Weaver et al. 2007), the spacecraft system (Fountain et al. 2007),

Table 1 New Horizons Pluto–Charon and KBO measurement objectives

Group 1: Required
Characterize the global geology and morphology of Pluto and Charon
Map the surface composition of Pluto and Charon
Characterize the neutral atmosphere of Pluto and its escape rate
Group 2: Highly desired
Characterize the time variability of Pluto's surface and atmosphere
Image Pluto and Charon in stereo
Map the terminators of Pluto and Charon with high resolution
Map the surface composition of selected areas of Pluto and Charon at high resolution
Characterize Pluto's ionosphere and solar wind interaction
Search for neutral atmospheric species including H, H_2, HCN, and C_xH_y, and other hydrocarbons and nitriles upper atmosphere
Search for an atmosphere around Charon
Determine bolometric Bond albedos for Pluto and Charon
Map the surface temperatures of Pluto and Charon
Group 3: Desirable
Characterize the energetic particle environment of Pluto and Charon
Refine bulk parameters (radii, masses, densities) and orbits of Pluto and Charon
Search for additional satellites and rings

the mission design (Guo et al. 2007), the Ralph visible/infrared imager/imaging spectrometer instrument (Reuter et al. 2007), the Alice ultraviolet spectrometer instrument (Stern et al. 2007), the LORRI high resolution imager instrument (Cheng et al. 2007), the REX radio science instrument (Tyler et al. 2007), the SWAP plasma instrument (McComas et al. 2007), the PEPSSI high energy particle spectrometer instrument (McNutt et al. 2007), and the Venetia Burney Student Dust Counter instrument (Horanyi et al. 2007).

2 Pluto Mission Background Studies

In this section I briefly recapitulate the relevant history of Pluto mission studies. I begin with NASA's Voyager mission and work forward monotonically in time through the many studies of the 1990s; more details can be found in the book, *Pluto and Charon* (Stern and Mitton 2005). In Sect. 3, I will describe the call for competed Pluto–Kuiper Belt (PKB) mission proposals in early 2001 and the selection of New Horizons at the end of that year.

2.1 Voyager Pluto

NASA's Voyager 1 and 2 outer planets reconnaissance flyby missions included an option for Voyager 1 to fly from Saturn in 1980 to a late 1980s Pluto flyby. This option, however, was mutually exclusive with Voyager 1 making a close flyby of Saturn's large and complex, atmosphere-laden moon Titan during its late-1980 exploration of the Saturn system. Owing in part to the lower risk of the Titan flyby than a long cruise to Pluto, and also the higher scientific priority at the time of Titan, the Pluto option was not exercised. Of course, at the time this decision was made, Pluto's atmosphere, its small satellites, its complex surface composition, and the entire Kuiper belt all remained undiscovered, perhaps rationalizing the

Titan choice from today's perspective. By the time of the 1989 Voyager 2 flyby of Pluto-analog Triton, however, Pluto's richness and context were beginning to be understood. That, combined with the fascinating results of Voyager 2's Triton flyby, including a pathologically young surface, active geysers, and an atmosphere, motivated interest, particularly in a handful of young planetary scientists, to successfully appeal to NASA in 1989 to begin Pluto mission studies.

2.2 Dedicated Pluto Mission Studies

Owing to the scientific interest and pressure resulting from Voyager's results at Triton and the burgeoning suspicion in the late 1980s that a Kuiper belt existed beyond Neptune, NASA began studying dedicated Pluto flyby reconnaissance missions. The first such study (eventually dubbed "Pluto-350") was undertaken as a part of the Discovery Program Science Working Group (DPSWG) in 1989–1990. The study scientists for this effort were S.A. Stern and F. Bagenal; the study manager was R. Farquhar. The concept for this study was to send a "minimalist" scientific payload to Pluto–Charon for a bare bones reconnaissance flyby; the Kuiper belt was then undiscovered and not a part of the mission study. The resulting spacecraft (Farquhar and Stern 1990), was a 350-kg, RTG-powered vehicle with four instruments (imager, ultraviolet spectrometer, radio science, and a plasma package). Pluto-350 was to launch on a Delta II launch vehicle in 1999, perform several Earth and Venus gravity assists, and then use Jupiter for a final gravity assist in 2006 so as to arrive at Pluto ca. 2015. At the time of this study, a four-instrument spacecraft weighing half what Voyager did, and much lighter still than the Galileo, Magellan, and Cassini planetary spacecraft of the day, was considered controversial in terms of its small scope and its perceived high risk.

Shortly after the Pluto-350 study, NASA began studying flying a much larger, Cassini-class Mariner Mark II mission to Pluto. This mission, though much more costly, was perceived to have lower risk and a broader scientific potential. It would also provide a logical follow on for the RTG-powered Mariner Mark II line that Cassini was then starting. Notably, this Pluto mission would have replaced the Cassini Huygens Titan entry probe with a short-lived, deployable second flyby spacecraft designed to fly over Pluto's far hemisphere some 3.2 days (one Pluto half-rotation) before or after the mother ship. This mission, along with a Mariner Mark II mission to orbit Neptune was adopted as a high priority in the Solar System Exploration Subcommittee (SSES) 1990s planetary exploration plan derived in a "community roadmap shoot out" meeting held in February 1991. Following this, NASA's Solar System Exploration Division (then under the direction of W. Huntress) formed the Outer Planets Science Working Group (OPSWG; S.A. Stern, chair) to shape the Pluto mission's scientific content, document its rationale, and prepare for an instrument selection process by the mid-1990s. By 1992, OPSWG had completed most of its assigned mission study support tasks. However, owing to tightening budgets at NASA, OPSWG also was asked to debate the large Mariner Mark II versus the much smaller Pluto-350 mission concepts. In early 1992, OPSWG selected Pluto-350 as the more pragmatic choice. It is worth noting that by this time, Mars Pathfinder and NEAR, also small spacecraft, were being started in NASA's Discovery program, so smaller missions were becoming more accepted.

However, in early 1992, a new and radical mission concept called Pluto Fast Flyby (PFF) was introduced by the Jet Propulsion Laboratory's (JPL's) R. Staehle as a "faster, better, cheaper" alternative to the Mariner Mark II and Pluto-350 Pluto mission concepts. As initially conceived, PFF was to weigh just 35–50 kg and carry only 7 kg of highly miniaturized (then non-existent) instruments, and fly two spacecraft to Pluto for <$500M, excluding launch costs. PFF caught the attention of then NASA Administrator D. Goldin, who directed all Pluto-350 and Mariner Mark II work to cease in favor of PFF. PFF would have launched its two flyby spacecraft on Titan IV-Centaur launchers; these low-mass spacecraft would have shaved the Pluto-350 and Mariner Mark II flight times from 12–16 years down to 7 or 8 years. Like Mariner Mark II and Pluto-350, PFF involved RTG power and JGAs. The heavier missions also involved Earth and Venus gravity assists on the way to Jupiter. All of these mission concepts were developed by JPL mission study teams.

Shortly after PFF was introduced, however, it ran into problems. One was mass growth, which quickly escalated the flight system to the 140 kg class with no increase in mission payload mass. A second issue involved cost increases, largely due to a broad move within NASA to include launch vehicle costs in mission cost estimates. Because two Titan IV launchers alone cost over \$800M, this pushed PFF to well over \$1B. A third issue was the turmoil introduced into NASA's planetary program by the loss of the Mars Observer in 1993. These various events began to sour then NASA Administrator Goldin on PFF. Cost concerns subsequently caused PFF to be cut back to one spacecraft, but even this was too expensive for Administrator Goldin.

Following this, OPSWG chair Stern attempted to gain European and Russian collaboration in the mission to reduce cost so that a new start could be afforded. European interest was generally lukewarm. However, Russian interest was stronger. A concept emerged between Stern and Russia's director A. Galeev of the IKI space research center in Moscow that a Russian Proton launch vehicle would loft PFF, saving NASA the ~\$400M cost of the Titan IV launch. The incentive for Russia would be a probe, called a Drop Zond, which would enter Pluto's atmosphere to obtain mass spectroscopy and imagery before an impact on Pluto, as well as their first entrée to outer planets exploration. However, when Russia later asked in 1995 to be paid for this launch, rather than accepting the Drop Zond as a quid pro quo, W. Ip and I. Axford at Germany's Max Plank Institute for Planetary Physics offered to pursue German national funding for the Russian launch; the plan of the German scientists was to pay Russia for the Proton launch (~\$30M at that time) in exchange for NASA accommodating a second probe on PFF which would impact Jupiter's moon Io during the JGA encounter.

Even with such innovative arrangements, however, PFF was never started into development owing to higher priorities within NASA for then Administrator Goldin. During 1994–1995, Mr. Goldin directed a series of studies to determine if PFF could fly without any foreign participation, without nuclear power (to Pluto!), and also whether it could be launched on a small launcher (i.e., a Delta II). These studies were widely considered in OPSWG to be diversionary tactics by Mr. Goldin, who was perceived as not being able to cancel the Pluto effort but was unwilling to start it. Nonetheless, JPL carried the requested studies over a period of about a year; they concluded that although a slow (12–15 year) Delta II launched mission was feasible (something previously established for Pluto-350), non-nuclear Pluto missions were either too risky (e.g., using battery power alone) or beyond the cost or technological capability of the era. During this same period, however, PFF did solicit, select, and fund the breadboard/brassboard development of a breakthrough suite of competitively miniaturized imagers, spectrometers, and radio science, and plasma instruments suitable for PFF.

Following on the rapidly expanding interest in the Kuiper belt by the mid-1990s, NASA next directed JPL to reinvent PFF as Pluto Express (later named and more commonly known as Pluto–Kuiper Express, or PKE). PKE was a single spacecraft PFF mission with a 175 kg spacecraft, a 9-kg science payload, and a 2 Gbit solid state memory. It would have launched in the 2001–2006 JGA launch window. A Science Definition Team (SDT) chaired by J.I. Lunine was constituted in 1995 and delivered its report in 1996 for an anticipated instrument selection in 1996–1997. However, in late 1996 PKE mission studies were drastically cut back by Administrator Goldin and no instrument selection was initiated.

By 1999, however, continued interest and pressure by the scientific community caused NASA to release a solicitation for PKE instruments; proposals were due in March 2000. Many of the proposals, including a radio science investigation led by L. Tyler, an energetic particle spectrometer led by R. McNutt, and a remote sensing investigations suite led by A. Stern, resulted from the PFF miniaturized instrument development program. These proposals were evaluated and ranked, but never selected. By September 2000, NASA cancelled PKE, still in Phase A, owing to mission cost increases which had once again pushed the projected mission cost well over the $1B mark.

3 The Birth of New Horizons

Following the cancellation of PKE, intense scientific and public pressure caused then NASA Associate Administrator for Space Science E. Weiler to solicit mission proposals for a "Pluto Kuiper Belt" (PKB) flyby reconnaissance mission. That early 2001 solicitation and the resulting late 2001 selection of New Horizons are discussed in this section. For additional details about early Pluto mission studies the following references are recommended: Stern (1993), Terrile et al. (1997), and Stern and Mitton (2005).

3.1 PKB Mission AO and Selection Process

The PKB AO was announced in a NASA Press Conference on 20 December 2000 and released on 19 January 2001. The AO (NASA 2001) mandated a two-step selection process with initial proposals due 20 March 2001 (later extended to 6 April 2001). Following a down-select to two teams, Phase A studies would be performed with due dates in the August-September timeframe. Because no PI-led mission to the outer planets, nor any PI-led mission involving RTGs, had ever been selected, the AO was termed experimental by NASA, which made clear it was not obligated to select any proposals at all.

The PKB AO required responders to propose an entire PKB mission (i.e., not just the science payload or science investigation), to meet at least the detailed specifications of the Group 1 measurement objectives, to complete the Pluto flyby before the end of 2020, to launch aboard a U.S. Atlas V or Delta IV launch vehicle, and to do so within a complete mission cost cap of $506M FY2001 dollars. Launch vehicle selection between the Atlas V and Delta IV was planned for 2002. Two spare Cassini-Galileo RTGs were made available for use to proposal teams, with associated costs of $50M and $90M (the latter with higher power).

Shortly after the 19 January 2001 AO release, on 6 February 2001, the then-new Administration released its first budget, which cancelled PKB by not funding it in FY02 and future years. Within days, NASA announced the suspension of the PKB AO as well. However, intensive work on Capital Hill by the science community resulted in less than a week in a directive from the U.S. Senate to NASA to proceed with the AO so as not to limit Congressional authority to override the PKB cancellation decision.

Five proposals were turned in to NASA. The contenders included two proposals from JPL (L. Soderblom and L. Esposito, PIs) and one from APL (S.A. Stern, PI). The Soderblom et al. proposal cleverly involved ion propulsion in order to remove the 2004–2006 JGA launch window constraint. The Esposito et al. and Stern et al. proposals both involved conventional JGA trajectories and no ion propulsion. I will now summarize the New Horizons mission as proposed.

The New Horizons team was formed by an agreement between PI Stern and APL Space Department Head Dr. Stamatios ("Tom") Krimigis that was made on 22 December 2000.

Fig. 1 New Horizons depicted over Pluto by planetary scientist and space artist Dan Durda

The science team was formed from Stern's PKE PERSI instrument proposal team and Dr. Lenard Tyler's PKE radio science proposal team, plus about five other scientists added from APL and other institutions to add scientific breadth for a full mission proposal. Dr. Andrew Cheng was named the New Horizons project scientist. The Tyler et al. radio science team had been the only radio science proposal for the 1999–2000 PKE AO, and he considered their participation to be a key strategic element of a winning PKB proposal.

The first face-to-face meeting of the New Horizons science and spacecraft teams took place at APL on 8 January 2001. Mission payload selection was largely complete by 22 January, just three days after the PKB AO was released. The mission concept was to launch a small (400 kg class) flyby spacecraft based on heritage from APL's CONTOUR multi-comet flyby mission, then in development for launch in 2002. The PKB spacecraft would be able to fly about 30 kg of instruments—far more than the 7 to 9 kg PKE would have been able to. It also would include substantial avionics and propulsion system redundancy for the long voyage, and it would use the lower-power (and lower cost) of the two RTGs that NASA offered in the AO.

In the proposal, strong emphasis was placed on reducing programmatic (i.e., cost and schedule) risk because (i) APL was viewed as a new entrant to outer planet missions and (ii) it was important to convincingly avoid the repeated cost escalations of the 1990s Pluto study and mission development attempts at JPL. A very large, 48 Gbit solid state memory was proposed for the mission in order to allow the spacecraft to take maximum advantage of its time in the Pluto system (by contrast, the PKE mission planned a 2 Gbit memory). Finally, every effort was made to propose the earliest feasible launch and arrival; so we proposed that launch would be in December 2004 toward a JGA, with a January 2006 backup JGA. The December 2004 launch would target a July 2012 arrival. After a long process of winnowing, on 5 February 2001, our PKB proposal was named New Horizons. The name was meant to symbolize both the new scientific horizons inherent in the exploration of the Pluto system and the Kuiper belt, as well as the programmatic new horizons of PI-led outer planet missions. PI Stern commissioned planetary scientist and artist D. Durda to provide a "2001-esque" Pluto flyby graphic that evoked a sense of new horizons. That image, with an as-launched New Horizons substituted for the 2001-era concept, is shown in Fig. 1.

The proposed New Horizons payload consisted of the following four instrument packages:

- PERSI, a PKE-proposed instrument package consisting of the Alice ultraviolet spectrometer and the Ralph multi-color imager/infrared imaging spectrometer.

Table 2 Proposed New Horizons payload

Instrument	Type	Sensor characteristics	Builders
PERSI	Remote sensing suite	MVIC (panchromatic and four-color CCD imager, 0.4–1.0 microns, 20 microradians/pixel), LEISA (near infrared imaging spectrometer, wedged filter, 1.25–2.5 μm, $R = 600$ for 2.1–2.25 microns and $R = 300$ otherwise, 62 microradians/pixel), and Alice (ultraviolet imaging spectrometer, 500–1850 Å, spectral resolution 3 Å, 5 milli-radians/pixel).	Ball, SwRI, NASA/Goddard Space Flight Center (GSFC)
REX	Uplink radio science, passive radiometry	Signal/noise power spectral density 55 db-Hz; ultrastable oscillator stability 1×10^{-13} in 1 second samples. Disk-averaged radiometry to ± 0.1 K.	Stanford, JHU/APL
PAM	Plasma and high energy particle spectrometers	SWAP (solar wind plasmas up to 6.5 keV, toroidal electrostatic analyzer and retarding potential analyzer), and PEPSSI (ions 1–5000 keV and electrons 20–700 keV, time-of-flight by energy to separate pickup ions).	SwRI, JHU/ APL
LORRI	High resolution imager	Panchromatic, narrow angle CCD imager, 0.30–0.95 microns, 5 microradians/pixel.	JHU/APL

- REX, an uplink radio science instrument with radiometer capabilities.
- LORRI, a long focal length panchromatic Charged Couple Device (CCD) camera to provide 4x higher resolution imaging than Ralph could accomplish.
- PAM, a plasma package consisting of both the high-sensitivity Solar Wind Around Pluto (SWAP) solar wind monitor to address Pluto atmospheric escape objectives and the Pluto Energetic Particle Sensor Spectrometer Investigation (PEPSSI) energetic particle spectrometer adapted from the Energetic Plasma Sensor (EPS) sensor then in development for NASA's MESSENGER Mercury orbiter.

Table 2 provides some additional details on the payload as proposed. The article in this volume by Weaver et al. (2007) provides a more detailed overview of the as-launched scientific payload, which differs primarily in terminology (i.e., instrument names) and a few minor technical aspects from that described here.

The objectives of this payload were to significantly exceed the minimum mission requirements laid out by the AO, to significantly exceed what PKE would have accomplished, but not to overburden the mission with a costly "Christmas tree" array of instruments incompatible with a highly cost constrained outer planets mission. Other instruments considered but not included in this payload for various reasons were a magnetometer, a plasma wave sensor, a dust instrument for cruise science in the deep outer solar system and the Kuiper belt, bolometers, and a mass spectrometer.

PERSI and REX were termed the New Horizons "core payload," because they were sufficient to accomplish all of the Group 1 science that the PKB AO required proposers to meet. LORRI and PAM were termed the "supplementary payload" and were included to add depth and breadth to what the core instruments could do; however, the supplementary payload was clearly stated to be descopable should technical or programmatic considerations force cutbacks during development.

I now return to the proposal process. Proposals were turned in on 6 April 2001. After a two month technical and programmatic review process, on 6 June 2001, NASA announced the selection of JPL's POSSE (Pluto Outer Solar System Explorer; L. Esposito, PI) and APL's New Horizons for Phase A studies and further competition. PI Stern was at a Kuiper belt meeting in Paris and was informed by a phone message to call home to "Dr. Yung" (meaning Co-Investigator Leslie Young), who relayed to him that NASA had called with the selection news somewhat earlier in the day. A kickoff meeting for the two Phase A studies was sponsored by NASA Headquarters on 18 June 2001.

Both POSSE and New Horizons were funded by NASA at the $500,000 level for Phase A studies that were to be due on 18 September 2001. Both teams contributed substantial internal funds to supplement the NASA funding they received. The ground rules of the Phase A study were that the proposal teams could not augment their proposed science payloads or science teams, but were instead to provide additional engineering, cost, and schedule study to further flesh out their mission concepts. The 11 September 2001 terrorist attacks on New York and Washington, D.C. interceded in the final days of proposal preparation. Owing to the nationwide stoppage of air transport (including overnight mail), a shut down of government activities in central Washington, D.C. for several days, and the general national paralysis that temporarily ensued, NASA extended the final proposal deadline to 25 September 2001.

Formal oral briefings on the proposals to a NASA Concept Study Evaluation Review Board were held for New Horizons and POSSE on 16 and 18 October, 2001, respectively.

In parallel with the Phase A and proposal activities described above, the scientific community and the New Horizons team also undertook a difficult effort to put funding in place in the NASA budget for FY2002's needed Phase B development. Had this not been done, any selection of a mission would have been moot, because no contract could be let to begin work, thereby ensuring that the 2004–2006 JGA launch window would not be met and no mission would be built (the next JGA window would not open until 2015). Ultimately, after much work and some intrigue, this effort succeeded with the Senate passage and House-Senate conference agreement of a NASA FY2002 budget in early 2002 that included $30M in supplementary funding for the PKB mission to initiate spacecraft and science instrument development as well as work toward launch vehicle procurement.

NASA selected New Horizons in mid-November. However, the formal announcement of this award was held up until 29 November. PI Stern was informed of the selection of New Horizons for Phase B development by a phone call from NASA's PKB Program Scientist, Dr. Denis Bogan, while he was at the annual AAS Division for Planetary Sciences meeting, which was held in New Orleans that year. A win party was held on Bourbon Street that night in the New Orleans French Quarter, but the details remain (perhaps understandably) fuzzy.

4 Mission Development Overview

Initiating the development of New Horizons was difficult for a variety of reasons. To begin, NASA's selection letter to PI Stern pointed out the numerous obstacles the mission faced before it could be confirmed. Among these were a lack of funding or a plan to fund after Phase B; the lack of a nuclear qualified launch vehicle; the short time to launch; and the lack of sufficient fuel to power a flight RTG. The award letter also postponed launch from December 2004 to January 2006, which implied a 5 year delay in arrival date from mid-2012 to mid-2017. NASA also soon insisted on New Horizons using the more expensive RTG of the two in inventory. Also complicating matters was the tragic loss of two key APL engineers responsible for REX ultrastable oscillator development, who died in a small aircraft accident at the end of 2001.

The New Horizons team nonetheless began work in January 2002, initially focusing on the requirements development and documentation phase that would lead to a May 2002

Fig. 2 New Horizons spacecraft concept as originally proposed. Contrast to Fig. 1 with artwork by Dan Durda

System Requirements Review (SRR). PI Stern and the mission design team worked hard to shorten the flight time and move the arrival date earlier than 2017, ultimately achieving a mid-2015 arrival date. Stern and Cheng (2002) and Stern (2002) summarize the mission at this early development stage.

The New Horizons science team, the larger planetary science community, APL, SwRI, and others worked to see funds included in the FY2003 budget for mission development after Phase B. A key aspect of this battle was meeting Space Science Associate Administrator E. Weiler's challenge that NASA and the Administration would support New Horizons if the soon-to-be finalized NRC Decadal Report in Planetary Sciences (Belton et al. 2002) ranked PKB as *the* highest priority new start for solar system exploration. Owing to the scientific significance of the Kuiper belt exploration in general, and Pluto system exploration in particular, this key milestone was accomplished in the summer of 2002, thereby largely ending funding battles over the mission (though severe cash flow difficulties persisted into FY2003).

Figure 2 depicts the spacecraft as designed. Figure 3 shows the project organization during spacecraft construction. Figure 4 depicts the mission trajectory. Figure 5 shows the assembled spacecraft during checkout at the launch site.

Major milestones in the development of New Horizons were as follows:

May 2002: Systems Requirements Review
October 2002: Mission Preliminary Design Review
July 2002: Selection of the Boeing STAR-48 upper stage
March 2003: Non-Advocate Review and Authorization for Phases C and D
July 2003: Selection of the Atlas V 501 launch vehicle
October 2003: Mission Critical Design Review
September 2004: First instrument payload delivery
January 2005: Spacecraft structure complete
March 2005: Final instrument payload delivery
April 2005: Spacecraft integration complete
May 2005: Beginning of spacecraft environmental testing
September 2005: Spacecraft shipment to the launch site in Florida
December 2005: Spacecraft mating with its launch vehicle
January 2006: Launch.

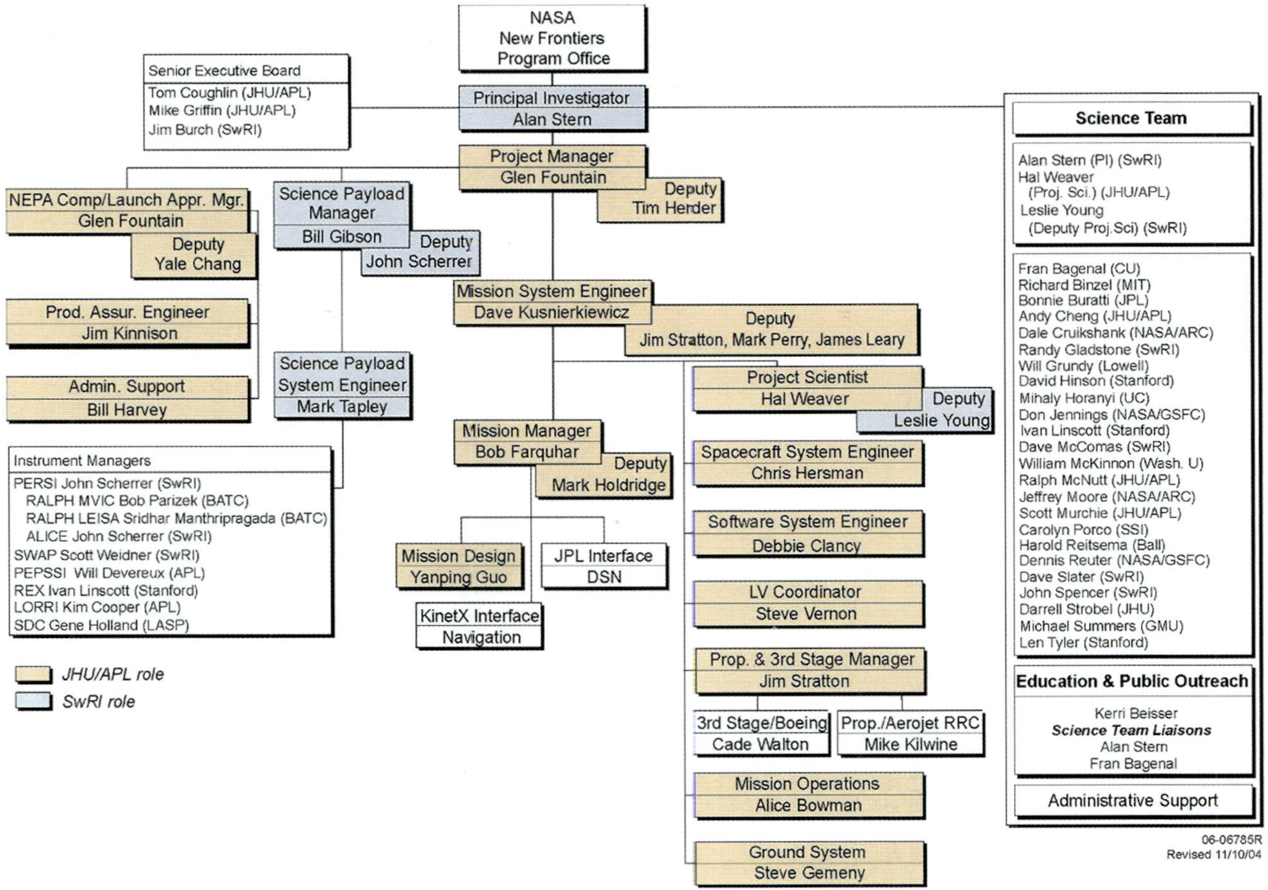

Fig. 3 New Horizons spacecraft-payload team project organization chart during spacecraft construction

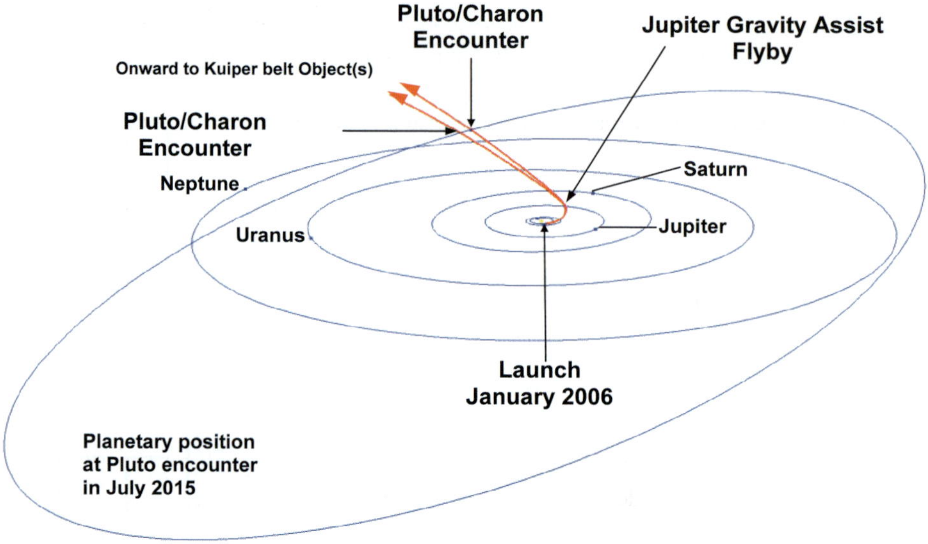

Fig. 4 New Horizons trajectory depiction. The two red trajectory lines show the range of possible encounter dates (2015–2020) that applied for all possible launch dates in the 35-day long 2006 launch window. Planetary positions are shown at the time of Pluto encounter in 2015

Fig. 5 The New Horizons spacecraft is shown here in a clean room, a few weeks before launch

During the course of the development of New Horizons, both the spacecraft and instrument payload designs evolved in many ways. The most important spacecraft changes during development included:

- RTG fuel production difficulties that resulted in a 30 watt (15%) power decrease at Pluto.
- RTG mount and spacecraft balance issues that added over 50 kg in dry mass.
- Downsizing of the telecom high gain antenna from 3.0 to 2.1 m to save mass.
- A 25% increase in the power system capacitor bank capacity to source load transients up to 33 milli-farads.
- Removing corners on the triangular spacecraft structure to save mass.
- Increasing the onboard solid state memory to 64 Gigabits.
- Substituting heavier sun trackers when advanced development units stalled in production.

- Substituting traveling wave tubes for solid state power amplifiers in the telecom system.
- Changes in thruster positioning to accommodate plume impingement and fuel line routing concerns.
- Added telecom redundancy through cross strapping of the antenna and receiver/transmitter networks.

The most important instrument payload changes during development included:

- Addition of the Education and Public Outreach (EPO) Student Dust Counter to the payload.
- Separating the PERSI instrument into distinct Ralph (visible/infrared) and Alice (ultraviolet spectrometer) instruments.
- Separating the PAM instrument into distinct SWAP (low energy) and PEPSSI (high-energy) instruments.
- Adding launch doors to PEPSSI, SWAP, and LORRI.

Some notable scientific developments that occurred during mission development included the following:

- The discovery of Kuiper belt satellites in 2001.
- The discovery of factor of two increases in pressure and changes in the vertical structure of Pluto's atmosphere between 1988 and 2002.
- The discovery of ammonium hydrates on Charon in 2004.
- The discovery of high albedos and Pluto-like surface compositions on some KBOs by 2005.
- The discovery of Pluto's satellites Nix and Hydra in 2005.
- The discovery of objects roughly as large or larger than Pluto in the KB and inner Oort Cloud by 2005.

The as-flown New Horizons payload is summarized in Table 3; many more details can be found in the accompanying article describing the New Horizons payload by Weaver et al. (2007).

During its development, over 2500 individuals worked directly on spacecraft, payload, ground system, RTG, and launch vehicle/upper stage development of New Horizons. Also during development, numerous personnel and programmatic changes also took place. The initial New Horizons project manager, Mr. Thomas Coughlin, retired. Tom was replaced by APL's highly qualified Mr. Glen Fountain at the start of 2004. The initial project scientist, Dr. Andrew Cheng stepped down during development and was replaced by Dr. Harold Weaver, in order to permit Dr. Cheng to place more emphasis on his critical role as the LORRI instrument PI. It is worth noting, to APL's credit, that in both of these cases, APL accepted the first choices of the PI for the replacement personnel. Since the project's inception, Dr. Leslie Young of SwRI served as the Deputy Project Scientist. The NASA Marshall Space Flight Center Discovery-New Frontiers Program office came into being in late 2004, following the dissolution of the JPL Discovery-New Frontiers Program office that operated from late 2003 to mid-2004.

These and other details of the development of New Horizons could easily fill an entire book; and indeed such a book might even be written some day. For now, the above listed summary will suffice as an introduction to the papers that follow in this volume of Space Science Reviews.

Table 3 New Horizons Payload

Instrument, PI	Measurement objectives	Characteristics
UV imaging spectrometer (Alice), S.A. Stern	• Upper atmospheric temperature and pressure profiles of Pluto • Temperature and vertical temperature gradient should be measured to ~10% at a vertical resolution of ~100 km for atmospheric densities greater than $\sim 10^9$ cm^{-3} • Search for atmospheric haze at a vertical resolution <5 km • Mole fractions of N_2, CO, CH_4 and Ar in Pluto's upper atmosphere • Atmospheric escape rate from Pluto • Minor atmospheric species at Pluto • Search for an atmosphere of Charon • Constrain escape rate from upper atmospheric structure	Bandpass: 465–1881 Å; FOV: $4° \times 0.1°$ plus $2° \times 2°$; Spectral resolution: 1.8 Å/spectral element, Spatial resolution: 5 mrad/pixel
Ralph/Multicolor Visible Imaging Camera (Ralph/MVIC), S.A. Stern	• Hemispheric panchromatic maps of Pluto and Charon at best resolution exceeding 0.5 km/pixel • Hemispheric 4-color maps of Pluto and Charon at best resolution exceeding 5 km/pixel • Search for/map atmospheric hazes at a vertical resolution <5 km • High resolution panchromatic maps of the terminator region • Panchromatic, wide phase angle coverage of Pluto and satellites • Panchromatic stereo images of Pluto and Charon, Nix, and Hydra • Orbital parameters, bulk parameters of Pluto and satellites • Search for rings • Search for additional satellites	Bandpasses: 400–975 nm (panchromatic), plus 4 color filters (Blue, Red, 890 nm Methane, Near-IR); FOV: $5.7° \times 0.15°$ (stare, pan), or $5.7° \times$ TBD (scan); IFOV: 20 μrad/pixel
Ralph/Linear Etalon Imaging Spectral Array (Ralph/LEISA), D. Jennings	• Hemispheric near-infrared spectral maps of Pluto and Charon at best resolution exceeding 10 km/pixel • Hemispheric distributions of N_2, CO, CH_4 on Pluto at a best resolution exceeding 10 km/pixel • Surface temperature mapping of Pluto and Charon • Phase-angle-dependent spectral maps of Pluto and Charon	Bandpass: 1.25–2.50 μm, $\lambda/\delta\lambda \approx 240$; 2.10–2.25 μm, $\lambda/\delta\lambda \approx 550$; FOV: $0.9° \times 0.9°$; IFOV: 62 μrad/pixel
Radio Science Experiment (REX), L. Tyler	• Temperature and pressure profiles of Pluto's atmosphere • Surface number density to ±1.5%, surface temperature to ±2.2 K and surface pressure to ±0.3 μbar • Surface brightness temperatures on Pluto and Charon • Masses and chords of Pluto and Charon; detect or constrain Pluto's J_2 • Detect, or place limits on, an ionosphere for Pluto	X-band (7.182 GHz uplink, 8.438 GHz downlink); Radiometry $T_{Noise} < 150$ K; Ultra-Stable Oscillator (USO) frequency stability requirement: $\delta f/f = 3 \times 10^{-13}$ over 1 sec
Long Range Reconnaissance Imager (LORRI), A. Cheng	• Hemispheric panchromatic maps of Pluto and Charon at best resolution exceeding 0.5 km/pixel • Search for atmospheric haze at a vertical resolution <5 km • Panchromatic maps of the far-side hemisphere • High resolution panchromatic maps of the terminator region • Panchromatic, wide phase angle coverage of Pluto and satellites • Panchromatic stereo images of Pluto, Charon, Nix, and Hydra • Orbital parameters, bulk parameters of Pluto and satellites • Search for satellites and rings	Bandpass: 350–850 nm; FOV: $0.29° \times 0.29°$; IFOV: 5 μrad/pixel

Table 3 (*Continued*)

Instrument, PI	Measurement objectives	Characteristics
Solar Wind At Pluto (SWAP), D. McComas	• Atmospheric escape rate from Pluto • Solar wind velocity and density, low energy plasma fluxes and angular distributions, and energetic particle fluxes at Pluto–Charon • Solar wind interaction with Pluto and Charon	FOV: 200° × 10° Energy range: 0.25–7.5 keV Energy resolution: RPA: 0.5 V (< 1.5 keV) ESA: 0.4 $\Delta E/E$ (>1.4 keV)
Pluto Energetic Particle Spectrometer Science Investigation (PEPSSI), R. McNutt	• Composition and density of pick-up ions from Pluto, which indirectly addresses the atmospheric escape rate • Solar wind velocity and density, low energy plasma fluxes and angular distributions, and energetic particle fluxes in the Pluto system	Energy range: 1 keV–1 MeV FOV: 160° × 12° Resolution: 25° × 12°
Venetia Burney Student Dust Counter (SDC), M. Horanyi	• Trace the density of dust in the Solar System along the New Horizons trajectory from Earth to Pluto and beyond	12 PVF panels to detect dust impacts and 2 control panels shielded from impacts

5 Launch and Early Flight

On schedule as directed at selection in November 2001, New Horizons took flight in January 2006. Initial launch attempts on 17 and 18 January 2006 had been foiled by a weather front that adversely affected the launch site near Cape Canaveral, Florida on 17 January and the APL mission operations site in Columbia, Maryland on 18 January. On its third attempt, New Horizons launched, at 1900 UT on 19 January 2006. This date, coincidentally, was five years to the day since the PKB AO was released, 10 years to the week since the death of Pluto's discoverer, Clyde Tombaugh, and 76 years to the week since the discovery images of Pluto were obtained. It is worth noting that 19 January was the first date on which the Atlas V actually attempted to count to zero, and when it did, it launched.

At launch, New Horizons carried all of the instruments proposed for its scientific payload in 2001 as well as the Venetia Burney Student Dust Counter (SDC) added by PI Stern as an EPO enhancement in 2002; no payload descoping had occurred. When launched, New Horizons carried 78 kg of fuel and pressurant; virtually all of the 80 kg load it could possibly have carried in its fuel tank, a testament to the spacecraft and payload team's ability to control mass gain during development. At launch, the spacecraft also carried nine mementos to its target of the ninth planet. These were: two U.S. flags, the state quarters of Maryland and Florida, a small piece of the first private manned spacecraft, SpaceShip 1, a CDROM with over 100,000 names being sent to Pluto, another CDROM with numerous pictures of the spacecraft and spacecraft-mission development teams, a 1990 US postage stamp ("Pluto: Not Yet Explored"), and a small amount of the ashes of Pluto's discoverer, Clyde Tombaugh, whose remains have become, with the launch of New Horizons, the first of a human being launched beyond the solar system, to the stars.

The New Horizons Atlas V launch vehicle and STAR-48 upper stage both performed flawlessly, releasing the spacecraft on the proper trajectory some 50 minutes after launch. The spacecraft was contacted three minutes later and was found to be in good health. Subsequent tracking revealed the spacecraft to have received a highly accurate injection onto its

Jupiter course, with velocity errors of only about 18 meters/second. This was far less than the ~100 meters/second fuel budget New Horizons carried for the purpose of post-launch trajectory correction. As a result, the spacecraft fuel supply available for mission science at Pluto and to explore KBOs is almost twice as large as nominal preflight predictions.

New Horizons was the fastest spacecraft ever launched. It crossed the orbit of the moon in ~9 hours and reached Jupiter in record time—just 13 months. It will reach Pluto 9.5 years after launch and will then continue across the Kuiper belt on a hyperbolic trajectory that will escape to interstellar space.

During the first 10 weeks of flight, the spacecraft was spun down to its nominal cruise 5 RPM spin rate, and a thorough series of subsystem checkouts was conducted. These subsystem tests revealed very good performance in all subsystems. No significant hardware problems were revealed on the spacecraft. Also during this period a series of three trajectory correction maneuvers were carried out to refine the course to Jupiter; 20 meters/second of fuel were expended to carry out this sequence of maneuvers.

During the period from late March to late September 2006, all seven of the payload instruments were turned on, checked out, and calibrated. All of the instruments are working well. Six of the seven instrument door deployments were completed in this phase; the last, opening the Alice UVS solar occultation port to space, took place in July 2007 near 6.5 AU. No instrument problems were found to exist that will compromise their scientific capabilities at the Pluto system or KBOs, though minor issues—a PEPSSI mounting error on the spacecraft and a very small light leak with Ralph—have been documented.

Also during March–September 2006, several significant spacecraft software upgrades were performed. These added significant new capability to the onboard fault detection and correction ("autonomy") system, and corrected various software bugs and idiosyncrasies in both the Command and Data Handling (C&DH) and Guidance and Control (G&C) software package that had been detected after launch.

On 13 June 2006, New Horizons serendipitously flew past a small (~4 km diameter), S-type asteroid called 2002 JF56 (later named "APL" by the IAU after nomination by PI Stern) at a fortuitously close range of just 104,000 kilometers. This "encounter," while distant from a scientifically important range, did result in some insights into this small body (Olkin et al. 2006). More importantly, however, the flyby was used to test instrument pointing and image motion compensation capabilities with a moving target. These tests were all successful.

New Horizons encountered Jupiter on 28 February 2007. Surrounding this event, from January to June 2007, New Horizons conducted an extensive series of over 700 Jupiter system observations. The Jupiter encounter was planned out to further calibrate instruments, test spacecraft and ground system procedures and capabilities as a risk reduction in advance of the Pluto encounter, and obtain new science to follow up on discoveries made by the most recent missions to Jupiter: Galileo and Cassini. More data was generated and downlinked from Jupiter than planned for Pluto, thanks to the higher New Horizons telemetry rates available at ~5 AU than at 32 AU. The Jupiter encounter included a traverse far down Jupiter's magnetotail, which the spacecraft exited only after 2500 R_J; no previous spacecraft had ventured more than ~100 R_J in this magnetotail.

During the latter half of 2007, New Horizons began its hibernation-wakeup checkout-hibernation cycling. A small (2.4 m/s) course correction was applied in September 2007, correcting aim point errors at Pluto from over 500,000 km to ~50,000 km. Also in 2007, the science team reanalyzed the optimal arrival date and closest approach distance to Pluto, settling on 14 July 2015 and 12,500 km from Pluto's surface, respectively. Plans are in place for further flight software upgrades in 2008–2009 based on lessons learned at Jupiter.

New Horizons will require almost 8 years of cruise to fly from the end of its Jupiter encounter to the start of its Pluto encounter. During this time the spacecraft will spend about 10 months of each year in hibernation, preserving avionics lifetimes by shutting down much of its equipment and reducing costs by relaxing the need for constant monitoring and commanding. Once per week, the spacecraft operators will check a beacon tone indicating general health status from the spacecraft. Once each month, engineering telemetry will be collected to assess spacecraft health and subsystem trends in more detail. For two months each year, the spacecraft will be awakened for thorough check outs, instrument calibrations, cruise science, and trajectory corrections (if necessary). Cruise science includes interplanetary plasma and dust measurements, phase curve studies of Uranus, Neptune, Pluto, and bright KBOs, interplanetary Lyα measurements, and other studies. In 2010 and 2014, the spacecraft and ground team will additionally conduct full-scale Pluto encounter rehearsals in flight.

The Pluto system encounter will span $\sim$200 days beginning in early 2015 and lasting until about 30 days after Pluto closest approach. Complete data transmission, however, will require some 4 to 9 months after the encounter owing to the 1,000 bit/sec downlink rate at Pluto.

Near closest approach some 12,500 km over Pluto, Ralph will obtain maps of Pluto and Charon with kilometer-scale resolution; at closest approach, LORRI images at scales as high as 75 m/pixel may be achieved (depending on the final flyby distance selected). In addition, the Group 1 objectives call for mapping the surface composition and distributions of major volatile species, for which Ralph will obtain: (i) four-color global (dayside) maps at 1.6 km resolution, and (ii) diagnostic, hyper-spectral near-infrared maps at 7 km/pixel resolution globally (dayside), and at higher resolution for selected areas. Maps of surface-lying CH_4, N_2, CO, CO_2, and H_2O abundances will be obtained. Surface temperatures will be mapped by Ralph using temperature-sensitive infrared spectral features; these maps are expected to have resolutions as good as 2 deg K and 10 km; hemispheric-averaged surface brightness temperature will also be measured by the REX radiometer mode.

Characterization of Pluto's neutral atmosphere and its escape rate will be accomplished by: (i) Alice ultraviolet airglow and solar occultation spectra to determine the mole fractions of N_2, CH_4, CO and Ar to 1% in total mixing ratio and to determine the temperature structure in the upper atmosphere, (ii) REX radio occultations at both Pluto and Charon, measuring the density/temperature structure of Pluto's neutral atmosphere to the surface, (iii) SWAP and PEPSSI in situ determination of the atmospheric escape rate by measuring Pluto pickup ions, and (iv) Alice H Lyα mapping of Pluto and Charon in order to determine the rate of Roche-lobe flow of atmosphere from Pluto to Charon. Searches for atmospheres around Charon and KBOs will be made using Alice with both airglow and solar occultation techniques.

REX-derived Doppler tracking will also be used to measure the masses of Pluto and Charon, and to attempt J2 determinations; together with imagery-derived 3-D volumes, these data will be used to obtain improved densities. SDC will measure the density and masses of dust particles in the solar system from 1 AU to at least 40 AU, far surpassing the 18 AU boundary beyond which any dust detector has as yet penetrated.

Approximately three weeks after the Pluto closest approach, New Horizons will perform a trajectory correction maneuver to target its first KBO encounter. The target KBO will be selected after a deep search for candidates to be made in 2011–2013 after Pluto exits the dense background star fields of Sagittarius. Monte Carlo simulations indicate that 4 to perhaps 12 candidate KBOs may be detected within reach of the spacecraft's available fuel supply. These same simulations indicate a typical 2 to 3 year cruise time to reach the first KBO flyby. A second KBO flyby may also be possible given the available fuel. It is important to note, however, that any KBO encounter will require the approval of an extended mission budget from NASA.

6 Concluding Remarks

New Horizons is now safely in flight to make the first reconnaissance in 2015 of the fascinating Pluto-system: the farthest of the classical planets, the most accessible example of ice dwarf planets, and the most well known KBO.

If all goes well, New Horizons will then go on to make a flyby reconnaissance of a 30 kilometer to 50 kilometer diameter KBO some two to four years later. A second KBO encounter two to four years later still may even be possible, depending on spacecraft health, fuel status, our ability to detect sufficient KBOs within reach of the vehicle before it passes out of the classical KB at about 50 AU, and NASA funding.

Beyond completing the initial reconnaissance of the nine classical planets, New Horizons has already demonstrated the ability of PI-led missions to be mounted to the outer solar system, opening up a wide range of future possibilities. As well, the mission has proven the ability of non-traditional entities, like APL and SwRI to explore the outer solar system, giving NASA new programmatic flexibility and enhancing the competitive options available to NASA when selecting outer planet missions.

If New Horizons is successful, it will represent a watershed development in the scientific exploration of dwarf planets—an entirely new class of bodies in the solar system, of worlds with exotic volatiles on their surfaces, of rapidly (possibly hydrodynamically) escaping atmospheres, and of giant impact derived satellite systems. It will also provide the first dust density measurements beyond 18 AU, cratering records that shed light on both the ancient and present-day KB impactor population down to tens of meters, and a key comparator to the puzzlingly active, former dwarf planet (now satellite of Neptune) called Triton, which is as large as Eris and Pluto.

Following this article is a series of other articles describing in significantly more detail the mission trajectory, the mission science, the spacecraft, and the science payload and its individual instruments.

Acknowledgements I thank the entire New Horizons team, including the NASA Headquarters and NASA Marshall Space Flight Center program offices, the Johns Hopkins Applied Physics Laboratory, the Southwest Research Institute, all of the mission subcontractors, Lockheed-Martin and Boeing, the New Horizons science team, the Department of Energy, the Jet Propulsion Laboratory (JPL), and NASA Kennedy Space Flight Center for their dedication, commitment, and drive throughout the development of this incredible mission of discovery. I also thank New Horizons Program Scientist D. Bogan, New Horizons Project Scientist H. Weaver, Deputy New Horizons Project Scientist Leslie Young, two anonymous referees, and Space Science Reviews editor C. Russell for helpful comments on this manuscript.

References

M.J. Belton et al., New frontiers in the solar system. An integrated exploration strategy. National Research Council, 2002, 145 pp.

Cheng et al., Space Sci. Rev. (2007, this issue). doi:10.1007/s11214-007-9271-6

R. Farquhar, A. Stern, Pushing back the frontier: A mission to the Pluto–Charon system. Planet. Rep. **10**(4), 18–23 (1990)

Fountain et al., Space Sci. Rev. (2007, this issue)

Guo et al., Space Sci. Rev. (2007, this issue). doi:10.1007/s11214-007-9242-y

Horanyi et al., Space Sci. Rev. (2007, this issue)

J.I. Lunine et al., Report of the Pluto–Kuiper Express Science Definition Team, NASA, 1996

McComas et al., Space Sci. Rev. (2007, this issue). doi:10.1007/s11214-007-9205-3

McNutt et al., Space Sci. Rev. (2007, this issue)

NASA, Pluto Kuiper belt mission announcement of opportunity. Announcement of Opportunity 01-OSS-01, 2001

C.B. Olkin, D. Reuter, A. Lunsford, R.P. Binzel, S.A. Stern, The New Horizons distant flyby of asteroid 2002 JF56. American Astronomical Society, Division of Planetary Science meeting 38, 2006, abstract 9.22

Reuter et al., Space Sci. Rev. (2007, this issue)

S.A. Stern, The Pluto reconnaissance flyby mission. EOS **74**, 73–75 (1993)

S.A. Stern, Journey to the farthest planet. Sci. Am. **286**, 56–59 (2002)

S.A. Stern, A. Cheng, NASA plans Pluto–Kuiper belt mission. EOS **83**, 101–106 (2002)

S.A. Stern, J. Mitton, *Pluto and Charon: Ice Worlds on the Ragged Edge of the Solar System* (Wiley-VCH, 2005), 244 pp.

Stern et al., Space Sci. Rev. (2007, this issue)

R.J. Terrile, S.A. Stern, R.L. Staehle, S.C. Brewster, J.B. Carraway, P.K. Henry, H. Price, S.S. Weinstein, Spacecraft missions to the Pluto and Charon system, in *Pluto and Charon*, ed. by S.A. Stern, D.J. Tholen (University of Arizona Press, Tucson, 1997), pp. 103–136

Tyler et al., Space Sci. Rev. (2007, this issue)

H.A. Weaver, S.A. Stern, M. Mutchler, A.J. Steffl, M.W. Buie, W.J. Merline, J.R. Spencer, E.F. Young, L.A. Young, The discovery of two new satellites of Pluto. Nature **439**, 943–945 (2006)

Weaver et al., Space Sci. Rev. (2007, this issue)

Young et al., Space Sci. Rev. (2007, this issue)

The New Horizons Spacecraft

Glen H. Fountain · David Y. Kusnierkiewicz · Christopher B. Hersman · Timothy S. Herder · Thomas B. Coughlin · William C. Gibson · Deborah A. Clancy · Christopher C. DeBoy · T. Adrian Hill · James D. Kinnison · Douglas S. Mehoke · Geffrey K. Ottman · Gabe D. Rogers · S. Alan Stern · James M. Stratton · Steven R. Vernon · Stephen P. Williams

Originally published in the journal Space Science Reviews, Volume 140, Nos 1–4, 23–47.
DOI: 10.1007/s11214-008-9374-8

Abstract The New Horizons spacecraft was launched on 19 January 2006. The spacecraft was designed to provide a platform for seven instruments designated by the science team to collect and return data from Pluto in 2015. The design meets the requirements established by the National Aeronautics and Space Administration (NASA) Announcement of Opportunity AO-OSS-01. The design drew on heritage from previous missions developed at The Johns Hopkins University Applied Physics Laboratory (APL) and other missions such as Ulysses. The trajectory design imposed constraints on mass and structural strength to meet the high launch acceleration consistent with meeting the AO requirement of returning data prior to the year 2020. The spacecraft subsystems were designed to meet tight resource allocations (mass and power) yet provide the necessary control and data handling finesse to support data collection and return when the one-way light time during the Pluto fly-by is 4.5 hours. Missions to the outer regions of the solar system (where the solar irradiance is 1/1000 of the level near the Earth) require a radioisotope thermoelectric generator (RTG) to supply electrical power. One RTG was available for use by New Horizons. To accommodate this constraint, the spacecraft electronics were designed to operate on approximately 200 W. The travel time to Pluto put additional demands on system reliability. Only after a flight time of approximately 10 years would the desired data be collected and returned to Earth. This represents the longest flight duration prior to the return of primary science data for any mission by NASA. The spacecraft system architecture provides sufficient redundancy to meet this requirement with a probability of mission success of greater than 0.85. The

G.H. Fountain (✉) · D.Y. Kusnierkiewicz · C.B. Hersman · T.S. Herder · T.B. Coughlin · D.A. Clancy · C.C. DeBoy · T.A. Hill · J.D. Kinnison · D.S. Mehoke · G.K. Ottman · G.D. Rogers · J.M. Stratton · S.R. Vernon · S.P. Williams
Applied Physics Laboratory, The Johns Hopkins University, 11100 Johns Hopkins Rd., Laurel, MD 20723, USA
e-mail: glen.fountain@jhuapl.edu

W.C. Gibson
Southwest Research Institute, 6220 Culebra Rd., San Antonio, TX 78238, USA

S.A. Stern
Southwest Research Institute, 1050 Walnut St., Suite 400, Boulder, CO 80302, USA

spacecraft is now on its way to Pluto, with an arrival date of 14 July 2015. Initial in-flight tests have verified that the spacecraft will meet the design requirements.

Keywords Pluto · New Horizons · Spacecraft · Radioisotope thermoelectric generator

1 Introduction

A reconnaissance mission to Pluto and the Kuiper Belt immediately brings a number of issues to mind for those charged with the development of the spacecraft. Pluto's orbital distance from the Sun ranges from 28 to 40 astronomical units (AU) and has an orbital period of 248 years. Pluto's most recent closest approach to the Sun occurred in 1989, and it is now receding such that in 2015 the solar distance will be 32 AU. At those distances the travel time to Pluto is significant, and a spacecraft must be designed to be highly reliable, which implies both simplicity in design and sufficient redundancy to guard against single-point failures. The solar energy at Pluto is on the order of $1/1000$ of the irradiance received in Earth orbit; therefore, the spacecraft must carry its own energy source. The only currently available technology is the radioisotope thermoelectric generator (RTG), which uses the thermal energy created by the decay of plutonium 238 to produce electrical energy. A single RTG was available to the development team, which limited the total electrical energy available to operate the spacecraft to approximately 200 W at the time of the Pluto encounter. The limitation in electrical energy also required a very efficient thermal design that did not require significant electrical energy to supply makeup heat. The communications system is affected by the power limitations as well as the necessity for managing variable data rates over these great distances (on the order of 500 bits/s to kilobits per second). The command and data handling system architecture, is dominated by the low data rates and the need to operate the spacecraft safely with large delays between the receipt of data and the ability to command the spacecraft, given the roundtrip light time at Pluto is 9 hours. Thus the spacecraft is required to operate autonomously throughout the close encounter with Pluto, and it will take months to return the entire data set to Earth.

With the limited power of a single RTG, both the power dissipation of each subsystem and the operation of the spacecraft were carefully considered. Design choices provided some relief, such as the design of the communication system described in Sect. 6 of this paper. The remaining degrees of freedom in power management consist of the power cycling of instruments and other subsystems and the ability to minimize the power required for thermal management. The thermal management issues are described in Sect. 8. Though it is possible to operate with various subsystem elements and instruments on at the same time, the plan for the Pluto encounter is to operate with only those units essential to a particular observation turned on at any one time.

The guidance and control system design is capable of pointing the spacecraft with sufficient accuracy for the science instruments to observe Pluto, its moons Charon, Hydra, and Nix, and other objects of interest during its journey. In addition, the attitude control of the spacecraft is capable of pointing the antenna toward Earth while requiring very little control activity during the journey to Pluto. For New Horizons, the guidance and control system operates in both spinning and 3-axis modes to meet these two requirements. Finally, the system provides the ability to make adjustments in the spacecraft trajectory to correct the residual launch trajectory errors, to correct for small disturbances encountered along the way to Pluto, and to modify the trajectory after Pluto to allow the reconnaissance of Kuiper Belt objects if NASA approves an extended mission.

The cost of flight operations for such a long-duration mission also influenced the system design. The first factor (already discussed) was to keep the spacecraft design simple. A second strategy invoked for New Horizons was to develop operational modes sufficiently robust to allow a very low level of ground monitoring and control during the long cruise periods from launch to Jupiter, and especially the nearly 8-year period from operations at Jupiter to Pluto. The operational approach (designated the "hibernation" mode) utilizes the robustness of operating the spacecraft in a passive spinning attitude control state. Special command and autonomy modes provide a logical path to transition into various safe states in case a spacecraft fault occurs during the periods of hibernation. And, the mode include a very-low-rate communication "beacon" to alert the ground should intervention be required.

During hibernation only the critical systems to assure health and safety are powered on. The operational scenario during hibernation is to monitor the beacon tone once a week. The beacon tone provides the signal for mission operations to determine if everything is OK (a green beacon) or if ground intervention is required (a red beacon). Housekeeping data are transmitted once a month to gain additional insight into spacecraft health. The spacecraft is returned to normal operational status for a short time each year for an annual checkout to confirm that all systems are operating to the required specification. The hibernation mode allows the spacecraft to be operated with a very small (less than 10) mission operation staff for the long cruise to Pluto, thus minimizing the operations cost of the mission. The impacts of the hibernation mode on the various subsystems are described in the sections below.

The primary purpose of the New Horizons spacecraft is to bring a selected set of science instruments sufficiently close to Pluto to gather information to meet the science requirements established by NASA in the Announcement of Opportunity AO-OSS-01, released on 19 January 2001. These requirements are described by Stern (2008b) and Weaver et al. (2008) in their papers in this volume. The New Horizons science team chose seven instruments for the New Horizons mission:

1. Alice, an ultraviolet spectrometer (Stern 2008a)
2. Ralph, a multicolor imager/infrared imaging spectrometer (Reuter et al. 2008)
3. REX, an uplink radio science instrument with radiometer capabilities (Tyler et al. 2008)
4. LORRI, a long-focal-length panchromatic charge-coupled device (CCD) camera (Cheng et al. 2008)
5. SWAP, a solar wind monitor to address Pluto atmospheric escape rates (McComas et al. 2008)
6. PEPSSI, an energetic particle spectrometer to measure the composition of the ion species existing in region of Pluto (McNutt et al. 2008)
7. Venita Burney (VB) dust counter, which measures the density of fine dust particles in the solar system along New Horizons' trajectory from Earth to Pluto and beyond (Horanyi et al. 2008)

The placement of these instruments on the spacecraft is shown in Fig. 1. The optical instruments (Alice, Ralph, and LORRI) are approximately co-aligned. The Alice instrument has a secondary optical aperture aligned with the high-gain antenna (HGA) used by REX, thereby allowing both instruments to make simultaneous measurements of Pluto's atmosphere during that portion of the mission trajectory when Pluto and Charon pass between the spacecraft and the Sun and the Earth, respectively. SWAP and PEPSSI are oriented to use both spinning and 3-axis control modes to collect data. The orientations are also set to minimize solar glint into the PEPSSI aperture and maximize the ability of SWAP to collect pickup ions during

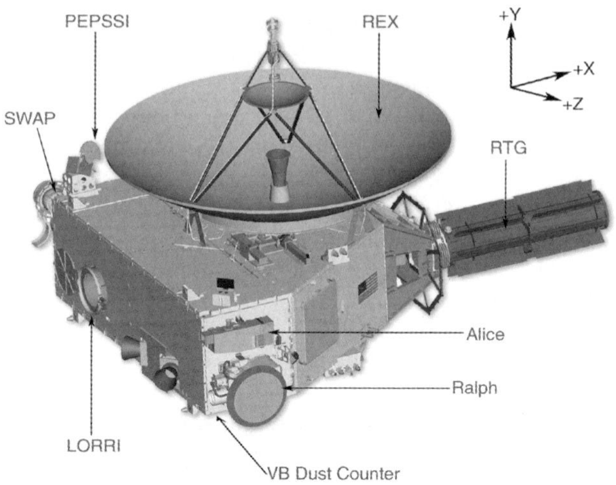

Fig. 1 The New Horizons spacecraft supports seven science instruments and is powered by a single RTG

Fig. 2 The New Horizons spacecraft operates either in spin mode or under 3-axis control. For the majority of its journey to Pluto, it spins about the +Y axis to which the antenna assembly is aligned (+Y low gain, medium gain, and high gain)

the approach and close encounter at Pluto. The VB dust counter is located on the side of the spacecraft pointing in the direction of the spacecraft velocity vector (the so-called "ram" direction) for the majority of the trip to Pluto.

The development team used these design constraints while transforming the concept of Fig. 1 into the spacecraft shown in Fig. 2. The following sections provide additional insight about design choices, principally focusing on how the spacecraft operates.

2 Spacecraft Configuration

2.1 Mechanical Configuration

The mechanical configuration of the spacecraft was driven by the need to spin the spacecraft about the axis defined by the antenna assembly; maintain spin axis alignment during launch; place instruments and propulsion system thrusters to meet observation requirements without interfering with one another; provide space for the spacecraft subsystems; and accommodate the launch loads. The Ulysses spacecraft (Wenzel et al. 1992) had similar design requirements. The New Horizons design team used the general configuration of Ulysses as a starting point for the spacecraft mechanical configuration, including the following:

1. A configuration that aligns the principal moment of inertia axis with the HGA, the +Y axis of Fig. 1
2. Placement of the single RTG in the X-Z plane of the spacecraft to increase the angular momentum and maximize the distance of this source of radiation from the electronics and instruments

This configuration provides a highly stable platform to point the spinning spacecraft in a desired direction, usually toward the Earth.

The foundation of the spacecraft structure is a central cylinder located within the spacecraft main body. This single-piece, machined aluminum ring forging is the focal point for the load path and, as such, is the primary load-bearing member of the spacecraft. The central cylinder also incorporates the separation interface adaptor to the third stage. Directly attached to the central cylinder are the propulsion system propellant tank (centered within the cylinder) and the top, aft, and main bulkhead decks. The bulkhead decks provide flat mounting surfaces and interfaces for subsystems. Five additional outer panels attach to the aft, top, and main bulkhead decks. The decks are of an aluminum honeycomb-type sandwich structure in which areas subject to high stress include magnesium edge members.

The titanium RTG support structure is attached to the decks on the opposite side of the cylinder from the instruments. The support structure is made primarily of titanium to provide high stiffness, low mass, and low thermal conductivity. Since the RTG support structure also had to provide electrical isolation, an all-metallic electrical isolation system was developed that didn't compromise structural stiffness. The isolation system employs multiple layers of non-conductive surface coatings applied to metallic plates located at the structure joints. To meet RTG interface requirements, an aluminum interface flange is attached to the titanium structure on the RTG side. Additional requirements to ensure proper stiffness and prevent deterioration of material properties due to RTG induced heating necessitated the addition of cooling fins to the RTG adaptor collar. The cooling fins assured that the strength of the RTG interface flange which is made of aluminum had sufficient margin during the 3rd stage burn when the acceleration peaked at 10.8g's and the interface temperature reached 260°C.

The propulsion system (hydrazine) propellant tank is centered on the axis defined by the principal axis and the center of mass of the spacecraft. At this location the variation in the fill fraction as the propellant is used has minimal impact on changes in the alignment of the spin axis. This ensures that when the spacecraft is spinning the HGA remains pointing in a fixed direction throughout mission life. The tank has a capacity of 90 kg of hydrazine. The total mass at launch was 77 kg of hydrazine and helium pressurant to meet the maximum allowable launch mass and to ensure stability of the combined spacecraft and third stage stack.

The placement of the propellant tank has two further advantages. It is surrounded by the cylinder that carries the launch loads directly into the third stage attach fitting. Since the tank is between the instruments and most of the electronics and RTG, the propellant further reduces the radiation effects of the RTG. The modest radiation output of the RTG and its placement guarantees that total dose received by the electronics is less than 5 krads (from all sources) throughout the primary phase of the mission. The majority of this dose is received at Jupiter when the spacecraft passes at a distance of 32 Jupiter radii (R_J). In addition, waste heat from the RTG and the electronics is used to keep the hydrazine sufficiently above the freezing point without significant use of makeup heater power.

2.2 System Configuration

Figure 3 provides an overview of the major functional system elements and their connectivity (Kusnierkiewicz et al. 2005).

To meet the overall reliability requirements there is significant block redundancy. There are two integrated electronic modules (IEMs). Each IEM contains a command and data handling (C&DH) processor; a guidance and control (G&C) processor; RF electronics, which are the heart of the communication system; an instrument interface card that provides connectivity to each instrument; and a 64-Gbit solid state recorder (SSR). In addition there is block redundancy in most of the remaining system elements, including the traveling wave tube amplifiers (TWTAs), star trackers, and inertial measurement units (IMUs). Other system elements (such as the power distribution unit, PDU) include redundant boards or have redundancy at the circuit level. For example, the shunt regulator unit has triply redundant control elements (Carlsson 2005).

To improve system reliability, significant cross-strapping below the block level is included in the design. For example, a C&DH processor in IEM 1 can be used as the controlling unit and the G&C system can operate using the G&C processor in IEM 2. The instruments' interfaces and much of their electronics (up to the focal plane detectors) are redundant, and much of the redundant circuitry can be accessed from the instrument interface card in each IEM. This cross-strapping is continued with each C&DH processor being able to access each of the redundant 1553 buses that provide the data pathways to the IMUs, star trackers, and sun sensors.

Only a few elements of the system are not redundant. These include the RTG; the propulsion system tank, line filter, and orifice; the RF hybrid coupler; and the HGA. Each of these units has a very robust design and a long history of failure-free service.

3 Propulsion Subsystem

The propulsion system (Stratton 2004) includes twelve 0.8-N thrusters, four 4.4-N thrusters, and the hydrazine propellant tank and associated control valves, as shown in Fig. 3. The titanium propellant/pressurant tank feeds the thrusters through a system filter, a flow control orifice, and a set of latch valves that prevent flow of the propellant until commanded to the open position after launch. Helium was selected as the tank pressurant instead of nitrogen to allow the loading of an additional kilogram of hydrazine. Blowdown curves for the system are shown in Fig. 4. Measurements of tank pressure and temperatures at various points in the system allow the mission operations team to monitor system performance and the amount of propellant remaining in the tank.

The 16 rocket engine assemblies (REAs) are organized into 8 sets and placed on the spacecraft as shown in Fig. 5. Pairs of the 0.8-N thrusters (each thruster from a different set) are usually fired to produce torques and control rotation about one of the three spacecraft

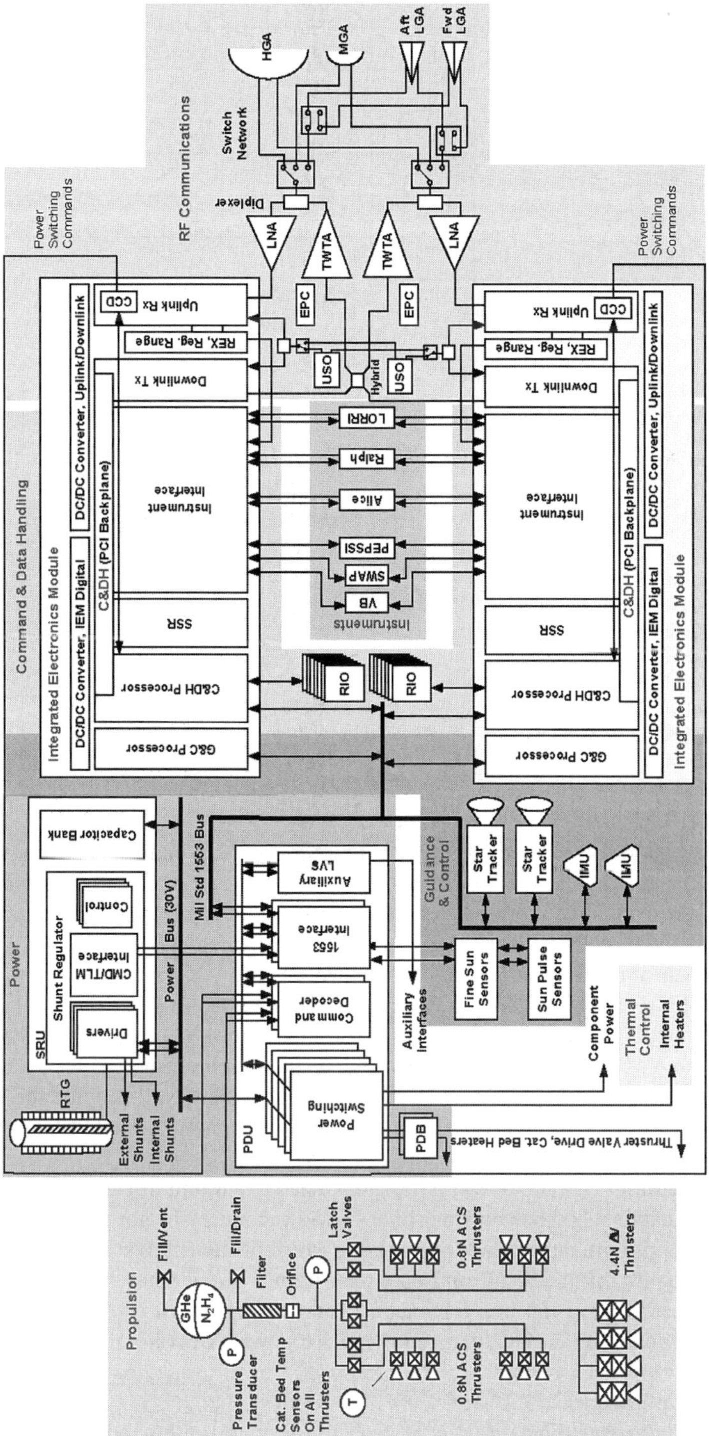

Fig. 3 The New Horizons spacecraft is designed with significant redundancy to ensure that the mission requirements are met after a 9.5-year flight to Pluto

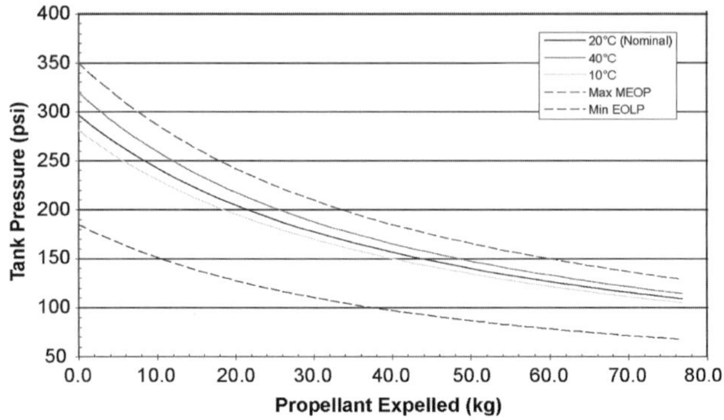

Fig. 4 New Horizons propulsion system blowdown profile

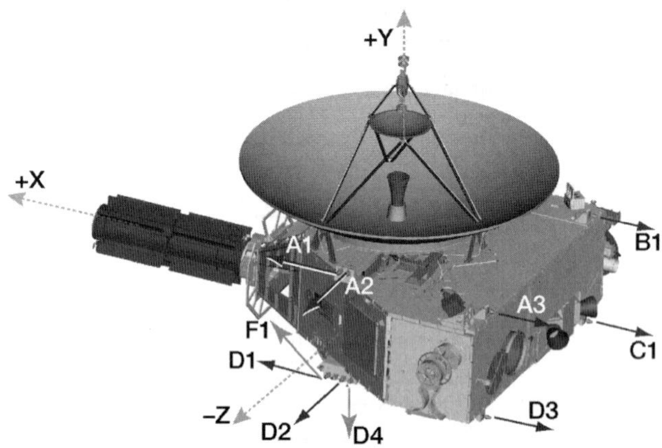

Fig. 5 The propulsion system thrusters are organized in eight sets (six of which are shown in the figure). The sets are distributed such that firing of pairs of 0.8-N thrusters located on opposite sides of the spacecraft produces couples for attitude control. The 4.4-N thrusters are paired in two sets on the $-Y/+Z$ spacecraft edge (F1 & D4 shown) and the $-Y/-Z$ spacecraft edge (not shown) and are used when larger ΔV maneuvers are required

axes. The one exception to the use of coupled thruster firings to control spacecraft rates is that of controlling rates about the spacecraft $\pm X$ axis during science observations, where uncoupled thruster firings are required to meet the maximum spacecraft drift rates allowed during this operational mode. Control rates for each of the spacecraft axes are shown below in Table 2. One pair of the 4.4-N thrusters is aligned along the $-Y$ spacecraft axis to provide ΔV for large propulsive events such as trajectory correction maneuvers (TCMs). The second pair of 4.4-N thrusters is aligned to produce thrust along the $+Y$ axis. These thrusters are rotated 45° in the Y-Z plane to minimize the plume impingement on the HGA dish. The net propulsive effect of these thrusters is therefore reduced. They still provide the required redundancy and the ability to generate thrust in both directions without a 180° rotation of the spacecraft.

Each thruster requires a heater to warm its catalyst bed to a minimum temperature prior to use. Each thruster catalyst bed has both a primary and a secondary heater element, with each element drawing approximately 2.2 W of power. Control of the catalyst bed heater circuits is grouped functionally by pairs (to minimize the number of switches required), so that a total of 16 switches control the heater elements, allowing great flexibility to operate the spacecraft safely while drawing the minimum required power. Operational requirements for catalyst bed heater use are enveloped by the spacecraft power usage, discussed in Sect. 7.3 and shown below in Fig. 10).

The pulse duration and total on-time of each thruster are commanded very precisely, providing accurate control of the total impulse generated during a maneuver. The 0.8-N

Table 1 New Horizons propellant budget allocations at launch

	ΔV (m/s)	Propellant usage (kg)
Primary mission TCM	110	22.3
Altitude control		29.3
Primary mission margin	132	25.2
Original margin allocation	91	17.5
Additional margin obtained from unused spacecraft dry mass allocation	41	7.7
Total navigation ΔV	242	
Total propellant load		76.8

thrusters can be turned on for periods as short as 5 ms, as further described in Sect. 4.2. The initial propellant load was allocated between primary mission TCMs, attitude control (including science and communication operations), and primary mission margin. At the end of the primary mission, sufficient margin may allow for an extended mission to one or more objects in the Kuiper Belt. This allocation is given in Table 1. The original margin was augmented during the final mission preparations when the unused dry mass margin was converted to additional propellant. Due to the excellent performance of the launch vehicle, the total margin has nearly doubled over that budgeted in Table 1.

Nutation is controlled (damped) both passively and actively on New Horizons, depending upon the spacecraft state and the current event executing (e.g. TCM, precession maneuver, etc.). If the spacecraft is performing a TCM or if the spacecraft is in Sun Acquisition state, then nutation damping is exclusively passive. If the spacecraft is precessing, then active nutation control is used only if the nutation angles grows exceedingly large. The size of this threshold is determined by onboard tunable parameters. Near the end of a precession maneuver (last few tenths of a degree), special onboard logic will fire thrusters so that (1) precessing the spacecraft reduces nutation angle and (2) reducing nutation angle will precess the spacecraft in the proper direction. Because the rotational frequency and nutation frequency are different values, it could take several additional minutes to complete a precession maneuver. Since the threshold to actively control nutation angle is a tunable parameter, the threshold could be decreased as propellant is depleted to ensure the nutation angle drops below a pre-determined value in a set amount of time. There should always be some small residual passive nutation damping through the life of the mission. Thus the propellant required for nutation control is very small and subsumed in the value for attitude control.

4 Guidance and Control

The guidance and control system uses a suite of sensors to determine the attitude of the spacecraft, the propulsion system thrusters as actuators, and one of the redundant guidance and control processors to compute the required control actions necessary to meet the commanded attitude state (Rogers et al. 2006). The attitude sensors (see Fig. 3) include the IMUs, the star trackers, and the sun sensors. The guidance and control system is capable of providing spin axis attitude knowledge of the spacecraft to better than $\pm 0.027^\circ$ (3σ) and spin phase angle knowledge within $\pm 0.30^\circ$ (3σ). The same $\pm 0.027^\circ$ knowledge (3σ) is provided for all axes when the spacecraft is in 3-axis mode. The control algorithms must

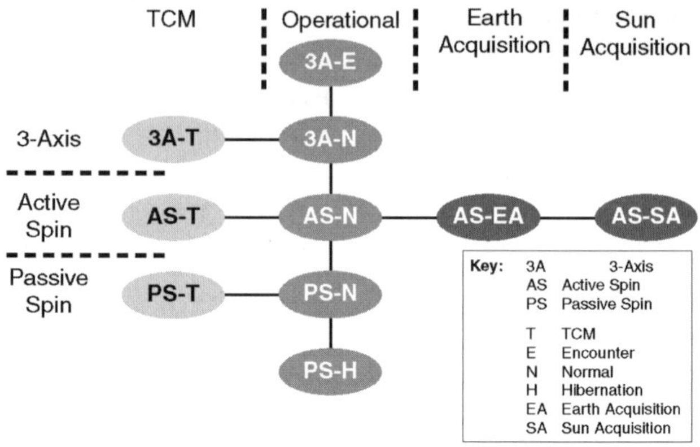

Fig. 6 The guidance and control system provides three operating modes to support TCMs, nominal operations, and two safe states (Earth and Sun acquisition)

maintain the spacecraft attitude to within $\pm 0.059°$ (3σ) and the spacecraft rotational rate to within $\pm 0.0019°/\mathrm{s}$ (3σ).

4.1 Attitude Control Modes

Figure 6 illustrates the various attitude control modes (3-axis, active spin, and passive spin) and the four nominal state classes (TCM, operational, Earth acquisition, and Sun acquisition). TCMs can be conducted in any of the three attitude modes. The choice of mode depends on factors such as the size of the ΔV to be imparted to the spacecraft, the desire to maintain the telemetry link while executing the burn, etc. The various operational states depend on spacecraft activity and other constraints such as the duration of the mode, the need for conservation of propellant, the level of ground monitoring, etc.

At launch, the spacecraft was put in the Passive Spin Hibernation (PS-H) state. This state precludes any onboard attitude control and minimizes the electrical power demand. As the name denotes, it is the primary mode used during the long cruise from Jupiter to Pluto, when there will be a low level of monitoring from the ground and the spacecraft is put into "hibernation". The Passive Spin Normal (PS-N) state also precludes any onboard attitude control activity, but does not limit other power demands. This mode is used during significant portions of the cruise from Earth to Jupiter and at other times in the mission. The Active Spin Normal (AS-N) state is used when an attitude maneuver is required and the spacecraft is spinning. This state is used to maintain the spin rate at its nominal 5 rpm value or to precess the spacecraft to a new orientation. The 3-axis mode states allow the spacecraft to be rotated about any spacecraft axis or desired axis to point in a particular orientation. These modes are used to point a particular instrument to a desired target or to scan a particular instrument field of view across a particular target (see Fig. 7).

The 3-Axis Normal (3A-N) state is used during most instrument activities, including instrument commissioning, engineering tests, and the Jupiter science campaign. The 3-Axis Encounter (3A-E) state is used during the period of closest approach at Pluto where acquiring data has priority. The difference between 3A-N and 3A-E is how the onboard autonomy system reacts to critical faults (e.g., a reboot of the C&DH processor). In the 3A-E state, the autonomy system attempts to correct the fault while keeping the spacecraft in 3A-E state to maximize science data acquisition. Conversely, critical faults in 3A-N state (and other states) result in the autonomy system exercising one of its two "go-safe" chains, Earth Acquisition and Sun Acquisition.

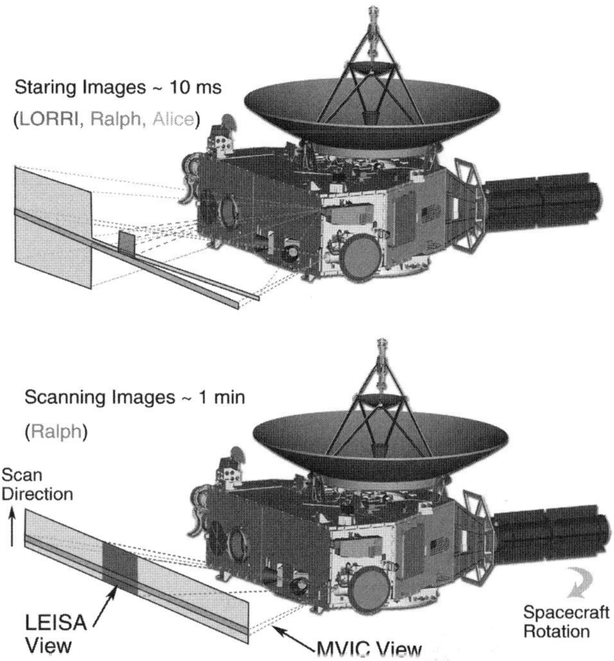

Fig. 7 (**A**) The optical instruments (Ralph, Alice, and LORRI) are aligned such that their fields of view nearly overlap. Only small motions of the spacecraft are needed to image the same area. (**B**) Precise control of the spacecraft attitude is needed both for staring and to provide smooth scanning motion about the +Z axis for Ralph

The Earth Acquisition Go-Safe chain transitions the spacecraft from its current mode and state into Active Spin Earth Acquisition (AS-EA) following the state transitions shown in Fig. 6. Once in Earth Acquisition mode, the autonomy system configures the spacecraft with the HGA pointed to Earth using emergency uplink (7.8125 bits/s) and downlink (10 bits/s) rates.

The second "go-safe" chain, Sun Acquisition Go-Safe, is exercised by the autonomy system as a last resort in the event that communication from the ground cannot be established with the spacecraft while in Earth Acquisition mode. This can be the result of a corrupted onboard ephemeris (which is needed by the spacecraft to determine Earth position), failed star trackers (which are needed for inertial reference), prolonged period of command inactivity (command loss timeout), etc. In Sun Acquisition (AS-EA) mode, the autonomy system configures the spacecraft to orient the HGA towards the Sun and to send a radio signal that a critical fault has occurred on the spacecraft (a so-called "red" beacon tone). Since Sun Acquisition Mode assumes that any Earth knowledge has been lost, the Sun provides the only "guidepost" for the spacecraft pointing. Additional information on the autonomy system is provided in Sect. 9.

4.2 Attitude Control Requirements and Performance

The thrusters generate the control torques that change the attitude motion required to execute the slews to targets and to point or scan the desired instrument toward or across the target. There are two major strategies of observation by the optical instruments:

1. To first point the appropriate instrument field of view (see Fig. 7) and "stare" with the spacecraft rotation limited to values of less than one image pixel during the maximum expected exposure time. This rate limit was determined by the size of the LORRI image pixel and the expected maximum exposure time of 10 ms.

Table 2 The spacecraft G&C system controls the body rates using small torques provided by thruster pairs on the spacecraft. The resulting minimum control body rates, as measured after launch, are provided by the thruster pulse duration of 5 ms

Thruster combination	Average rate change from 5-ms pulse (°/s)
+X axis (A2)	0.0013
−X axis (B2)	0.00092
+Y axis (C1/D1)	0.0014
Y axis (C3/D3)	0.0013
+Z axis (A1/D3)	0.00086
−Z axis (A3/D1)	0.00092

2. To scan about the spacecraft Z axis with a rate control sufficiently precise to allow the Ralph instrument to perform time delay integration of faint images.

Both of these requirements are met if the body rates are controlled with an accuracy of 0.0019°/s.

The thrusters are used in a pulse width modulation mode to achieve the desired control precision, with the IMUs used as the fine rate control sensors. The minimum pulse width available in the system of 5 ms is a value set by design based on the actuation time constants of the thruster control valves. The actual performance of the guidance and control system, as measured during the commissioning tests in the spring of 2006, are given in Table 2. These numbers improve (decrease) as the feed pressure of the spacecraft propulsion system decreases.

5 Command and Data Handling

The Command and Data Handling (C&DH) functions—command management, science and engineering data management, timekeeping, and the capability to provide autonomous system recovery and safing—are implemented with the resources within the two redundant IEMs (see Fig. 3):

- C&DH processor card
- SSR card
- Instrument interface card
- Critical command decoder on the uplink card
- Downlink formatter on the downlink card

Communication between cards within the IEM is over a peripheral component interconnect (PCI) backplane. Remote input/output (RIO) units provide temperature and voltage measurements for monitoring the health and safety of the spacecraft. These RIO units communicate with the C&DH using an inter-integrated circuit (I^2C) bus. A MIL-STD-1553 serial bus interface is used between the IEMs to allow for redundancy and cross-strapping as well as the transfer of commands and data between G&C components and the G&C processors.

The C&DH system consists of these hardware resources and the software running on the C&DH processor. The processor in IEM 1 is designated C&DH 1 and the processor in IEM 2 is designated C&DH 2. During normal operations, the C&DH section of one IEM is fully powered and designated as prime, the so-called "bus controller", while the second unit is powered in a standby mode. During the long cruise between Jupiter and Pluto, this second unit will be powered off. The fault protection system monitors, from multiple locations, a pulse-per-second "heartbeat" from the bus controller. If the heartbeat is interrupted for a period of 180 seconds, the redundant C&DH will be powered on and take command as the bus controller. In addition, other safing actions will be taken as described in the section on fault protection and autonomy.

5.1 Command Management

Ground commands are received via the RF uplink card. A critical command decoder on the uplink card provides a simple, robust mechanism to receive and execute commands without the operation of the C&DH processor. These include C&DH processor resets, switching of bus controllers, and all other power switching commands. The C&DH software receives CCSDS (Consultative Committee for Space Data Systems) telecommand transfer frames from the ground via the RF system uplink card. The C&DH software extracts telecommand packets from the transfer frames. These telecommand packets are delivered to other subsystems on the spacecraft or to the C&DH processor itself. The telecommand packets that the C&DH processor receives can contain commands to be acted upon in real time or commands to be stored for later use. Commands intended for later use are designated as command macros. Command macros can consist of a single simple action (such as commanding power on or off to a subsystem) or a string of switching commands and information (data) for use by the various subsystems. Each C&DH subsystem has 0.75 Mbytes of storage for command macros.

The C&DH subsystem supports time-tagged rules. These trigger and execute one or more of the command macros when the value of the onboard mission elapsed time (MET) is greater than or equal to the value of the MET specified in the rule. Mission Operations uses time-tagged rules to implement almost all onboard activities, including science activities at Jupiter and Pluto. There is storage for 512 time-tagged rules.

The C&DH software implements the onboard autonomy system. An autonomy rule is a postfix expression whose inputs can include any spacecraft telemetry point. An autonomy rule is triggered when it has evaluated to "true" for the required number of times in a given interval. A triggered autonomy rule also executes one or more of the command macros. There is storage for 512 autonomy rules. The autonomy system is further described in Sect. 9.

5.2 Time Management

The accuracy of the correlation of MET to Universal Time (UT) is of major importance to support navigation, G&C activities, and the collection of science data. The ultra-stable oscillator (USO) used as the onboard source of a 1 pulse per second (PPS) signal maintains the spacecraft time base. Careful design of the ground to MET clock register in the C&DH system ensures onboard correlation of better than ±4 seconds when the spacecraft is at Pluto (a 9-hour round trip light time delay). Science instrument time correlation (post-facto) requirements are ±10 ms for REX, LORRI, Ralph, and Alice. Post-launch measurements in July 2006 verified that the post-facto timing correlation is well within this requirement.

5.3 Data Management

The instrument interface card communicates to each of the instruments. Communication consists of:

1. Providing commands and MET from the IEM to each instrument

2. Collecting housekeeping data and low-rate science data from the instruments to the IEM
3. Receiving and formatting instrument high-rate science data and IEM-generated header data
4. Distributing spacecraft time markers to each instrument
5. Processing and transferring IEM analog telemetry voltages (data input from the RIO devices using the I^2C bus)

Each instrument interface card (in IEM 1 and IEM 2) provides connections to both sides of each instrument's redundant electronics, thereby providing significant cross-coupling and increased reliability.

The SSR consists of 64 Gbits of non-volatile storage organized in 16 independently addressable segments corresponding to the 16 physical memory stacks on the SSR card. Raw science data can be streamed onto the recorder at rates up to 13 Mbits/s. Once a segment is filled, the recorder switches to the next segment. Raw science data is read off the SSR, compressed, and written back to the recorder for later transmission. Erasing of the SSR is done on a segment basis, after all data in that segment are either transmitted or compressed and stored to another segment. Once a segment is erased it is available for storage of new data. The C&DH software provides several mechanisms to control the writing and playback of data onto the recorder. The recorder uses the concept of data types to allow differentiation of, and thereby control, the data. There are a total of 51 data types, including raw (non-packetized) high-speed science data; compressed versions of the science data, with each type of compression of a particular type of science data resulting in a different data type; data from instruments that produce low-speed, packetized science data; and instrument and spacecraft housekeeping data.

The C&DH software also implements SSR bookmarks. Bookmarks allow access of data on the SSR associated with a specific activity. Bookmarks are opened at the start of the activity and closed at the end of the activity. Data types are included in the bookmark. Bookmarks are a substitution for using MET to define the data to be compressed or played back. All commands to compress or play back data can be specified using either bookmarks or MET. The priority of the data types to be played back from the SSR can be specified. This applies both to a playback by bookmark or a playback by MET.

The New Horizons software implements both lossless and lossy compression. Non-packetized science data are read off of the SSR, compressed and formed into CCSDS packets, and written back to the SSR. There is also the option to read the non-packetized science data off of the SSR, form the data into CCSDS packets without doing any sort of compression, and write the data back to the SSR.

Lossless compression can be combined with subframing, or windowing. Rather than performing the lossless compression on the entire image, it is possible to specify up to eight subframes of the image and then perform the lossless compression on the data within these subframes.

A special method of subframing is applied to images collected by the Linear Etalon Imager Spectral Array (LEISA) infrared sensor within the Ralph instrument. This is known as dark-sky editing. A LEISA frame consists of one spatial dimension (X) and one wavelength dimension (Y). Observations are made with the spacecraft oriented such that a rotation causes the observed object to pan across the focal plane at a rate of approximately one spatial pixel per frame. For each frame, a subframe can be defined such that the number of pixels that most likely contain the object is determined and stored on the SSR. A single subframe "walks" across a sequence of LEISA images. The original Y offset of this sub-frame, the direction it moves from image to image, and the pixel rate it moves from image to image can be specified, as well as the size of the subframe.

6 Communication System

The New Horizons RF telecommunication system (DeBoy et al. 2005) provides the command uplink, telemetry downlink, and essential elements of both the REX instrument and radio navigation capabilities. The system is designed to provide communications with the Earth using the Deep Space Network (DSN) in both spinning and 3-axis stabilized attitude control modes. The system (see Fig. 3) consists of an antenna assembly, RF switch network, hybrid coupler, redundant TWTAs, redundant USOs, and the uplink and downlink cards located in the redundant IEMs.

The antenna system includes both a forward antenna system (aligned with the +Y axis of Fig. 1) and a hemispherical low-gain antenna (LGA) mounted below the spacecraft (not shown in Fig. 1) and aligned along the −Y axis. The forward antenna assembly (Schulze and Hill 2004) consists of the +Y hemispherical-coverage LGA, and two co-aligned parabolic antennas. The 2.1-m HGA was designed to meet a requirement for a minimum of 600 bits/s downlink telemetry rate at 36 AU to return the Pluto data set. The HGA provides better than 42 dBic gain for angles within 0.3° of the +Y axis. The secondary reflector assembly consists of a medium-gain antenna (MGA) and the HGA subreflector. The MGA allows communication at larger angles between the +Y axis and Earth (up to 4°), and specifically allows commands to be received by the spacecraft at ranges up to 50 AU. The two LGAs provide communication with Earth at any attitude orientation early in the mission. The LGAs are capable of maintaining two-way communications up to distances of approximately 1 AU from Earth. Beyond that distance (reached in the spring of 2006), only the MGA and HGA can be used.

6.1 Command Reception and Tracking

The uplink card provides the command reception capability as well as a fixed down-conversion mode for the uplink radio science experiment (REX). The uplink command receiver utilizes a low-power digital design that significantly reduces the power consumption of this critical system. Previous deep space command receivers consume ~12 W of primary power; the digital receiver developed for New Horizons consumes approximately 4 W. Since both receivers are typically powered on, the total power savings is 16 W, a mission-enabling achievement for New Horizons, with its very limited total power budget of ~200 W. The uplink card also provides critical command decoding, ranging tone demodulation, X-band carrier tracking, and a regenerative ranging subsystem. New Horizons also flies non-coherent Doppler tracking and ranging developed by APL (Jensen and Bokulic 2000), implemented largely on the uplink card and capable of providing Doppler velocity measurements of better than 0.1 mm/s throughout the mission.

Range tracking of interplanetary spacecraft is normally accomplished by sending tones phase-modulated on an RF carrier from the DSN to the spacecraft, receiving the signal with a wide-bandwidth spacecraft receiver, and retransmitting that signal back to the DSN stations. There the retransmitted signal is processed to determine two-way round trip light time and thus the distance to the spacecraft. The primary error source in this system is the received uplink noise accompanying the ranging modulation signal on the spacecraft, which is also amplified and retransmitted back to the DSN. In addition to supporting standard tone ranging, the New Horizons communication system has incorporated a regenerative ranging circuit (RRC) to limit this turnaround uplink noise. This implementation uses a delay-locked loop (DLL) that generates an onboard replica of the received ranging signal and adjusts the timing of the onboard signal to align it with what the spacecraft receives. The signal transmitted to the ground is free of wideband uplink noise, significantly reducing the primary error source contribution. This will enable New Horizons to determine spacecraft range to a precision of less than 10 m (1σ) from near-Earth ranges to beyond 50 AU, or achieve mission range measurement accuracy for integration times orders of magnitude shorter than those for sequential ranging.

6.2 Ultra-Stable Oscillator Performance

The New Horizons USO is a critical component of the RF telecommunication system, providing the precision reference frequency (30 MHz) for the uplink, downlink, and the radio science experiment (REX). The USO is built at APL, and is based on heritage of systems developed over the last 30 years for missions such as Mars Observer, Cassini, GRACE, and Gravity Probe B. New Horizons carries two USOs. Each USO is a pristine version of an ovenized crystal oscillator. Short-term frequency stability (Allan deviation) at 1-second and 10-second intervals is better than 3×10^{-13} and 2×10^{-13}, respectively. This stability, and the USO low-output phase noise (<-125 dBc/Hz at 100 Hz), is crucial to REX.

6.3 Downlink System Performance

The downlink card in each IEM is the exciter for the TWTAs and encodes frame data from the spacecraft C&DH system into rate 1/6, CCSDS turbo-coded blocks. It also calculates and inserts navigation counts into the frame data to support the noncoherent Doppler tracking capability. In addition, it is used to transmit beacon tones during the hibernation cruise period. The redundant TWTAs are the high-power amplifiers for the downlink signal. The hybrid coupler connects the downlink exciter outputs and the TWTA RF inputs. This allows either TWTA to be connected to either downlink card. The two TWTA outputs are connected via the RF switch assembly to the antennas. The network allows both TWTAs to simultaneously transmit through the HGA (if there is sufficient spacecraft power), with one transmitting a right hand circular (RHC) polarized signal and the other a left hand circular (LHC) polarized signal. The DSN has the capability to receive both signals and combine them on the ground to enhance the received signal-to-noise ratio and thereby increase the data rate by approximately 1.9 times above that using a single TWTA. The resulting downlink data rate (which includes 2 dB of margin in the link calculation) is shown in Fig. 8.

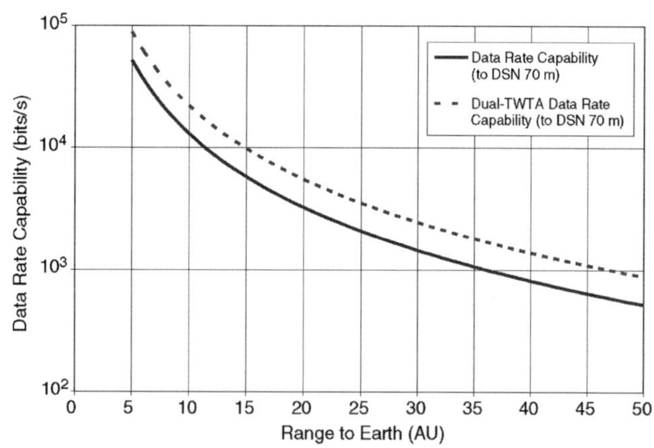

Fig. 8 The RF system provides downlink telemetry above 1 kbit/s at Pluto and above 700 bits/s at 50 AU. A mode using both the redundant TWTAs at once and combining the signals on the ground increases the data rate such that over 900 bits/s can be transmitted at 50 AU

Table 3 Spacecraft power system summary. Numbers in parentheses are the spare services within the power system at launch

RTG beginning of mission power	245.7 We
RTG minimum power at Pluto	180.0 We
Power shunt rating	312 W
Regulated bus voltage	+30 V ± 1 V
Switched loads	86 (12)
Thruster/latch valve drivers	56 (4)
Pulsed loads	148 (16)
Opto-isolated contact closures	8 (8)
Spacecraft configuration relays	24 (1)
Digital telemetry	1.22 kBytes
Analog telemetry	0.58 kBytes

The downlink system will guarantee that the entire Pluto data set (estimated to be 5 Gbits after compression) in 172 days with one 8-hour pass per day using the DSN 70-m antennas. If there is sufficient power, such that both TWTAs can be used, the time to downlink the data set can be reduced to less than 88 days.

7 Power System

The power system supplies and distributes power to the spacecraft subsystems and the instruments. It also provides hardware redundancy and fault protection. The key components of the power system (see Fig. 3) are the general purpose heat source radioisotope thermoelectric generator (GPHS-RTG), the shunt regulator unit (SRU), the external power dissipation shunts, the PDU, and the propulsion diode box (PDB). Table 3 gives a functional summary of the power system.

7.1 Radioisotope Thermoelectric Generator Performance

The "F8" GPHS-RTG supplied by the U. S. Department of Energy (Harmon and Bohne 2007) is the latest in a series of RTGs of the same design supplied for NASA missions beginning in the late 1980s. The unit converts heat generated by radioactive decay of the plutonium heat source into electricity using silicon-germanium (SiGe) thermocouples (designated "unicouples" because of the close physical proximity of the "hot" and "cold" junctions of the thermocouple). The plutonium heat source is made up of 72 individual pellets of plutonium dioxide. Each pellet is encased in an iridium cladding and assembled (four to a module) into the 18 heat source modules that make up the total RTG inventory. The RTG assembly is illustrated in Fig. 9. The fully fueled RTG had 132,465 curies of plutonium at launch with a half-life of 87.7 years (NASA 2005). The thermal output of the heat modules was 3,948 W at launch (Cockfield 2006), which is radiated to space during flight. Results of spacecraft RTG integration and early mission performance tests have been flawless, with the F8 normalized power output tracking above its contemporaries (Ottman and Hersman 2006).

7.2 Shunt Regulator Unit

The SRU is the power interface with the RTG. It is designed to maintain the bus voltage to 30 V by dissipating the excess RTG power in either external resistive shunts (waste

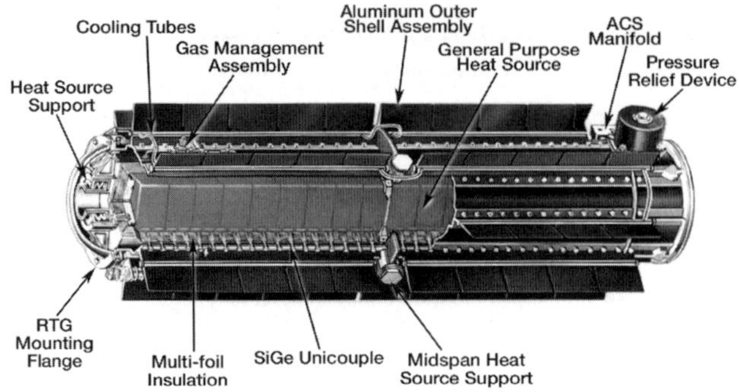

Fig. 9 The GPHS-RTG has 9.75 kg of plutonium dioxide distributed in 18 GPHS modules. The modules are designed to contain the plutonium in case of a launch accident. The aluminum outer assembly radiates the heat from the modules into space. The unicouples generate electricity based on temperature differential between the modules and the outer assembly. Prior to launch, the assembly is filled with an inert gas to keep the unicouples from oxidizing. Once in space, the gas pressure relief device allows the gas to escape and the RTG then operates at maximum efficiency (Image courtesy of the Department of Energy)

heat is radiated to space) or internal spacecraft heaters, depending on the thermal control needs. The three SRU loop controllers use a majority-voted design to control the 16 sequential analog shunts that maintain a constant load on the RTG during spacecraft operational mode changes. Each full shunt is capable of dissipating 19.5 W, with the first two shunts steerable to internal heaters in half-power increments. The shunts were designed to be $n + 1$ redundant with a fully fueled RTG and can be individually disabled if a fault is detected. Relays within the SRU maintain the configuration of the shunts, internal heaters, or external dissipaters. The spacecraft bus voltage is set at 30.25 V, which is slightly above the maximum power transfer level of the RTG, to account for voltage drops as the power is distributed to the individual loads. The SRU also contains a 33.6-mF capacitor bank that provides for short duration current surges at load turn-on or fault condition.

7.3 Power Distribution Unit and Propulsion Diode Box

The PDU contains fully redundant solid-state power switching, pulsing, and monitoring functions and hardware-based fault protection features for the spacecraft. The communication interface to the IEM is through the redundant 1553 interfaces with redundant universal asynchronous receiver/transmitter (UART) serial links for the passage of critical commands and telemetry. Loads such as the IEM, command receiver, USO, and the PDU 1553 board are configured as critical loads with both the primary and the redundant units powered on by default. Software and hardware enabling is required to turn off any of the spare units. Dual-level low-voltage sensing sheds non-critical loads if the bus voltage drops below the set points (nominally 28.5 V and 27.0 V, respectively). Additional bus protection features within the PDU also address the power-limited nature of the RTG power source. The circuit breakers have selectable levels for tight load monitoring and allow multiple attempts at power cycling the loads. Loads are fused as a final protection measure. The PDU also contains spacecraft configuration relays, sun sensor interface, sensing of load telemetry (both voltage and current) and temperature telemetry from the power system. The configuration relays provide a non-volatile means of storing spacecraft state information. They are used by the autonomy system in certain fault conditions (e.g., temporary loss of power or a switch

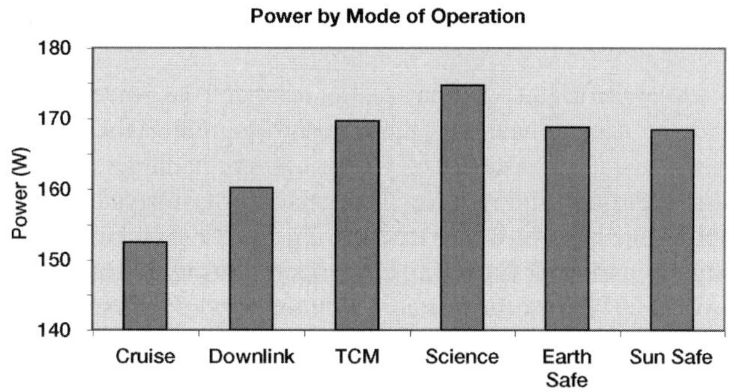

Fig. 10 Spacecraft and instrument power for each of the operating states were tailored to ensure safe power margins through the primary mission

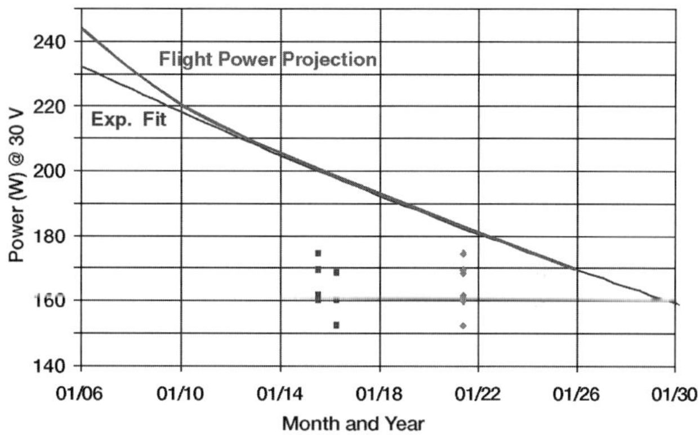

Fig. 11 (Color online) The RTG power output decreases with time, primarily as a result of the plutonium decay half-life of 87 years. The generator supplies sufficient power even at 50 AU (~ 2021) to support the power requirements for all necessary spacecraft states for an encounter and data playback (*blue squares* indicate Pluto and *gray diamonds* indicate a Kuiper Belt object at 50 AU)

to the redundant C&DH system) to assure that critical spacecraft state information is maintain onboard. The PDB is the power system interface between the PDU and the propulsion thrusters, latch valves, and the catalyst bed heaters.

Figure 10 shows the power for each of the nominal operating modes. The spacecraft power requirements vary as different subsystems are powered to perform various functions. The requirement is smallest during the cruise mode, when most instruments and some subsystems such as the transmitter and IMUs are powered off. Data return after the Pluto encounter is performed in a spinning mode. The downlink power assumes that only one transmitter is on. If the second TWTA is powered to increase the data rate downlink, as discussed in Sect. 6, the spacecraft must be in a passive spin state so that other components, such as the catalyst bed heaters, can be powered off to accommodate the additional 31 W. TCM operations can be performed in several ways (see Sect. 4.1); the minimal power mode shown here is the active spin mode. Science operations at Pluto are performed in the 3-axis control mode, with only those instruments powered that support a particular observation (the maximum is shown here). The two safe modes (Earth Safe and Sun Safe) require the propulsion system's catalyst bed heaters to be powered, thus requiring the same power as a TCM. Figure 11 illustrates the expected power provided by the RTG over the lifetime of the mission and the required power for each of these modes relative to that power. Thus, New Horizons can expect to operate with significant power margin for the Pluto encounter and adequate margin for a Kuiper Belt object encounter at up to 50 AU.

8 Thermal Management

The New Horizons thermal design balances the power and waste heat provided by the RTG and the natural heat loss through the thermal blankets, instrument apertures, and control mechanisms to ensure each of the system elements remains within safe operating temperatures through the mission. The spacecraft subsystems are contained within the thermos bottle-like core of the spacecraft. The average internal spacecraft temperature varies from slightly under 40°C (during early operations with the lower deck facing the Sun at 1 AU) to around 20°C over the mission. Key features of the New Horizons thermal design are shown in Fig. 12.

The "thermos-bottle" design ties all the subsystems and components to the bulk spacecraft temperature. All the propulsion system components, including the rocket engine assemblies, are thermally tied to the spacecraft bus and are kept warm through thermal contact with the structure. The instruments are all thermally controlled based on conduction to the bus. Only the HGA and the star trackers are thermally isolated from the spacecraft bus.

The multilayer insulation blankets surrounding the spacecraft provide the dominant heat leak. The design uses the internal electronics dissipation and approximately 15 W of waste heat from the RTG to support the internal temperatures. The minimum allowable spacecraft heat leak was measured during spacecraft thermal-vacuum testing to be about 105 W with a 10°C bus temperature. Early in the mission, louvers actuated when the internal box temperatures exceeded 25°C to keep the internal temperature from getting too high. These events occurred when periods of the high internal dissipation combined with spacecraft attitudes that produced large solar loading. Since then, the bus has been running at nearly room temperature based on the onboard thermal management system.

The thermos-bottle design approach allows the spacecraft heater system to be shifted away from temperature control to control based on managing the total power level in the spacecraft. Because of the 9-hour round trip light time to Pluto, and because available electrical power is limited, it was necessary to combine the thermal and power management functions to simplify the overall design. The onboard heater control software monitors the power dissipated in the spacecraft and turns on various heaters to keep the total power close

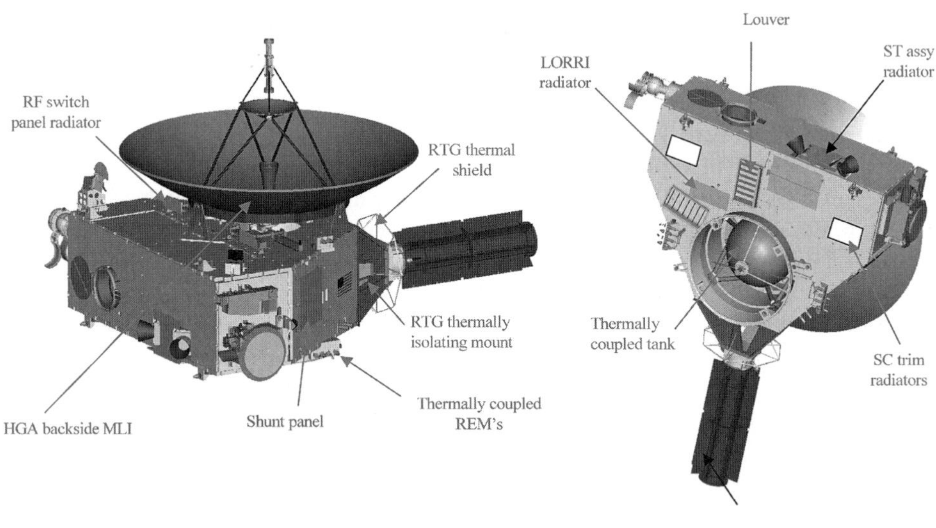

Fig. 12 Key thermal design features of the New Horizons spacecraft

to a commanded, target value. There are 2.5-W, 5-W, 10-W, and 20-W heaters spread around the central spacecraft core. The locations of the specific heaters are less important than the total power input to the system.

Because some components, such as the TWTAs and catalyst bed heaters, do not dissipate all their power inside the spacecraft, an allowance was made in the software to apply only a fraction of those loads to the measured internal dissipation. The correct functioning of this heater control software was verified in thermal vacuum testing. During simulated mission operations, the total power dissipation goes through a defined profile, as the various boxes are turned on and off. The onboard heater software finds the total internal power dissipation by adding the different loads, including the reductions for those loads where the power is dissipated outside the thermal envelope. Then, heaters are turned on to make up the difference between the measured and targeted value. The total power dissipated within the spacecraft is the combination of the box dissipation and the heater power.

9 Autonomy and Fault Protection

The New Horizons mission has the longest primary mission duration to date. The primary science goal can only be achieved after a 9.5-year journey culminating in a complex set of observations requiring significant time to transmit the data to Earth. During spacecraft development, much thought and energy were devoted to fault protection. This effort continues as the operations team evaluates in-flight mission performance. The fault protection architecture uses the redundancy of the spacecraft system (shown in Fig. 3) if off-nominal operation is detected. Basic elements of fault protection are resident in redundant elements of the PDU. The PDU monitors C&DH bus traffic and will automatically switch to the alternate C&DH system if it detects that nominal C&DH processor activity of the controlling system has stopped. The major elements of fault protection are implemented by software running on the controlling C&DH processor. This software is the principal component of the autonomy subsystem. The software evaluates telemetry data in real time and, based on the evaluation, takes one or more of the following actions:

1. Execute a set of commands to correct a detected fault.
2. Generate a "beacon tone" to alert operators that an event on the spacecraft requiring attention has occurred.
3. Execute one of two "go safe chain" command sets, which puts the spacecraft into either an Earth Safe or a Sun Safe state, as described in Sect. 4 (in the event of a critical fault).

The evaluation of onboard data is performed by a set of rules that check for data that exceed defined limits for a period of time. The time period (or persistence) of the exceedance varies from rule to rule. The persistence length minimizes the chance of a rule "firing" on noisy data, or on transient data that occur during a commanded change in spacecraft pointing. Processors (other than the C&DH processor, whose activity is monitored by the PDU circuitry) are monitored via a set of "heartbeat" rules that use a telemetry point to determine if the processor is stuck at either a "one" or a "zero" state. The persistence of each of these heartbeat rules is adjusted as appropriate to match the nominal operation of the specific processor. The autonomy software can also compute dynamic limits. For example, the autonomy system monitors the propulsion system for potential propellant leaks. The system monitors the propellant as a function of both the pressure and the temperature of the propellant tank using the ideal gas law to compute a current volume and compares it to an initial value set at a previous time appropriate to the phase or mode of spacecraft operation. At the time of launch, the autonomy system used 126 rules to determine the state of health of the spacecraft.

The command sets are organized as user-defined macros and stored in memory space defined by the C&DH system. The macros can include any allowable C&DH command and can be used to power units on or off, change spacecraft modes, enable or disable autonomy rules, or execute other macros. These macros can be executed by either real-time commands or by the autonomy subsystem. The macros can also be executed by time-tagged commands, allowing the commands in the macros to execute at a specific time in the future. The autonomy subsystem used 132 macros at launch. This set has been modified as the spacecraft position along its trajectory has changed and will continue to be modified as different phases of the mission occur, system performance changes, and operational experience dictates.

The capabilities of the autonomy system are used to support a number of mission operation tasks as well as providing fault protection. For example, the "command load" sequences generated by the mission operations team are loaded into one of two memory segments. Upon the completion of one sequence, an autonomy rule is used to switch to the next sequence. The autonomy rules also check to see that an appropriate sequence has been loaded into the second memory segment, and if it has not, a rule fires causing the system to enter the "go safe" chain and point to Earth. And, as discussed in Sect. 4.1 Attitude Control Modes, in the Encounter mode, the autonomy system will attempt to correct any onboard fault and continue the observation timeline.

10 Performance and Lifetime

For mission success, the New Horizons spacecraft must continue to function through post-Pluto data playback. As discussed above, the system design included numerous element redundancies to increase the probability of proper operation beyond that of a single-string design. A probabilistic risk assessment (PRA) was carried out to guide overall system architecture design and to support trade studies, with initial results given at the concept study review and final results several months prior to launch. While the result is not intended to be a true statistical measure of the probability of success for the New Horizons mission, setting a design goal of 85% and tracking the PRA result through the design cycle gave assurance that risk was appropriately managed. The results of the final PRA (Fretz 2005) are shown in Fig. 13 for the best- and worst-case launch dates and mission lifetimes for minimum science and full science. The bands in Fig. 13 give the 90% confidence limits for each calculation, with the mean shown as the vertical line in each distribution. The mean probability of success remains above the design goal of 85% for the actual mission duration, and is acceptable for all possible mission durations based on changes in launch date.

Prior to launch, the sensitivity of the mean probability of success to changes in failure rates for individual system elements was examined, in part to understand the effect on the probability of success of unresolved anomalies from integration, test, and pre-launch activities. In addition, this sensitivity analysis studied how the probability of success estimates depend on the accuracy of the failure rates for individual system elements. The results from altering the failure rates fall well within the uncertainty range for the PRA results in Fig. 13, indicating that the impact of unresolved issues or inaccurate element failure rate estimates on the overall mission reliability was negligible.

Management of both limited-life system elements and use of consumable resources onboard the spacecraft is essential to achieving these lifetime probabilities. Limited-life items are listed in Table 4. The limited-life items are discussed in three categories:

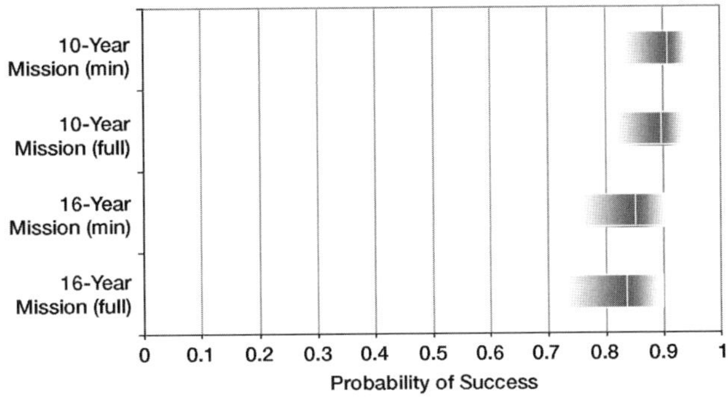

Fig. 13 An analysis was performed to determine the probability of the New Horizons system design to successfully perform the mission and meet the science objectives. With the successful launch early in the 2006 launch window, the probability of meeting the full mission requirements are well above the mission requirements and the probability of mission success in an extended mission (through the Kuiper Belt to at least 50 AU) is also very high

Table 4 New Horizons limited-life items

Latch valve cycles
Thruster valve cycles
Telecom system RF switch cycles
PDU configuration relay cycles
Alice instrument aperture door cycles
Ralph instrument side select relay cycles
Instrument processor non-volatile memory programming cycles
TWTA filament on-time
IMU leak rate
IMU on-time/laser intensity
IEM Processor flash memory programming cycles
SSR EEPROM/flash memory erase cycles

1. Items exercised very few times under normal operations (e.g., latch valves, RF switches, PDU configuration relays, instrument door, relay, and non-volatile memory programming cycles)
2. Items that are purely time dependent (e.g., IMU leak rate)
3. Items used regularly under normal operations that must be carefully managed (thruster valve cycles, TWTA and IMU on-time/laser intensity, and IEM and SSR EEPROM/flash memory cycles)

The latch valves that supply attitude control thrusters were commanded to the open state shortly after launch and there are no plans to operate them again. The latch valves that supply ΔV thrusters are opened only for TCMs, resulting in a very small number of cycles. RF switches are operated to check out the redundant elements of the RF system. Some operations remain as the communication link uses the MGA or HGA as the mission continues, but these operations are few in comparison to the lifetime capabilities of the switches. The same is true for the operation of the PDU configuration relays, instrument door, instrument relays, and non-volatile memory programming cycles, where the operational limit exceeds the expected usage by orders of magnitude.

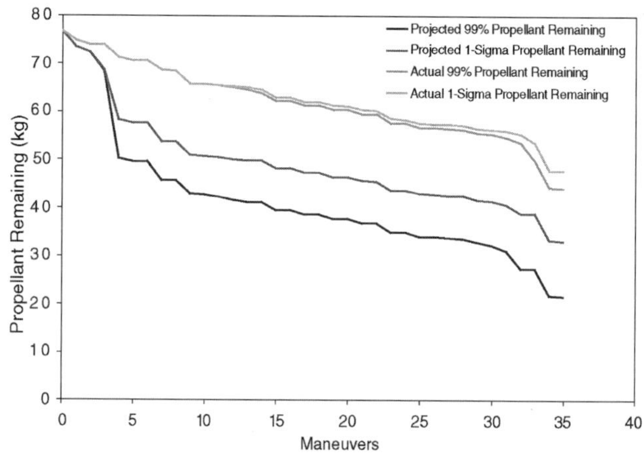

Fig. 14 The hydrazine propellant is the only onboard consumable for the mission. Consumption analysis included 35 events through Pluto data return that required expenditure of hydrazine. The pre-launch prediction indicated that a good margin would remain available (>20 kg with a 99% probability). Because of the excellent trajectory provided by the launch vehicle that number has increased to over 40 kg (figure courtesy of Stewart Bushman)

The IMU leak rate was measured before launch and was determined to be substantially small, such that it provided margin greater than 10 times the requirement. Information on gross leak rates is available via telemetry No change in the rate has been observed during the first year of flight.

The TWTA filaments are rated at 260,000 hours. The planned operating time through the Pluto downlink is less than 45,000 hours. Thus there is sufficient margin in the use of a single TWTA to meet mission requirements.

The IMU operating hours are limited by the operating life of the gyro lasers, due to degradation in the laser intensity. These units were carefully selected during manufacturing to ensure lifetimes for all 3-axis measurements exceeding 42,000 hours. Pre-launch estimates of the projected lifetime actually exceeded 52,000 hours. These units operate a significant amount of time during the early cruse to Jupiter and during the Jupiter encounter. After Jupiter, the spacecraft spends most of the next 7 years in the hibernation mode, during which the IMUs are off. By this means, the planned operating time on each IMU is less than 23,000 hours.

The thruster valve cycle limit of 273,000 was adopted from the limit used on the Cassini program. This limit was derived from qualification tests, which demonstrated performance beyond 409,000 cycles. New Horizons uses the same thruster as the Cassini and Voyager missions. The Voyager spacecraft, which experienced its only failure at over 500,000 cycles, provides valuable in-flight data to increase confidence in the thruster valve cycle performance. New Horizons plans to cycle each thruster valve approximately 200,000 times, well under the 273,000 cycle limit.

The New Horizons spacecraft is fully functional with significant margins for both consumables (see Fig. 14) and limited-life resources to support the full mission requirements at Pluto. The initial cruise phase and Jupiter encounter provide many opportunities to fully exercise the system elements and to allow the mission operations team to demonstrate the kind of operational activities planed for the Pluto reconnaissance. With the excellent launch and the power supplied by the RTG, the spacecraft is expected to operate productively well into the Kuiper Belt, providing NASA and the Science Community the opportunity to explore this region after the 2015 Pluto rendezvous is completed.

Acknowledgements The New Horizons spacecraft is an accomplishment that represents the work of many hundreds of individuals and many organizations. The New Horizons spacecraft development was performed at The Johns Hopkins University Applied Physics Laboratory under NASA Contract NAS5-97271, under the leadership of Dr. S. Alan Stern, the New Horizons Principal Investigator. In addition, Dr. Stamatios M. Krimigis provided essential leadership during the early phases of the program, when the mission was vulnerable to cancellation, until its value was re-enforced by the National Academy study of 2002.

References

U. Carlsson, in *Seventh European Space Power Conf., SP-589*, vol. I, 2005

A. Cheng et al., Space Sci. Rev. (2008, this issue). doi:10.1007/s11214-007-9271-6

R.D. Cockfield, in *4th Int. Energy Conversion Eng. Conf. Exhibit (IECEC)*, San Diego, CA, 26–29 June 2006

C.C. DeBoy et al., Acta Astron. **57**, 540–553 (2005). doi:10.1016/j.actaastro.2005.03.051

K. Fretz, Internal Memorandum SEA-2005-075, The Johns Hopkins University Applied Physics Laboratory, Laurel, MD, 2005

B.A. Harmon, W.A. Bohne, Space Technol. Applications Int. Forum, Albuquerque, NM, 11–15 February 2007

M. Horanyi et al., Space Sci. Rev. (2008, this issue)

J.F. Jensen, R.S. Bokulic, IEEE Trans. Aerosp. Electron. Syst. **36**(4), 1401–1406 (2000). doi:10.1109/7.892689

D. Kusnierkiewicz et al., Acta Astron. **57**, 135–144 (2005). doi:10.1016/j.actaastro.2005.03.030

D. McComas et al., Space Sci. Rev. (2008, this issue). doi:10.1007/s11214-007-9205-3

R.E. McNutt et al., Space Sci. Rev. (2008, this issue)

NASA: Final Environmental Impact Statement for the New Horizons Mission, National Aeronautics and Space Administration, July 2005

G.K. Ottman, C.B. Hersman, in *4th Int. Energy Conversion Eng. Conf. Exhibit (IECEC)*, San Diego, CA, 26–29 June 2006

D. Reuter et al., Space Sci. Rev. (2008, this issue)

G.D. Rogers et al., *Autonomous Star Tracker Performance*. Valencia Papers, 2006

R.C. Schulze, S. Hill, in *IEEE Aerospace Conf.*, Big Sky, Montana, vol. 2 (2004), pp. 966–974

S.A. Stern, Space Sci. Rev. (2008a, this issue)

S.A. Stern, Space Sci. Rev. (2008b, this issue). doi:10.1007/s11214-007-9295-y

J. Stratton, in *55th Int. Astronautical Congress of the Int. Astronautical Federation*, Int. Academy of Astronautics, Int. Institute of Space Law, Vancouver, Canada, 4–8 October 2004

L. Tyler et al., Space Sci. Rev. (2008, this issue). doi:10.1007/s11214-007-9302-3

H.A. Weaver et al., Space Sci. Rev. (2008, this issue)

K.-P. Wenzel, R.G. Marsden, D.E. Page, E.J. Smith, Astron. Astrophys. Suppl. Ser. **92**(2), 207–215 (1992)

New Horizons Mission Design

Yanping Guo · Robert W. Farquhar

Originally published in the journal Space Science Reviews, Volume 140, Nos 1–4, 49–74.
DOI: 10.1007/s11214-007-9242-y © Springer Science+Business Media B.V. 2007

Abstract In the first mission to Pluto, the New Horizons spacecraft was launched on January 19, 2006, and flew by Jupiter on February 28, 2007, gaining a significant speed boost from Jupiter's gravity assist. After a 9.5-year journey, the spacecraft will encounter Pluto on July 14, 2015, followed by an extended mission to the Kuiper Belt objects for the first time. The mission design for New Horizons went through more than five years of numerous revisions and updates, as various mission scenarios regarding routes to Pluto and launch opportunities were investigated in order to meet the New Horizons mission's objectives, requirements, and goals. Great efforts have been made to optimize the mission design under various constraints in each of the key aspects, including launch window, interplanetary trajectory, Jupiter gravity-assist flyby, Pluto–Charon encounter with science measurement requirements, and extended mission to the Kuiper Belt and beyond. Favorable encounter geometry, flyby trajectory, and arrival time for the Pluto–Charon encounter were found in the baseline design to enable all of the desired science measurements for the mission. The New Horizons mission trajectory was designed as a ballistic flight from Earth to Pluto, and all energy and the associated orbit state required for arriving at Pluto at the desired time and encounter geometry were computed and specified in the launch targets. The spacecraft's flight thus far has been extremely efficient, with the actual trajectory error correction ΔV being much less than the budgeted amount.

Keywords Mission design · Trajectory design · New Horizons mission · Space mission · PKB mission · Pluto · Kuiper Belt objects

Y. Guo (✉) · R.W. Farquhar
The Johns Hopkins University Applied Physics Laboratory, 11100 Johns Hopkins Road, Laurel, MD 20723-6099, USA
e-mail: yanping.guo@jhuapl.edu

R.W. Farquhar
e-mail: robert.farquhar@jhuapl.edu

1 Introduction

The early mission design work for the New Horizons mission began in late 2000, shortly after NASA terminated the "Pluto–Kuiper Express" program. A team at The Johns Hopkins University Applied Physics Laboratory (JHU/APL), which was wrapping up NASA's Near-Earth Asteroid Rendezvous (NEAR) mission with an unprecedented soft landing on an asteroid, was assembled to put together a feasible mission implementation plan, including the early mission design concept. The team hoped to save the long-sought mission to Pluto, the only remaining planet not yet visited (at the time Pluto was still the ninth planet). Urged on by the science community, NASA issued an Announcement of Opportunity (AO) in January 2001 to solicit proposals for the so-called "Pluto–Kuiper Belt (PKB) Mission," the first mission of NASA's New Frontiers program. Later, the early mission design concept evolved and became a part of the New Horizons mission proposal, led by Principal Investigator Alan Stern of the Southwest Research Institute (San Antonio, TX). The proposal was submitted to NASA and was selected for a three-month concept study (Phase A), and on November 19, 2001, NASA concluded its rigorous evaluations on two final proposals and selected the New Horizons proposal for the PKB mission.

The mission design underwent numerous revisions and updates before the New Horizons spacecraft was launched successfully on January 19, 2006, aboard an Atlas V 551 with a Star 48B third stage, in accordance with an updated baseline mission design using a Jupiter gravity-assist (JGA) trajectory to Pluto. The spacecraft flew by Jupiter on February 28, 2007, to gain a needed speed boost and will encounter Pluto on July 14, 2015, after a 9.5-year journey from launch, followed by an extended mission to the Kuiper Belt objects (KBOs). This paper discusses the objectives, requirements, and goals of the New Horizons mission design and reviews various mission design scenarios regarding routes to Pluto and launch opportunities. The baseline mission design is described in detail, covering the key aspects of launch window, interplanetary trajectory, JGA flyby, Pluto–Charon encounter with science measurement requirements, and extended mission to the Kuiper Belt and beyond. This paper also presents analyses of the launch data and the early post-launch flight results.

2 Mission Design Requirements

The scope of the PKB mission, the science requirements, and the program schedule and constraints were defined in the NASA AO (NASA 2001), which also specified the candidate launch vehicles for the PKB mission. The mission objectives were classified further as either requirements (objectives that must be accomplished by the mission) or goals (objectives that are desirable but not required to be accomplished by the mission). The tasks of the New Horizons mission design—especially the selection of the launch dates, the design of the interplanetary trajectory and Pluto encounter, and the ΔV budgeting—in principle were guided and bound by the mission objectives, program requirements, and constraints identified in the NASA AO.

2.1 Mission Scope and Objectives

According to the NASA AO, the principal goal of the PKB mission is to perform high-quality scientific investigations of the PKB region of the solar system (NASA 2001). Spacecraft had been sent to the other eight planets but not to Pluto, although planning for a mission to Pluto dates as far back as the 1960s (Long 1969; Keller 1971; Farquhar and Stern 1990; Weinstein 1992; Staehle et al. 1994; Minovitch 1994; Stern and Mitton 1998). The Pluto mission certainly is one of the most challenging deep-space missions, because it requires extremely high launch energy, long flight time, observation of multiple bodies in a brief flyby with high flyby velocity, communications to Earth from great distances, and long light-time delay.

Pluto was discovered by Clyde Tombaugh in 1930 and currently is located more than 31 astronomical units (AU) from the Sun. Our current knowledge about Pluto is based on observations taken from the ground and from orbits around the Earth. Pluto orbits the Sun in 248 Earth years in one revolution in an elliptical orbit of perihelion at 29.7 AU and aphelion at 49.4 AU. Its orbit is inclined 17° from the ecliptic plane, in contrast to the other planets, which reside within a few degrees of the ecliptic plane. Pluto has a half-sized moon, Charon, discovered in 1978, and two recently discovered small moons, Nix and Hydra. Before the two new moons were discovered, Pluto often was regarded as a binary system because the system's center of mass is outside Pluto. Charon does not move around Pluto; instead, Pluto and Charon move around the center of mass of the Pluto system, the Pluto barycenter. The PKB mission is to carry out the first scientific reconnaissance of the Pluto system and accomplish the specified science objectives and goals through a close flyby of Pluto and Charon.

The outer space beyond the orbit of Neptune is referred to as the Kuiper Belt, named after Gerald Kuiper, who hypothesized in 1951 that the short-period comets originate from a collection of material left over from the formation of the solar system. Kuiper's theory was validated with the discovery of the first KBO by David Jewitt and Jane Luu (Jewitt and Luu 1993) in 1992. Since then, numerous KBOs have been discovered each year. So far, the number of KBOs identified is over 1000, which is believed to be only a very small fraction of the total number of KBOs. The PKB mission aims to explore the Kuiper Belt region for the first time by visiting one or more KBOs in an extended mission after the Pluto–Charon encounter. The KBO encounter is a highly desired mission goal but not one of the NASA AO's mission requirements.

2.2 Science Requirements

The science objectives at Pluto and Charon were identified by the NASA Science Definition Teams and categorized into three groups, listed in Table 1, according to their priorities (NASA 2001). The group 1 objectives have the highest priority and are required to be fully accomplished by the PKB mission, the group 2 objectives are desirable, and the group 3 objectives are optional. All group 1 objectives are requirements, and the group 2 and group 3 objectives are goals.

2.3 Program Requirements and Constraints

The PKB mission is divided into two mission phases in terms of program requirements:

(i) the primary mission to Pluto, a mission requirement, and
(ii) the extended mission to the KBOs, a mission goal.

The total cost for the mission was required to be capped at 500 million FY01 dollars, including launch vehicle and launch services, spacecraft and science instruments, full mission development, and flight operations for the primary mission to Pluto. Flight operations for the extended mission to the Kuiper Belt (Phase F) were excluded from the capped funding. Candidate launch vehicles suggested in the NASA AO for the PKB mission were the new Evolved Expendable Launch Vehicle (EELV) classes, either Atlas V or Delta IV. The upper kick stage was not included in the launch vehicle package but was to be chosen by the mission implementation team.

Table 1 Science objectives at Pluto and Charon

Group	Objectives
1	Characterize the global geology and morphology of Pluto and Charon
	Map surface composition of Pluto and Charon
	Characterize the neutral atmosphere of Pluto and its escape rate
2	Characterize the time variability of Pluto's surface and atmosphere
	Image Pluto and Charon in stereo
	Map the terminators of Pluto and Charon with high resolution
	Map the surface composition of selected areas of Pluto and Charon with high resolution
	Characterize Pluto's ionosphere and solar wind interaction
	Search for neutral species, including H, H_2, HCN, and CxHy, and other hydrocarbons and nitriles in Pluto's upper atmosphere and obtain isotopic discrimination where possible
	Search for an atmosphere around Charon
	Determine bolometric Bond albedos for Pluto and Charon
	Map the surface temperatures of Pluto and Charon
3	Characterize the energetic particle environment of Pluto and Charon
	Refine bulk parameters (radii, masses, densities, etc.) and orbits of Pluto and Charon
	Search for magnetic fields from Pluto and Charon
	Search for additional satellites and rings

The NASA AO set a firm deadline for the time of the Pluto–Charon encounter: the mission is required to arrive at Pluto as early as possible but no later than 2020. The deadline was driven mainly by the concern that Pluto's atmosphere may collapse after 2020. Since passing the perihelion in 1989, Pluto has been continuously moving farther away from the Sun, and the planet's highly eccentric and inclined orbit causes its environment to change with time. Scientists predict that Pluto's thin atmosphere will be frozen onto its surface around 2020. After that, the atmosphere will not reappear until two centuries later when Pluto returns from the aphelion and approaches perihelion. In addition, if the arrival at Pluto is too late, more of Pluto's northern polar regions will fall into shadow as its north pole tilts farther away from the Sun. More surface area will fall into the shadow and, consequently, less surface area will be able to be imaged.

3 Mission Design Scenarios

Various trajectory options to get to Pluto and the associated launch opportunities were analyzed to determine and select the best mission design that will not only meet the NASA AO requirements but also maximize the mission accomplishments under the program constraints. During the mission development phase, as the program progressed and design constraints evolved, the New Horizons mission design was revised many times, and several design scenarios were investigated and considered (Guo and Farquhar, 2002, 2005, 2006).

3.1 Routes to Pluto

Sending a spacecraft to Pluto requires extremely high launch energy and so far is one of the most demanding launches of all the interplanetary missions. Of the nine planets, Pluto is located outermost from the Sun and also is the most distant one from Earth. For a direct Earth-to-Pluto flight, the required launch energy is higher than any of the past missions to the other eight planets. This imposes a significant challenge to the capability of the launch vehicle. The most powerful launch vehicles available are the EELV classes, but none of them can provide the needed launch energy without adding an additional kick stage. Even with a kick stage, the spacecraft has to be light at greatly reduced launch mass. In order to ease the high launch energy demand, alternative routes that require lower launch energy always are preferable over the direct route.

For the combination of lower launch energy and required arrival time, the best route to Pluto is via a JGA flyby instead of flying directly from Earth to Pluto. The gravity assist received at the Jupiter swingby acts like a slingshot, accelerating the spacecraft to reach Pluto faster, and allows for lower launch energy compared to a direct flight with the same flight time. The launch energy savings provided by the JGA trajectory is indispensable and sometimes crucial for mission feasibility, especially when the performance of the launch vehicle is insufficient for a direct flight.

Besides the JGA trajectory (Minovitch 1994) that proceeds directly from Earth to Jupiter and then to Pluto, there are other indirect JGA trajectories, such as the three-year ΔV-Earth–Jupiter gravity-assist approach (Farquhar and Stern 1990) and the Venus–Venus–Earth–Jupiter gravity-assist trajectory (Weinstein 1992). These indirect JGA trajectories include additional Earth or Venus-and-Earth flybys before approaching Jupiter, further reducing the launch energy to a level such that a small launch vehicle would be sufficient. However, the indirect JGA trajectories were rejected because the further reduction of the launch energy comes at the cost of a longer flight time, necessary for completing the loops for the Earth and Venus flybys, and a sizable deep space maneuver, as for the ΔV-Earth–Jupiter gravity-assist trajectory type.

In general, there are other options of trajectories using flybys of other planets. However, the inner planets, notably the Earth, cannot provide a sufficient gravity assist, and a powered swingby of a significant ΔV would be required. As for the outer planets, no feasible flyby trajectories exist within the PKB mission schedule. Among the other mission options analyzed, the one closest to the required PKB mission schedule is the Saturn gravity-assist (SGA) flyby trajectory, but the earliest launch opportunity for a SGA trajectory is in 2009 with a Pluto arrival time no earlier than 2022.

3.2 Launch Opportunities

With either the direct or indirect JGA trajectories, Jupiter must be in the right phase with Earth and Pluto at the time of launch. Additional phase-matching is required if more planetary flybys are involved. An excellent launch opportunity for a JGA trajectory was found to exist in December 2004, when Earth, Jupiter, and Pluto formed an almost perfect phase, allowing a very powerful gravity assist at the Jupiter swingby while maintaining a reasonable distance from Jupiter to avoid high doses of radiation.

The JGA launch opportunity occurs every 13 months (one Earth–Jupiter synodic period), as long as Jupiter does not advance so far as to render its gravity assist unusable for reaching Pluto. The next JGA launch opportunity was found to be in January 2006, which also was the last chance for a launch onto a JGA trajectory to reach Pluto by 2020. However, the velocity boost gained from the Jupiter flyby in 2006 would not be as great as that of the 2004 launch because Jupiter was moving gradually out of phase.

Because of the extra flight time needed for completing the Earth or Venus flybys, the launch opportunity for an indirect JGA trajectory has to occur at least two to three years prior to the time of the direct JGA launch opportunity, assuming that the phasing for the Earth or Venus flybys are right (this would be in the time frame of 2001–2003). Given the

PKB mission schedule of starting Phase B in 2002, it was infeasible to consider any indirect JGA trajectories. Furthermore, the use of a radioisotope thermoelectric generator (RTG) as the onboard power supply also disfavors the indirect JGA trajectories that would include an Earth flyby.

Launches scheduled after 2006 with arrival at Pluto before 2020 required use of the Pluto-direct trajectory. Launch opportunities for the Pluto-direct trajectory occur once every 12 months, and because there is no gravity-assist flyby to gain an extra boost, the direct trajectory requires more launch energy.

3.3 New Horizons Approach

Multiple launch opportunities and trajectory options, including launches in 2004, 2006, and 2007, were considered during the mission planning and development phase. The first mission design developed in 2001 in the initial proposal and concept study was to launch in December 2004. New Horizons would arrive at Pluto in July 2014 through the JGA trajectory, completing the extended mission to the KBOs by 2019.

Because of insufficient funding, in early 2002 NASA directed that the PKB mission could not be ready for launch in 2004. The baseline mission then was revised to launch in January 2006, the last launch opportunity for the JGA trajectory, pushing the earliest Pluto arrival to late 2015. Because the speed boost by Jupiter is much less than that of the 2004 launch case, much higher launch energy was required for the 2006 launch. At the time, the launch vehicle had not been selected yet, and the mission design was required to accommodate whichever launch vehicle NASA would select. The most capable launch vehicles from the two candidate EELV launch vehicle classes, Delta IV Heavy and Atlas V 551, were considered as a reference for designing the mission. These two launch vehicles, however, have significant differences in launch capability, according to the estimated contract performance released from NASA. In order to take advantage of the full potential of each vehicle, two baseline mission designs tailored to the specific performances of each vehicle were developed (Guo and Farquhar 2002). For the Delta IV Heavy vehicle, the baseline mission was to launch in January 2006 and arrive at Pluto in 2015–2016, whereas for the Atlas V 551 vehicle, the arrival time was one year later in 2016–2017.

In July 2003, NASA selected the Atlas V 551 as the launch vehicle for the New Horizons mission. Outfitted with several enhancements tailored specifically to the New Horizons payload, the performance of the Atlas V 551 was improved significantly, and an updated launch vehicle performance curve was provided by NASA's Kennedy Space Center (Cape Canaveral, FL). Based on the latest Atlas performance data, the baseline mission design as well as the backup mission design were determined in October 2003. After that, there were times when alternative Pluto arrival times were studied and considered as possible options in response to concerns of possible lower RTG power and/or the possible situation in which a new KBO is discovered near the predicted New Horizons trajectory path. Eventually, however, the mission was implemented with the baseline mission design developed in October 2003, with some minor adjustments of the Pluto arrival time in response to an update of the new Pluto satellite ephemerides released in March 2005.

A design for a backup mission option was developed alongside the primary launch design, given the potential uncertainties regarding launch and the critical Pluto arrival time constraint. The backup design was planned for launch in February 2007 during a 14-day launch period: arrival at Pluto in 2019 for the first 12 launch days and arrival at Pluto in 2020 for the last 2 launch days, all using the Pluto-direct trajectory. The 2006 baseline, 2007 backup, and other mission scenarios analyzed in detail are listed in Table 2.

Table 2 Mission scenarios to Pluto and KBO

Mission scenario	Launch		Encounter	
	Period	C3 (km^2/s^2)	Body	Year
2006 Baseline JGA or Pluto-direct → Pluto → KBOs	35 days (January 11–February 14)	164	Pluto	2015–2020
2007 Backup Pluto-direct → Pluto → KBOs	14 days (February 2–15)	166.2	Pluto	2019–2020
2006 Launch JGA → Pluto → KBOs	20 days (January 10–29) 20 days (January 9–28)	166 156.7	Pluto Pluto	2015 2016
2006 Launch (extended launch period) Pluto-direct → Pluto → KBOs	16 days (January 30–February 14) 4 days (February 5–8)	166 156.7	Pluto Pluto	2019 2020
2006 Launch 2+ year ΔV EGA → Pluto → KBOs	20 days (January 7–26) 20 days (January 3–22) 20 days (December 24–January 13)	28.2 28.4 28.8	Pluto Pluto Pluto	2015 2016 2020
2006 Launch 3+ year ΔV EGA → Pluto → KBOs	20 days (January 18–February 6) 20 days (January 13–February 1) 20 days (January 4–23)	50.4 50.6 51	Pluto Pluto Pluto	2015 2016 2020
2006 Launch 4+ year ΔV EGA → Pluto → KBOs	20 days (January 27–February 15) 20 days (January 22–February 10) 20 days (January 10–29)	65.1 65.3 65.8	Pluto Pluto Pluto	2015 2016 2020
2007 Launch Pluto-direct → Pluto → KBOs	10 days (February 4–13) 10 days (February 4–13)	165 162.3	Pluto Pluto	2019 2020
2008 Launch Pluto-direct → Pluto & KBOs	10 days (February 7–16)	168.5	Pluto	2020
2008 Launch JGA → Neptune → KBOs	20 days (March 15–April 3)	161	Neptune	2018
2008 Launch JGA → Uranus → KBOs	20 days (March 9–28)	109	Uranus	2015
2008 Launch JGA → KBO (1992 QB1)	20 days (March 8–27)	99	1992 QB1	2025
2009 Launch SGA → Pluto → KBOs	20 days (November 18–December 7)	148	Pluto	2022
2010 Launch SGA → Pluto → KBOs	20 days (November 30–December 19)	143	Pluto	2024

Notes: JGA, Jupiter gravity assist; SGA, Saturn gravity assist; 2+, 3+, 4+ year ΔV EGA, deep space burn-Earth gravity-assist trajectory with time of flight more than 2, 3, or 4 years of the Earth return orbit

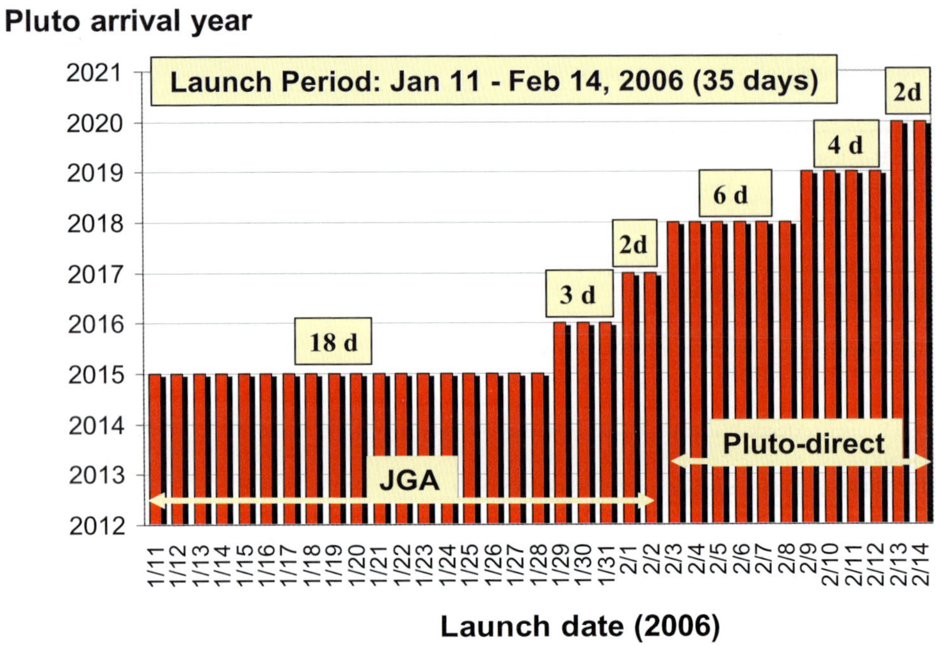

Fig. 1 New Horizons baseline mission design

4 Baseline Mission Design

The goal for the baseline mission design was to reach Pluto at the earliest time possible with the given launch vehicle performance and the required spacecraft launch mass as well as to have a long launch period to maintain a very high launch probability. The baseline mission design had a 35-day launch period starting on January 11, 2006, for the earliest Pluto arrival in 2015 via a JGA flyby trajectory, and ending on February 14, 2006, for the latest Pluto arrival in 2020 via a Pluto-direct trajectory, as shown in Fig. 1. All arrival times for the different arrival years were chosen at the favorable solar opposition seasons in the summer for the best science observations at Pluto–Charon flyby, achieving both Earth and solar occultations by both Pluto and Charon with the desired Pluto–Charon encounter geometry. The design requires maximum launch energy, C3, of 164 km^2/s^2 and allows for a spacecraft wet mass of 478 kg. The mission is divided into seven distinct phases: launch and early operations, first cruise to Jupiter, the Jupiter flyby, second cruise to Pluto, the Pluto–Charon encounter, post-encounter of science data playback, and the extended mission to KBOs.

4.1 Launch

A prolonged launch period of 35 days from January 11, 2006, to February 14, 2006, was selected for New Horizons, which is almost twice as long as a typical launch period (21 days) for interplanetary missions in order to ensure a very high probability of being launched within the 2006 launch opportunity. The first part of the launch period (January 11–February 2) used the JGA trajectories that get to Pluto as early as 2015, and the later part of the launch period (February 3–14) used the Pluto-direct trajectories with the Pluto arrival time in 2020 as the latest. Experiences with past mission launches indicate that there

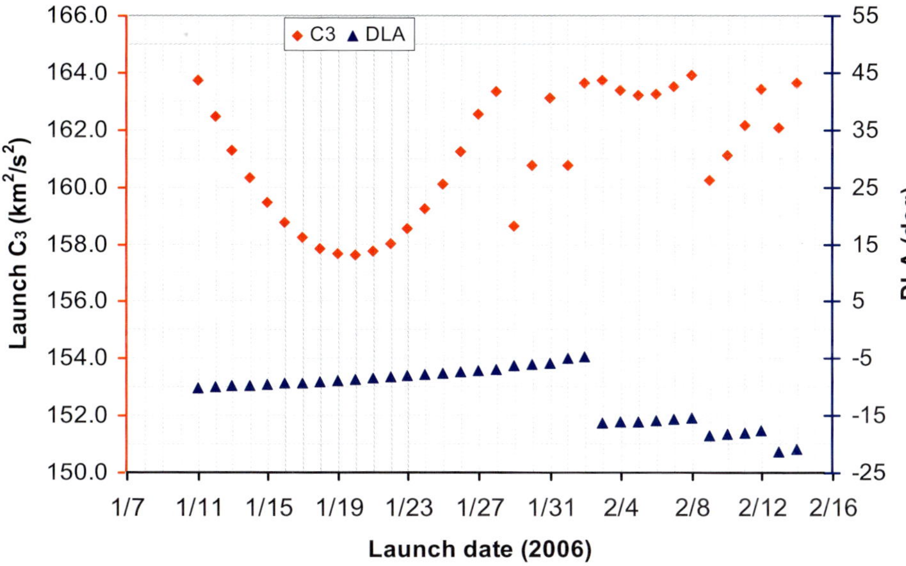

Fig. 2 Launch C3 and DLA requirements

is a much higher chance of launching in the early days of the launch period, which is reflected within the launch period design strategy that started with the earliest Pluto arrival and ended with the latest.

The unusually long launch period did not require extra launch energy and was made possible by

(i) combining the launch opportunities of two different types of trajectories together and
(ii) not fixing but varying the Pluto arrival time, as illustrated in Fig. 1.

The dominant Pluto arrival time is 2015, obtained in the first 18 days of the primary launch period, from January 11 to January 28, 2006, by means of the JGA trajectory. When the 2015 arrival window closed after January 28, the later Pluto arrival times were considered. Continuing with the JGA trajectory, five more days were added to the launch period: three days for the 2016 arrival and two days for the 2017 arrival. When the window for all JGA trajectories closed, the Pluto-direct trajectory was considered to further extend the launch period. This results in 12 extra launch days until the Pluto arrival year reached 2020, which still met the mission requirement for the latest Pluto arrival time. The Pluto arrival time in Fig. 1 is not a continuous curve but jumps from year to year because of certain science geometry requirements at the Pluto flyby. More on the selection of the Pluto arrival time is described in Sect. 4.4.2.

The launch energy requirement for New Horizons is the highest of all space launches to date, about 10 times more than a typical mission to Mars. The launch energy, C3, is defined as the square of the hyperbolic excess velocity (V_∞) of the spacecraft with respect to Earth, a measure related to how much velocity increase must be supplied to the spacecraft by the launch vehicle at launch. The C3 requirements for each launch day for the designed trajectory, whether it is a JGA trajectory or a Pluto-direct trajectory, are shown in Fig. 2. Also included in Fig. 2 are the values of the declination of launch asymptote (DLA) for all launch days. All of the DLA angles are less than the launch site latitude of 28.5°, indicating there is no additional launch penalty from having to perform a plane change.

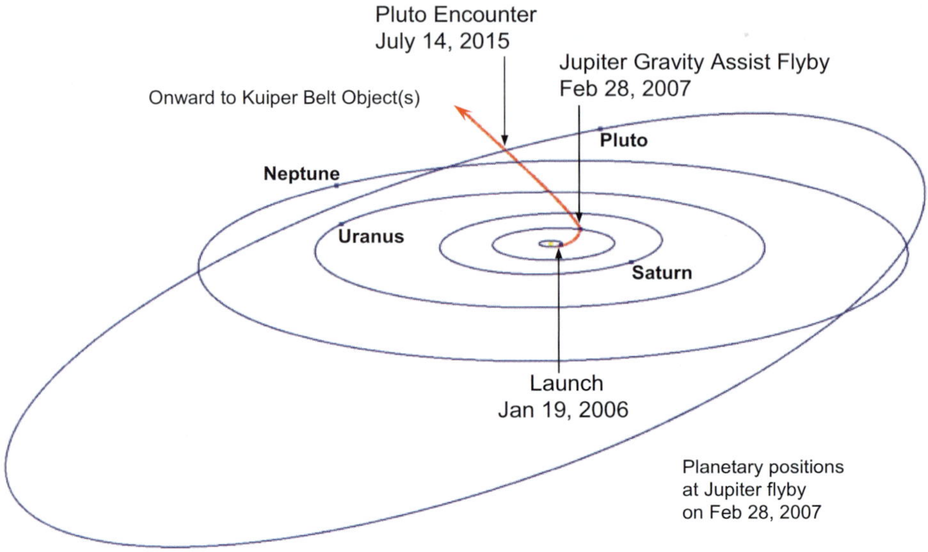

Fig. 3 Interplanetary trajectory

The C3, DLA, and the right ascension of the launch asymptote (RLA)—the launch targets to which the spacecraft must be delivered by the launch vehicle—specify the New Horizons launch requirements. The New Horizons mission trajectory was designed as a ballistic flight from Earth to Pluto with or without the Jupiter flyby. All energy and the associated orbit state required for arriving at Pluto at the desired time and encounter geometry were computed and specified in the launch targets that were provided to the launch vehicle provider. The New Horizons launch required a three-stage rocket consisting of the Atlas V 551 EELV launch vehicle and the Star 48B third stage. The Atlas V 551 is a two-stage rocket supplied by Lockheed Martin. The first stage consists of a common core booster and five strap-on solid rocket boosters, and the second stage is a powerful Centaur booster that has restart capability. The third stage Star 48B is a spin-stabilized solid rocket made by Boeing and customized for the New Horizons mission. The New Horizons spacecraft was placed into an Earth parking orbit by the first stage and the Centaur's first burn. It was then injected into the specified heliocentric trajectory through the combined injection burn supplied by the Centaur (second burn) and the Star 48B after a short coasting in the parking orbit.

4.2 Interplanetary Trajectory

The baseline mission design considered two interplanetary trajectories:

(i) the JGA trajectory for the primary launch period for a fast flight to Pluto and
(ii) the Pluto-direct trajectory for the extended launch period.

The January 19, 2006, launch put the spacecraft into the favorable early Pluto arrival JGA trajectory that flew by Jupiter on February 28, 2007, and will encounter Pluto and Charon on July 14, 2015, as shown in Fig. 3. The flight from Earth to Jupiter only took 404 days; no spacecraft has ever reached Jupiter from Earth in such a short time with such a fast speed.

However, by nature, New Horizons cannot maintain this high speed for its entire flight. During the interplanetary flight toward Pluto, the spacecraft is immersed in the Sun's gravitational field, which slows down the spacecraft along its path as it moves away from the

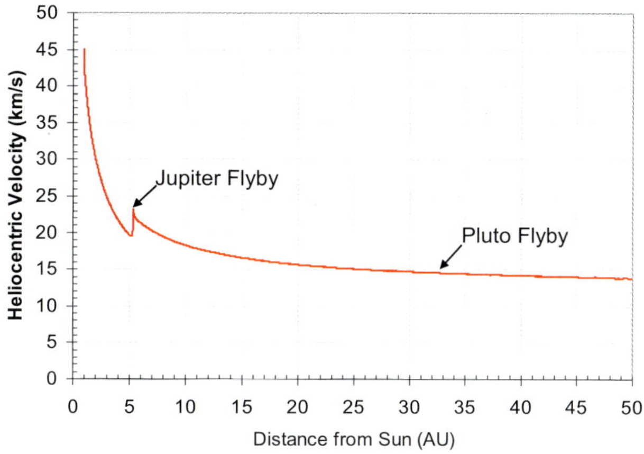

Fig. 4 Heliocentric velocity of the New Horizons spacecraft over the flight from Earth to Pluto

Sun. The close flyby of Jupiter was designed to inject a speed boost from the appropriate body motions relative to Jupiter. The heliocentric speed of the spacecraft (the magnitude of the spacecraft velocity vector with respect to the Sun) is plotted in Fig. 4 against the solar distance during the flight from Earth to Pluto and beyond. The highest heliocentric speed, as shown in Fig. 4, is at the beginning when the spacecraft was injected into the heliocentric trajectory at launch. The speed then decreased until it reached Jupiter. The speed "jump" was clearly observed at the Jupiter flyby in February and March 2007. An acceleration of 3.83 km/s was gained at the JGA flyby. After that, the speed decreased again because of the Sun's gravity. When the spacecraft reaches Pluto, the heliocentric speed will go down to 14.5 km/s. The speed increase at the Pluto flyby is only a few meters per second, which is not visible from the plot in Fig. 4. After Pluto, the speed will continue to decrease as the spacecraft moves into the Kuiper Belt region and beyond.

The plot in Fig. 5 provides a mission profile showing the spacecraft's distances from the Sun and Earth as a function of time over the mission, along with the Sun–Earth–Probe (SEP) and Sun–Probe–Earth (SPE) angles. The solar distance increases monotonically, as expected, while the distance from Earth oscillates because of the periodic motion of the Earth around the Sun. Both the SEP and SPE angles are periodic. The predicted solar conjunction periods are defined to be when the SEP angle is less than three degrees and are listed in Table 3. There will be no communications expected with the spacecraft during the solar conjunction periods, which occur at the minima of the SEP angle. The solar opposition occurs at the maxima of SEP angle; the predicted solar opposition dates are listed in Table 4.

On the way to Pluto, the spacecraft crosses almost the entire solar system on a path within close proximity of the ecliptic plane. Although it has or will cross the orbits of five planets (Table 5), there will be no close flyby of the planets except for Jupiter. In most of the cases, the planet is very distant when the spacecraft crosses its orbit.

4.3 Jupiter Gravity-Assist Flyby

The Jupiter flyby was aimed at a point slightly below Jupiter's equatorial plane and more than 2.3 million km from Jupiter's center, as illustrated in Fig. 6. The Jupiter flyby trajectory design used the JGA to accelerate the spacecraft and change the trajectory inclination to achieve the desired Pluto encounter. The closest approach (C/A) to Jupiter took place on February 28, 2007. Figure 7 displays the Jupiter flyby geometry at the C/A as observed from above Jupiter's equatorial plane. The spacecraft flew by Jupiter outside the orbits of

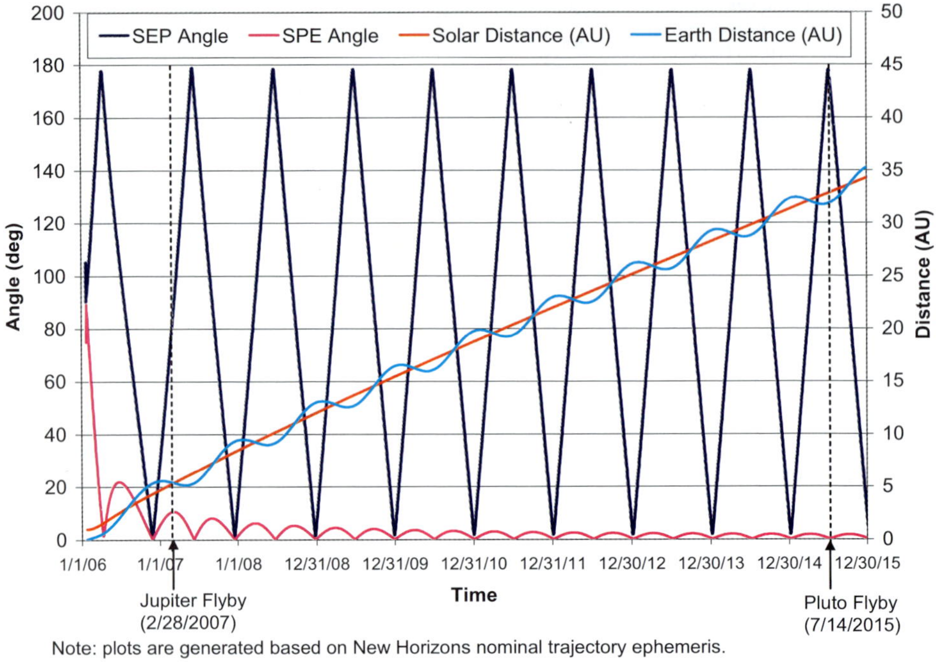

Fig. 5 New Horizons mission profiles of the distances from the Sun and Earth and the SEP and SPE angles

Table 3 Time of solar conjunction

Number of days	Start of conjunction (UTC)	End of conjunction (UTC)
8	11/19/2006 20:00	11/27/2006 12:00
6	12/11/2007 4:00	12/17/2007 11:00
6	12/19/2008 5:00	12/24/2008 21:00
5	12/24/2009 3:00	12/29/2009 12:00
5	12/27/2010 9:00	1/1/2011 13:00
5	12/29/2011 17:00	1/3/2012 17:00
5	12/30/2012 12:00	1/4/2013 10:00
5	12/31/2013 23:00	1/5/2014 19:00
5	1/2/2015 4:00	1/6/2015 23:00
5	1/3/2016 5:00	1/7/2016 23:00

Notes:

1. The listed New Horizons solar conjunction dates are predicted dates when communications between the spacecraft and Earth are blocked by the Sun
2. The conjunction time is computed down to hours with the SEP angle less than three degrees

the Galilean satellites at a speed of 21.2 km/s with respect to Jupiter and at a relatively large distance of 32.25 Jupiter radii (R_J). The radiation doses experienced by the spacecraft are very low at such a great distance.

Besides the large Galilean satellites, there are numerous small irregular Jovian satellites, for a total of 63 satellites discovered in the Jupiter system so far. Many of the small irregular

Table 4 Dates of solar opposition

Solar opposition time (UTC) (Max. SEP angle)	SEP angle (deg)	Spacecraft–Earth distance (AU)
2006-04-06 02:00	177.77	0.61
2007-06-04 12:00	178.87	5.34
2008-06-17 20:00	178.42	9.14
2009-06-24 13:00	178.26	12.68
2010-06-28 14:00	178.17	16.06
2011-07-01 09:00	178.12	19.34
2012-07-02 12:00	178.09	22.54
2013-07-04 04:00	178.06	25.68
2014-07-05 12:00	178.04	28.77
2015-07-06 16:00	178.03	31.83

Table 5 Planet orbit passing dates

Planet	Date	Days from launch
Mars	April 7, 2006	78
Jupiter	February 28, 2007	404
Saturn	June 8, 2008	871
Uranus	March 18, 2011	1884
Neptune	August 24, 2014	3139
Pluto	July 14, 2015	3463

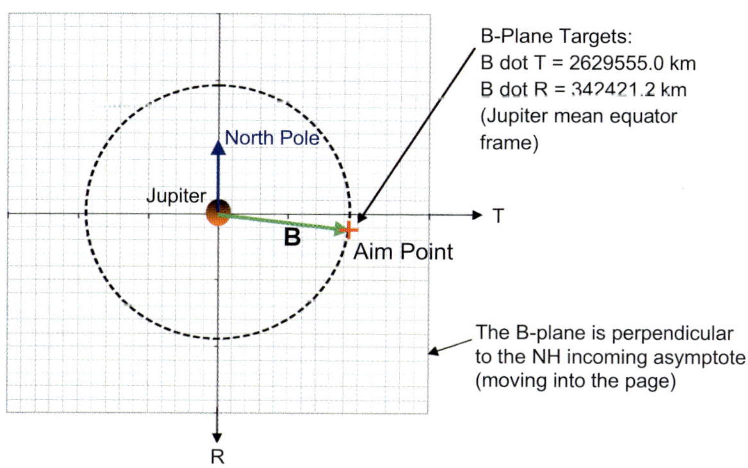

Fig. 6 Jupiter B-plane targeting

ones were discovered in recent years, and knowledge about them is very limited. We always have been interested in and hoped to find close-encounter opportunities with the Jovian satellites during the New Horizons Jupiter flyby. However, the trajectory analyses indicated there are no good close encounters with the Jovian satellites unless a trajectory adjustment

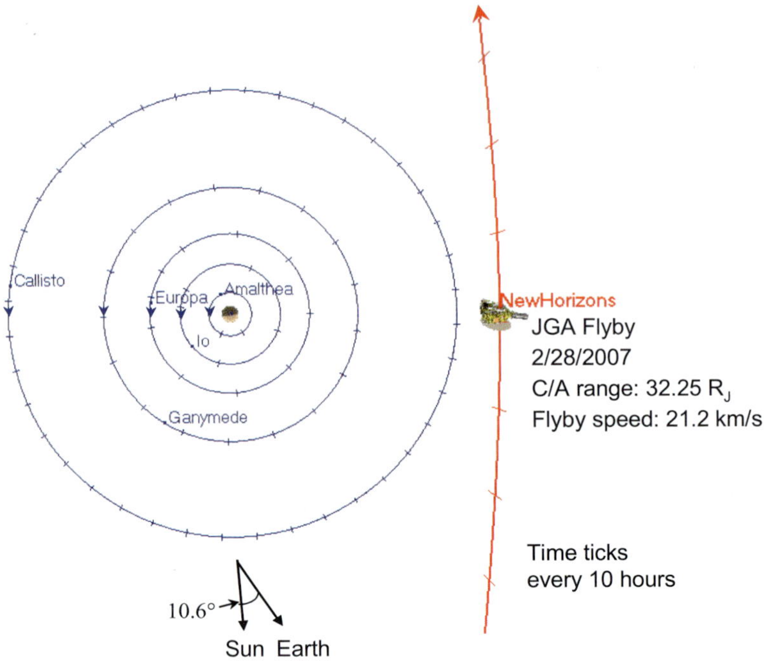

Fig. 7 Jupiter flyby geometry

is performed. The project decided not to expend more ΔV and to save it for the KBOs. Figure 8 shows the C/A distances of the New Horizons spacecraft to the 63 Jovian satellites.

4.4 Pluto–Charon Encounter

The design of the Pluto–Charon encounter trajectory in combination with the selection of the Pluto arrival date and C/A time is a critical part of the mission design because it directly affects how many of the mission's science objectives and goals can be accomplished in the brief Pluto flyby. There are 16 itemized Pluto/Charon science objectives (Table 1) identified in the original NASA PKB mission AO. The New Horizons mission seeks to accomplish 15 of the 16 set objectives (all except the search of magnetic fields) in this first Pluto reconnaissance investigation with seven onboard instruments: ultraviolet (UV) imaging spectrograph ("Alice"), visible and infrared (IR) imager ("Ralph"), Radio science EXperiment (REX), LOng Range Reconnaissance Imager (LORRI), Solar Wind Around Pluto (SWAP), Pluto Energetic Particle Spectrometer Science Investigation (PEPSSI), and Student Dust Counter (SDC).

The Pluto–Charon encounter was challenged by the many science requirements and goals and required careful planning and compromise. The encounter design uses multiple instruments to observe two bodies in a single flyby while conducting coordinated measurements involving four bodies: the Earth, the Sun, Pluto, and Charon. Based on the NASA AO guidelines and the capability and characteristics of New Horizons instruments, the mission design team and the science team worked out the derived science measurement requirements.

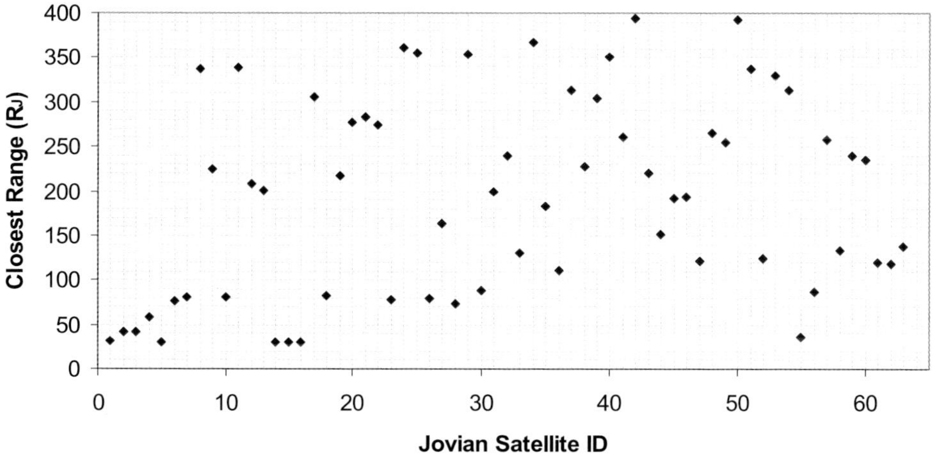

ID	Name	ID	Name	ID	Name	ID	Name	ID	Name
1	Io	14	Thebe	27	Chaldene (S/2000_J10)	40	S2002_J1	53	S2003_J13
2	Europa	15	Adrastea	28	S/2000_J11	41	S2003_J1	54	S2003_J14
3	Ganymede	16	Metis	29	Autonoe (S2001_J1)	42	S2003_J2	55	S2003_J15
4	Callisto	17	S/1999_J1	30	Thyone (S2001_J2)	43	S2003_J3	56	S2003_J16
5	Amalthea	18	S/1975_J1	31	Hermippe (S2001_J3)	44	S2003_J4	57	S2003_J17
6	Himalia	19	Kalyke (S/2000_J2)	32	S2001_J4	45	S2003_J5	58	S2003_J18
7	Elara	20	Iocaste (S/2000_J3)	33	S2001_J5	46	S2003_J6	59	S2003_J19
8	Pasiphae	21	Erinome (S/2000_J4)	34	S2001_J6	47	S2003_J7	60	S2003_J20
9	Sinope	22	Harpalyke (S/2000_J5)	35	S2001_J7	48	S2003_J8	61	S2003_J21
10	Lysithea	23	Isonoe (S/2000_J6)	36	S2001_J8	49	S2003_J9	62	S2003_J22
11	Carme	24	Praxidike (S/2000_J7)	37	S2001_J9	50	S2003_J10	63	S2003_J23
12	Ananke	25	Megaclite (S/2000_J8)	38	S2001_J10	51	S2003_J11		
13	Leda	26	Taygete (S/2000_J9)	39	S2001_J11	52	S2003_J12		

Fig. 8 Jovian satellite encounter profile

4.4.1 Science Measurement Requirements

4.4.1.1 Priority Ranking In general, measurements associated with Pluto have higher priority than those associated with Charon. Among the Pluto measurements, group 1 science objectives have higher priority than those of group 2, and group 2 science objectives have higher priority than those of group 3; this also applies to the Charon measurements. When not all requirements can be achieved, the following priority ranking takes place:

A. Pluto Earth occultation
B. Pluto solar occultation
C. Two Deep Space Network (DSN) station coverage during Pluto Earth occultation
D. Charon solar occultation
E. Charon Earth occultation

In the encounter design, the observation selection follows the same priority order as defined for the science objectives. The highest priority is given to the observations for accomplishing the group 1 science objectives: the atmosphere of Pluto and global geology, morphology, and surface composition of Pluto and Charon. In the same measurement category, Pluto is considered the primary observation body. Because Charon also holds essential information for understanding the Pluto–Charon binary system, the mission seeks to take as many measurements of Charon as possible without undermining the fulfillment of the Pluto objectives.

4.4.1.2 Requirements for Remote Sensing The three onboard imaging instruments, Ralph, LORRI, and Alice, are equipped with visible imaging, IR spectral mapping, and UV measurements. They are responsible for implementing the investigation of the global geology, morphology, and surface composition of Pluto and Charon. The critical measurement conditions for the remote sensing are the solar phase angle and the flyby distance. The Pluto arrival condition determines the solar phase angle at encounter, while the appropriate flyby distance depends on the field of view (FOV) of the sensors. Distance from the surface must be less than 25,000 km for the visible imaging to achieve a resolution better than 1 km per pixel, and less than 161,300 km for the IR mapping to achieve a resolution better than 10 km per pixel. The C/A distance to the surface of Pluto is required to be no greater than 25,000 km. To match the instruments' performances and capabilities for imaging a fast-moving body, the desired C/A distance from Pluto's surface currently is selected at about 10,000 km to strike a balance between high spatial resolution and a lack of smear. An overall goal for remote sensing is to cover as much of the surfaces of Pluto and Charon as possible.

4.4.1.3 Requirements for Atmosphere Investigation The atmosphere investigation is carried out primarily by the REX experiments and supported by the Alice measurements. The REX experiments conduct the radiometric measurements of the atmosphere by analyzing the variation of the RF signals passing through the atmosphere that are received by the spacecraft. Alice measures the received UV signals emitted from the Sun and passing through the atmosphere. Both measurements must be performed when occultation takes place. For the Pluto atmosphere, the Earth–Pluto occultation is required for REX measurements and the Sun–Pluto occultation for Alice measurements. In addition, the REX experiments are uplink-based and transmit high-powered signals from the DSN to the spacecraft. Thus, it is highly desirable to have the RF signals transmitted to the spacecraft simultaneously from two DSN complexes during the Earth occultation. This improves the signal-to-noise ratio and provides redundancy, as the spacecraft is at a great distance of 32 AU away from Earth at the Pluto encounter. The search for atmosphere around Charon requires similar REX and Alice measurements at the Earth and Sun occultation by Charon.

In summary, the science measurements require a Pluto–Charon encounter trajectory that has the desired C/A distance to Pluto and enables the occurrences of Earth–Pluto occultation, Sun–Pluto occultation, Earth–Charon occultation, and Sun–Charon occultation as well as the existence of simultaneous uplinks to the spacecraft from two DSN complexes during the Earth occultation. These are very stringent constraints for a single flyby of two bodies. The goal for the Pluto–Charon encounter design is to optimize the encounter geometry and flyby trajectory under the arrival constraints to enable all of the desired science measurements.

4.4.2 Selection of Pluto Arrival Time

The Pluto arrival time is an encounter design parameter. Theoretically, the Pluto arrival can be at any time as long as launch energy permits it. However, to enable the required science measurements described in the previous section, the time of Pluto arrival must be selected when the Earth, the Sun, and Pluto are positioned in such a geometry that can support the formation of the desired Earth and solar occultations by Pluto as well as by Charon. The earliest year to reach Pluto depended on the launch date and the trajectory taken, as shown in Fig. 1. For each year, as the Earth orbits the Sun once, there are two opportunities for the desired occultation geometry when the Earth and Sun are about in a line with Pluto (one in the summer and the other in the winter, as illustrated in Fig. 9), making it possible to achieve both the Earth and the Sun occultations during the flyby of Pluto and Charon. The summer

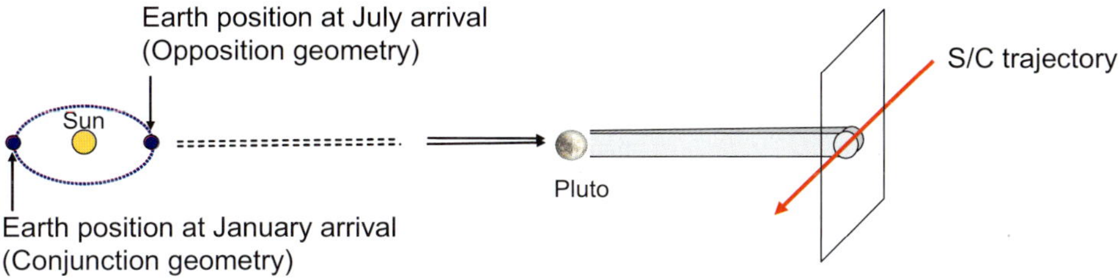

Fig. 9 Options for Pluto arrival time selection

opportunity corresponds to the solar opposition geometry with Earth in between the Sun and Pluto, and the winter opportunity corresponds to the solar conjunction geometry with the Sun positioned in the center between Earth and Pluto. The summer arrival time was selected for the solar opposition geometry, which is more favorable for communications and the REX measurements.

Inclusion of the Charon occultation in the flyby in addition to the Pluto occultation further constrains the time of arrival. There are only two possible times when the Charon occultation can occur during each Charon orbit (6.4 days), once before the spacecraft passes Pluto and the other after passing Pluto. The time of the Charon occultation after passing Pluto was selected for the preferred flyby sequence of approaching Pluto first and Charon second. This flyby sequence results in a flyby geometry of Pluto in front of Charon. The large disk of Charon, about half the size of Pluto, is believed to be able to shed adequate light for imaging the dark surface of Pluto.

Among the potential arrival time options separated by the Charon orbit period within the summer opportunity, only those that result in an Earth–Pluto occultation supported by uplinks from two DSN complexes became candidates for the final encounter trajectory. The time of the Pluto arrival eventually was selected in accordance with the encounter trajectory design that maximizes the overall science accomplishments.

4.4.3 Pluto at Approach

The heliocentric transfer orbit determines the conditions upon arrival at Pluto, such as the solar phase angle and the direction of the incoming trajectory asymptote with respect to Pluto and Charon. The spacecraft is to arrive at Pluto from a heliocentric transfer trajectory inclined 2.34° above the ecliptic plane and to approach Pluto from its southern hemisphere (as shown in Fig. 10) at a solar phase angle of 15°, an excellent illumination condition for a full-spectrum survey of Pluto and Charon on the approaching hemisphere. The subsolar position is at the latitude of 49° south, showing that the southern hemisphere is sunlit and the north portion is in permanent Sun shade. As Pluto rotates at a rate of about 6.4 Earth days, different portion of its surface will be imaged.

4.4.4 Pluto Flyby Trajectory and Geometry

The goal of the Pluto flyby trajectory design is to maximize the required and desired science measurements at Pluto and Charon in accordance with the science measurement priority ranking and requirements as described in Section 4.4.1, providing the necessary supporting geometry and conditions for science measurements during the Pluto flyby. Prior to the delivery of the final launch targets to the launch vehicle provider, the Pluto–Charon encounter

Fig. 10 Pluto at approach

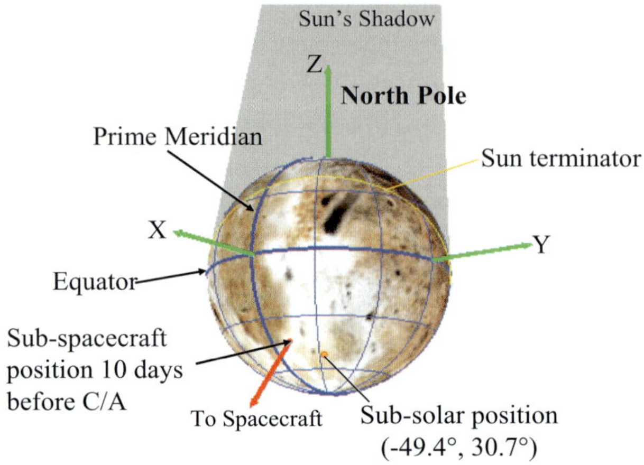

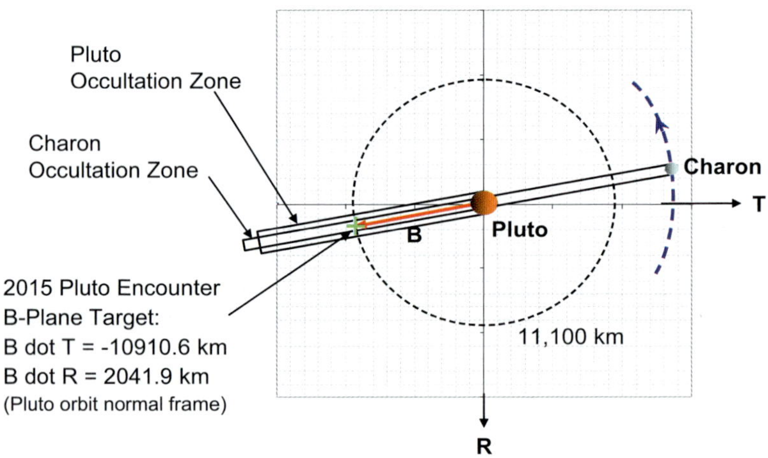

Fig. 11 Pluto B-plane targeting

trajectory design was revised further based on the latest updated planetary and Pluto/Charon ephemerides. Figure 11 illustrates the B-plane targeting at Pluto arrival, which defines the Pluto flyby trajectory. Figure 12 shows the Pluto flyby trajectory and close-encounter geometry, as viewed from the direction perpendicular to the Pluto–Sun line.

New Horizons passes by Pluto and Charon inside Charon's orbit from the same side, which is convenient for switching the observation target from Pluto to Charon for imaging. Charon orbits Pluto in a circular retrograde orbit at a rate synchronized with Pluto's rotation period of 6.387 Earth days, with a mean radius of 19,600 km. The considerable size of Charon (593-km radius) relative to Pluto (1195-km radius) causes the center of mass of the system to lie outside of Pluto, a unique situation in the planetary system. The trajectory crosses Charon's orbital plane at about 43°, and the angle between Charon's orbit normal and the trajectory outgoing asymptote is about 133°.

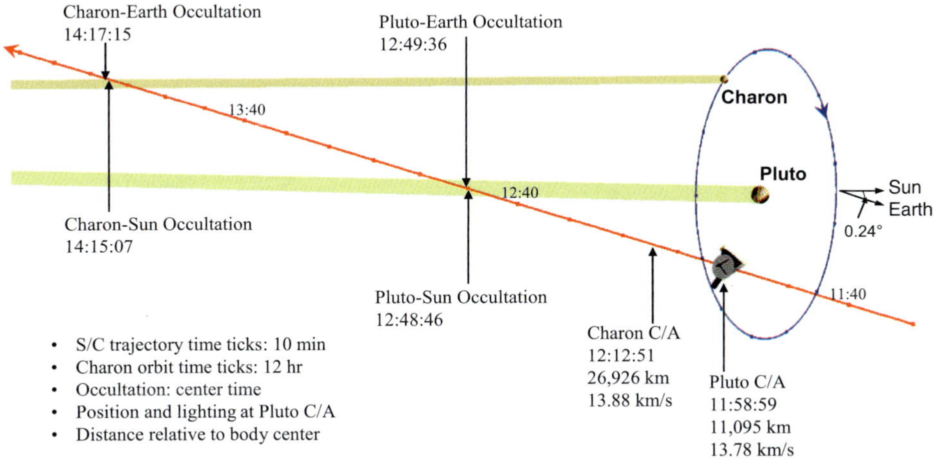

Fig. 12 Pluto–Charon encounter geometry

4.4.5 Encounter Sequence and Event Timeline

As shown in Fig. 12, the flyby proceeds in a sequence of encounters, first with Pluto and then followed by Charon. The major flyby events start with the C/A to Pluto on July 14, 2015, at 11:58:59 UTC of spacecraft time, at a distance of 11,095 km from the center of Pluto and a flyby speed of 13.78 km/s. It is followed shortly by the C/A to Charon at 12:12:51 UTC at a C/A range of 26,926 km. Within the next two hours, New Horizons travels through the solar and Earth occultation zones of Pluto and Charon, passing by behind Pluto and Charon. The Pluto occultation occurs first at about 36 minutes after Charon's C/A, and the Charon occultation takes place 1 hour and 26 minutes later. In both cases, the solar occultation starts before the Earth occultation but with a short time separation. During the Earth and solar occultations, the two responsible instruments, REX (which measures the radial signals from Earth) and Alice (which measures the UV signals from Sun), are configured to be capable of handling the measurements simultaneously. The atmosphere investigation will carry on continuously from Pluto through Charon. More detailed encounter parameters and a timeline of the Pluto–Charon flyby events are listed in Table 6.

4.4.6 Deep Space Network Access Profile

During the Earth occultation, two DSN complexes, Canberra and Goldstone, will be in view of the spacecraft and able to transmit RF signals to the spacecraft from both complexes simultaneously. Their access profile is shown in Fig. 13, in which the spacecraft elevation angle from the three DSN complexes, Goldstone, Canberra, and Madrid, are plotted over a period of 24 hours on the day of the Pluto–Charon encounter. The ground transmission time shown in Fig. 13 is 4 hours, 25 minutes, and 19 seconds earlier than the occultation time to account for the light's propagation time.

The desired elevation angle is above 15° to assure adequate transmission of the RF signal from the complexes, although lower elevation angles also may work. The shaded region in Fig. 13 is for elevation angles less than 15°. As the elevation angle profile indicates, the overlapping period of about three hours between Goldstone and Canberra is the only time period when elevation angles are above 15° from two DSN complexes. The Earth occultation

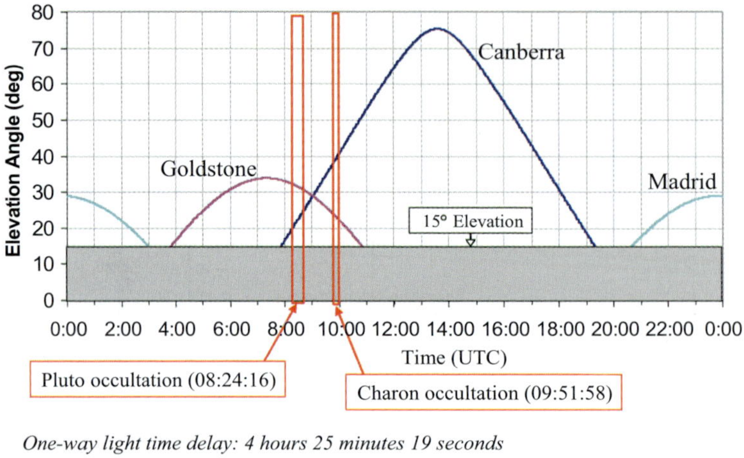

Fig. 13 DSN access profile

by Pluto and Charon is targeted to take place within this time period. As Fig. 13 indicates, the spacecraft is accessible simultaneously from Goldstone and Canberra for the time period from before Earth occultation by Pluto through after the Earth occultation by Charon, with sufficient ingress and egress time margins at elevation above 15 degrees.

4.5 Extended Mission to the Kuiper Belt and Beyond

After the flyby of the Pluto system, New Horizons will continue its journey to explore the Kuiper Belt as an extended mission. The plan for the Kuiper Belt exploration is to conduct similar science investigations as carried out at Pluto and Charon, using the same onboard instruments built for Pluto investigations, through a close flyby of one or more KBOs with a size of 50-km diameter or greater.

4.5.1 Plans for Kuiper Belt Object Encounter

The KBO flyby targets will be selected just prior to the Pluto encounter, because the trajectory to Pluto will not be changed regardless of the chosen KBO flyby target. Delaying the selection of the KBO target(s) until 2015 allows for many more years of searching for new KBOs. Plans and resources have been in place to conduct a series of KBO searches from near-Earth orbit (Hubble Space Telescope) and Earth-based observatories in the region of the sky where the New Horizons trajectory is predicted. No candidate KBOs have been identified yet.

One of the preparations for the KBO mission is to develop the strategy and plans to target KBOs with the available onboard resources. The spacecraft, including the communications system, is designed for the KBO mission to go as far as 50 AU from the Sun. However, the onboard power supply is expected to decrease in output as a function of time, so it may impose limitations on spacecraft and instrument operations at a later time. Onboard propellant is a key element that determines how many of the KBOs are accessible to the spacecraft. Current estimates show that there will be as much as 250 m/s of ΔV capacity left after the Pluto flyby, which is attributed mainly to the accurate orbit injection at launch.

Because of the small mass possessed by Pluto, the gravity assist to be gained from a Pluto flyby is negligible. During the Pluto encounter design, analyses were performed to

Table 6 Pluto–Charon encounter parameters

Pluto encounter date		2015-07-14
Pluto	C/A time	11:58:59
	C/A dist (km)	11095
	C/A vel (km/s)	13.78
Charon	C/A time	12:12:51
	C/A dist (km)	26926
	C/A vel (km/s)	13.88
Pluto–Sun occultation	Start time	12:43:12
	End time	12:54:18
	S/C dist (km)	42609
Pluto–Earth occultation	Start time	12:43:57
	End time	12:55:13
	S/C dist (km)	43272
Charon–Sun occultation	Start time	14:13:36
	End time	14:16:37
	S/C dist (km)	105307
Charon–Earth occultation	Start time	14:15:11
	End time	14:19:17
	S/C dist (km)	107032
Sun–Pluto–Earth angle		0.24°
Earth distance (AU)		31.9
Sun distance (AU)		32.9

Note: Time is spacecraft time in UTC. C/A distances are relative to object center

investigate whether the direction of the outgoing trajectory of the spacecraft could be altered by the Pluto flyby, through adjusting the B-plane aiming point so that the spacecraft could fly toward the first KBO target. The calculations indicated that Pluto can hardly bend the spacecraft flyby trajectory because of its low mass and the relatively high spacecraft flyby speed. This feature clearly is displayed in Fig. 13, where the flyby trajectory is almost a straight line, implying the Pluto flyby cannot help tune the spacecraft's trajectory toward the selected KBO target. Instead, trajectory change maneuvers must be applied for targeting the KBOs.

As soon as two weeks after the Pluto flyby, with the key science data transmitted back to Earth at the 24-hour-per-day continuous playback, a trajectory correction maneuver (TCM) can be applied to alter the spacecraft's trajectory toward the first KBO target. The more time between the execution of the TCM and the encounter date, the less ΔV is required for the needed trajectory adjustment. Limited by the available onboard ΔV capacity, the accessible KBO targets will be in the region near the extended post-Pluto trajectory path, as illustrated in Fig. 14. Large ephemeris uncertainties are expected for the KBO targets because of the short observation time and the targets' great distances from Earth. The plan is to acquire OpNav images of the KBO target by using onboard imagers Multispectral Visible

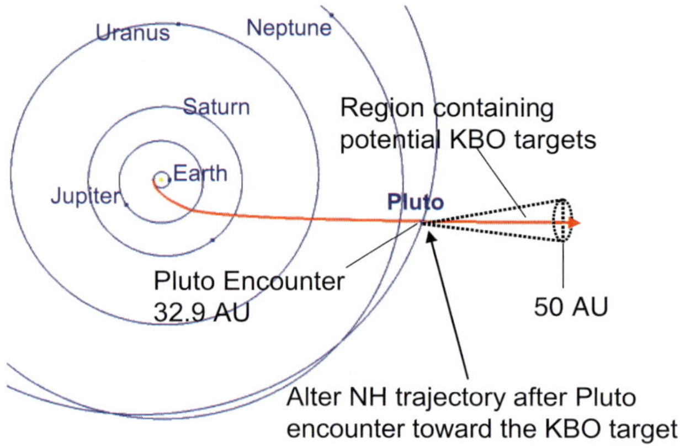

Fig. 14 Mission to the KBOs

Imaging Camera (MVIC) and LORRI to refine the spacecraft trajectory relative to the KBO target. The KBO OpNav images are desired as early as possible so that the trajectory can be corrected with minimum ΔV. Based on the current estimate, the high-resolution imager LORRI is capable of detecting a KBO target (visual magnitude 17.4) as far as 43 days out. Once the KBO OpNav images are obtained, a trim TCM will be executed to correct the KBO position errors. It will be followed with a cleanup TCM to refine the encounter targeting a few days prior to the encounter.

The most likely KBO flyby is estimated to occur in 2018 when New Horizons' heliocentric distance reaches 42 AU, where the distribution of KBO objects over heliocentric distance peaks (Spencer et al. 2003). Exploration of the KBOs is planned to go as far as 50 AU from the sun. The spacecraft will reach the 50-AU distance in 2021 and is expected to encounter one or more KBOs by then.

4.5.2 Departing the Solar System

After completing the primary mission to Pluto and the extended mission to the Kuiper Belt, the spacecraft will continue to move out of the solar system in a Sun-escape trajectory. Right after the Pluto flyby, its asymptotic solar system excess velocity is 12.5 km/s, in the direction of right ascension of 293° and declination of 2.1° in the Sun-centered mean ecliptic of J2000 reference frame. The trajectory adjustments to be performed for targeting the KBOs may alter the trajectory slightly, but the onboard propellant is insufficient to stop the spacecraft from escaping from the solar system.

5 Flight Results

New Horizons was launched on January 19, 2006, and successfully injected into the desired heliocentric trajectory as designed. The flight so far has been extremely smooth, and the needed trajectory maintenance has been less than what was planned. After three trajectory corrections that took out the small injection errors, the spacecraft flew by Jupiter on February 28, 2007, and will encounter Pluto on July 14, 2015, as planned.

5.1 Launch and Orbit Injection

At 19:00 UTC (2:00 p.m. EST) on January 19, 2006, New Horizons lifted off from Launch Complex 41 at Florida's Cape Canaveral Air Force Station atop the Star 48B aboard the

Table 7 Launch targets: achieved versus designed

Launch targets	A—Designed	B—Achieved	Injection error (B − A)	Predicted 3σ injection error*
C3 (km^2/s^2)	157.6561	157.7502	0.0941	0.4245
DLA (deg)	−8.8407	−8.8683	−0.0276	0.3307
RLA (deg)	209.3855	209.3124	−0.0731	0.3603

Notes: A, required launch target specified by the Mission Design Team; B, launch target derived from the determined trajectory (OD005 solution) provided by the Navigation Team based on post-launch DSN tracking data

*Based on Boeing Trajectory Cycle 3 report

Atlas V launch vehicle. It first was inserted into an elliptical Earth parking orbit of perigee altitude 165 km and apogee altitude 215 km. After a short coast in the parking orbit, the spacecraft then was injected into the desired heliocentric orbit by the Centaur second stage and Star 48B third stage. At the Star 48B burnout, the New Horizons spacecraft reached the highest Earth departure speed, estimated at 16.2 km/s, becoming the fastest spacecraft ever launched from Earth. In less than nine hours, it passed by the Moon at a distance of 184,700 km from Moon center.

The conditions for injection into the heliocentric orbit were defined as launch targets specified in C3, DLA, and RLA at the target interface point (TIP) defined as 10 minutes after Star 48B ignition. The designed and achieved launch targets along with the actual injection errors and the 3σ values are presented in Table 7. The orbit injection was remarkably accurate, with the orbit injection errors much less than 1σ.

By targeting New Horizons to the designed Pluto B-plane aim point with an optimized Jupiter flyby from the injected TIP state, the ΔV required for correcting the injection errors at TIP was determined at 18.2 m/s. Most of the ΔV is for adjusting the velocity direction by 0.069°, or 1.2 mrad, and a small portion of it is for reducing the 4-m/s overburn. The ΔV budgeted for injection error correction was 92 m/s with 99% probability. Consequently, a significant amount of propellant now is available and can be used for targeting the KBOs.

5.2 Summary of Trajectory Corrections

Since launch, a total of three TCMs have been performed thus far. These trajectory corrections removed all the launch errors associated with orbit injection and placed New Horizons into the designed trajectory to Jupiter.

The first two TCMs, TCM-1A and TCM-1B, were designed together in a pair and implemented in a manner that minimized mission risk. The very first trajectory maneuver of the spacecraft was decided to be executed using the passive spin TCM (PS-TCM) mode, which is an open-loop axial ΔV execution without the guidance and control (G&C) system in control. TCM-1A served as a calibration burn to verify that the misalignment or unbalanced thruster performance of the paired thrusters would not make the spacecraft unstable and was limited to a magnitude of 5 m/s to avoid any chance of destabilizing the spacecraft. If the TCM was executed as planned, TCM-1B would complete the needed ΔV by carrying out the remaining part of the ΔV implemented in the same PS-TCM mode and the same pair of thrusters as used for TCM-1A. TCM-1A, with a nominal magnitude of 5 m/s, was executed successfully on January 28, 2006, nine days after launch, and TCM-1B, with a nominal magnitude of 13.32 m/s, was applied successfully two days later on January 30, 2006.

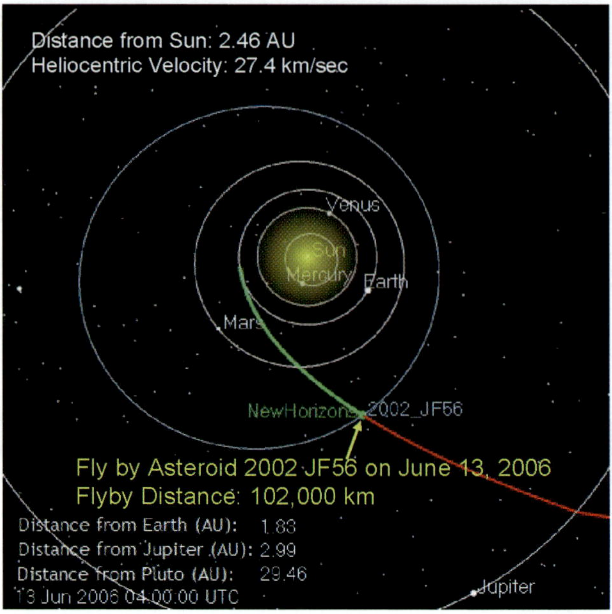

Fig. 15 Flyby of asteroid 2002 JF56 on June 13, 2006

Both TCM-1A and TCM-1B were terminated by timing. There was about 6% underburn attributed to thruster performance that did not match the expected values obtained from pre-launch thruster tests. TCM-2, originally scheduled on February 15, 2006, was canceled because the orbit solution at the time had an uncertainty that was comparable to the ΔV values. On March 9, 2006, TCM-3 was applied with a nominal magnitude of 1.16 m/s to make up the 6% underburn from the previous two TCMs. TCM-3 was implemented with the three-axis TCM (3A-TCM) mode, which is a closed-loop ΔV execution with G&C system in control of the burn. TCM-3 was very accurate, with only a small execution error (0.01%). Because the trajectory errors are so small, the project has decided not to perform any TCMs before the Jupiter flyby. TCMs-4, -5, -6, and -7, which were scheduled at 76, 20, and 5 days before the Jupiter flyby and 15 days after the Jupiter flyby, respectively, were canceled. The next trajectory correction (TCM-8) is planned in September 2007, after the Jupiter flyby.

5.3 Flyby of Asteroid 2002 JF56

On June 13, 2006, New Horizons flew by a small asteroid designated 2002 JF56 at a C/A distance of 102,000 km, as shown in Fig. 15. This unexpected close encounter offered a great opportunity for the New Horizons spacecraft to perform a Pluto-like pointing and tracking exercise to test both the G&C ability of attitude control of scanning and pointing and the instrument performance of Ralph imager. The high-resolution imager, LORRI, was unable to participate in this exercise because New Horizons was still too close (<3 AU) to the Sun, and door opening still was restricted.

5.4 ΔV Status

At launch, a total of 76.85 kg of propellant was loaded on the spacecraft. The propellant is for TCMs and for spacecraft attitude maneuvers. The spacecraft has a blow-down monopropellant hydrazine propulsion system consisting of a central tank, 12 0.8-N attitude control system (ACS) thrusters, and 4 4.4-N TCM thrusters. The propellant usage for the trajectory corrections has been much less than the pre-launch budgeted amount. So far, about 9 kg of propellant has been used. Furthermore, it is estimated that there will be as much as 47 kg of propellant remaining after the Pluto flyby for the Kuiper Belt mission, corresponding to a ΔV capacity of 250 m/s.

6 Conclusion

After five years of planning and revisions, the New Horizons mission design was finalized, and the spacecraft was launched successfully on January 19, 2006. The spacecraft was injected into the favorable JGA trajectory to Pluto, flew by Jupiter on February 28, 2007, and gained a significant speed boost well on its way to encounter Pluto on July 14, 2015. Using Jupiter's gravity assist in the trajectory design shortened the time of flight to Pluto by three years and provided a great opportunity for preparing the team for the upcoming Pluto encounter and for acquiring a great deal of Jupiter science data as a bonus to the mission.

The launch of New Horizons used a strategy that combined the launch opportunities of two different types of trajectories and allowed the Pluto arrival time to vary from an early arrival to a late arrival. With this strategy, the launch energy good for an 18-day launch period also was adequate for a 35-day launch period, thus maintaining the early arrival preference and extending the launch period without increasing the required launch energy.

The high accuracy in trajectory simulation was maintained throughout the trajectory design by using a fully integrated trajectory with high-fidelity models updated with the latest planetary and Pluto ephemerides. High-precision computations of the trajectory for each of the 35 launch days, integrated from TIP to Jupiter flyby and to Pluto encounter and beyond, ensured minimum trajectory adjustments and corrections post-launch. Precisely specifying the launch targets for the orbit injection conditions and accurate delivery of the spacecraft by the launch vehicle and third stage resulted in a remarkable orbit insertion that saved significant onboard propellant. Since launch, New Horizons' flight has been extremely efficient, closely following the designed trajectory. Five of the eight planned TCMs from launch through post-Jupiter flyby were canceled because they were not needed. Many of the lessons learned and strategies used for the New Horizons mission design can be applied to the design of future missions.

Acknowledgements The authors thank New Horizons Principal Investigator, Alan Stern; the science team for their support and review of the Pluto–Charon encounter design; and the Jet Propulsion Laboratory for providing the planetary and satellite ephemerides. The authors also thank Tom Strikwerda for review and helpful edits on this manuscript. This work was supported by NASA under contract no. NAS5-97271.

References

R.W. Farquhar, S.A. Stern, Planet. Rep. **10**, 18–23 (1990)

Y. Guo, R.W. Farquhar, in *New Horizons Mission Design for the Pluto–Kuiper Belt Mission*, AIAA-2002-4722, AIAA/AAS Astrodynamics Specialist Conference and Exhibit, Monterey, CA, August 5–8, 2002

Y. Guo, R.W. Farquhar, Acta Astronaut. **56**, 421–429 (2005)

Y. Guo, R.W. Farquhar, Acta Astronaut. **58**, 550–559 (2006)

D.C. Jewitt, J.X. Luu, Nature **362**, 730–732 (1993)

J.W. Keller, in *Grand Tour Surveys of the Outer Solar System, Advances in the Astronautical Sciences*, vol. 29, part I, ed. by J. Vagners (1971), pp. 37–47

E.J. Long, Astron. Aeronaut. **7**, 32–47 (1969)

M.A. Minovitch, J. Spacecr. Rockets **31**, 1029–1037 (1994)

NASA, *Announcements of Opportunity, Pluto–Kuiper Belt Mission*, AO 01-OSS-01, January 19, 2001
J. Spencer, M. Buie, L. Young, Y. Guo, A. Stern, Earth Moon Planets **92**, 483–491 (2003)
R.L. Staehle, R.J. Terrile, S.S. Weinstein, Planet. Rep. **14**, 4–11 (1994)
S.A. Stern, J. Mitton, *Pluto and Charon* (Wiley, New York, 1998)
S. Weinstein, *Pluto Flyby Mission Design Concepts for Very Small and Moderate Spacecraft*, AIAA-92-4372, AIAA/AAS Astrodynamics Conference, Hilton Head Island, SC, August 10–12, 1992

Overview of the New Horizons Science Payload

H.A. Weaver · W.C. Gibson · M.B. Tapley · L.A. Young · S.A. Stern

Originally published in the journal Space Science Reviews, Volume 140, Nos 1–4, 75–91.
DOI: 10.1007/s11214-008-9376-6

Abstract The New Horizons mission was launched on 2006 January 19, and the spacecraft is heading for a flyby encounter with the Pluto system in the summer of 2015. The challenges associated with sending a spacecraft to Pluto in less than 10 years and performing an ambitious suite of scientific investigations at such large heliocentric distances (>32 AU) are formidable and required the development of lightweight, low power, and highly sensitive instruments. This paper provides an overview of the New Horizons science payload, which is comprised of seven instruments. *Alice* provides moderate resolution (∼3–10 Å FWHM), spatially resolved ultraviolet (∼465–1880 Å) spectroscopy, and includes the ability to perform stellar and solar occultation measurements. The *Ralph* instrument has two components: the *Multicolor Visible Imaging Camera (MVIC)*, which performs panchromatic (400–975 nm) and color imaging in four spectral bands (Blue, Red, CH_4, and NIR) at a moderate spatial resolution of 20 μrad/pixel, and the *Linear Etalon Imaging Spectral Array (LEISA)*, which provides spatially resolved (62 μrad/pixel), near-infrared (1.25–2.5 μm), moderate resolution ($\lambda/\delta\lambda \sim 240$–550) spectroscopic mapping capabilities. The *Radio Experiment (REX)* is a component of the New Horizons telecommunications system that provides both radio (X-band) solar occultation and radiometry capabilities. The *Long Range Reconnaissance Imager (LORRI)* provides high sensitivity ($V < 18$), high spatial resolution (5 μrad/pixel) panchromatic optical (350–850 nm) imaging capabilities that serve both scientific and optical navigation requirements. The *Solar Wind at Pluto (SWAP)* instrument measures the density and speed of solar wind particles with a resolution $\Delta E/E < 0.4$ for energies between 25 eV and 7.5 keV. The *Pluto Energetic Particle Spectrometer Science Investigation (PEPSSI)* measures energetic particles (protons and CNO ions) in 12 energy

H.A. Weaver (✉)
Johns Hopkins University Applied Physics Laboratory, 11100 Johns Hopkins Road, Laurel, MD 20723, USA
e-mail: hal.weaver@jhuapl.edu

W.C. Gibson · M.B. Tapley
Southwest Research Institute, 6220 Culebra Road, San Antonio, TX 78238, USA

L.A. Young · S.A. Stern
Southwest Research Institute, 1050 Walnut St., Suite 400, Boulder, CO 80302, USA

channels spanning 1–1000 keV. Finally, an instrument designed and built by students, the *Venetia Burney Student Dust Counter (VB-SDC)*, uses polarized polyvinylidene fluoride panels to record dust particle impacts during the cruise phases of the mission.

Keywords New Horizons mission · Pluto · Kuiper belt

1 Introduction

New Horizons was the first mission selected in NASA's New Frontiers series of mid-sized planetary exploration programs. The New Horizons spacecraft was launched on 2006 January 19 and is now on a 3 billion mile journey to provide the first detailed reconnaissance of the Pluto system during the summer of 2015. Assuming that this primary objective is successful, NASA may authorize an extended mission phase that will permit a flyby of another Kuiper belt object (KBO), as yet unidentified, probably within 3 years of the Pluto encounter. The genesis and development of the New Horizons mission is described by Stern (2007). The scientific objectives of the mission are discussed by Young et al. (2007). Here we provide a high level overview of the scientific payload. Detailed descriptions of individual instruments are given elsewhere in this volume, as referenced below.

The New Horizons mission is an ambitious undertaking that required the development of lightweight, low power, and highly sensitive instruments. Pluto will be nearly 33 AU from the sun at the time of the encounter in 2015, and a launch energy (C3) of nearly 170 $km^2\ s^{-2}$ was needed to reach this distance within the 9.5 year transit to the Pluto system. Even using the powerful Lockheed-Martin Atlas 551 launcher in tandem with its Centaur second stage and a Boeing Star48 third stage, the entire spacecraft mass had to be kept below 480 kg, of which less than 50 kg was allocated to the science payload. At Pluto's large heliocentric distance, the use of solar photovoltaic cells was not an option, so the New Horizons mission relies on a radioisotope thermoelectric generator (RTG) for all of its power needs. The mission requirement on the total power available at the Pluto encounter is only 180 W, of which less than 12 W can be used at any one time by the scientific instruments. The solar output (light and particle) at Pluto is approximately 1000 times smaller than at the Earth, which means that the instruments attempting to measure reflected sunlight or the solar wind during the Pluto encounter must be extremely sensitive. Finally, we note that the long mission duration imposes strict reliability requirements, as the spacecraft and science payload must meet their performance specifications at least 10 years after launch.

Fortunately, all of the New Horizons instruments successfully met these daunting technical challenges without compromising any of the mission's original scientific objectives. Below we provide a high-level description of all the instruments on New Horizons, discuss their primary measurement objectives, and summarize their observed performance, which has now been verified during in-flight testing. But first we begin by briefly describing the spacecraft pointing control system as it relates to the science payload.

2 Payload Pointing Control

The New Horizons spacecraft does not have enough power to support a reaction wheel based pointing control system and instead relies on hydrazine thrusters to provide slewing capability and attitude control. The positions of stars measured by one of two star trackers (the second star tracker provides redundancy) are used to determine the absolute orientation of the spacecraft (i.e., the RA and DEC locations of some reference axis on the spacecraft), and the drift rate is monitored by a laser-ring gyro system (the inertial measurement unit, or IMU). The attitude data from the star tracker and IMU are used in a feedback loop to set the pointing within prescribed limits in both absolute position and drift rate. The spacecraft IMUs, star trackers, sun sensors, and guidance computers are all redundant.

The New Horizons spacecraft spends much of its time spinning at $\sim$5 RPM around the Y-axis. In this mode, useful data can be obtained by REX, SWAP, PEPSSI, and the VB-SDC, but typically not by any of the other instruments.

For virtually all observations made by the imaging instruments, 3-axis pointing control mode is required. In 3-axis mode, the spacecraft can be slewed to a targeted location to an accuracy of $\pm$1024 μrad (3σ) and controlled to that location within a typical "deadband" of $\pm$500 μrad. For some Alice observations, when the target must be kept near the center of its narrow slit, the deadband can be reduced to $\pm$250 μrad. The drift rate is controlled to within $\pm$34 μrad/sec (3σ) for both fixed and scanning observations. The post-processing knowledge of the attitude and drift rate derived from the star tracker and IMU data are $\pm$350 μrad (3σ) and $\pm$7.5 μrad/sec (3σ), respectively. Ralph observations usually require the spacecraft to scan about its Z-axis. The nominal scan rate for Ralph/MVIC is 1.1 mrad/sec, and the nominal scan rate for Ralph/LEISA is 0.12 mrad/sec.

Further details about the New Horizons guidance and control system can be found in Rogers et al. (2006).

3 Science Payload

3.1 Overview

All of the fundamental ("Group 1") scientific objectives for the New Horizons mission (Stern 2007; Young et al. 2007) can be achieved with the *core* payload comprised of: (i) the *Alice* ultraviolet (UV) imaging spectroscopy remote sensing package, (ii) the *Ralph* visible and infrared imaging and spectroscopy remote sensing package, and (iii) the *Radio Experiment (REX)* radio science package. The *supplemental* payload, which both deepens and broadens the mission science, is comprised of the *Long Range Reconnaissance Imager (LORRI)*, which is a long-focal-length optical imaging instrument, and two plasma-sensing instruments: the *Solar Wind Around Pluto (SWAP)* and the *Pluto Energetic Particle Spectrometer Science Investigation (PEPSSI)*. The supplemental payload is not required to achieve minimum mission success, but these instruments provide functional redundancy across scientific objectives and enhance the scientific return by providing additional capabilities not present in the core payload. The *Venetia Burney Student Dust Counter (VB-SDC)*, which was a late addition to the supplemental payload approved by NASA as an Education and Public Outreach (EPO) initiative, also provided a new capability to New Horizons, namely, an interplanetary dust detection and mass characterization experiment.

Pictures of all seven instruments are displayed in Fig. 1, which also lists the mass and power consumption of each instrument. The locations of the instruments on the New Horizons spacecraft are displayed in Fig. 2.

As discussed further below, Ralph is essentially two instruments rolled into a single package: the *Multispectral Visible Imaging Camera (MVIC)* is an optical panchromatic and color imager; the *Linear Etalon Imaging Spectral Array (LEISA)* is an infrared imaging spectrometer. The boresights of MVIC, LEISA, LORRI, and the Alice airglow channel are aligned with the spacecraft $-X$ axis (Fig. 2) except for minor tolerancing errors. The projections of the fields of view of those instruments onto the sky plane are depicted in Fig. 3.

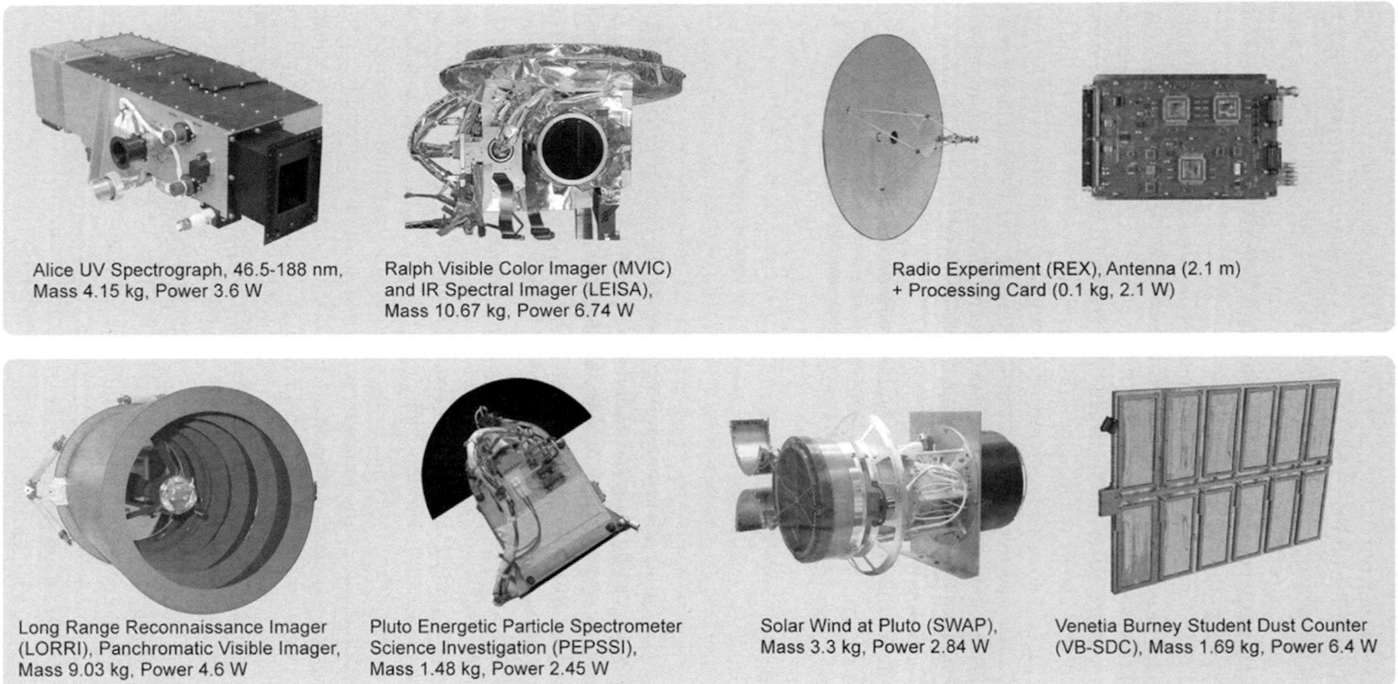

Fig. 1 The three instruments comprising the New Horizons core payload are shown along the *top row*, and the instruments comprising the supplemental payload are displayed along the *bottom row*. The approximate mass and power consumption are shown just below the picture of each instrument. The total mass of the entire science payload is approximately 30 kg, excluding the antenna, and the total power drawn by all the instruments is approximately 29 W

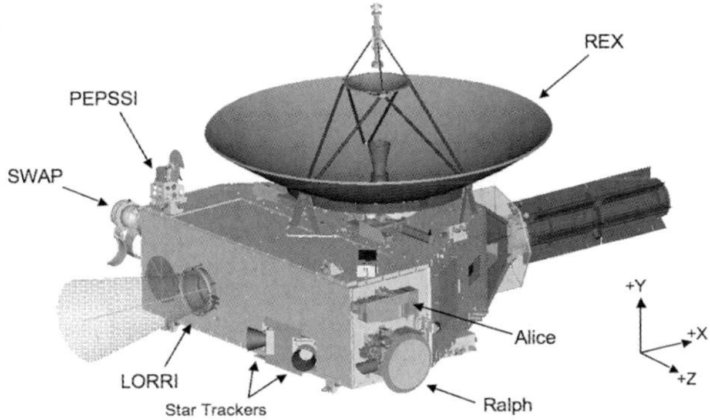

Fig. 2 This drawing shows the locations of the instruments on the New Horizons spacecraft. The VB-SDC is mounted on the bottom panel, which is hidden from view. The boresights of LORRI (sketched in figure), Ralph, and the Alice airglow channel are all approximately along the $-X$ direction. The boresights of the Alice solar occultation channel and the antenna are approximately along the $+Y$ direction. SWAP covers a swath that is $\sim 200°$ in the XY plane and $\sim 10°$ in the YZ plane. PEPSSI's field-of-view is a $\sim 160°$ by $\sim 12°$ swath whose central axis is canted with respect to the principal spacecraft axes to avoid obstruction by the backside of the antenna. The black structure with fins located at $+X$ is the RTG, which supplies power to the observatory. The star trackers, which are used to determine the attitude, can also be seen. The antenna diameter is 2.1 m, which provides a scale for the figure

The types of observations performed by the New Horizons instruments are depicted in Fig. 4. None of the instruments have their own scanning platforms, so the entire spacecraft must be maneuvered to achieve the desired pointings. As described above, the guidance and control system uses hydrazine thrusters to point the spacecraft at the desired target.

The principal measurement objectives and the key characteristics of the New Horizons science payload are summarized in Table 1, which also includes the names and affiliations of the instrument Principal Investigators (PIs) and the primary builder organization for each instrument. The measurement objectives that are directly related to the mission Group 1 scientific objectives are highlighted in boldface.

In the following subsections, we provide further discussion of each of the New Horizons instruments. We attempt to provide a high-level summary of the instruments' capabilities, with detailed descriptions left to the individual instrument papers, which are referenced in each subsection.

3.2 Alice

The Alice instrument aboard New Horizons is an ultraviolet (UV) imaging spectrometer that provides moderate spectral and spatial resolution capabilities over the wavelength range ~465–1880 Å with a peak effective area (actual input aperture area times the system transmittance) of $\sim 0.3\ \text{cm}^2$. Light enters Alice's $f/3$ telescope via either the main entrance aperture (called the Airglow Aperture, co-aligned with the Ralph and LORRI apertures), or, via a small, fixed pickoff mirror, through the Solar Occultation Channel (SOCC, co-aligned with the New Horizons high-gain antenna). Light from either aperture is reflected off the 4 cm $\times$ 4 cm primary mirror, passes through a single slit, is reflected off a holographic grating, and finally is detected using a photon-counting, microchannel plate double delay line device, read out as a 32 $\times$ 1024 element digital array. The SOCC aperture is stopped down by a factor of 6400 relative to the Airglow Aperture to allow Alice to look directly at the

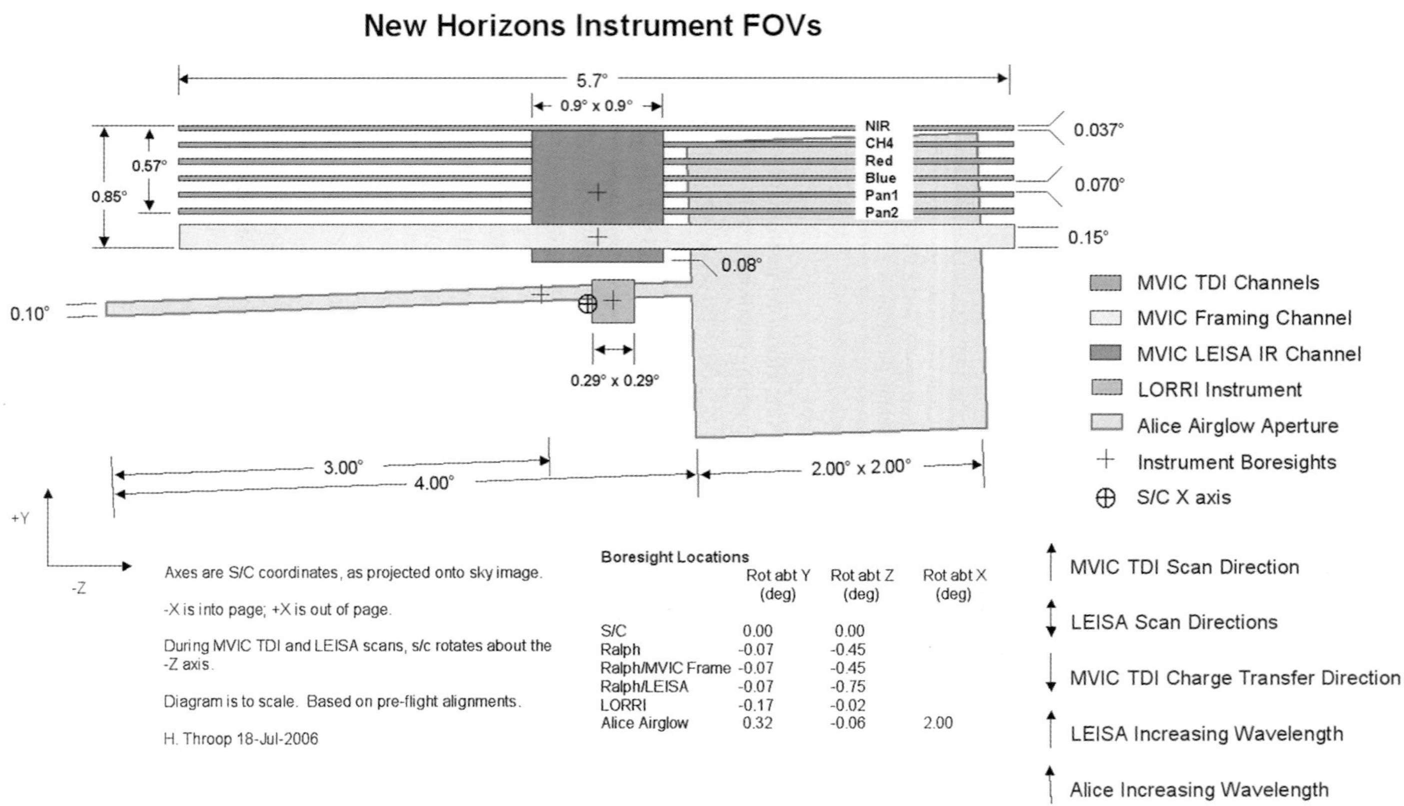

Fig. 3 The fields of view (FOVs) of the MVIC, Ralph, Alice airglow, and LORRI instruments are projected onto the sky plane; the listed boresights are measured in-flight values. The angular extent of each instrument's FOV is also listed. The spacecraft $+X$ direction is out of the page, the $+Y$ direction is up, and the $+Z$ direction is to the left. The LORRI field FOV overlaps the narrow portion of the Alice airglow channel, and the MVIC FOV overlaps the wide portion. The LEISA FOV overlaps the MVIC FOV

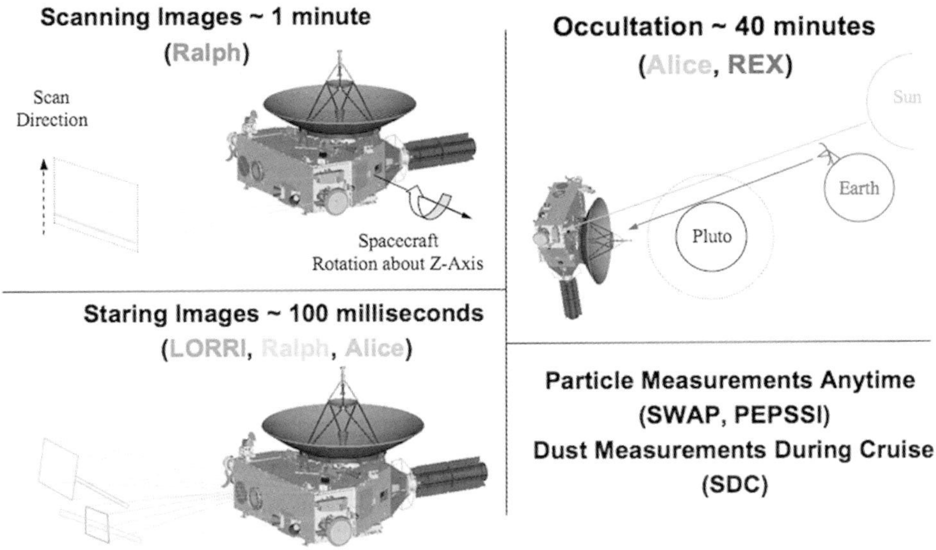

Fig. 4 Types of New Horizons observations. Typical Ralph MVIC Time Delay Integration (TDI) and LEISA observations (*upper left*) are performed by rotating the spacecraft about the Z-axis. Typical Ralph MVIC frame, LORRI, and Alice airglow observations (*lower left*) are made with the spacecraft staring in a particular direction. The Alice and REX occultation observations (*upper right*) are performed by pointing the antenna at the Earth and the Alice occultation channel at the Sun, so that radio signals from the DSN on Earth can be received by REX at the same time that Alice observes the Sun. Observations by the particle instruments (SWAP, PEPSSI, and VB-SDC; lower right) can occur essentially anytime, in either spinning or 3-axis mode. However, most of the VB-SDC data will be collected during cruise mode, when the other instruments are in hibernation mode and the spacecraft is passively spinning, because thruster firings add a large background noise level to the VB-SDC's data

Sun for solar occultations of Pluto's and Charon's atmospheres. The Alice entrance slit is a "lollipop" (see Fig. 3) with a $0.1^\circ \times 4^\circ$ "slot" used primarily for airglow observations and a $2^\circ \times 2^\circ$ "box" used mainly during solar occultation observations. The point source spectral resolution is 3–6 Å, depending on wavelength, and the plate scale in the spatial dimension is 0.27° per pixel. During the Pluto and Charon occultation observations, the Sun has an apparent diameter of $\sim 1'$, and the spectral resolution is 3–3.5 Å. During filled-slit airglow aperture observations, the spectral resolution is ~9–10 Å.

Alice is a name, not an acronym, taken from one of the main characters of the American television show *The Honeymooners*. Alice is sometimes called Pluto-Alice (P-Alice) to distinguish it from its predecessor, Rosetta-Alice (R-Alice), which is a similar instrument being flown on the European Space Agency (ESA) Rosetta mission to comet 67P/Churyumov-Gerasimenko. Compared to R-Alice, P-Alice has a somewhat different bandpass and various enhancements to improve reliability. P-Alice also includes a separate solar occultation channel, which is not available on R-Alice. Both P-Alice and R-Alice are significantly improved versions of the Pluto mission "HIPPS" UV spectrograph (HIPPS/UVSC), which was developed at Southwest Research Institute (SwRI) in the mid-1990s with funds from NASA, JPL, and SwRI.

Alice's principal measurement objectives and its key characteristics are summarized in Table 1. Alice was designed to measure Pluto's upper atmospheric composition and temperature, which is a New Horizons Group 1 scientific objective. Alice will also obtain model-dependent escape rate measurements from Pluto's atmosphere, and it will provide some

Table 1 New Horizons Instruments: Pluto system measurement objectives and characteristics (PI = principal investigator; instrument characteristics are summary values with details provided in the individual instrument papers)

Instrument, PI	Measurement objectives	Instrument characteristics
UV imaging spectrometer (Alice), S.A. Stern (SwRI), SwRI	• Upper atmospheric temperature and pressure profiles of Pluto • Temperature and vertical temperature gradient should be measured to ~10% at a vertical resolution of ~100 km for atmospheric densities greater than $\sim 10^9$ cm^{-3} • Search for atmospheric haze at a vertical resolution <5 km • Mole fractions of N_2, CO, CH_4 and Ar in Pluto's upper atmosphere • Atmospheric escape rate from Pluto • Minor atmospheric species at Pluto • Search for an atmosphere of Charon • Constrain escape rate from upper atmospheric structure	UV spectral imaging; 465–1880 Å; FOV $4° \times 0.1°$ plus $2° \times 2°$; Resolution 1.8 Å/spectral element, 5 mrad/pixel; Airglow and solar occultation channels
Multispectral Visible Imaging Camera (Ralph/MVIC), S.A. Stern (SwRI), Ball and SwRI	• Hemispheric panchromatic maps of Pluto and Charon at best resolution exceeding 0.5 km/pixel • Hemispheric 4-color maps of Pluto and Charon at best resolution exceeding 5 km/pixel • Search for/map atmospheric hazes at a vertical resolution <5 km • High resolution panchromatic maps of the terminator region • Panchromatic, wide phase angle coverage of Pluto, Charon, Nix, and Hydra • Panchromatic stereo images of Pluto and Charon, Nix, and Hydra • Orbital parameters, bulk parameters of Pluto, Charon, Nix, and Hydra • Search for rings • Search for additional satellites	Visible imaging; 400–975 nm (panchromatic); 4 color filters (Blue, Red, Methane, Near-IR); FOV $5.7° \times 0.15°$ (stare, pan), or $5.7° \times$ arbitrary (scan); IFOV 20 μrad/pixel
Linear Etalon Imaging Spectral Array (Ralph/LEISA), D. Jennings (GSFC), GSFC, Ball, and SwRI	• Hemispheric near-infrared spectral maps of Pluto and Charon at best resolution exceeding 10 km/pixel • Hemispheric distributions of N_2, CO, CH_4 on Pluto at a best resolution exceeding 10 km/pixel • Surface temperature mapping of Pluto and Charon • Phase-angle-dependent spectral maps of Pluto and Charon	IR spectral imaging; 1.25 to 2.5 μm; 1.25–2.50 μm, $\lambda/\delta\lambda \approx 240$; 2.10–2.25 μm, $\lambda/\delta\lambda \approx 550$; FOV $0.9° \times 0.9°$; IFOV 62 μrad/pixel
Radio Science Experiment (REX), L. Tyler (Stanford), Stanford and JHU/APL	• Temperature and pressure profiles of Pluto's atmosphere to the surface • Surface number density to ±1.5%, surface temperature to ±2.2°K and surface pressure to ±0.3 μbar. • Surface brightness temperatures on Pluto and Charon (give wavelength) • Masses and chords of Pluto and Charon; detect or constrain J2s. • Detect, or place limits on, an ionosphere for Pluto	X-band (7.182 GHz uplink, 8.438 GHz downlink); Radiometry $T_{Noise} < 150$ K; Ultra-Stable Oscillator (USO) frequency stability: $\delta f/f = 3 \times 10^{-13}$ over 1 sec

Table 1 (*Continued*)

Instrument, PI	Measurement objectives	Instrument characteristics
Long Range Reconnaissance Imager (LORRI), A. Cheng (JHU/APL), JHU/APL and SSG	• Hemispheric panchromatic maps of Pluto and Charon at best resolution exceeding 0.5 km/pixel. • Search for atmospheric haze at a vertical resolution <5 km • Long time base of observations, extending over 10 to 12 Pluto rotations • Panchromatic maps of the far-side hemisphere • High resolution panchromatic maps of the terminator region • Panchromatic, wide phase angle coverage of Pluto, Charon, Nix, and Hydra • Panchromatic stereo images of Pluto, Charon, Nix, and Hydra • Orbital parameters, bulk parameters of Pluto, Charon, Nix, and Hydra • Search for satellites and rings	Visible panchromatic images; 350–850 nm; FOV $0.29° \times 0.29°$; IFOV 5 µrad/pixel; Optical Navigation
Solar Wind At Pluto (SWAP), D. McComas (SwRI), SwRI	• Atmospheric escape rate from Pluto • Solar wind velocity and density, low energy plasma fluxes and angular distributions, and energetic particle fluxes at Pluto-Charon • Solar wind interaction of Pluto and Charon	Solar wind detector; FOV $200° \times 10°$; Energy Range 0.25–7.5 keV; Energy Resolution RPA: 0.5 V (<1.5 keV) ESA: 0.4 $\Delta E/E$ (>1.4keV)
Pluto Energetic Particle Spectrometer Science Investigation (PEPSSI), R. McNutt (JHU/APL), JHU/APL	• Composition and density of pick-up ions from Pluto, which indirectly addresses the atmospheric escape rate • Solar wind velocity and density, low energy plasma fluxes and angular distributions, and energetic particle fluxes in the Pluto system	Energetic particle detector; Energy Range 1 keV–1 MeV; FOV $160° \times 12°$; Resolution $25° \times 12°$
Venetia Burney Student Dust Counter (VB-SDC), M. Horanyi (U. Colorado), LASP/Colorado	• Trace the density of dust in the Solar System along the New Horizons trajectory from Earth to Pluto and beyond	12 PVF panels to detect dust impacts and 2 control panels shielded from impacts

limited surface mapping and surface composition capabilities in the UV. Alice's spectral bandpass includes lines of CO, atomic H, Ar, and Ne, which may be detectable as airglow, and the electronic bands of N_2, CH_4, and other hydrocarbons and nitriles, which are detectable during solar and stellar occultation observations. Young et al. (2007) provide a detailed discussion of Alice's scientific objectives. Stern et al. (2007) should be consulted for further details on Alice's design and performance.

3.3 Ralph: MVIC and LEISA

Ralph and Alice together comprise the primary remote sensing payload on New Horizons. *Ralph* is named after Alice's husband in *The Honeymooners*. It is a combined visible/NIR imager (called MVIC) and imaging IR spectrograph (called LEISA). Both of these two focal planes are fed by a single telescope assembly, and there is a windowed door (with a transmittance of ~80%) at the input aperture to provide protection against damage and contamination. *MVIC (Multi-spectral Visible Imaging Camera)* is an optical imager employing CCDs with panchromatic and color filters. *LEISA (Linear Etalon Imaging Spectral Array)* is a near infrared (IR) imaging spectrograph employing a 256 × 256 mercury cadmium telluride (HgCdTe) array. In addition to its scientific capabilities, MVIC also serves as an Optical Navigation camera for New Horizons.

The common telescope assembly for Ralph has a three-mirror, off-axis anastigmat design with a 7.5 cm primary mirror. A dichroic reflects the optical light to the MVIC focal plane and transmits the IR light to the LEISA focal plane. Only one focal plane is active at a time, with a relay used to select either MVIC or LEISA. Owing to Ralph's critical role in achieving the New Horizons Group 1 scientific objectives, all of its electronics and some of its focal plane CCDs are redundant.

The MVIC focal plane has seven independent CCD arrays mounted on a single substrate. Figure 3 shows the relative positions of the arrays, as projected on the sky. Six of the arrays have 5000 (columns) × 32 (rows) photosensitive pixels and operate in time-delay integration (TDI) mode. Two of the TDI arrays provide panchromatic (400–975 nm) images, and the other four TDI arrays provide, respectively, color images in blue (400–550 nm), red (540–700 nm), near IR (780–975 nm), and narrow band methane (860–910 nm) channels. The frame transfer array has 5000 (columns) × 128 (rows) pixels and provides panchromatic images (400–975 nm). All of the MVIC arrays have square pixels that are 20 μrad on a side. Thus, the TDI arrays have a field of view of 5.7° × 0.037°, and the frame transfer array has a field of view of 5.7° × 0.15°. To obtain MVIC TDI images, the spacecraft scans the TDI arrays across the target (Fig. 4) at the same rate that charge is shifted from one row to the next, so that the effective exposure time is 32 times the row transfer time. The two TDI panchromatic arrays are sized to meet the 0.5 km/pixel Group 1 mapping requirement near closest approach when Pluto's diameter subtends ~5000 pixels. Each panchromatic array can be operated independently, for redundancy. The four color arrays are operated in tandem. The primary measurement objectives and key characteristics of MVIC are summarized in Table 1.

MVIC images in the three broadband colors will provide information on spectral slopes of Pluto's surface and on its atmospheric properties. The narrow band filter permits mapping of the surface methane abundance, as the well-known 890 nm absorption band is the strongest methane feature available at optical wavelengths. The near IR filter doubles as the continuum comparison for this methane mapping. MVIC's framing array is operated in stare, not scanning, mode, and is used when geometrical fidelity is important (e.g., for optical navigation) or when scanning is not practical (e.g., observing Pluto at closest approach when the apparent motion is too fast). The 700–780 nm gap between the red and near IR bandpasses overlaps another methane band at 740 nm; combining data from the panchromatic, blue, red, and near IR filters can provide some information about band depth in this "virtual" filter. Young et al. (2007) discuss MVIC's scientific objectives in more detail. Further details on MVIC and its performance can be found in Reuter et al. (2007).

LEISA's dispersive capability is provided by its wedged etalon (a linear variable filter, or LVF), which is mounted ~100 μm above its 256 × 256 pixel HgCdTe PICNIC array. The etalon covers 1.25–2.5 μm, a spectral region populated with many absorption features of N_2, CH_4, H_2O, NH_3, CO, and other molecules, at a resolving power of ~250. A higher-resolution sub-segment, covering 2.10–2.25 μm at a resolving power of ~560, will be used to discern grain sizes, mixing states, and pure versus solid-solution abundances (Quirico et al. 1999). The higher-resolution segment is also critical for taking advantage of the temperature-sensitive N_2 bands (Grundy et al. 1993, 1999), and the symmetric, doubled $\upsilon 2 + \upsilon 3$ CH_4 band that is diagnostic of pure versus diluted CH4 abundances (Quirico and Schmitt 1997). As was the case for MVIC, LEISA images are obtained by scanning its field of view across the target (Fig. 4) with the frame transfer rate synchronized with the scan rate. The LVF is oriented so that wavelength varies along the scan direction. Thus, scanning LEISA over a target produces images at different wavelengths (this is unlike the case for MVIC where the scanning simply increases the signal 32-fold). LEISA builds up a conventional spatial-spectral data cube (256 monochromatic images) by scanning the FOV across all portions of the target at a nominal scan rate of 125 μrad/sec. A nominal framing rate of 2 Hz is currently planned to maintain <1 pixel attitude smear and provide good signal-to-noise ratio in the Pluto system. The primary measurement objectives and key characteristics of LEISA are summarized in Table 1. Reuter et al. (2007) provide further details on LEISA's design and performance, and Young et al. (2007) provides a more in-depth discussion of LEISA's scientific objectives.

3.4 REX

REX is the radio science package on New Horizons. *REX* stands for Radio EXperiment. The REX instrument is unique among the suite of instruments comprising the New Horizons payload in that it is physically and functionally incorporated within the spacecraft telecommunications subsystem. Because this system is entirely redundant, two REX's are carried on New Horizons. They can be used simultaneously to increase SNR.

The REX principle of operations for radio science is as follows: the 2.1 m High Gain Antenna aboard New Horizons (see Fig. 1) receives radio signals from NASA's Deep Space Network (DSN) at a carrier frequency of 7.182 GHz. New Horizons transmits radio signals via the antenna to the DSN at a carrier frequency of 8.438 GHz. By measuring phase delays in the received signal as a function of time, the instrument allows one to invert a radio occultation profile into a temperature, number density profile of the intervening atmosphere, if it is sufficiently dense. REX can also operate in a passive radiometry mode to measure radio brightness temperatures at its receiver frequency.

The heart of the REX instrument is an ultra-stable oscillator (USO), which operates at 30 MHz for the down-conversion to an intermediate frequency (IF). An Actel Field Programmable Gate Array (FPGA) takes samples of the IF receiver output and generates wideband radiometer and narrowband sampled signal data products. The REX hardware also includes an analog-to-digital converter (ADC) and other electronics interface components in the telecommunications system. As noted above, there are two copies of the entire telecommunications system (except for the High Gain Antenna), which means that there is full system redundancy in the REX capabilities. The primary measurement objectives and key characteristics of REX are summarized in Table 1. Tyler et al. (2007) discuss REX and its performance in much greater detail.

REX addresses the Group 1 scientific objective of obtaining Pluto's atmospheric temperature and pressure profiles down to the surface using a unique uplink radio occultation technique. REX detects the changes induced by Pluto's atmosphere in the radio signal transmitted to the spacecraft from the DSN. This differs from the typically used downlink method, in which the spacecraft transmits to receivers on Earth. REX will also address Group 2 and Group 3 scientific objectives, including probing Pluto's ionospheric density, searching for Charon's atmosphere, refining bulk parameters like mass and radius, and measuring the surface emission brightness at a wavelength of 4.2 cm, which permits the determination of both the dayside and nightside brightness temperatures with an angular resolution of $\sim 1.2^\circ$ (full-width between the 3 dB points). Young et al. (2007) provide further discussion of the REX's scientific objectives.

3.5 LORRI

The *Long Range Reconnaissance Imager (LORRI)* is a narrow angle (field of view $= 0.29^\circ \times 0.29^\circ$), high resolution (4.96 μrad/pixel), panchromatic (350–850 nm) imaging system. It was placed on New Horizons to augment and also back up Ralph's panchromatic imaging capabilities. LORRI's primary function is to provide higher resolution imagery.

LORRI's input aperture is 20.8 cm in diameter, making LORRI one of the largest telescopes flown on an interplanetary spacecraft. The large aperture, in combination with a high throughput ($QE_{peak} \approx 60\%$) and wide bandpass, will allow LORRI to achieve a signal-to-noise ratio exceeding 100 during disk-resolved observations of Pluto, even though exposure times must be kept below 100 ms to prevent smearing from pointing drift. A frame transfer 1024×1024 pixel (optically active region), thinned, backside-illuminated charge-coupled device (CCD) detector records the image in the telescope focal plane. The CCD output is digitized to 12 bits and stored on the spacecraft's solid state recorder (SSR).

Raw images can be downlinked, but typically the images will be either losslessly or lossy compressed before transmission to the ground in order to minimize the use of DSN resources. LORRI image exposure times can be varied from 0 ms to 29,967 ms in 1 ms steps, and images can be accumulated at a maximum rate of 1 image per second. LORRI's large dynamic range allows it to be an imaging workhorse during the Jupiter encounter, when saturation limits MVIC observations to relatively large solar phase angles.

LORRI operates in an extreme thermal environment, mounted inside the warm spacecraft and viewing cold space, but the telescope's monolithic, silicon carbide construction allows the focus to be maintained over a large temperature range (-120 to 50°C) without any focus adjustment mechanisms. Indeed, LORRI has no moving parts making it a relatively simple, reliable instrument that is easy to operate. A one-time deploy aperture door, mounted on the spacecraft structure, protected LORRI from the harsh launch environment. Cheng et al. (2007) provide a detailed description of LORRI and its performance.

Owing to its higher spatial resolution, higher sensitivity, and lower geometrical distortion (< 0.5 pixel across the entire field of view) compared to Ralph/MVIC, LORRI is also serving as the prime optical navigation instrument on New Horizons. During a typical 100 ms exposure using the full format (1024×1024) mode, LORRI can achieve a signal-to-noise ratio of ~ 5 on $V = 13$ stars. On-chip 4×4 binning, used in conjunction with a special pointing control mode that permits exposing up to 10 s while keeping the target within a single rebinned pixel, allows imaging of point sources as faint as $V \approx 18$, which will permit LORRI to detect a 50 km diameter KBO ~ 7 weeks prior to encounter, thereby enabling accurate targeting to the KBO.

LORRI's primary measurement objectives and key characteristics are summarized in Table 1. LORRI has first successfully detected Pluto (on September 21, 2006 at a distance of 28 AU), and its resolution at Pluto will start exceeding that available from the Hubble Space Telescope approximately 3 months prior to closest approach. En route to Pluto, LORRI will obtain rotationally resolved phase curves of Pluto and later Charon, once the two can be separately resolved. LORRI will obtain panchromatic maps over at least 10 Pluto rotations during approach, with the final complete map of the sunlit hemisphere exceeding a resolution of 0.5 km/pixel. LORRI will map small regions near Pluto's terminator with a resolution of ~ 50 m/pixel near the time of closest approach, depending on the closest approach distance selected. LORRI will also be heavily used for studies requiring high geometrical fidelity, such as the determining the shapes of Pluto, Charon, Nix, and Hydra and refining the orbits of all these objects relative to the system barycenter. LORRI observations at high phase angles will provide a sensitive search for any particulate hazes in Pluto's atmosphere. Young et al. (2007) provides a more detailed discussion of the scientific objectives addressed by LORRI observations.

3.6 SWAP

The Solar Wind Around Pluto (SWAP) instrument is one of two particle detection in situ instruments aboard New Horizons. It is comprised of a retarding potential analyzer (RPA), a deflector (DFL), and an electrostatic analyzer (ESA). Collectively, these elements are used to select the angles and energies of solar wind ions entering the instrument. The selected ions are directed through a thin foil into a coincidence detection system: the ions themselves are detected by one channel electron multiplier (CEM), and secondary electrons produced from the foil are detected by another CEM. SWAP can measure solar wind particles in the energy range from 25 eV up to 7.5 keV with a resolution of $\Delta E/E < 0.4$. SWAP has a fan-shaped field of view that extends $>200°$ in the XY-plane of the spacecraft by $>10°$ out of that plane (see Fig. 2). For typical observations, SWAP measures solar wind speed and density over a 64 sec measurement cycle. The principal measurement objectives and key characteristics of SWAP are summarized in Table 1. Further details on SWAP and its performance can be found in McComas et al. (2007).

SWAP was designed to measure the interaction of the solar wind with Pluto, which addresses the Group 1 scientific objective of measuring Pluto's atmospheric escape rate. Additionally, SWAP has a specific goal of characterizing the solar wind interaction with Pluto as a Group 2 objective. SWAP also addresses the Group 3 objectives of characterizing the energetic particle environment of Pluto and searching for magnetic fields, which it does indirectly. For more details on SWAP's scientific objectives, see McComas et al. (2007) and Young et al. (2007).

3.7 PEPSSI

The *Pluto Energetic Particle Spectrometer Science Investigation (PEPSSI)* is the other in situ particle measurement instrument aboard New Horizons. It is a compact, radiation-hardened particle instrument comprised of a time-of-flight (TOF) section feeding a solid-state silicon detector (SSD) array. Each SSD has 4 pixels, 2 dedicated to ions, and 2 for electrons. PEPSSI's field of view (FOV) is fan-like and measures $160° \times 12°$, divided into six angular sectors of $25° \times 12°$ each. Ions entering the PEPSSI FOV generate secondary electrons as they pass through entrance and exit foils in the TOF section, providing "start" and "stop" signals detected by a microchannel plate (MCP). Particle energy information, measured by the SSD, is combined with TOF information to identify the particle's composition. Each particle's direction is determined by the particular 25° sector in which it is detected. Event classification electronics determine incident mass and energy, with 12 channels of energy resolution. Protons can be detected in the energy range 40–1000 keV, electrons in the range 25–500 keV, and CNO ions in the range 150–1000 keV. TOF-only measurements extend to <1 keV for protons, to 15 keV for CNO ions, and to 30 keV for N_2^+. TOF measurements are possible in the range 1–250 ns to an accuracy of ± 1 ns. The geometrical factor for ions is slightly larger than 0.1 cm^2 sr. A typical measurement includes 8-point spectra for protons and electrons and reduced resolution energy spectra for heavier ions for all six look directions. The mass resolution of PEPSSI varies with energy: for CNO ions, it is <5 AMU for

>1.7 keV/nucleon, and <2 AMU for >5 keV/nucleon. The principal measurement objectives and key instrument characteristics of PEPSSI are summarized in Table 1. McNutt et al. (2007) provide a detailed discussion of PEPSSI and its performance.

The PEPSSI design is derived from that of the Energetic Particle Spectrometer (EPS), which is flying on the MESSENGER mission to Mercury. PEPSSI has thinner foils than EPS, which enables measurements down to smaller energy ranges. PEPSSI also has a slightly increased geometric factor and draws less power than EPS. Both EPS and PEPSSI trace back their heritage to a NASA PIDDP program in the 1990s to develop a particle instrument for use on a Pluto flyby mission.

By measuring energetic pickup ions from Pluto's atmosphere, PEPSSI provides information related to the atmospheric escape rate on Pluto, which is a New Horizons Group 1 scientific objective. PEPSSI's primary role, however, is to address the Group 3 objective of characterizing the energetic particle environment of Pluto. Fluxes of energetic pickup ions may be measured as far as several million kilometers from Pluto (see Bagenal et al. 1997), and PEPSSI observations will be used to determine the mass, energy spectra, and directional distributions of these energetic particles (Bagenal and McNutt 1989). Secondarily, PEPSSI will also provide low-resolution, supporting measurements of the solar wind flux, complementing SWAP. Young et al. (2007) provides a more detailed discussion of PEPSSI's scientific objectives.

3.8 VB-SDC

The Student Dust Counter (SDC), also known as the Venetia Burney SDC in honor of the student who named Pluto in 1930, is an impact dust detector that will be used to map the spatial and size distribution of interplanetary dust along the trajectory of the New Horizons spacecraft from the inner solar system to and through the Kuiper Belt.

Unlike all of the other instruments, the VB-SDC was not part of the original New Horizons proposal and was added by NASA as an Education and Public Outreach (EPO) experiment. For the first time ever, students were given the opportunity to design, build, and operate an instrument for an interplanetary mission. (NASA-certified personnel performed all quality assurance inspections and supervised the final assembly.) Approximately 20 undergraduate physics and engineering students at the University of Colorado worked on the VB-SDC and, despite getting a rather late start, their instrument was the first to be delivered to the New Horizons spacecraft.

The VB-SDC's sensors are thin, permanently polarized polyvinylidene fluoride (PVDF) plastic films that generate an electrical signal when dust particles penetrate their surface. The SDC has a total sensitive surface area of $\sim$0.1 m^2, comprised of 12 separate film patches, each 14.2 cm $\times$ 6.5 cm, mounted onto the top surface of a support panel. In addition, there are two reference sensor patches mounted on the backside of the detector support panel, protected from any dust impacts. These reference sensors, identical to the top surface sensors, are used to monitor the various background noise levels, from mechanical vibrations or cosmic ray hits.

The entire support panel is mounted on the exterior of the New Horizons spacecraft, outside the spacecraft multi-layer insulating (MLI) blanket, facing the ram ($-Y$) direction. The VB-SDC observations are most useful during the cruise phases of the mission, when the spacecraft is spinning and the other instruments are turned off. Thruster firings during 3-axis operations generate large VB-SDC background signals, which make it very difficult to detect true IDP impacts.

The VB-SDC was designed to resolve, to within a factor of $\sim$2, the masses of interplanetary dust particles (IDPs) in the range of $10^{-12} < m < 10^{-9}$ g, which corresponds roughly to a size range of 1 – 10 μm in particle radius. Bigger grains are also recorded, but their masses cannot be resolved. With the characteristic spacecraft speed during cruise of ~13 km/s, current models of the dust density in the solar system (Divine 1993) suggest that the VB-SDC should record approximately 1 IDP hit per week.

The principal measurement objectives and key instrument characteristics of the VB-SDC are summarized in Table 1. Horanyi et al. (2007) provide a detailed discussion of the VB-SDC and its performance.

4 Science Payload Commissioning Overview

The New Horizons instrument commissioning activities began shortly after the nominal performance of the spacecraft subsystems was verified; this was approximately 1 month after launch. Over a period of about 8 months, each instrument team developed a detailed set of science activity plans (SAPs) to characterize the in-flight performance and functionality, and to verify that their measurement objectives could be achieved. Functional tests were executed first to demonstrate that critical engineering parameters (e.g., currents, voltages, temperatures, etc.) fell within their expected ranges. After nominal functional performance was verified, a series of performance tests were executed for each instrument. All of the commissioning tests discussed below took place during calendar year 2006. A small set of commissioning activities (about 10% of the total) were completed during and shortly after the Jupiter encounter in 2007.

The instruments completed their functional tests during February–March 2006. The first observations of an external target are termed "first light" observations, and these were staggered throughout the May to September 2006 period for the various instruments. Alice detected interplanetary hydrogen Lyman-α and Lyman-β emission during its first light observations on May 29. Alice then observed two UV calibration stars, γ Gruis and ρ Leonis, on August 31. Owing to safety reasons, the Alice SOCC door was not opened until July 2007, after the Jupiter encounter. Ralph/MVIC first light occurred during observations of its stellar calibration targets (the M6 and M7 galactic open clusters) through its windowed door on May 10, and then through the opened door on May 28. Both Ralph/MVIC and Ralph/LEISA observed the asteroid 2002 JF56 in a moving target tracking test during May 11–13. Ralph/LEISA made the first observations of its calibration star (Procyon) on June 29. LORRI first light occurred when it opened its aperture door on August 29 and observed M7. LORRI observed M7 for an extensive set of calibration observations on September 3, including a simultaneous observation with MVIC to measure the relative alignments of those two instruments. Both Ralph and LORRI observed Jupiter on September 4 as test observations in preparation for the Jupiter encounter in February 2007. Ralph and LORRI also observed Uranus and Neptune in September for optical navigation testing. LORRI observed Pluto during observations on September 21 and 23, and the Jovian irregular satellite Himalia on September 22, again as part of optical navigation testing. The first use of REX mode by the telecommunications system took place on April 19. REX scanning observations to measure the high gain antenna (HGA) beam pattern were performed on June 20. Two radio calibration sources (Cass A and Taurus A) and "cold sky" were observed on June 29 to measure the REX radiometry mode performance. All REX calibration observations in 2006 were performed on side-A; side-B calibration observations were executed in early January 2007. SWAP's door was opened on March 13, but the first solar wind observations started in late-September and continued through December. PEPSSI's door was opened on May 3, but its ability to measure particles was first tested in June. The VB-SDC attempted to take science data in early-March, but the spacecraft was in 3-axis mode and the high VB-SDC background rate produced by the nearly continual thruster firings made it essentially impossible to detect real dust particle events. The VB-SDC had its first real chance to detect dust particles while the spacecraft was in "passive" spin mode (thruster firings still occur during "active" spin mode) in April, but the relatively low count rate expected requires that the instrument be well-calibrated and the data carefully analyzed, Additional VB-SDC data was then taken from October through December 2006, while the spacecraft remained in spin mode.

5 In Flight Hibernation, Annual Checkouts, and Encounter Rehearsals

The New Horizons mission is exceptionally long in duration, with the primary mission objective not being completed until nearly 10 years after launch. Activities during the mission are generally either front-end or back-end loaded, with the first 14 months busy with instrument commissioning and the Jupiter encounter, and the last year of the mission devoted to intensive observations of the Pluto system. For most of the time during the 8 years between the encounter phases (2007–2014, inclusive), the spacecraft will be placed into a "hibernation" mode, with all non-essential subsystems, including the scientific payload, powered off. This preserves component life.

During the hibernation period, beacon radio tones are sent periodically from the spacecraft to the Earth that allow flight controllers to verify the basic health and safety of the spacecraft. Additionally, monthly telemetry passes are scheduled to collect engineering trend data.

Although the spacecraft is kept in hibernation to reduce component use prior to the Pluto-system encounter, it is also important to verify periodically the performance of the spacecraft subsystems and instruments, and to keep the mission operations team well trained and prepared for the Pluto encounter activities. Therefore, the spacecraft will be brought out of hibernation each year for roughly 60 days, called "annual checkouts" (ACOs), during which time the performance of the spacecraft subsystems and instruments can be verified. Generally, the ACO instrument activities are comprised of a subset of the commissioning activities that focus on the instrument's performance (e.g., stellar calibration observations). ACOs are also the opportunity for annual cruise science observations to be collected, such as interplanetary charged particle measurements, studies of the hydrogen distribution in the interplanetary medium, and extensive phase curve studies of Pluto, Charon, Uranus, Neptune, Centaurs, and KBOs, none of which can be obtained from spacecraft near Earth.

In addition, two full rehearsals of the Pluto encounter will be conducted, during the summers of 2009 and 2014, respectively, that will serve both to verify that the Pluto encounter sequence will work and to provide essential training for the mission operations team in preparation for the actual encounter.

6 Current Status of the Science Payload

All seven of the instruments comprising the New Horizons science payload have essentially completed their in-flight commissioning activities. In all cases, the in-flight performance verifies that the science payload can meet its measurement objectives, thereby accomplishing all of the scientific objectives of the New Horizons mission.

The Ralph instrument was used to observe asteroid 2002 JF56 during a serendipitous flyby at a closest approach distance of 100,000 km on 2006 June 13 (Olkin et al. 2006), which verified the spacecraft's ability to track reliably a fast-moving target. The SWAP, PEPSSI, and VB-SDC instruments also began taking scientific data, in addition to commissioning data, during 2006.

All of the instruments participated in the Jupiter encounter (with closest approach on 2007 February 28), which was considerably more ambitious than any of the previous activities. In fact, the Jupiter encounter had roughly twice the number of observations, and double the data volume, of what was originally planned for the Pluto encounter in 2015. Most importantly, however, the Jupiter encounter provided an invaluable and unique opportunity to test the mission's capabilities. Even though a few of the observations taken in the Jovian system failed, the lessons learned from that encounter undoubtedly improved the prospects for a successful Pluto system encounter in 2015, which is the most important activity of the New Horizons mission.

Acknowledgements We thank all of the New Horizons Instrument Teams for their extraordinary efforts in designing, developing, testing, and delivering a highly capable science payload that promises to revolutionize our understanding of the Pluto system and the Kuiper belt. We also thank the numerous contractors who partnered with the Instrument Teams for their outstanding work and dedication. Partial financial support for this work was provided by NASA contract NAS5-97271 to the Johns Hopkins University Applied Physics Laboratory.

References

F. Bagenal R.L. McNutt Jr., Pluto's interaction with the solar wind. Geophys. Res. Lett. **16**, 1229–1232 (1989)

F. Bagenal, T.E. Cravens, J.G. Luhmann, R.L. McNutt Jr., A.F. Cheng, Pluto's interaction with the solar wind, in *Pluto and Charon*, ed. by S.A. Stern, D.J. Tholen (Univ. of Arizona, Tucson, 1997), pp. 523–555

A.F. Cheng et al., Long Range Reconnaissance Imager on New Horizons. Space Sci. Rev. (2007, this issue). doi:10.1007/s11214-007-9271-6

N. Divine, Five populations of interplanetary meteoroids. J. Geophys. Res. **98**, 17029–17051 (1993). doi:10.1029/93JE01203

W.M. Grundy, B. Schmitt, E. Quirico, The temperature dependent spectra of alpha and beta nitrogen ice with application to Triton. Icarus **105**, 254–258 (1993). doi:10.1006/icar.1993.1122

W.M. Grundy, M.W. Buie, J.A. Stansberry, J.R. Spencer, B. Schmitt, Near-infrared spectra of icy outer solar system surfaces: Remote determination of H_2O ice temperatures. Icarus **142**, 536–549 (1999). doi:10.1006/icar.1999.6216

M. Horanyi et al., The Student Dust Counter on the New Horizons mission. Space Sci. Rev. (2007, this issue). doi:10.1007/s11214-007-9250-y

D. McComas et al., The Solar Wind Around Pluto (SWAP) instrument aboard New Horizons. Space Sci. Rev. (2007, this issue). doi:10.1007/s11214-007-9205-3

R.E. McNutt et al., The Pluto Energetic Particle Spectrometer Science Investigation (PEPSSI) on New Horizons. Space Sci. Rev. (2007, this issue)

C.B. Olkin, D. Reuter, A. Lunsford, R.P. Binzel, S.A. Stern, The New Horizons distant flyby of asteroid 2002 JF56. DPS Meeting #38 (2006) abstract #59.22

E. Quirico, B. Schmitt, A spectroscopic study of CO diluted in N_2 ice: Applications for Triton and Pluto. Icarus **128**, 181–188 (1997). doi:10.1006/icar.1997.5710

E. Quirico, S. Doute, B. Schmitt, C. de Bergh, D.P. Cruikshank, T.C. Owen, T.R. Geballe, T.L. Roush, Composition, physical state, and distribution of ices at the surface of Triton. Icarus **139**, 159–178 (1999).

D. Reuter et al., Ralph: A visible/infrared imager for the New Horizons mission. Space Sci. Rev. (2007, this issue). doi:10.1007/s11214-008-9375-7

G.D. Rogers, M.R. Schwinger, J.T. Kaidy, T.E. Strikwerda, R. Casini, A. Landi et al., Autonomous star tracker performance, in *Proc. The 57th IAC Congress*, Valenica, Spain, 2006

S.A. Stern, The New Horizons Pluto Kuiper belt Mission: An overview with historical context. Space Sci. Rev. (2007, this issue). doi:10.1007/s11214-007-9295y

S.A. Stern et al., Alice: The ultraviolet imaging spectrometer aboard the New Horizons Pluto-Kuiper belt mission. Space Sci. Rev. (2007, this issue) doi:10.1007/s11214-007-9295-y

L. Tyler et al., The Radio EXperiment (REX) on New Horizons. Space Sci. Rev. (2007, this issue) doi:10.1007/s11214-007-9302-3

L.A. Young et al., New Horizons: Anticipated scientific investigations at the Pluto system. Space Sci. Rev. (2007, this issue)

New Horizons: Anticipated Scientific Investigations at the Pluto System

Leslie A. Young · S. Alan Stern · Harold A. Weaver · Fran Bagenal · Richard P. Binzel · Bonnie Buratti · Andrew F. Cheng · Dale Cruikshank · G. Randall Gladstone · William M. Grundy · David P. Hinson · Mihaly Horanyi · Donald E. Jennings · Ivan R. Linscott · David J. McComas · William B. McKinnon · Ralph McNutt · Jeffery M. Moore · Scott Murchie · Catherine B. Olkin · Carolyn C. Porco · Harold Reitsema · Dennis C. Reuter · John R. Spencer · David C. Slater · Darrell Strobel · Michael E. Summers · G. Leonard Tyler

Originally published in the journal Space Science Reviews, Volume 140, Nos 1–4, 93–127.
DOI: 10.1007/s11214-008-9462-9

L.A. Young (✉) · S.A. Stern · C.B. Olkin · J.R. Spencer
Southwest Research Institute, Boulder, CO, USA
e-mail: layoung@boulder.swri.edu

H.A. Weaver · A.F. Cheng · R. McNutt · S. Murchie
Johns Hopkins University Applied Physics Lab., Laurel, MD, USA

F. Bagenal · M. Horanyi
University of Colorado, Boulder, CO, USA

R.P. Binzel
Massachusetts Institute of Technology, Cambridge, MA, USA

B. Buratti
Jet Propulsion Laboratory, Pasadena, CA, USA

D. Cruikshank · J.M. Moore
NASA Ames Research Center, Moffett Field, CA, USA

G.R. Gladstone · D.J. McComas · D.C. Slater
Southwest Research Institute, San Antonio, TX, USA

W.M. Grundy
Lowell Observatory, Flagstaff, AZ, USA

D.P. Hinson · I.R. Linscott · G.L. Tyler
Stanford University, Stanford, CA, USA

D.E. Jennings · D.C. Reuter
NASA Goddard Space Flight Center, Greenbelt, MD, USA

Abstract The New Horizons spacecraft will achieve a wide range of measurement objectives at the Pluto system, including color and panchromatic maps, 1.25–2.50 micron spectral images for studying surface compositions, and measurements of Pluto's atmosphere (temperatures, composition, hazes, and the escape rate). Additional measurement objectives include topography, surface temperatures, and the solar wind interaction. The fulfillment of these measurement objectives will broaden our understanding of the Pluto system, such as the origin of the Pluto system, the processes operating on the surface, the volatile transport cycle, and the energetics and chemistry of the atmosphere. The mission, payload, and strawman observing sequences have been designed to achieve the NASA-specified measurement objectives and maximize the science return. The planned observations at the Pluto system will extend our knowledge of other objects formed by giant impact (such as the Earth–moon), other objects formed in the outer solar system (such as comets and other icy dwarf planets), other bodies with surfaces in vapor-pressure equilibrium (such as Triton and Mars), and other bodies with N_2:CH_4 atmospheres (such as Titan, Triton, and the early Earth).

Keywords Pluto · Charon · Nix · Hydra · New Horizons

1 Introduction

The New Horizons spacecraft was launched on 19 January 2006 on its 9.5-year journey to make the first reconnaissance of the Pluto system, including its satellites Charon, Nix, and Hydra, and a possible extended mission to one or more Kuiper Belt Objects (KBOs). While spacecraft have visited each of the terrestrial and jovian planets, this will be the first spacecraft to explore any of the icy dwarf planets that dominate the third, outer portion of our solar system. By improving our knowledge of these distant bodies, we will extend our understanding of ice dwarfs and Kuiper Belt Objects and of the origin and evolution of our solar system.

The New Horizons science payload is comprised of Alice, an ultraviolet spectrometer; Ralph/MVIC, a visible panchromatic and color imager; Ralph/LEISA, an infrared imaging spectrometer; REX, a radio science experiment; LORRI, a high-resolution panchromatic imager; PEPSSI, an energetic particle detector; SWAP, a solar wind analyzer; and SDC, a student-built dust counter.

In this article, we describe the scientific investigations that the New Horizons science team (Table 1) intends to perform at the Pluto system. In Sect. 2, we describe the science goals for which the New Horizons mission was designed. In Sect. 3, we describe the planned

W.B. McKinnon
Washington University, Saint Louis, MO, USA

C.C. Porco
Space Science Institute, Boulder, CO, USA

H. Reitsema
Ball Aerospace and Technologies Corporation, Boulder, CO, USA

D. Strobel
Johns Hopkins University, Baltimore, MD, USA

M.E. Summers
George Mason University, Fairfax, VA, USA

Table 1 New Horizons science team

Name	Role
Alan Stern (SwRI)	New Horizons PI, Alice PI, Ralph PI
Hal Weaver (JHU/APL)	New Horizons PS, LORRI IS, GGI Team Member, ATM Team Member
Leslie Young (SwRI)	New Horizons Deputy PS, ATM Team Member
Fran Bagenal (CU)	P&P Lead, EPO Lead (2001–2007)
Richard Binzel (MIT)	GGI Team Member
Bonnie Buratti (JPL)	GGI Team Member
Andrew Cheng (JHU/APL)	LORRI PI, P&P Team Member
Dale Cruikshank (NASA/ARC)	COMP Team Member
Randall Gladstone (SwRI)	ATM Lead
William Grundy (Lowell)	COMP Lead
David Hinson (Stanford)	ATM Team Member
Mihaly Horanyi (CU)	SDC PI, P&P Team Member
Donald Jennings (NASA/GSFC)	Ralph/LEISA PI, COMP Team Member
Ivan Linscott (Stanford)	REX IS, ATM Team Member
Dave McComas (SwRI)	SWAP PI, P&P Team Member
William McKinnon (Wash. U)	GGI Team Member
Ralph McNutt (JHU/APL)	PEPSSI PI, P&P Team Member, ATM Team Member
Jeffrey Moore (NASA/ARC)	GGI Lead
Scott Murchie (JHU/APL)	GGI Team Member
Catherine Olkin (SwRI)	EPO Lead (2007–), GGI Team Member
Carolyn Porco (SSI)	GGI Team Member
Harold Reitsema (Ball)	GGI Team Member
Dennis Reuter (NASA/GSFC)	Ralph IS, GGI Team Member, COMP Team Member
David Slater (SwRI)	Alice IS, ATM Team Member
John Spencer (SwRI)	GGI Deputy, COMP Team Member
Darrell Strobel (JHU)	ATM Team Member
Michael Summers (GMU)	ATM deputy
Leonard Tyler (Stanford)	REX PI, ATM Team Member

PI = Principal Investigator. PS = Project Scientist. IS = Instrument Scientist. ATM = Atmospheres Theme Team. GGI = Geology, Geophysics and Imaging Theme Team. COMP = Composition Theme Team. P&P = Particles and Plasmas Theme Team. EPO = Education & Public Outreach

observations at Pluto and the Pluto system to fulfill these goals. This section includes a summary of the instrument suite (Sect. 3.1), the mission design (Sect. 3.2), the encounters with asteroid 2002 JF_{56} and Jupiter (Sects. 3.3 and 3.4), the planned encounter with the Pluto system (Sect. 3.5) and possible encounters with KBOs (Sect. 3.6) based on pre-launch strawman sequences. In this paper, we describe the payload and mission design in the context of our science goals. Further details on the payload and mission design, including in-flight observations, are given in companion papers in this issue.

In Sect. 4, we show the mapping of the observations to the specific mission goals, and in Sect. 5, we discuss some of the ways in which the planned observations will further planetary science in general.

2 Mission Science Goals

The Pluto system (Table 2) consists of Pluto, discovered in 1930, its large satellite Charon, discovered in 1978, and two smaller moons Nix and Hydra, discovered in 2005. Pluto's heliocentric orbit has an inclination of 17.2 deg and an eccentricity of 0.25, with a heliocentric distance that ranges from 29.7 AU at perihelion (1989 Sep 5) to 49.3 AU (2114 Feb 19), leading to a factor of 2.8 in received insolation over its orbit. Pluto has a high obliquity of 119 deg (Tholen and Buie 1997); the sub-solar latitude ranges from 0 deg at equinox (1987 Dec 11) to 57.5 deg at the solstices (−57.5 deg on 2029 Jul 10). At the time of the New Horizons closest approach, Pluto will be 32.9 AU from the Sun at an ecliptic latitude of only 1.9 deg, with a sub-solar latitude of −49.5 deg. (Here and in the remainder of the paper, as is the convention for the New Horizons mission, we use the International Astrophysical Union (IAU) definition of Pluto's North pole, which is equivalent to Pluto's south rotational pole.)

Pluto and Charon form a double planet, with the center of mass of the system exterior to Pluto. Despite their proximity, they are vastly different worlds (Stern 1992). Pluto is red in color, with a surface covered with volatiles (N_2, CH_4, CO) that support a seasonally variable atmosphere. The sublimation/condensation cycle, plus photochemical processing, leads to an overall high albedo, and large albedo variations. Charon, in contrast, is neutral in color, and has a water-ice dominated surface with little spatial variation; no Charonian atmosphere has been detected. Little is currently known about the physical properties of Nix and Hydra beyond their orbits, magnitudes, and gross colors (Buie et al. 2006; Stern et al. 2007), with sizes and masses that have been inferred from likely ranges of albedos and densities.

Based on the remarkable growth in our knowledge of Pluto and Charon in the 1980's and early 1990's, and guided by the Voyager encounter of Triton, in 1992 NASA's Outer Planet Science Working Groups (OPSWG, S.A. Stern, chair) described three high priority, nine second priority, and four third priority science objectives for Pluto system flyby reconnaissance. These objectives were adopted, essentially verbatim, in the October 1995 Pluto–Kuiper Express Science Definition Team Report (J.I. Lunine, chair), the September 1999 Announcement of Opportunity for the Pluto–Kuiper Express (PKE-MPD AO: 99-OSS-04), and the January 2001 Announcement of Opportunity for the Pluto-Kuiper Belt Mission (PKB-AO: AO:01-OSS-01) (Table 3). More detail on the history of the mission is given by Stern (2008).

Important new results for Pluto and Charon since the Mission Science Goals were first proposed include the discovery of the moons Nix and Hydra (Weaver et al. 2006), the measurement of Pluto's post-perihelion atmospheric expansion (Elliot et al. 2003; Sicardy et al. 2003), new limits on Charon's atmosphere and measurement of its radius (Gulbis et al. 2006; Person et al. 2006), rotationally or spatially resolved measurements of Pluto's temperature (Lellouch et al. 2000), surface composition (Grundy and Buie 2001), and color (Young et al. 2001a). These have upheld, rather than challenged, the importance of the science goals to which the New Horizons mission was designed.

2.1 Geology and Morphology (Goals 1.1, 2.1, 2.2, 2.3, 2.8)

From rotational photometric and compositional lightcurves, mutual events (Pluto and Charon transiting and eclipsing each other), and the first post-repair HST images, it is clear that Pluto had extreme albedo variations, suggesting a complex geological and geomorphological surface (Buie et al. 1997b; Stern et al. 1997a; Young et al. 1999, 2001a; see Fig. 1). Panchromatic images of Pluto, Charon, Nix, and Hydra will provide a spatial context for the interpretation of New Horizons and observatory-based (ground-based or near-Earth space-based) data. Broad-band colors can be used with albedo to differentiate the surfaces into

Table 2 Inventory of the Pluto system

Object	Semi-major axis, a (km)[a]	Absolute magnitude, H[c]	Radius (km)	Mass (10^{22} kg)	Geometric albedos, V	Surface constituents	Atmosphere
Pluto	2042 ± 86[b]	-0.5368 ± 0.0022[d]	1140–1190[e]	1.305 ± 0.006[d]	0.52[h]	N_2:CH_4, CH_4, CO, possible tholins[i]	N_2, CH_4, CO[k]
Charon	17529 ± 86[b]	1.3488 ± 0.0868[d]	606.0 ± 1.5[f]	0.152 ± 0.006[d]	0.35[h]	H_2O, possible NH_3, $NH_3 \cdot 2H_2O$[j]	None detected[f]
Nix (2005/P2)	48675 ± 121[b]	8.58 ± 0.11[g]	20–70[g]	< 0.0004[g]	0.04–0.35[g]	Unknown	Unlikely
Hydra (2005/P1)	64780 ± 88[b]	8.13 ± 0.12[g]	23–83[g]	< 0.0004[g]	0.04–0.35[g]	Unknown	Unlikely

[a] Semi-major axis of orbit, relative to system barycenter. Charon semimajor axis (relative to Pluto) is 19571.4 ± 4.0 (Buie et al. 2006)

[b] Buie et al. (2006)

[c] V magnitude at geocentric distance = 1 AU, heliocentric distance = 1 AU, phase angle = 0 degrees

[d] Average over a rotation period from observations in 1992 and 1993 (Buie et al. 1997a)

[e] Tholen and Buie (1997)

[f] Young et al. (2005), Gulbis et al. (2006), Sicardy et al. (2006)

[g] Weaver et al. (2006), Stern et al. (2007). Albedo ranges and densities are assumed

[h] Based on rotationally averaged magnitudes. Radius of 1180 km used for Pluto albedo

[i] Owen et al. (1993), Olkin et al. (2007)

[j] Brown and Calvin (2000), Brown (2002)

[k] Owen et al. (1993), Young et al. (1997), Young et al. (2001b)

Table 3 New Horizons Pluto System Science Objectives as stated in PKB-AO: AO:01-OSS-01

Group 1 (required)	
Goal No.	Goal
1.1	**Characterize the global geology and morphology of Pluto and Charon**
1.1a	*Hemispheric panchromatic maps:* Obtain panchromatic viewable disk coverage of both Pluto and Charon at a resolution of 1 kilometer per line pair (1 km/lp), or equivalent. Viewable disk means the entire lit and visible surface of the target body viewed from the spacecraft at a single point in time during the approach to the target. The 1 km/lp objective applies to the subspacecraft point; it is understood that a combination of image projection effects and spacecraft data storage limitations may degrade resolution away from the subspacecraft point.
1.1b	*Color maps:* Obtain viewable disk coverage of both Pluto and Charon in 2 to 5 color bands at a resolution of 3–10 km/lp (or equivalent). The resolution objective applies to the subspacecraft point; it is understood that a combination of image projection effects and spacecraft data storage limitations may degrade resolution away from the subspacecraft point.
1.1c	*Phase angle coverage:* Obtain sufficient imaging at moderate and high phase angles to specify the phase integrals of Pluto and Charon.
1.1d	*Image dynamic range and signal-to-noise ratio (S/N):* For all imaging, provide sufficient dynamic range to cover brightness contrasts of up to 30 (i.e., normal albedo between 0.03 and 1) with an average S/N goal of about 100, but somewhat lower S/N in the darkest regions.
1.2	**Map surface composition of Pluto and Charon**
1.2a	*Hemispheric infrared spectroscopic maps:* Obtain infrared spectroscopic maps of one hemisphere of both Pluto and Charon with approximately 10 km/pixel resolution at disk center with the ability to detect a <0.02 change in albedo everywhere in the spectrum.
1.2b	*Goal for compositional determination:* Determine the spatial distribution of frozen N_2 and secondary constituents such as CO, CH_4. Determine quantitatively the presence of such additional major exposed volatiles, hydrocarbons, and minerals (or rocks) as may exist, all at spatial resolution of 5–10 km/pixel or equivalent.
1.2c	*Spectral coverage and resolution:* For each spatial resolution element, obtain a spectral resolution ($\lambda/\Delta\lambda$) of at least 250 over all or part of the 1–5 micron region (or beyond, if relevant).
1.3	**Characterize the neutral atmosphere of Pluto and its escape rate**
1.3a	*Composition:* Determine the mole fractions of N_2, CO, CH_4 and Ar in Pluto's atmosphere to at least the 1% level of the total mixing ratio.
1.3b	*Thermospheric thermal structure:* Measure T and dT/dz at 100 km vertical resolution to 10% accuracy at gas densities of 10^9 cm^{-3} and higher.
1.3c	*Aerosols:* Characterize the optical depth and distribution of near-surface haze layers over Pluto's limb at a vertical resolution of 5 km or better.
1.3d	*Lower atmospheric thermal structure:* Measure temperature and pressure at the base of the atmosphere to accuracies of ± 1 K and 0.1 microbar.
1.3e	*Evolution:* Determine the escape rate.

distinct color/albedo units, presumably related to geological units, as was done for Triton (Brown et al. 1995). On Pluto, global albedo maps are also critical for understanding the seasonal volatile transport cycle, since the temperatures of both volatiles and substrates depend on their bond albedos. While the requirements (Goals 1.1a and 1.1b) are to image one hemisphere of Pluto and Charon, the New Horizons mission will map as much of the surfaces of Pluto, Charon, Nix and Hydra as possible. Because the frosts on Pluto's surface are expected to migrate, New Horizons will also image Pluto over time (Goal 2.1).

Both hemispheric maps (Goal 1.1) and maps at higher resolution (Goal 2.3) will elucidate the underlying geological and geomorphological processes. For example, visible imaging will detect craters, tectonic features, and changes in reflectivity indicative of various com-

Table 3 (*continued*)

Group 2 (strongly desired)	
Goal No.	Goal
2.1	Characterize the time variability of Pluto's surface and atmosphere
2.2	Image Pluto and Charon in stereo
2.3	Map the terminators of Pluto and Charon with high resolution
2.4	Map the surface composition of selected areas of Pluto and Charon with high resolution
2.5	Characterize Pluto's ionosphere and solar wind interaction
2.6	Search for neutral species including H, H_2, HCN, and C_xH_y, and other hydrocarbons and nitriles in Pluto's upper atmosphere, and obtain isotopic discrimination where possible
2.7	Search for an atmosphere around Charon
2.8	Determine bolometric Bond albedos for Pluto and Charon
2.9	Map the surface temperatures of Pluto and Charon.
Group 3 (desired)	
Goal No.	Goal
3.1	Characterize the energetic particle environment of Pluto and Charon
3.2	Refine bulk parameters (radii, masses, densities) and orbits of Pluto and Charon
3.3	Search for magnetic fields of Pluto and Charon (indirectly addressed by New Horizons)
3.4	Search for additional satellites and rings.

Fig. 1 False-color B-V image of Pluto (Young et al. 2001a) rotated so that the IAU North pole (rotational South pole, and current winter pole) is up

positions, origins, and evolution, while observations of the same areas at different viewing geometries can yield bond albedo (Goal 2.8, related to the phase integrals, Goal 1.1c), and grain size. Height and terrain information are also important, and can be derived from stereo imaging (Goal 2.2) and from shadows and shading (especially effective at high phase angle, Goal 2.3, and combined with a photometric phase function, Goal 1.1c).

New Horizons flies two cameras for panchromatic imaging (LORRI and Ralph/MVIC), one of which (Ralph/MVIC) also has three broad color filters and a filter designed to detect CH_4 ice.

2.2 Surface Composition (Goals 1.2, 2.1, 2.4, 2.9)

Infrared spectra of surfaces in the outer solar system (Goal 1.2c) are rich in spectral features. These spectra are particularly diagnostic at wavelengths shorter than 2.5 microns, because the intensity of sunlight (and consequently signal-to-noise) drops at longer wavelengths. Spectroscopy of Pluto's surface shows spectral features of frozen N_2 (2.15 micron), CH_4 (weak and strong bands that dominate Pluto's spectrum) and CO (1.58 and 2.35 micron) (Owen et al. 1993; Douté et al. 1999; Cruikshank et al. 1997; see Fig. 2a). Charon's infrared spectrum shows H_2O ice (including the 1.65 micron feature indicative of crystalline H_2O ice) and probably NH_3 or its hydrate $NH_3 \cdot 2H_2O$ near 2.2 micron (Brown and Calvin 2000; Brown 2002), although other species, such as aluminum-bearing phyllosilicates, also have absorption bands at 2.215 micron (Buie and Grundy 2000).

N_2, CH_4, and CO are all volatile at Pluto's surface temperature, and therefore support Pluto's atmosphere and participate in Pluto's seasonal cycle (Spencer et al. 1997). These ices are not uniformly distributed over Pluto's surface (Grundy and Buie 2001). An understanding of the spatial distribution of frozen N_2, CH_4, and CO (Goal 1.2b) is vital for understanding the surface energy balance, since energy is transported across Pluto's surface through latent heat of sublimation (Spencer et al. 1997). Spatially resolved spectra will also allow a more sensitive search for additional surface species (Goal 1.2b). On Pluto, for example, exposed H_2O ice can be more easily identified without the interference of CH_4 bands (Grundy and Buie 2002). On Charon, NH_3 may be confirmed in spatially resolved spectra by the detection of the 1.99 micron band of NH_3, which may be suppressed by the presence of the broad H_2O band at 2.0 micron in disk-averaged spectra (Buie and Grundy 2000).

With sufficient signal-to-noise ratio and spectral resolution (Goals 1.2a and 1.2c), infrared (IR) spectra can be used for much more than the mere determination of the presence of surface species. The IR spectra of N_2, CH_4, and H_2O are all temperature dependent (Goal 2.9; Tryka et al. 1995; Grundy et al. 1993, 1999, 2002), especially when the simultaneous observation of both strong and weak bands of CH_4 can break the grainsize/temperature ambiguity. The bands of frozen CH_4 shift according to whether or not the CH_4 ice is pure or in a solid solution (i.e., whether a CH_4 molecule neighbors other CH_4 molecules or different host, or matrix, molecules).

Observations over a long timebase may detect the migration of volatiles (Goal 2.1). High spatial resolution composition maps (Goal 2.9) allow the connection between geologic processes and composition, illuminating questions such as: what are the chromophores causing Pluto's red color? If New Horizons finds sublimation scarps, will they be associated with areas of volatiles? Do craters reveal a change in composition with depth?

For composition mapping, New Horizons flies a 1.25–2.5 micron spectral imager (Ralph/LEISA), augmented by mapping of CH_4 through its visible-wavelength 0.89 micron band (Ralph/MVIC).

2.3 Atmospheric and Particles/Plasma (Goals 1.3, 2.1, 2.5, 2.6, 2.7, 3.1)

Pluto's atmosphere is believed to be in vapor pressure equilibrium with its surface ices. Of the three volatiles detected on Pluto (N_2, CO, and CH_4), N_2 has the highest vapor pressure, and so dominates the atmosphere, with CH_4 and CO as trace components (Owen et al.

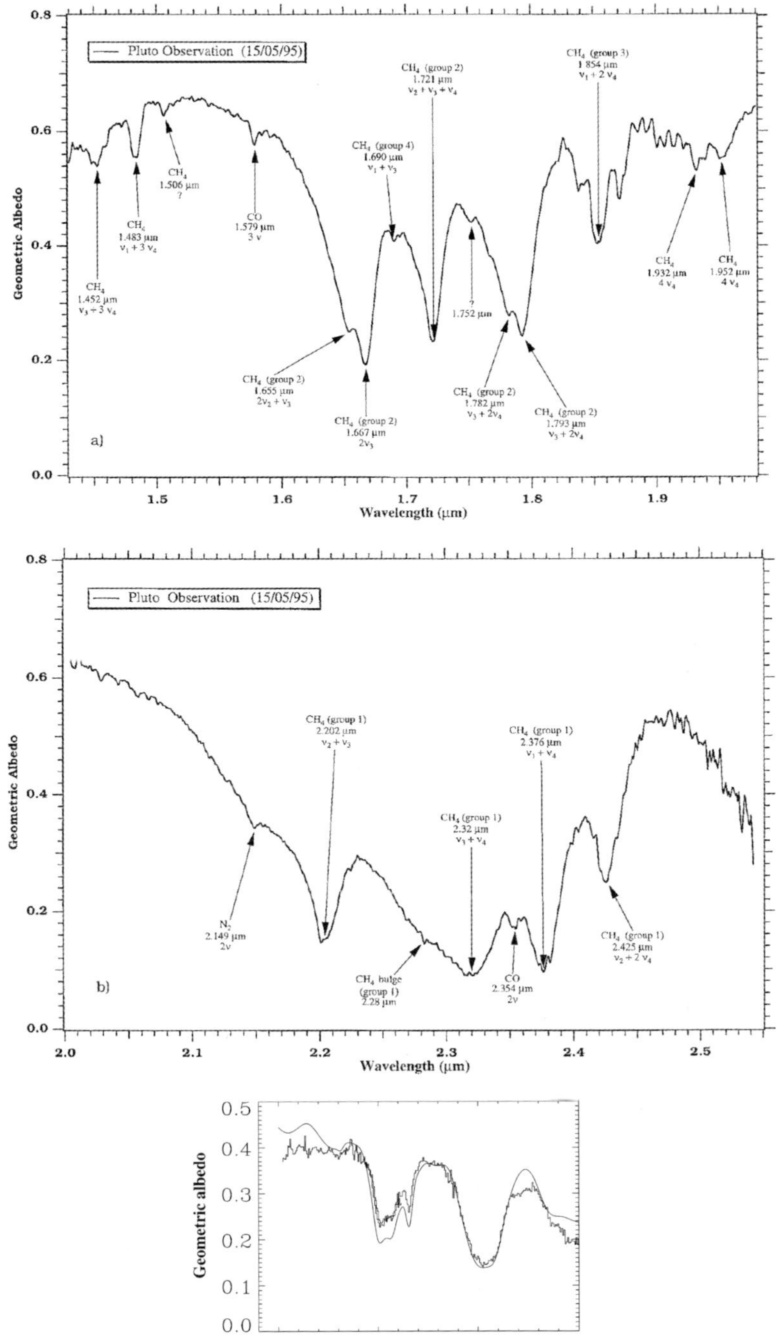

Fig. 2 *Top* and *middle*: Pluto's near-infrared spectrum (Douté et al. 1999), showing absorption bands due to N_2, CH_4, and CO ices. *Bottom*: Charon's near-infrared spectrum (Brown 2002), with the best-fit water ice model as a smooth line, showing H_2O ice, with the 1.65 micron feature indicative of crystalline ice, and an additional absorber at 2.2 micron (most likely NH_3 or its hydrate)

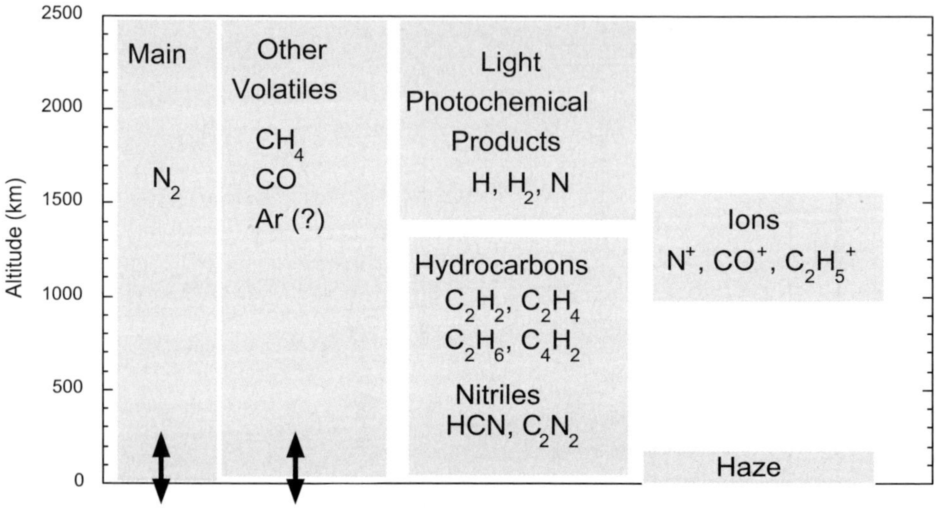

Fig. 3 Summary of atmospheric structure and composition, based on the models of Summers et al. (1997)

1993). Although the total column abundance of CH_4 has been measured (Young et al. 1997), its vertical profile and mixing ratio are unknown. Only upper limits on gaseous CO exist (Young et al. 2001b; Bockelée-Morvan et al. 2001). Models of Pluto's atmosphere (Yelle and Elliot 1997; Summers et al. 1997; Krasnopolsky and Cruikshank 1999; see Fig. 3) depend critically on the bulk composition (Goal 1.3a) for energetics and thermal structure (Goal 1.3b), chemistry (Goal 2.6), condensation products (Goal 1.3c), escape rates (Goal 1.3e), and the formation and composition of an ionosphere (Goal 2.5).

Pluto has an atmosphere that is thick enough to globally transport volatiles (e.g., Yelle and Elliot 1997). In theory, the surface pressure is a simple function of N_2 ice temperature. In practice, even if surface temperatures were measured with perfect accuracy, much is uncertain about the relation between the N_2 surface pressure (Goal 1.3d) and the temperature of N_2:CH_4 ice (Trafton et al. 1998), since the less-volatile CH_4 may form a barrier to N_2 sublimation, or a near-surface turbulent layer may affect the vertical transport of energy, CH_4, and CO, affecting the atmospheric composition (Goal 1.3a) and near-surface temperature structure (Goal 1.3d).

Pluto's atmosphere is likely to be in hydrodynamic escape (Goal 1.3e) with escape rates of a few 10^{10} $cm^{-2}\,s^{-1}$ (Trafton et al. 1997; Krasnopolsky 1999), equivalent to the loss of up to roughly a km of frost over the age of the solar system. Light photolysis products (H, H_2, N) escape, and heavier products may escape at higher N_2 flux rates (Trafton et al. 1997). Thus, Pluto's escaping atmosphere should leave its mark on Pluto's surface in the sublimation record, and possibly the amount of photochemical products that have rained onto the surface. Pluto's atmosphere is expected to extend past the orbit of Charon, leading to a possible secondary atmosphere on Charon that depends on Pluto's escape rate (Goal 2.7). Pluto's ionosphere and extended atmosphere should interact with the solar wind (Goal 2.5), with high fluxes leading to a large region (tens of Pluto radii) of comet-like interaction, and low fluxes leading to a smaller region with Venus-like interaction (Bagenal et al. 1997). Neutrals escaping from Pluto's atmosphere are ionized by solar EUV photons or by charge transfer, leading to an energetic particle environment of pickup ions (Goal 3.1).

New Horizons addresses the atmospheric goals primarily through UV occultations and airglow observations (Alice) and uplink radio occultations (REX). The haze observations are made by imaging (Ralph/MVIC, LORRI) and UV occultations, with SWAP and PEPSSI addressing the solar wind, energetic particle, and atmospheric escape rate goals.

2.4 Interiors, Environment, Origin and Evolution (Goals 3.2, 3.3, 3.4)

Clues to the formation and evolution of Pluto, Charon, Nix, and Hydra (presumably by giant impact, Stern et al. 1997b; Canup 2005) lie in their orbits (Dobrovolskis et al. 1997; Malhotra and Williams 1997; Goal 3.2) and bulk properties (Tholen and Buie 1997; Goal 3.2). Further information can be found from the discovery of (or limits on) additional rings or satellites (Goal 3.4) interior to the Hill sphere (6.0×10^6 km; Steffl et al. 2007). The heliocentric orbit of Pluto points to its formation in the outer solar nebula and its subsequent resonance capture by Neptune. The obliquity of each body in the system, the orbits of each body about the system barycenter, the state of tidal lock, and the interior structure of each body give additional constraints on the formation of the system. The interior structure gives external expression in the masses, radii, and shapes (and, possibly, the magnetic fields, Goal 3.3) of each body.

New Horizons will measure the orbits and bulk properties by a combination of imaging (LORRI, Ralph/MVIC) and the radio system (REX). If Pluto has a substantial magnetic field, this may be detectable by its effect on the solar wind (SWAP). Finally, SDC will measure the dust environment of Pluto.

2.5 Kuiper Belt Investigation

The Kuiper belt has emerged in recent years as a fundamental architectural element of the outer solar system. In addition to containing a rich reservoir of nearly pristine relics from the era of planet formation, the Kuiper belt provides context for understanding the origin and evolutionary environment of the Pluto–Charon system (e.g., Stern et al. 1997b). The Kuiper Belt and scattered disk are currently considered the source region of the short-period Jupiter family comets (Farinella et al. 2000; Weissman and Levison 1997) and provide a link between our solar system and the debris disks seen around main sequence stars such as β Pic. A goal of New Horizons is to investigate one or more Kuiper belt object (KBO) in an extended mission, fulfilling as many of the Group 1, 2 and 3 goals as possible.

3 Observations

3.1 Instrument Suite

The New Horizons payload (Table 4) addresses all of the Group 1, 2, and 3 goals listed in Table 3 except the Group 3 goal of a search for magnetic fields, which is only addressed indirectly. An overview of the science payload is given by Weaver et al. (2006), and detailed discussions of the individual instruments are provided by Cheng et al. (2008) (LORRI), Horanyi et al. (2008) (SDC), McNutt et al. (2008) (PEPSSI), McComas et al. (2008) (SWAP), Reuter et al. (2008) (Ralph), Stern et al. (2008) (Alice), and Tyler et al. (2008) (REX). In this section, the instruments are described with an emphasis on the relationship between the instrument specifications and the New Horizons science goals.

Table 4 New Horizons payload

Instrument	Description	Comments
Ralph/MVIC (Multi-spectral Visible Imaging Camera)	Visible panchromatic and color imager	Panchromatic and color imaging
Ralph/LEISA (Linear Etalon Imaging Spectral Array)	IR imaging spectrometer	Surface composition maps
Alice (not an acronym)	UV spectrometer	Upper atmospheric temperature, composition and escape rate
REX (Radio EXperiment)	Radio science experiment (uplink and radiometry)	Lower atmospheric temperature, pressure, and density; radiometry
LORRI (LOng Range Reconnaissance Imager)	High-resolution panchromatic imager	Panchromatic imaging
SWAP (Solar Wind Around Pluto)	Solar wind analyzer	Solar wind interaction
PEPSSI (Pluto Energetic Particle Spectrometer Science Investigation)	Energetic particle detector	Energetic particle environment
SDC (Student Dust Counter)[a]	Dust counter	Interplanetary dust environment

[a]Renamed VBSDC (Venetia Burney Student Dust Counter) after launch

3.1.1 Ralph: MVIC and LEISA

Ralph has two focal planes, MVIC and LEISA. MVIC's prime purpose is to obtain the Group 1 panchromatic and color maps. It also supports the Group 1 goal of characterizing of haze properties, the Group 1 composition maps, and the Group 2 and 3 goals of time variability, stereo imaging, terminator imaging, bond albedos, bulk parameters, and satellite/ring search. Both MVIC and LORRI take panchromatic images, with complementary observing techniques. Since MVIC operates its time-delay integration (TDI) panchromatic array in a scanning mode, it can obtain the Group 1 0.5 km/pixel hemispheric map of Pluto at a distance of 25,000 km in a single scan, observing the whole of Pluto's visible hemisphere in roughly 60 seconds without seams, and a change in scale of only $\sim 3\%$. Furthermore, MVIC was specifically designed to take observations of Pluto's atmosphere after closest approach, to search for hazes in forward scattering.

MVIC three broad-band colors, blue (400–550 nm), red (540–700 nm), and near-IR (780–975 nm) were chosen to give information on color slopes, as well as atmospheric properties. The narrow band CH_4 filter (860–910 nm) allows mapping of CH_4 abundance through the well known 0.89 micron CH_4 feature seen on Pluto's surface, the strongest methane feature at CCD wavelengths (Fig. 4). The NIR filter acts as the continuum comparison for this methane mapping. The 700–780 nm gap between the red and NIR includes another CH_4 band at 740 nm; combining information from panchromatic, blue, red, and NIR filters can give some information about band depth in this "virtual" filter.

LEISA's prime purpose is to obtain the Group 1 surface composition maps, and to support the Group 2 and 3 goals of time variability, high-resolution surface composition, and surface temperatures. LEISA's full range of 1.25–2.5 micron with a resolving power ($\lambda/\Delta\lambda$) of

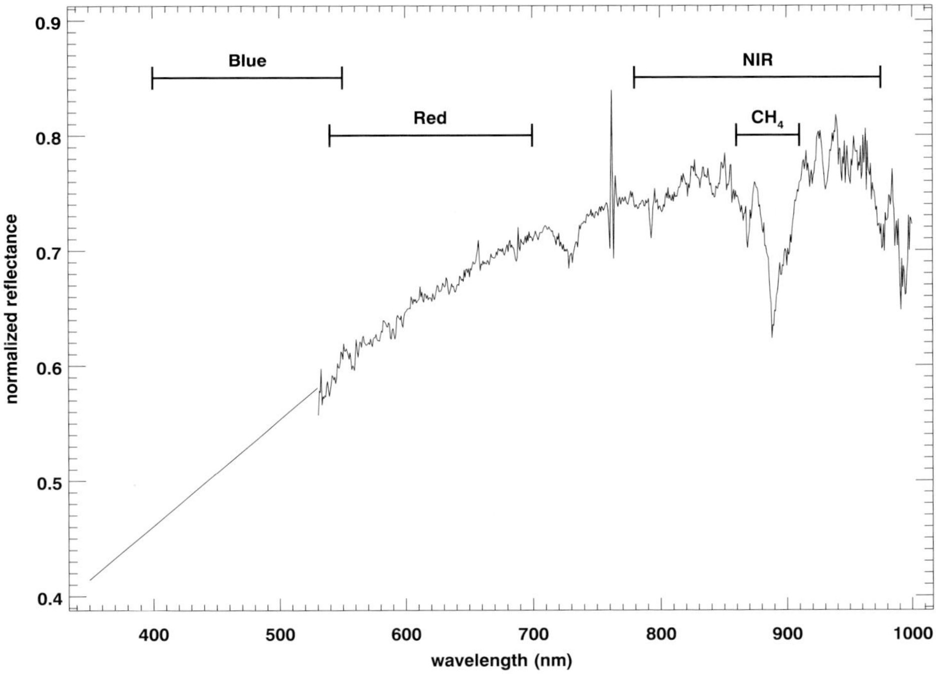

Fig. 4 Spectrum of Pluto (Grundy 1995; Buie et al. 1997a), and the MVIC color filters for comparison

roughly 250 was chosen to include absorption features of species previously detected on Pluto (N_2, CH_4, and CO) and Charon (H_2O, NH_3) as well as other species relevant to the outer solar system (Fig. 5). Because of the importance of the N_2 feature at 2.15 microns, as well as the $\upsilon_2 + \upsilon_3$ CH_4 band at 2.20 microns (diagnostic of pure vs. diluted CH_4 abundances; Quirico and Schmitt 1997), LEISA also measures the region covering 2.10–2.25 micron with a resolving power of roughly 560. LEISA operates by scanning its 256×256 pixel array across a target to build up a complete spectral image (Reuter et al. 2008).

3.1.2 Alice

Alice's prime purpose is to obtain the Group 1 atmospheric goals of composition, thermospheric temperature structure, aerosols, and evolution (escape rate). It also supports the Group 2 and 3 goals of time variability, minor atmospheric species, and Charon atmosphere search. Alice observes occultations of stars or the sun in a time-tagged mode with a spectral resolution of 3.5 Å. Longer integrations, for airglow or surface studies, are made in histogram mode. These can either be observed through Alice's 2×2 deg "box" portion of the slit or through the 0.1×2 deg "slot" portion, at resolutions of 3–4.5 Å for point sources or 9.0 Å for objects that fill the 0.1 deg $\times 4$ deg "slot." The spatial resolution is 0.27 deg pixel^{-1} along the slit. Alice's 465–1881 Å bandpass includes the electronic cutoffs of N_2, Ar, CH_4, and other hydrocarbons and nitriles that may be detectable by solar and stellar occultations (Fig. 6), and emission lines of CO, atomic H, Ar, and Ne that may be detectable as airglow (Fig. 7). Because N_2 dominates Pluto's atmosphere, the thermal profile can be derived from the N_2 density profile. The temperatures at high altitudes serve as lower boundary conditions for atmospheric escape; Alice can also map the cloud of escaping H by Ly-α against the interstellar background.

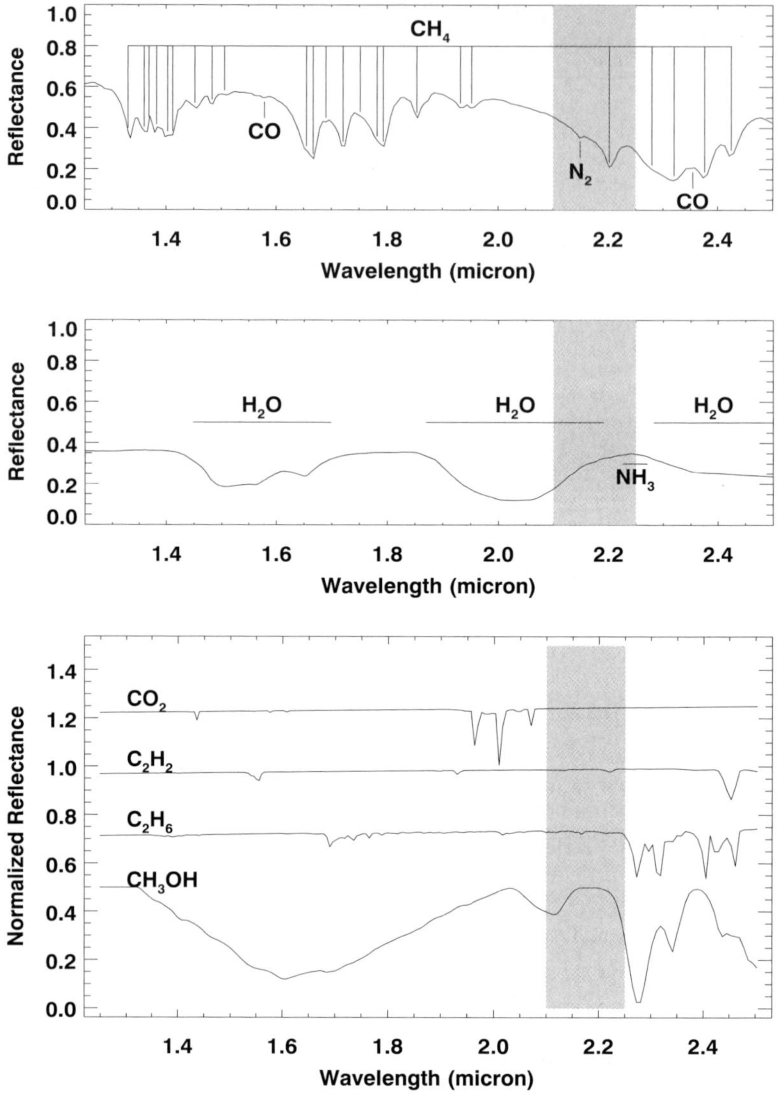

Fig. 5 Spectrum of Pluto (*top*), Charon (*middle*), and other significant solids (*bottom*) over the LEISA spectral range, at a resolving power of 250 over 1.25–2.50 micron, and a resolving power of 560 over 2.10–2.25 micron (indicated in *gray*)

3.1.3 REX

REX's prime purpose is to obtain the Group 1 atmospheric goals of measuring the lower atmospheric pressure and temperature; during Earth occultations, REX will detect the delay of radio signals uplinked from multiple DSN antennas. REX will also address Group 2 and Group 3 objectives, including probing Pluto's electron density within its ionosphere, searching for Charon's atmosphere, refining bulk parameters like mass and radius, and measuring surface emission at 4.2 cm.

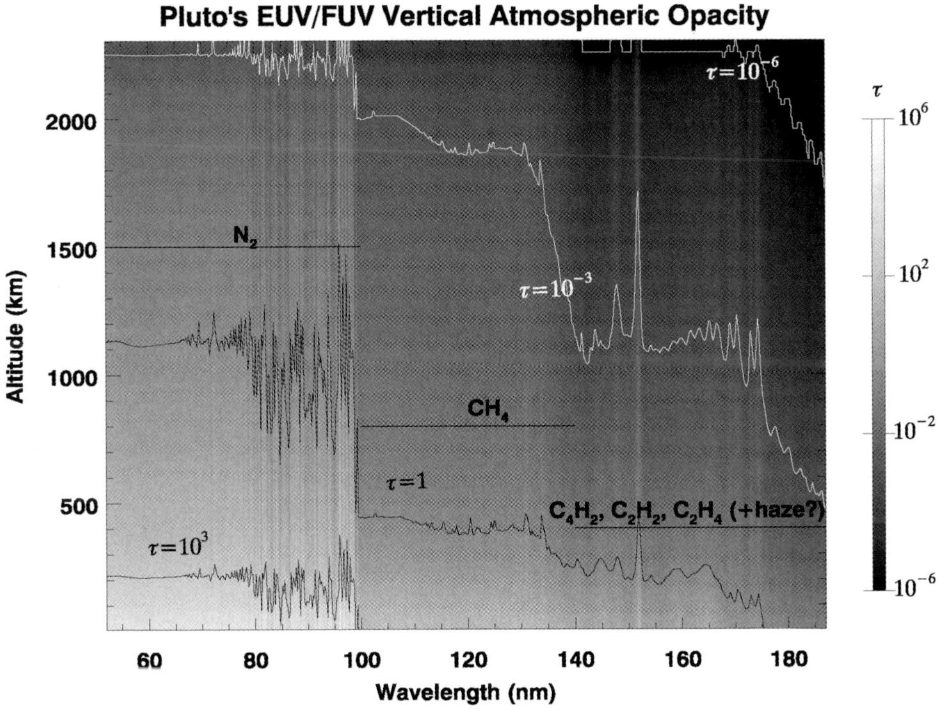

Fig. 6 Simulated line-of-sight optical depth (τ) for the M2 model atmosphere of Krasnopolsky and Cruikshank (1999), with an added 5% Ar

3.1.4 LORRI

LORRI augments most of the MVIC panchromatic imaging goals, including the Group 1 panchromatic mapping goals. Compared to MVIC, LORRI has a larger aperture, a square field of view (1024 × 1024 vs. MVIC's 5000 × 32 framing array), and finer angular resolution (5 microrad pixel^{-1} vs. MVIC's 20 microrad pixel^{-1}). This makes LORRI especially valuable for long time-base observations of the Pluto system, for observing the "far side" hemisphere of Pluto during the last 6.4 days (one Pluto rotation) before closest approach, making observations where geometric fidelity is critical (which favors LORRI's stare mode over MVIC's scanning operation), and taking the highest resolution images near closest approach.

3.1.5 PEPSSI

PEPSSI's prime purpose is the Group 3 goal of characterizing the energetic particle environment. In particular, PEPSSI's primary objective is to determine the mass, energy spectra, and directional distributions of these energetic particles, with a resolution able to discriminate between the various types of species expected at Pluto. The bulk of the energetic particles are expected to be pickup ions, or particles from Pluto's tenuous cloud of escaping neutrals, some billion of km in extent, that are ionized by solar EUV photons or charge transfer collisions with solar wind protons (Bagenal et al. 1997). These new ions then feel and react to the solar wind, following roughly cycloidal trajectories. In Pluto's frame of reference, these

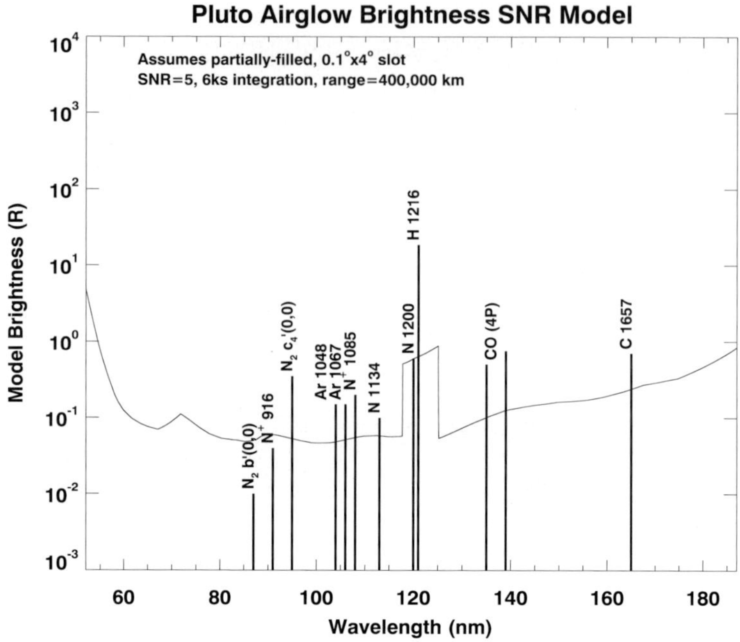

Fig. 7 Some of the airglow lines in the Alice bandpass

ions reach a maximum speed of twice the solar wind velocity, so that ions have a maximum energy of $m(2V_{\mathrm{sw}})^2/2$ (Bagenal et al. 1997), where m is the mass of the ion and V_{sw} is the solar wind velocity (approximately 450 km s^{-1}). Thus, N^+ will have energies near 35 keV, and N_2^+ will have energies near 70 keV. These are within PEPSSI's energy range, which can measure particles with energies up to 1 MeV. Since the bulk of the pickup ions originate from Pluto, PEPSSI supports the Group 1 objective of measuring the atmospheric escape rate by measuring the number and type of these ions. In addition, the energy range of PEPSSI supports the Group 2 objective of characterizing Pluto's solar wind interaction.

3.1.6 SWAP

The primary purpose of SWAP is the Group 3 objective of measuring Pluto's interaction with the solar wind. At Pluto, the unperturbed solar wind is expected to have a small proton temperature (i.e., a narrow energy spread), with typical solar wind energies of ~1000 eV. Within the region where the solar wind interacts with Pluto, the energy distribution of the solar wind particles will become less energetic with a wider spread. SWAP's energy range (35 eV to 7.5 keV for the centers of SWAP's electrostatic analyzer bins, McComas et al. 2008) will measure both the unperturbed solar wind, and the solar wind where it is deflected and decelerated due to mass loading by ionized atmospheric gases. Because the spatial scale of the interaction region is directly proportional to the atmospheric escape rate (Bagenal et al. 1997), SWAP supports the Group 1 objective of measuring Pluto's atmospheric escape rate. With an energy range extending to 7.5 keV, SWAP will also detect low energy pick-up ions, overlapping the energy range of the PEPSSI instrument. This allows cross-calibration of the two instruments, and supports the Group 3 objective of measuring Pluto's energetic particle environment.

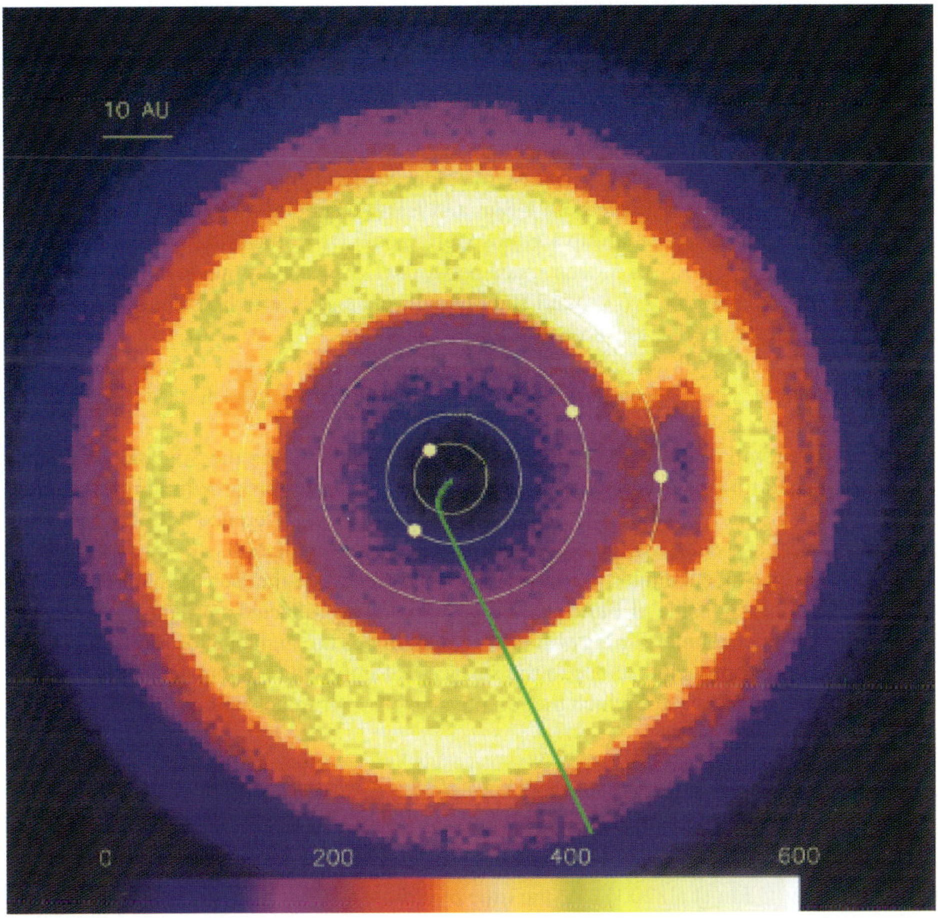

Fig. 8 Numerical simulation of the column density of 23 μm dust of Liou and Zook (1999, adapted from their Fig. 4a), with locations of Jupiter, Saturn, Uranus and Neptune plotted for 2015. The location of New Horizons relative to Neptune is plotted in *green*

3.1.7 SDC

The primary purpose of SDC is for Education and Public Outreach, allowing students the opportunity to design, build, fly, operate, and interpret results from a spacecraft instrument. SDC is designed to be the first instrument to measure the dust distribution in our solar system beyond 18 AU, where the dust may show the signature of resonances with Neptune (Fig. 8). Dust measurements are not specified in the science objectives, but are similar to the Group 3 goal of searching for rings, as they relate to Pluto's environment.

3.2 Mission Design

The trajectory of the New Horizons spacecraft is designed to satisfy the science goals listed in Sect. 2. Mission design is covered in more detail in Guo and Farquhar (2008). The mission design satisfies the following requirements:

1. New Horizons will arrive at the Pluto system on 2015 July 14. This satisfies the AO-specified requirement that New Horizons arrive at the Pluto system as early as feasible, but before 2020. The arrival time is driven by the change in the sub-solar latitude on Pluto and Charon (and the corresponding decrease in the area on Pluto and Charon illuminated by sunlight), and by the increase in heliocentric distance (which increases the probability of a collapse of Pluto's atmosphere in response to colder ice temperatures).

2. New Horizons will achieve both a Sun and Earth occultation of Pluto. Both occultations are required to meet the Group 1 atmospheric goals. The time of day will allow simultaneous coverage by two NASA deep space network (DSN) stations at elevations >15 deg. Simultaneous coverage by two ground stations will enhance the radio science experiment and lower mission risk (due to weather, for example).

3. New Horizons will achieve a solar occultation by Charon, and, if possible, also an Earth occultation. The Charon solar occultation is required to meet Group 2 Charon atmospheric science objectives.

4. Encounter will occur near opposition, with the Earth between the Sun and New Horizons, at a Sun–Earth–spacecraft angle of 171 deg. This is greater than the minimum required angle of 45 deg. An opposition encounter will reduce solar interference with the radio science experiment and decrease the noise floor for the radio occultations.

5. At the time of encounter, Charon will be farther from the Sun than Pluto. In this configuration, the nightside of Pluto will be illuminated by reflected sunlight from Charon. This "Pluto-first" encounter timing will enable imaging of the nightside of Pluto.

6. The closest approach distance to the surface of Pluto will be about 10,000 km, which is the encounter geometry for which the remote sensing instruments are designed to achieve their resolution objectives.

3.3 Encounter with Asteroid 2002 JF_{56}

On 2006 June 13, New Horizons flew past the asteroid (132524 APL) 2002 JF_{56} at a distance of 102,000 km (Olkin et al. 2006). At the time of closest approach, the asteroid was moving relative to New Horizons at 54 arcsec sec^{-1}. This provided an excellent opportunity to test methods for tracking fast-moving objects such as Pluto, since the asteroid had the highest relative rate of any body that New Horizons could image before the encounter with the Pluto system. At the time of encounter, LORRI's door was not yet opened. We therefore carried out observations of the asteroid using both MVIC and LEISA on Ralph. Panchromatic observations were carried out at 35, 13 and 8 hours before closest approach and four-color filter data was collected at 60, 20 and 8 minutes before closest approach. These data include phase angles not achievable from the ground (spanning 5 deg to 89 deg). Because of the small size of this asteroid and the distant nature of this untargeted flyby, the asteroid was not resolved during any of the observations (at closest approach, an MVIC pixel subtends 2 km at the asteroid), but an unusual (non-stellar) pattern to the signal from the asteroid is consistent with either an irregularly shaped asteroid or a satellite of the asteroid.

3.4 Encounter with Jupiter

On 2007 Feb 28, the New Horizons spacecraft flew past Jupiter at a range of 2.5 million km, and conducted an extensive series of observations of the planet and its satellites. Space does not allow us to show the numerous and varied observations, which have been reported elsewhere. Highlights of the encounter include:

- Time-resolved near-IR image cubes of the Great Red Spot and its surroundings, and high-resolution CCD imaging of the "Little Red Spot" (Baines et al. 2007; Reuter et al. 2007; Cheng et al. 2008).
- UV imaging and spectra of the Jovian aurorae and airglow (Gladstone et al. 2007).
- UV stellar occultations observations of Jupiter's atmosphere (Greathouse et al. 2008).
- Global imaging of Io's plumes, post-Galileo surface changes, and high-temperature (0.4–2.5 micron) volcanic thermal emission from Io (Spencer et al. 2007).
- Observations of UV, and (in eclipse) visible and near-IR, atmospheric emissions from Io, Europa, and Ganymede (Retherford et al. 2007; Spencer et al. 2007).
- Near terminator imaging of large-scale topographic features on Europa (Grundy et al. 2007).
- 1.25–2.5 micron spectroscopy of Galilean satellite surface composition, with similar spatial resolution and better spectral resolution than Galileo NIMS global observations (Grundy et al. 2007).
- Extensive imaging of Jupiter's rings to search for embedded satellites and improve knowledge of the ring structure and ring particle phase functions (Showalter et al. 2007).
- Distant imaging of Himalia and Elara to determine shapes, sizes, and phase curves.
- Plasma observations of the magnetosphere, including an unprecedented 3-month flight down the magnetotail (McComas et al. 2007; McNutt et al. 2007).

3.5 Pluto

Plans for the Pluto system encounter are, at the time of writing (some 8 years before the encounter begins) still only preliminary. While encounter planning will be readdressed prior to the Pluto encounter rehearsals in 2012, a detailed plan was formed pre-launch in 2001–2003, to guide spacecraft design, instrument placement, pre-launch tests, commissioning, and checkout. In this paper, we present that prelaunch strawman sequence, updated for expected spacecraft performance (e.g., pointing accuracy, navigation error ellipses, and settle times) in October 2005.

In this plan, science observations of Pluto are divided into five phases. The first phase, the *cruise phase*, includes all observations of the Pluto system in the first nine years of flight. The second phase, the *approach observatory phase*, begins when LORRI can first resolve Pluto, in January 2015. The transition between this and the third phase, the *approach far encounter phase,* is defined by that time when downlink can no longer keep pace with the data collection. For the purposes of this paper, we describe these two phases together as the *approach phase*. Most of the group 1, 2 and 3 objectives are best satisfied in the *near encounter phase*, the period extending from 13 hours before to 5 hours after closest approach. After closest approach there will be a short *departure far encounter phase* of high phase angle observations, radio tracking, and in situ environmental measurements.

3.5.1 Cruise Phase

New Horizons observed Pluto with LORRI during instrument commissioning in 2006 (Fig. 9), and will continue to observe Pluto throughout the cruise period. Pluto and Charon are effectively point sources during the cruise phase. By 2013, LORRI will be also able to clearly separate Charon from Pluto. These observations are of particular benefit to the Group 1 goal of determining the photometric phase functions of Pluto and Charon, since the solar phase angle of Pluto as seen from New Horizons changes over this period, and since New Horizons and Earth observe Pluto from different angles. These observations will also

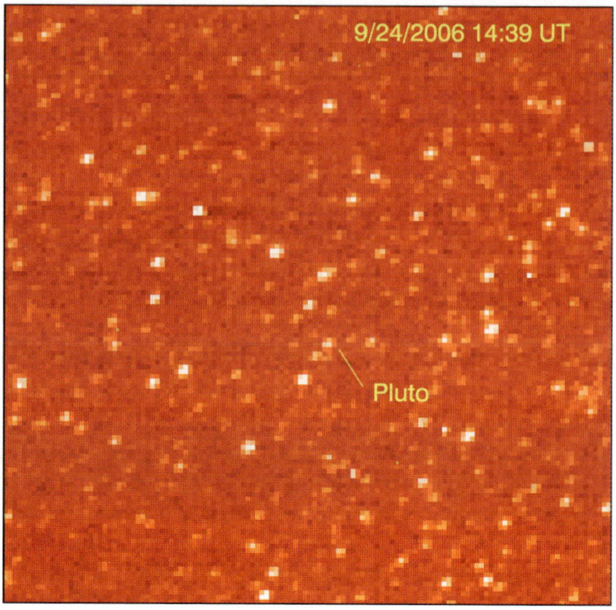

Fig. 9 One of six LORRI images of Pluto and the dense background of stars taken in September, 2006

address the Group 2 objective of time variability. As opportunity and spacecraft resources allow, New Horizons will accomplish other science observations as well. For example, New Horizons will be able to observe Uranus and Neptune at phase angles not achievable from Earth-based or near-Earth observatories (Fig. 10).

3.5.2 Approach Phase

During the approach phase, we will investigate Pluto, Charon, Nix, and Hydra over a long timespan. We will begin imaging the system with MVIC, LEISA, LORRI, and Alice 150 days before closest approach, when LORRI can first resolve Pluto's disk. At this distance, LORRI's resolution will be 900 km pixel^{-1}, similar to the to HST maps, MVIC and LEISA will separate Pluto and Charon, and LORRI will easily be able to detect Nix and Hydra and will begin phase curve studies on these satellites. We plan to observe Pluto and Charon every 12 hours from P-150 to P-144 days, covering a complete 6.4-day Pluto rotation period at 30-deg longitudinal spacing. This 6.5-day sequence will be repeated each time the resolution improves by 50%. The current plan is to take eight such sequences, each starting at 150, 100, 66, 44, 28, 19, 12, and 6 days before Pluto closest approach. These observations will address time variability, provide airglow spectra, obtain a series of maps at increasingly higher resolution, and refine the orbits (and hence indirectly refine the masses) of Pluto, Charon, Nix, and Hydra. We will also search the returned imagery for small satellites that had previously escaped detection from Earth. The final set of observations, starting at 6 days out, will give the highest spatial resolution images and spectra of the portions of the surface away from the closest-approach hemisphere (including the far-side hemisphere at 3.2 days before closest approach). One or two 27-day solar rotations before closest approach, SWAP will begin solar wind observations to allow interpolation of the unaffected solar wind upstream of Pluto. The PEPSSI campaign to detect pick-up ions from Pluto's atmosphere will begin during the final days before closest approach, possibly detecting Pluto ions as far as 1 million km upstream of Pluto.

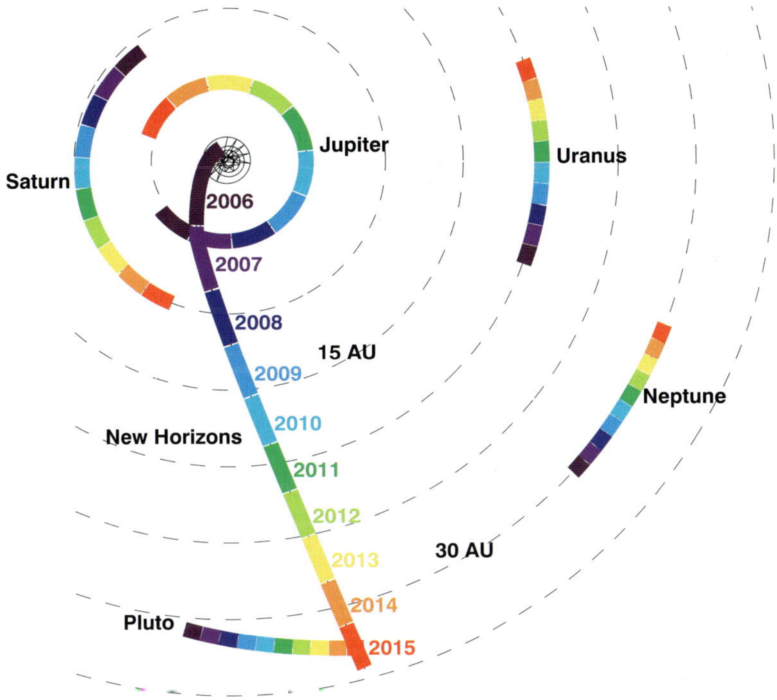

Fig. 10 Path of New Horizons and the outer planets from launch (2006) to Pluto encounter (2015)

3.5.3 Near Encounter Phase

The strawman near encounter phase was designed to fulfill the objectives outlined in Table 3. Near the start of the near encounter phase, roughly 13 hours before Pluto closest approach, LORRI will take images of Pluto and Charon where each body just fits within a single 1024×1024 pixel frame, giving high geometric fidelity of the entire approach hemisphere at 2–3 km pixel^{-1}. Much of the next seven hours will be spent taking approach airglow spectra with Alice, mainly for the Group 1 goal of atmospheric CO and Ar detection. Additional LORRI mosaics will be taken before LEISA takes the first set of Group 1 surface composition maps at about 3 hours before closest approach, at 10 km pixel^{-1}. Additional Alice observations will follow; Alice will resolve the surfaces of Pluto and Charon at this time. At two to 1.5 hours before closest approach, LEISA will take a second, redundant set of Group 1 surface composition maps at 5–7 km pixel^{-1}. New Horizons will then obtain the Group 1 geology (panchromatic and color) maps, with high-resolution visible and infrared images taken near Pluto and Charon closest approaches.

After closest approach, New Horizons will pass through the occultation zones of Pluto and Charon. Enough time is allocated for Alice to establish an unocculted solar spectrum and for REX to establish an upper baseline before and after Pluto's ionosphere is detectable. REX and Alice will observe the Earth and Sun occultations simultaneously. During the solid body occultation, REX will observe the thermal emission from Pluto or Charon. Immediately before and after the two occultations, high-phase angle images by MVIC will be used to search for hazes and rings in forward scattering, as well as attempt imaging of Pluto's night side in reflected Charon light. Although the particle instruments usually ride along

Table 5 Prelaunch strawman near encounter sequence

Start time re Pluto CA (min)	Target	Instrument	Description	Highest Group (1,2,3)	Start solar phase angle (deg)	Starting Resolution (km)	Start range to target center (km)
-800.0	Pluto	LORRI	P_pan_global_1x1_last	3	16.0	3.30	661,285
-589.7	Pluto	ALICE	PC_UV_airglow_spectrum_1	1	16.3	2431.5	487,484
-487.9	Charon	LORRI	C_pan_global_1x1_last	3	18.2	2.09	417,749
-486.1	Pluto	ALICE	PC_UV_airglow_spectrum_2	1	16.6	2003.6	401,902
-385.2	Pluto	ALICE	PC_UV_airglow_spectrum_3	1	17.0	1586.8	318,536
-284.2	Pluto	ALICE	PC_UV_airglow_spectrum_4	1	17.7	1169.8	235,131
-243.2	Pluto	LORRI	P_pan_global_3x3_1	2	18.2	1.00	201,289
-236.1	Pluto	ALICE	PC_UV_airglow_spectrum_5	1	18.2	971.3	195,444
-215.1	Pluto	LORRI	P_pan_global_3x3_2	2	18.6	0.88	178,118
-206.1	Pluto	LEISA	P_IR_global_scan_1	1	18.7	10.5	170,695
-198.6	Pluto	LEISA	P_IR_global_scan_2	1	18.9	10.1	164,510
-191.0	Pluto	LEISA	P_IR_global_scan_3	1	19.0	9.7	158,257
-183.3	Charon	LEISA	C_IR_global_scan_2	1	23.9	10.3	166,060
-177.2	Pluto	ALICE	PC_UV_airglow_spectrum_6	1	19.3	728.4	146,854
-165.5	Pluto	ALICE	P_UV_surface_spectrum_1	2	19.6	680.3	137,241
-153.7	Charon	ALICE	C_UV_surface_spectrum_1	2	25.5	706.2	141,839
-140.7	Pluto	LEISA	P_IR_global_doublescan_1	1	20.4	7.2	116,796
-131.8	Pluto	LEISA	P_IR_global_doublescan_2	1	20.8	6.7	109,530
-122.8	Pluto	LEISA	P_IR_global_doublescan_3	1	21.2	6.3	102,064
-113.3	Pluto	LEISA	P_IR_global_doublescan_4	1	21.8	5.8	94,263
-104.3	Charon	LORRI	C_pan_global_6x3_1	2	29.8	0.51	101,837
-88.7	Charon	LEISA	C_IR_global_doublescan_1	1	32.0	5.5	89,326
-78.0	Charon	LEISA	C_IR_global_doublescan_2	1	33.9	5.0	80,901
-66.3	Charon	MVIC	C_color_global_scan_1	1	36.5	1.42	71,809
-62.3	Charon	MVIC	C_color_global_scan_2	1	37.5	1.36	68,667
-59.1	Pluto	LORRI	P_pan_alongtrack_8x2_2	1	27.8	0.24	50,082
-44.8	Pluto	MVIC	P_color_global_scan_1	1	31.7	0.75	38,619
-39.6	Pluto	MVIC	P_color_global_scan_2	1	33.7	0.67	34,550
-34.1	Pluto	MVIC	P_pan_global_scan_1	1	36.5	0.58	30,281
-29.0	Pluto	MVIC	P_pan_global_scan_2	1	39.9	0.50	26,417
-23.5	Pluto	LEISA	P_IR_hires_scan_1	2	44.7	1.3	22,372
-14.2	Pluto	LORRI MVIC	P_pan_terminator_simultMVIC_1	2	58.5	0.07 0.30	16,139
-4.0	Charon	LEISA	C_IR_hires_scan_1	2	75.7	1.9	30,517
4.5	Charon	LORRI MVIC	C_pan_terminator_simultMVIC_1	2	88.6	0.14 0.54	27,785
13.1	Charon	MVIC	C_pan_global_scan_1	1	103.5	0.52	26,693
20.1	Pluto	MVIC	P_pan_GlobalHaze_scan_1	1	161.3	0.38	19,969
25.8	Pluto	ALICE	P_UV_occ_solar_1	1	167.6	7.6	24,067
25.8	Pluto	REX	P_radio_occ_simultALICE_1	1	167.6	1.0	24,067
76.1	Pluto	REX	P_radio_temperature_map_1	2	175.0	1338.5	63,860
83.3	Pluto	PEPSSI+SWAP	P_particles_roll_simultPAM_1	3	174.1		69,769
117.9	Pluto	ALICE	P_UV_nightglow_spectrum_2	2	171.4	484.6	98,094
129.8	Charon	ALICE	C_UV_occ_solar_1	2	179.3	31.3	100,262
129.8	Charon	REX	C_radio_occ_simultALICE_1	2	179.3	2.1	100,262
148.0	Pluto	MVIC	P_pan_GlobalRings_scan_1	3	170.1	2.43	122,831
155.0	Pluto	ALICE	P_UV_nightglow_spectrum_3	2	169.9	637.1	128,593
282.0	Pluto	LEISA	P_IR_hiphase_scan_1	2	167.7	14.4	233,342
292.0	Charon	LEISA	C_IR_hiphase_scan_1	2	171.8	14.4	233,328

Blue = Alice, Green = MVIC or LORRI, Red = LEISA, Gold = REX, Black = PEPSSI or SWAP. Grey shading = Charon. SWAP and PEPSSI on throughout the near encounter as power allows

with the pointing determined by the remote sensing instruments, a roll is planned between the two occultations specifically for SWAP and PEPSSI, when New Horizons will be downstream of Pluto and Charon. (The term "Pam" in the instrument description refers to both PEPSSI and SWAP; Stern and Cheng 2002.) Also after closest approach, Alice will look for nightglow emission from Pluto, and LESIA will take high phase angle observations of both Pluto and Charon.

3.5.4 Departure Phase

The departure phase is currently planned to extend for 30 days past Pluto closest approach, approximately four Pluto rotations. This period will allow SWAP and PEPSSI to measure in situ particle populations for one solar rotation downstream of Pluto. In departure, SDC will search for impact rate enhancements associated with dust from the Pluto system. REX will measure globally averaged temperatures for some days after closest approach. The strawman plan also calls for a series of remote sensing observations at the 165 deg phase angle asymptote during departure over one Pluto rotation.

3.6 Kuiper Belt Objects

The New Horizons mission is designed to make flybys of one or more KBOs after Pluto. The particular Kuiper Belt Objects for New Horizons are not yet selected. Thanks to a favorable launch, we expect sufficient fuel available to affect a 200 m s^{-1} change in velocity, giving a cone of accessibility with a full-width of 1.6 deg around New Horizon's nominal trajectory.

In this volume, there should be a 50% chance of reaching a KBO 72 km or larger, or 95% of reaching one 45 km or larger (Spencer et al. 2003). The probability for a KBO encounter as a function of distance along the New Horizons path is proportional to the number of KBOs per cubic AU and the width of the cone. This probability is strongly peaked at 42 AU. It is reasonable to expect that a post-Pluto KBO flyby to occur in 2018 or 2019.

The planned observations at KBOs are similar to those at Pluto. With MVIC, LORRI, and LEISA we can obtain visible color and panchromatic maps and IR composition maps of each flyby target, yielding shape, geological/geomorphological, photometric, cratering history, and compositional data, and we will search for evidence of volatiles, a past or present atmosphere, and potential satellites. In addition, LORRI can obtain long-range images to search for small satellites and globally map the object during approach, and high-resolution imagery at closest approach in order to study surface cratering and regolith processes, volatile transport, etc., at spatial scales down to 100 m or less (depending on the closest approach distance). Regarding any potential atmosphere, Alice can obtain airglow data like that obtained at Pluto and SWAP and PEPSSI can search for evidence of a solar wind interaction zone and upstream pickup ions (perhaps even sputtered from the surface). We also expect to have sufficient targeting accuracy to search for tenuous extended atmospheres using Alice for solar occultations, considering that any such atmosphere should be loosely bound. Doppler tracking of New Horizons will measure or place limits on the KBO mass. REX will measure brightness temperatures at 4.2 cm. Finally, SDC will measure the dust in the outer solar system, providing a different approach to understanding the collisional evolution of the Kuiper Belt.

4 Meeting the Mission Scientific Objectives at the Pluto System

The proposed mission trajectory, instruments, and strawman observing sequence fulfill the Group 1, 2 and 3 objectives listed in Table 3, as summarized in Table 6.

4.1 Group 1: Objectives Required for Mission Success

4.1.1 Characterize the Global Geology and Morphology of Pluto and Charon

Panchromatic Mapping Every panchromatic observation will be folded into the global maps, including images of Pluto and Charon, beginning one full rotation period (6.4 days)

Table 6 Mapping of science objectives to instruments

Objective		Measurement technique	
		Primary	Supporting
Group 1			
Geology/ Geophysics	pan	MVIC, LORRI	LEISA
	color	MVIC	LEISA
Surface composition		LEISA	MVIC, Alice
Neutral atmosphere		Alice, REX	MVIC, LORRI, PEPSSI, SWAP
Group 2			
Variability		MVIC, LEISA, Alice, LORRI, REX	PEPSSI, SWAP
Stereo		MVIC, LORRI	LEISA
Hi-res terminator maps		MVIC, LORRI	LEISA
Hi-res composition maps		LEISA	MVIC
Ionosphere/solar wind		REX, SWAP	PEPSSI
Other atmospheric species		Alice	PEPSSI, SWAP
Charon atmosphere		Alice	REX, LEISA
Bond albedos		MVIC, LORRI	LEISA
Surface temperatures		LEISA	REX
Group 3			
Energetic particles		PEPSSI	SWAP
Bulk parameters		REX, LORRI	MVIC, LEISA
Magnetic field search		not directly addressed	SWAP
Satellite and ring search		MVIC, LORRI	Alice, REX, SDC

before closest approach, when LORRI can resolve Pluto and its satellites at 36 km per pixel. We plan to take the main Group 1, 0.5 km pixel^{-1} maps of Pluto and Charon redundantly. For Charon, LORRI will make one mosaic map at low phase (C_pan_global_6x3_1) and MVIC will make another by scanning the pan TDI array near Charon closest approach (C_pan_global_scan_1). MVIC will take the Group 1 0.5-km pixel^{-1} map of Pluto with two redundant panchromatic TDI scans about 30 minutes before closest approach (P_pan_global_scan_1-2). In the current plan, LORRI will augment MVIC's 0.5 km pixel^{-1} hemispheric maps of Pluto with hemispheric mosaics at 1.0 and 0.9 km pixel^{-1}(P_pan_global_3x3_1-2), and a 400-km wide swath across the nadir of Pluto at 0.24 km pixel^{-1}(P_pan_alongtrack_8x8_2). Higher resolution hemispheric LORRI maps may be possible, depending on spacecraft pointing. As shown in Reuter et al. (2008) and Cheng et al. (2008), the signal-to-noise will meet the AO requirement of 50 or higher per pixel.

Color Mapping MVIC will make redundant four-color images of the Pluto system with color TDI scans. Pluto global color maps are taken at ~0.7 km pixel^{-1}(P_color_global_scan_1-2), while Charon global color maps are taken at 1.4 km pixel^{-1}(C_color_global_scan_1-2), exceeding the AO spatial resolution requirement of 1.5–5 km pixel^{-1}. As shown in Reuter et al. (2008), the signal-to-noise will meet the AO requirement of 50 or higher per pixel in two or more of the color channels.

Phase Angle Coverage The phase angles to Pluto and its satellites during the approach phase will increase slowly from 14.3 to 15 deg, then increase quickly during the near en-

Table 7 Ralph/LEISA pre-flight estimated IR detection limits for surface ices

Molecule	3-σ limit	Comments
N_2 (2.15 μm)	20%	Pure N_2 at 5 × 5 binning
N_2 detected by CH_4 band shifts	5%	For CH_4 abundance < 2%
CH_4 (pure or diluted 2.2 μm)	0.1%	Detect either mixing state
CH_4 vs. CH_4:N_2 (e.g., 2.2 μm)	10%	Distinguish mixing states
H_2O, crystalline (e.g., 1.65 μm)	10%	Quite model dependent
H_2O, amorphous (e.g., 2.0 μm)	10%	Quite model dependent
NH_3 & hydrates (2.20–2.25 μm)	5%	Quite model dependent
CO (2.35 μm)	0.1%	Probably dissolved in N_2
CO_2 (1.96 μm)	2%	Probably not in N_2
CH_3OH (2.28 μm)	25%	Better if dissolved in N_2
C_2H_6, C_2H_6:N_2 (1.68, 2.33 μm)	0.1%	Better if dissolved in N_2
C_2H_2, C_2H_2:N_2 (2.45, 2.42 μm)	0.1%	Better if dissolved in N_2
C_2H_4, C_2H_4:N_2 (2.22, 2.26 μm)	0.1%	Better if dissolved in N_2

Limits are conservative; we assumed 1 × 1 binning unless noted, intimate mixing with other expected species, and realistic albedos. We will also search for SO_2 (2.13 μm), H_2S (1.64 μm), H_2CO:N_2 (2.18 μm), HCN (1.91 μm), HC_3N (1.83 μm), pyroxene (1.79–2.33 μm), & kaolinite-serpentine clays (1.40 μm)

counter phase, as the spacecraft swings around Pluto and Charon. Phase angles for visible imaging before or near closest approach will be between 16 and 60 deg for Pluto and between 18 and 104 deg for Charon. Currently, we have observations of Pluto planned shortly after closest approach, with a phase angle near 160 deg, designed mainly for haze detection (see Sect. 4.1.3).

4.1.2 Map Surface Composition of Pluto and Charon

Ralph/LEISA will produce the main global spectral maps of Pluto and its satellites, producing spectral data cubes at a resolving power of ~250 in the interval 1.25–2.5 micron, and at ~560 in the interval 2.10–2.25 micron. LEISA will take hemispheric spectral images of both Pluto and Charon at 10 km pixel^{-1} about 3 hours before closest approach (P_IR_global_scan_1-3, C_IR_global_scan_2). At about 2 hours before closest approach, a redundant set of spectral images for Pluto will be made at resolutions of 5–7 km pixel^{-1} (P_IR_global_doublescan_1-4), and at about 1 1/2 hour before closest approach, the redundant hemispheric Charon observation are made at resolutions of 5 km pixel^{-1} (C_IR_global_doublescan_1-2). All spectral data cubes exceed the AO spatial resolution requirement of 5–10 km pixel^{-1}. Observations of Nix and Hydra will be planned after better orbit solutions are obtained. For each global map, LEISA scans over the target up to four times, both to increase the signal-to-noise ratio and to cover the entire target when it subtends more than 256 pixels. These hemispheric observations are made at phase angles of 15–35 deg. The hemispheric spectral images are augmented by regional spectral images near closest approach, at resolutions of 1–2 km pixel^{-1} and phase angles of 45 deg (Pluto) and 75 deg (Charon). If feasible, observations will be taken at high phase angle on departure, at resolutions near 14 km pixel^{-1}. At the expected SNR for the planned spectra (Reuter et al. 2008), LEISA will map the spatial distribution of N_2, CO, CH_4, and other species (Table 7).

4.1.3 Characterize the Neutral Atmosphere of Pluto and Its Escape Rate

Compositional Measurements Alice is the instrument New Horizons will use for measuring the neutral atmospheric composition at Pluto. N_2 and CH_4 will be measured through the solar UV occultation (P_UV_occ_solar_1). Stellar occultations are also being planned, but were not identified in the pre-launch sequence. These occultations will provide altitude-resolved abundance measurements at both the ingress and egress sites. N_2 densities are measured through continuum absorption and electronic transitions from 465–1000 Å. CH_4 will be detected by its continuum absorption at 1000–1300 Å. Airglow observations with Alice on approach (PC_UV_airglow_spectrum_1-6) will be the main technique for measuring CO and Ar, although both species may be detectable in the occultation spectra as well. The Fourth Positive Group (14,4) airglow band at 1356 Å, excited by resonance fluorescence of the (14,0) band by solar Lyα, should be the brightest CO feature above the level where CH_4 absorbs FUV. Ar will be searched for at its well-known EUV resonance lines (1048 and 1067 Å).

Upper Atmospheric Thermal Structure The primary data set for the Group 1 objective of measuring the thermal and pressure structure in Pluto's upper atmosphere comes from Alice solar and stellar occultations (P_UV_occ_solar_1). Since N_2 is expected to be the dominant species, we will derive temperatures from the line-of-sight number density of N_2 under the assumption of hydrostatic equilibrium. From our simulations (based on the M2 model atmosphere of Krasnopolsky and Cruikshank 1999), we will be able measure temperatures and thermal gradients to the AO-required accuracy (10% in 100 km bins) from altitudes of 1300 to 1800 km using only the N_2 continuum absorption shortward of 665 Å. By using the N_2 lines in addition to the N_2 continuum absorption, we will extend this range downward for several hundred km, until altitudes where CH_4 absorption becomes important.

Lower Atmospheric Thermal Structure Pluto's atmospheric pressure, number density and temperature near the surface will be measured with radio occultations (P_radio_occ_simultALICE_1). REX radio occultation profiles will measure atmospheric phase delays to better than 1 deg (or 0.05 of the estimated phase delay at 3 microbar). Pluto's atmospheric pressure at the time of arrival is uncertain, but the recent optical occultations of 2002 (Elliot et al. 2003; Sicardy et al. 2003) and 2006 (Young et al. 2008; Elliot et al. 2007) suggest that a 4 microbar surface pressure is a conservative estimate. For this case, pre-flight models estimate that REX will measure the base number density (n) to 1%, temperature (T) to 1.5 K, and pressure (p) to 0.2 microbar at 12 km vertical resolution (less than half a scale height) in a single occultation with only one REX operating and one DSN uplink. By using a second DSN station and combining ingress and egress occultations, REX will measure n to 0.5%, T to 0.75 K, and p to 0.1 μbar. See Tyler et al. (2008) for updated signal-to-noise estimates.

Evolution and Escape Rate Our objective here is to quantify Pluto's atmospheric loss rate (currently uncertain to over an order of magnitude; McNutt 1989; Trafton et al. 1997; Bagenal et al. 1997; Krasnopolsky 1999; Tian and Toon 2005), and determine whether the atmosphere is hydrodynamically escaping. Critical parameters to measure are the atmospheric scale heights of the escaping species and the total 500–1850 Å solar energy input that heats Pluto's thermosphere. Escape calculations suggest that Alice occultation profiles obtained by measuring these quantities upward from near the energy deposition peak (i.e., $\sim 3 \times 10^8$ cm^{-3} to $\sim 10^7$ cm^{-3}, perhaps 1300–2300 km altitude) can allow the determination of the species integrated escape flux to ~35%. The scale height of N_2 high in Pluto's atmosphere will be measured by solar occultation (P_UV_occ_solar_1). Also, Alice H Ly α airglow scale height profiles (from all Alice airglow measurements) can be used to derive an independent H-only escape flux (Clarke et al. 1992).

If the escape from Pluto's atmosphere is substantial (e.g. greater than about 1×10^{27} molecules per second) then ionization of the escaping neutrals will cause a measurable perturbation of the solar wind that extends well away from Pluto. The New Horizons plasma instruments will measure both the spatial extent and nature of the solar wind interaction with Pluto's escaping atmosphere. In particular, PEPSSI will measure the flux of Pluto pickup ions and SWAP will identify and measure the location of the atmosphere/solar wind interface (so long as we approach close enough to Pluto). Because pickup protons from ionized interstellar H are ubiquitous in the outer solar system, it is necessary to distinguish them from Pluto's pickup ions; PEPSSI energetic particle sensor has the mass resolution to separate interstellar H^+ from heavy CH_4^+, N_2^+, CO^+ and other molecular ions escaping from Pluto's atmosphere.

Aerosols and Haze Detection At Triton, Voyager 2 imaged both a uniform hydrocarbon haze (with a scale height of 10 km and a vertical optical depth of 0.001–0.01), and also patchy condensed N_2 clouds at altitudes <10 km with vertical optical depth near 0.1 (Rages and Pollack 1992). Based on Pluto's expected thermal profile, condensation clouds may be unlikely, but a hydrocarbon haze is expected by many. All low phase-angle approach maps of Pluto will be searched for clouds (e.g., P_pan_global_1x1_last, P_pan_global_scan_1). Additionally, the Alice solar occultation (P_UV_occ_solar_1) will achieve 15% accuracy in measuring haze with vertical optical depth of 0.005 in 5-km height bins above 500 km, using 1800–1850 Å wavelengths. After closest approach, MVIC images will be obtained at high phase angles to further characterize hazes by their forward scattering properties (P_pan_GlobalHaze_scan_1). Even if Pluto's haze is ten times less absorbing than Triton's, MVIC will be able to detect haze from the surface to 80 km above the surface at a phase angle of 135 deg. Higher phase angle measurements will also be attempted.

4.2 Group 2 Objectives

4.2.1 Characterize the Time Variability of Pluto's Surface and Atmosphere

For variability on timescales of a Pluto day (6.4-day rotation period), LEISA will use the temperature-sensitive N_2, CH_4, and H_2O bands to look for surface temperature changes as a function of local time of day and unit type, supported by REX radiometry of disk-averaged brightness temperatures at 4.2 cm on both day and night sides of Pluto. LEISA composition mapping and LORRI and MVIC imaging will be sensitive to differences in albedo and composition of areas at different times of day, perhaps showing the effects of ices condensing overnight. We will also use the combined entry and exit Earth radio occultation datasets (combined with imagery and IR spectroscopy of these locales) to search for variability at two local times of day in both the neutral atmosphere and the ionosphere. Spatially resolved UV airglow spectra of day and night-side atmospheres (and ingress and egress occultation profiles) will also be compared to search for diurnal and spatial variability in the neutral atmosphere.

New Horizons will begin observing Pluto 150 days before closest approach, at which time LORRI resolves Pluto is resolved with a diameter of 2.6 pixels across. Temporally resolved spectra and photometry during the approach phase will be searched for evidence of change. In the final three Pluto rotations, starting at 19 days before closest approach, the optical instruments will be able to track meteorology and cloud motions at rates as slow as $1\ \mathrm{m\,s^{-1}}$ and look for variations in the style or degree of geyser-like activity.

Observations during cruise may detect changes on longer, seasonal timescales. The powerful combination of surface composition maps, visible albedo maps, and atmospheric profiles will constrain seasonal models of Pluto's interacting atmosphere and volatile distribution.

MVIC and LORRI imaging will be used to study variations in surface unit age via stratigraphic relationships and crater density variations, and to search for wind streaks, dunes, or other aeolian features that can serve to suggest or place limits on past epochs of higher atmospheric bulk. Sublimation scarps may indicate how much frost escaped from Pluto over the age of the solar system.

4.2.2 Image Pluto and Charon in Stereo

Topography from stereo imaging will be inferred from comparison of images obtained from two vantage points, for which topographical height differences between two control points will have a more noticeable affect at larger horizontal separations. Throughout the ten hours before closest approach, large portions of Pluto and Charon are imaged at a range of viewing angles, and all such images will be used for stereo reconstruction. One such pair, P_pan_global_3x3_1 and P_pan_global_3x3_2, will allow height reconstruction of 1 km for features separated by 1000 km, or 5 km for features separated by 250 km. In addition to stereo reconstruction, topography can be deduced from shape-from-shading (photoclinometry), which relies on the photometric phase function (goal 1.1c).

4.2.3 Map the Terminators of Pluto and Charon with High Resolution

At roughly an hour before Pluto closest approach, LORRI will observe a strip crossing the terminator across Pluto roughly 1900 long by 450 km wide, at a phase angle of 28 deg with a resolution of 240 m $\mathrm{pixel^{-1}}$(P_pan_alongtrack_8x2_2). The highest resolution Pluto images will be taken fifteen minutes before Pluto closest approach, when LORRI and MVIC will take a series of simultaneous images of Pluto at 59 deg phase (P_pan_terminator_simultMVIC_1). Six to ten images will be made of Pluto by each instrument, depending on the size of the error ellipse, which is dominated by the time of flight uncertainty. With LORRI, the resolution will be 0.70 km $\mathrm{pixel^{-1}}$, so each image will cover 72×72 km. MVIC's resolution at the same distance is 0.30 km $\mathrm{pixel^{-1}}$, and each image will cover 38×1500 km rectangular areas.

As with Pluto, Charon will also be imaged with LORRI and MVIC for high-resolution terminator images. This is planned to occur an hour and a half before closest approach at a phase angle of 30 deg with LORRI (C_pan_global_6x3_1), and again simultaneously with MVIC and LORRI near closest approach at a phase angle of 89 deg (C_pan_terminator_simultMVIC_1).

4.2.4 Map the Surface Composition of Selected Areas of Pluto and Charon with High Resolution

Near Pluto closest approach, LEISA will take a high spatial resolution scan across Pluto at 1.3 km $\mathrm{pixel^{-1}}$, obtaining spectra on a strip measuring 824 km by 332 km (P_IR_hires_scan_1). The similar high spatial resolution LEISA scan of Charon will achieve a resolution of 1.9 km pixel^{-1}, over a strip measuring 1033 km by 486 km (C_IR_hires_scan_1). Additionally, MVIC will obtain approach hemisphere color maps of both bodies at a resolution (at the sub-spacecraft point) of 0.7 km pixel^{-1} for Pluto (P_color_globalscan_1-2) and 1.4 km pixel^{-1} for Charon (C_color_globalscan_1-2). These color maps will be obtained in all four MVIC color bands, including the CH_4 filter at 0.89 micron. Because the MVIC color maps and the high-resolution LEISA maps have resolutions comparable to the resolutions of the global panchromatic maps obtained as a Group 1 objective, they will allow us to relate compositional information with geology.

4.2.5 Characterize Pluto's Ionosphere and Solar Wind Interaction

REX will search for the signature of an ionosphere around Pluto during its Earth occultation (P_radio_occ_simultALICE_1), and will be sensitive to number densities of roughly 2000 e^{-} cm^{-3}. SWAP and PEPSSI also contribute to this goal. Even if the atmospheric escape rate from Pluto is small ($<1 \times 10^{27}$ molecules s^{-1}), the ionosphere forms a conducting barrier to the supersonic solar wind, with a bow shock forming upstream. The weak interplanetary magnetic field of the outer solar system means that the shock is likely to be very broad so the New Horizons' particle detectors should be able to detect the shock and perhaps measure structure within it. SWAP will also search for wake effects after closest approach (P_particles_roll_simultPAM_1, and during P_UV_occ_solar_1). By comparing low-energy data from SWAP with pickup ions detected by PEPSSI, we will be able to determine outflow rate from the planet's atmosphere.

4.2.6 Search for Neutral Species Including H, H_2, HCN, and C_xH_y, and Other Hydrocarbons and Nitriles in Pluto's Upper Atmosphere, and Obtain Isotopic Discrimination where Possible

Some minor species, such as H and Ne, will primarily be detected in Alice UV airglow measurements (PC_UV_airglow_spectrum_1-6). Other species will be detected from their opacity profiles in the Alice UV solar occultation data (P_UV_occ_solar_1). We should be able to detect HCN at mixing rations greater than 2×10^{-4}, C_2H_2 at 1×10^{-5}, C_2H_4 at 7×10^{-6}, and C_2H_6 at 3×10^{-4}.

4.2.7 Search for an Atmosphere Around Charon

The surface gravity of Charon is less than half that of Pluto's. As a result, its primordial atmosphere is expected to have escaped since Charon's formation. A secondary atmosphere on Charon from Pluto's escaping atmosphere, once considered unlikely (Trafton et al. 1997), could be diagnostic of Pluto's current escape rate. While analysis of a 1980 stellar occultation by Charon suggested the possibility of a tenuous atmosphere based on two anomalously low points in the upper baseline (Elliot and Young 1991), a stellar occultation in 2005, at much higher time resolution, showed no evidence for an atmosphere (Gulbis et al. 2006; Sicardy et al. 2006; Young et al. 2005), with upper limits near 2×10^{20} cm^{-2}. Our encounter plan includes Alice UV observations of a solar occultation by Charon (C_UV_occ_solar_1), which will be able to detect an atmosphere down to the nanobar range if it is present. While precedence is given to Alice in the targeting of the Charon occultation, a radio occultation may be observable as well (C_radio_occ_simulALICE_1).

4.2.8 Determine Bolometric Bond Albedos for Pluto and Charon

The bond albedo is the ratio of sunlight reflected in all directions to incident sunlight; it is relevant for calculating energy balance at the surface. The strict definition of bond albedo is a global quantity, derived from the disk-integrated brightness observed at a range of solar phase angles. Since Pluto is inhomogeneous in albedo, composition, and temperature, energy balance depends on the local version of bond albedo, namely normal reflectance and photometric phase functions. These rely on disk-resolved observations at a range of phase functions.

During the approach phase, New Horizons will make UV, visible, and infrared observations of Pluto, Charon, Nix, and Hydra at phase functions ranging from 14.1 to 14.9 deg. In the final week prior to closest approach, LORRI will observe Pluto, Charon, Nix and Hydra every 12 hours at resolutions better than 36 km. In the near encounter sequence, New Horizons will observe the full visible disks of Pluto and Charon with resolution better than 3 km pixel^{-1} at phase angles ranging from 16 to 58 deg (Pluto) and 16 to 104 deg (Charon). Higher phase observations are scheduled in the strawman sequence for phase angles ranging from 160 to 170 deg.

4.2.9 Map the Surface Temperatures of Pluto and Charon

We will measure the temperatures of Pluto, Charon, Nix, and Hydra using LEISA to observe IR spectral features that are diagnostic of temperature in all LEISA spectral images, at a resolution up to 1.3 km pixel^{-1} (P_IR_hires_scan_1). REX will measure the brightness temperature at 4.2 cm on a much coarser resolution scale (P_radio_temperature_map_1, but also passive radiometry during the solid body occultations of P_radio_occ_simultALICE_1 and C_radio_occ_simultALICE_1).

4.3 Group 3 Objectives

4.3.1 Characterize the Energetic Particle Environment of Pluto and Charon

For large atmospheric escape rates, the interaction may be best described as "comet-like," with significant mass-loading over an extensive region; for small escape rates the interaction is probably confined to a much smaller region, creating a more "Venus-like" interaction (Luhmann et al. 1991), where electrical currents in the gravitationally bound ionosphere deflect the solar wind flow. At aphelion (49 AU), should Pluto's atmosphere completely collapse and freeze onto the surface, then the interaction becomes "Moon-like" with the solar wind suffering minimal deflection and directly bombarding the bare, icy dayside surface. Not having a detectable atmosphere, Charon almost certainly has such a "Moon-like" interaction, remaining primarily in the solar wind if Pluto's interaction is weak but becoming totally engulfed if Pluto's interaction is strongly "comet-like" and extends beyond Charon's orbit at 17 R_P (where Pluto's radius, R_P, is 1150 km). The solar wind is supersonic so that when the flow impinges on an obstacle (such as the magnetosphere of the Earth or other planet) an upstream bow shock must form to slow and deflect the supersonic (actually superfast-mode magnetosonic) plasma. The weak interplanetary magnetic field (IMF) at 30 AU and heavy ions formed by photoionizing the heavy molecules of Pluto's escaping atmosphere have very large gyroradi ($\sim$500 R_P). The net results of these non-fluid or kinetic effects are to make the bow shock a thick transition region and to make the shape of the interaction region asymmetric where the direction of asymmetry is governed by the direction of the IMF. The New Horizons' SWAP instrument will measure the perturbation of the solar wind produced by the solar wind interaction with Pluto's atmosphere and both SWAP and PEPSSI can detect the energetic ions produced by the ionization of molecules from Pluto's atmosphere.

4.3.2 Refine Bulk Parameters (Radii, Masses, Densities) and Orbits of Pluto and Charon

The largest uncertainty in the heliocentric orbit of the Pluto system is its heliocentric distance. This will be improved by using Doppler ranging to determine the distance of the spacecraft, and optical navigation to determine the New Horizons trajectory relative to the Pluto system. Optical navigation is further aided by the differences in view angle to Pluto from New Horizons and the Earth.

LORRI will image the positions of Nix, Hydra, Charon, and Pluto over a 150-day timespan on approach, refining orbits of these bodies. Disk-resolved albedo maps will eliminate the persistent problem of center-of-light vs. center-of-mass offsets in Pluto-system astrometry (Buie et al. 2006). These orbits will improve measurements of the masses of Pluto and Charon, and possibly of Nix and Hydra. System masses will be augmented by REX Doppler measurements.

The radii and shapes of Pluto, Charon, Nix, and Hydra will be measured from imaging. Specifically, we will use LORRI for the highest geometric fidelity, obtaining best resolutions of order 3 km for all bodies (P_pan_global_1x1_last, C_pan_global_1x1_last). Alice and REX occultation measurements will augment measurements of the radii.

4.3.3 Search for Magnetic Fields of Pluto and Charon

Being a small body it is unlikely that Pluto has a substantial region of molten, convection iron inside and hence unlikely to have a magnetic dynamo. Nevertheless, the weak interplanetary magnetic field and tenuous solar wind at >30 AU mean that should Pluto have a weak magnetic field then it would produce a relatively large magnetosphere. Bagenal et al. (1997) estimate that a magnetic field >3700 nT would produce a magnetosphere that would hold off the solar wind upstream of the planet and produce a cavity (detected as a drop in solar wind flux by SWAP) extending beyond the flyby distance of the New Horizons spacecraft.

4.3.4 Search for Additional Satellites and Rings

Since the time of the formulation of the measurement goals for New Horizons, two new satellites, Nix and Hydra, have been discovered (Weaver et al. 2006). The entire Hill sphere will be searched by LORRI and MVIC for additional rings and satellites at low solar phase angle during the approach phase. MVIC will search for rings in forward scattering about two hours after Pluto closest approach (P_pan_GlobalRings_scan_1). Finally, SDC will search for dust from the Pluto system in the departure phase.

5 Broader Impact

By deepening our understanding of the Pluto system in particular, New Horizons will extend our knowledge of planetary science in general. We mention just a few examples here.

The surface composition of Pluto presumably reflects the nature of the solar nebula at large heliocentric distances 4.6 billion years ago. Does the dominant role played by N_2 on

Pluto's surface confirm the hypothesis that N_2 was the dominant nitrogen-bearing gas in the nebula? If N_2 was dominant in the nebula, why isn't that reflected in the composition of comets, which seem to be N_2 depleted? Are the N_2/CO and CO/CH_4 ratios on Pluto's surface consistent with current theories of the composition of the solar nebula? Solid CO_2 has been detected on Triton (Quirico et al. 1999; Grundy and Young 2004), but not yet on Pluto or Charon. What will a detection or upper limit tell us about Pluto and Charon's original volatile budget, cosmic-radiation-driven surface chemistry, and geologic histories (e.g., Shock and McKinnon 1993; Grundy et al. 2006)? Based on detailed composition measurements of Pluto and Charon, we will learn more about the origin and evolution of comets and icy dwarfs in the outer solar system.

While the concept of a surface and atmosphere in vapor-pressure equilibrium is well established (Spencer et al. 1997), many complications remain (Trafton et al. 1998). One is the relation between low-resolution albedo maps and volatile distribution. It is usually assumed that fresh ice is bright, and old ice, possibly chemically processed, is dark. However, new N_2 may be transparent if deposition rates are slow (Duxbury et al. 1997), or old frost might be brightened by the processes of subsidence of dark contaminants (sinking into pits created by their own local thermal heating; see Spencer et al. 1997). Moreover, the emissivity of N_2 frost, a factor in its equilibrium temperature, may depend on its crystalline phase (Stansberry and Yelle 1999). Another complication is the interaction of multi-component frosts with the atmosphere. On Pluto, the atmosphere is primarily N_2, with trace amounts of CO and CH_4, which are less volatile than N_2; all three species are seen as surface ices. Models for the multi-component interaction include the formation of a monolayer CH_4-rich cap above an N_2:CH_4 substrate, which chokes off communication of N_2 with the atmosphere, or the formation of warm, CH_4-rich patches, which leads to turbulent boundary layers that enhance CH_4 mixing ratios in the atmosphere. The powerful combinations of instruments on board New Horizons can investigate the relationships among geology, surface (and subsurface) energy balance, ice composition, mixing state, and temperature, along with the surface pressures and composition of the atmosphere. Of the eleven global atmospheres in our solar system, four are Jovian, three have atmospheres that do not condense on the surface (Venus, Earth, and Titan), and four (Pluto, Triton, Io, and Mars) have surface volatiles with significant vapor pressures at the surface temperatures. Thus, study of the behavior of Pluto's surface-atmosphere interaction leads to a greater understanding of a broad classes of atmospheres. As we learn more about the surfaces of other icy dwarf planets, the knowledge from Pluto can extend our understanding of the possibilities of other seasonal atmospheres in the Kuiper belt.

The thermal structure and composition of Pluto's atmosphere, measured with Alice and REX, will serve as the basic input for models investigating atmospheric processes. These models will be further constrained by other New Horizons observations. The results of atmospheric chemistry may be detectable by the presence of condensation products on the surface, by hazes in the atmosphere, by their UV signatures, or by heavier ions detected by the particle instruments. Pluto, Triton, and Titan are often thought of as an atmospheric triad, since all three have N_2 dominated atmospheres with CH_4; an improved understanding of chemistry and energetics on Pluto will improve our understanding of all three. Furthermore, a deeper understanding of atmospheric escape on Pluto has application to the early Earth, whose atmosphere is also thought to have been in hydrodynamic escape.

The anticipated science results from the New Horizons mission presented here are based on our current level of knowledge of the Pluto system. New Horizons is the first mission to the Pluto system, and the first mission to an icy dwarf planet. We should be prepared to be surprised and amazed.

Acknowledgements The authors acknowledge the work of the cast of scientists who have studied the Pluto system, and the numerous NASA and NSF grants and telescope allocations that have supported this work. We are grateful to be working with each of the engineers on the New Horizons project. Finally, the authors thank the public and the scientific community alike for their sustained interest in Pluto and the Pluto system that has made this mission possible.

References

F. Bagenal, T.E. Cravens, J.G. Luhmann, R.L. McNutt, A.F. Cheng, Pluto's interaction with the solar wind, in *Pluto and Charon*, ed. by S.A. Stern, D.J. Tholen (Univ. of Arizona Press, Tucson, 1997), pp. 523–555

K.H. Baines et al., Polar lightning and decadal-scale cloud variability on Jupiter. Science **318**, 226 (2007)

D. Bockelée-Morvan, E. Lellouch, N. Biver, G. Paubert, J. Bauer, P. Colom, D.C. Lis, Search for CO gas in Pluto, Centaurs and Kuiper belt objects at radio wavelengths. Astron. Astrophys. **377**, 343–353 (2001)

M.E. Brown, Pluto and Charon: formation, seasons, composition. Ann. Rev. Earth Planet. Sci. **30**, 307–345 (2002)

M.E. Brown, W.M. Calvin, Evidence for crystalline water and ammonia ices on Pluto's satellite Charon. Science **287**, 107–109 (2000)

R.H. Brown, D.P. Cruikshank, J. Veverka, P. Helfenstein, J. Eluszkiewicz, Surface composition and photometric properties of Triton, in *Neptune and Triton*, ed. by D.P. Cruikshank (Univ. of Arizona Press, Tucson, 1995), pp. 991–1030

M.W. Buie, W.M. Grundy, The distribution and physical state of H_2O on Charon. Icarus **148**, 324–329 (2000)

M.W. Buie, W.M. Grundy, E.F. Young, L.A. Young, S.A. Stern, Orbits and photometry of Pluto's satellites: Charon, S/2005 P1, and S/2005 P2. Astron. J. **132**, 290–298 (2006)

M.W. Buie, D.J. Tholen, L.H. Wasserman, Separate lightcurves of Pluto and Charon. Icarus **125**, 233–244 (1997a)

M.W. Buie, E.F. Young, R.P. Binzel, Surface appearance of Pluto and Charon, in *Pluto and Charon*, ed. by S.A. Stern, D.J. Tholen (Univ. of Arizona Press, Tucson, 1997b), pp. 269–293

R.M. Canup, A giant impact origin of Pluto–Charon. Science **307**, 546–550 (2005)

A.F. Cheng et al., Long range reconnaissance imager on New Horizons. Space Sci. Rev. (2008, this issue). doi:10.1007/s11214-007-9271-6

A.F. Cheng et al., Changing characteristics of Jupiter's little red spot. Astron. J. **135**, 2446–2452 (2008)

J.T. Clarke, S.A. Stern, L.M. Trafton, Pluto's extended atmosphere—an escape model and initial observations. Icarus **95m**, 173–179 (1992)

D.P. Cruikshank, T.L. Roush, J.M. Moore, M. Sykes, T.C. Owen, M.J. Bartholomew, R.H. Brown, K.A. Tryka, The surfaces of Pluto and Charon, in *Pluto and Charon*, ed. by S.A. Stern, D.J. Tholen (Univ. of Arizona Press, Tucson, 1997), pp. 221–268

A.R. Dobrovolskis, S.J. Peale, A.W. Harris, Dynamics of the Pluto–Charon binary, in *Pluto and Charon*, ed. by S.A. Stern, D.J. Tholen (Univ. of Arizona Press, Tucson, 1997), pp. 159–190

S. Douté, B. Schmitt, E. Quirico, T.C. Owen, D.P. Cruikshank, C. de Bergh, T.R. Geballe, T.L. Roush, Evidence for methane segregation at the surface of Pluto. Icarus **142**, 421–444 (1999)

N.S. Duxbury, R.H. Brown, V. Anicich, Condensation of nitrogen: Implications for Pluto and Triton. Icarus **129**, 202–206 (1997)

J.L. Elliot, L.A. Young, Limits to the radius and possible atmosphere of Charon from its 1980 stellar occultation. Icarus **89**, 244–254 (1991)

J.L. Elliot et al., The recent expansion of Pluto's atmosphere. Nature **424**, 165–168 (2003)

J.L. Elliot et al., Changes in Pluto's atmosphere: 1988–2006. Astron. J. **134**, 1–13 (2007)

P. Farinella, D.R. Davis, S.A. Stern, Formation and collisional evolution of the Edgeworth-Kuiper belt, in *Protostars and Planets IV*, ed. by V. Mannings, A.P. Boss, S.S. Russell (Univ. of Arizona Press, Tucson, 2000), pp. 1255–1282

G.R. Gladstone et al., Jupiter's nightside airglow and aurora. Science **318**, 229 (2007)

T.K. Greathouse et al., New Horizons Alice ultraviolet observations of a stellar occultation by Jupiter's atmosphere. Icarus (2008, submitted)

W.M. Grundy, Methane and nitrogen ices on Pluto and Triton: a combined laboratory and telescope investigation. Ph.D. Thesis, University of Arizona, 1995

W.M. Grundy, M.W. Buie, Distribution and evolution of CH_4, N_2, and CO ices on Pluto's surface: 1995 to 1998. Icarus **153**, 248–263 (2001)

W.M. Grundy, M.W. Buie, Spatial and compositional constraints on non-ice components and H_2O on Pluto's surface. Icarus **157**, 128–138 (2002)

W.M. Grundy, L.A. Young, Near infrared spectral monitoring of Triton with IRTF/SpeX I: Establishing a baseline. Icarus **172**, 455–465 (2004)

W.M. Grundy, B. Schmitt, E. Quirico, The temperature dependent spectra of alpha and beta nitrogen ice with application to Triton. Icarus **105**, 254–258 (1993)

W.M. Grundy, M.W. Buie, J.A. Stansberry, J.R. Spencer, B. Schmitt, Near-infrared spectra of icy outer solar system surfaces: Remote determination of H_2O ice temperatures. Icarus **142**, 536–549 (1999)

W.M. Grundy, B. Schmitt, E. Quirico, The temperature-dependent spectrum of methane ice I between 0.7 and 5 μm and opportunities for near-infrared remote thermometry. Icarus **155**, 486–496 (2002)

W.M. Grundy, L.A. Young, J.R. Spencer, R.E. Johnson, E.F. Young, M.W. Buie, Distributions of H_2O and CO_2 ices on Ariel, Umbriel, Titania, and Oberon from IRTF/SpeX observations. Icarus **184**, 543–555 (2006)

W.M. Grundy et al., New Horizons mapping of Europa and Ganymede. Science **318**, 234 (2007)

A.A.S. Gulbis, J.L. Elliot, M.J. Person, E.R. Adams, B.A. Babcock, M. Emilio, J.W. Gangestad, S.D. Kern, E.A. Kramer, D.J. Osip, J.M. Pasachoff, S.P. Souza, T. Tuvikene, Charon's radius and atmospheric constraints from observations of a stellar occultation. Nature **439**, 48–51 (2006)

Y. Guo, R. Farquhar, New Horizons mission design. Space Sci. Rev. (2008, this issue). doi:10.1007/s11214-007-9242-y

M. Horanyi et al., The Student Dust Counter on the New Horizons mission. Space Sci. Rev. (2008, this issue). doi:10.1007/s11214-007-9250-y

V.A. Krasnopolsky, Hydrodynamic flow of N_2 from Pluto. J. Geophys. Res. **104**, 5955–5962 (1999)

V.A. Krasnopolsky, D.P. Cruikshank, Photochemistry of Pluto's atmosphere and ionosphere near perihelion. J. Geophys. Res. **104**, 21979–21996 (1999)

E. Lellouch, R. Laureijs, B. Schmitt, E. Quirico, C. de Bergh, J. Crovisier, A. Coustenis, Pluto's non-isothermal surface. Icarus **147**, 220–250 (2000)

J.-C. Liou, H.A. Zook, Signatures of the giant planets imprinted on the Edgeworth-Kuiper belt dust disk. Astron. J. **118**, 580–590 (1999)

J.G. Luhmann, C.T. Russell, K. Schwingenschuh, E. Eroshenko, A comparison of induced magnetotails of planetary bodies – Venus, Mars, and Titan. J. Geophys. Res. **96**, 11,19911,208 (1991)

R. Malhotra, J.G. Williams, Pluto's heliocentric orbit, in *Pluto and Charon*, ed. by S.A. Stern, D.J. Tholen (Univ. of Arizona Press, Tucson, 1997), pp. 127–157

D.J. McComas, F. Allegrini, F. Bagenal, F. Crary, R.W. Ebert, H. Elliott, A. Stern, P. Valek, Diverse Plasma Populations and Structures in Jupiter’s Magnetotail. Science **318**, 217 (2007)

D. McComas et al., The Solar Wind Around Pluto (SWAP) instrument aboard New Horizons. Space Sci. Rev. (2008, this issue). doi:10.1007/s11214-007-9205-3

R.L. McNutt Jr., Models of Pluto's upper atmosphere. GRL **16**, 1225–1228 (1989)

R.L. McNutt et al., Energetic Particles in the Jovian Magnetotail. Science **318**, 220 (2007)

R.E. McNutt et al., The Pluto Energetic Particle Spectrometer Science Investigation (PEPSSI) on New Horizons. Space Sci. Rev. (2008, this issue). doi:10.1007/s11214-008-9436-y

C.B. Olkin, D. Reuter, A. Lunsford, R.P. Binzel, S.A. Stern, The New Horizons Distant Flyby of Asteroid 2002 JF56. AAS/Division for Planetary Sciences Meeting Abstracts **38** (2006) #59.22

C.B. Olkin, E.F. Young, L.A. Young, W. Grundy, B. Schmitt, A. Tokunaga, T. Owen, T. Roush, H. Terda, Pluto's spectrum from 1.0 to 4.2 μm: implications for surface properties. Astron. J. **133**, 420–431 (2007)

T.C. Owen, T.L. Roush, D.P. Cruikshank, J.L. Elliot, L.A. Young, C. de Bergh, B. Schmitt, T.R. Geballe, R.H. Brown, M.J. Bartholomew, Surface ices and atmospheric composition of Pluto. Science **261**, 745–748 (1993)

M.J. Person, J.L. Elliot, A.A.S. Gulbis, J.M. Pasachoff, B.A. Babcock, S.P. Souza, J. Gangestad, Charon's radius and density from the combined data sets of the 2005 July 11 occultation. Astron. J. **132**, 1575–1580 (2006)

E. Quirico, B. Schmitt, Near infrared spectroscopy of simple hydrocarbons and carbon oxides diluted in solid N_2 and pure ices: Implication for Triton and Pluto. Icarus **127**, 354–378 (1997)

E. Quirico, S. Douté, B. Schmitt, C. de Bergh, D.P. Cruikshank, T.C. Owen, T.R. Geballe, T.L. Roush, Composition, physical state and distribution of ices at the surface of Triton. Icarus **139**, 159–178 (1999)

K. Rages, J.B. Pollack, Voyager imaging of Triton's clouds and hazes. Icarus **99**, 289–301 (1992)

K.D. Retherford et al., Io's atmospheric response to eclipse: UV aurorae observations. Science **318**, 237 (2007)

D.C. Reuter et al., Jupiter cloud composition, stratification, convection, and wave motion: a view from New Horizons. Science **318**, 223 (2007)

D. Reuter et al., Ralph: A visible/infrared imager for the New Horizons mission. Space Sci. Rev. (2008, this issue). doi:10.1007/s11214-008-9375-7

E.L. Shock, W.B. McKinnon, Hydrothermal processing of cometary volatiles-applications to Triton. Icarus **106**, 464–477 (1993)

M.R. Showalter, A.F. Cheng, H.A. Weaver, S.A. Stern, J.R. Spencer, H.B. Throop, E.M. Birath, D. Rose, J.M. Moore, Clump detections and limits on moons in Jupiter’s ring system. Science **318**, 232 (2007)

B. Sicardy et al., Large changes in Pluto's atmosphere as revealed by recent stellar occultations. Nature **424**, 168–170 (2003)

B. Sicardy et al., Charon's size and an upper limit on its atmosphere from a stellar occultation. Nature **439**, 52–54 (2006)

J.R. Spencer, J.A. Stansberry, L.M. Trafton, E.F. Young, R.P. Binzel, S.K. Croft, Volatile transport, seasonal cycles, and atmospheric dynamics on Pluto, in *Pluto and Charon*, ed. by S.A. Stern, D.J. Tholen (Univ. of Arizona Press, Tucson, 1997), pp. 435–473

J. Spencer, M. Buie, L. Young, Y. Guo, A. Stern, Finding KBO flyby targets for New Horizons. Earth, Moon Planets **92**, 483–491 (2003)

J.R. Spencer et al., Io volcanism seen by New Horizons: A major eruption of the Tvashtar volcano. Science **318**, 240 (2007)

J.A. Stansberry, R.V. Yelle, Emissivity and the fate of Pluto's atmosphere. Icarus **141**, 299–306 (1999)

A.J. Steffl, M.J. Mutchler, H.A. Weaver, S.A. Stern, D.D. Durda, D. Terrell, W.J. Merline, L.A. Young, E.F. Young, M.W. Buie, J.R. Spencer, New constraints on additional satellites of the Pluto system. Astron. J. **132**, 614–619 (2007)

S.A. Stern, The Pluto–Charon system. Ann. Rev. Astron. Astrophys. **30**, 185–233 (1992)

S.A. Stern, The New Horizons Pluto Kuiper belt mission: an overview with historical context. Space Sci. Rev. (2008, this issue). doi:10.1007/s11214-007-9295-y

S.A. Stern, A. Cheng, NASA plans Pluto–Kuiper belt mission. EOS **83**, 101 (2002)

S.A. Stern, M.W. Buie, L.M. Trafton, HST high-resolution images and maps of Pluto. Astrophys. J. **113**, 827 (1997a)

S.A. Stern, W.B. McKinnon, J.L. Lunine, On the origin of Pluto, Charon, and the Pluto–Charon binary, in *Pluto and Charon*, ed. by S.A. Stern, D.J. Tholen (Univ. Arizona Press, Tucson, 1997b), pp. 605–663

S.A. Stern, M.J. Mutchler, H.A. Weaver, A.J. Steffl, The positions, colors, and photometric variability of Pluto's small satellites from HST observations: 2005–2006. Lunar and Planetary Sci. Conference **38**, no. 1722 (2007)

S.A. Stern et al., Alice: The ultraviolet imaging spectrometer aboard the New Horizons Pluto–Kuiper belt mission. Space Sci. Rev. (2008, this issue). doi:10.1007/s11214-008-9407-3

M.E. Summers, D.F. Strobel, G.R. Gladstone, Chemical models of Pluto's atmosphere, in *Pluto and Charon*, ed. by S.A. Stern, D.J. Tholen (Univ. Arizona Press, Tucson, 1997), pp. 391–434

D.T. Tholen, M.W. Buie, Bulk properties of Pluto and Charon, in *Pluto and Charon*, ed. by S.A. Stern, D.J. Tholen (Univ. of Arizona Press, Tucson, 1997), pp. 435–473

F. Tian, B. Toon, Hydrodynamic escape of nitrogen from Pluto. Geophys. Res. Lett. **32**, L18201 (2005)

L.M. Trafton, D.M. Hunten, K.J. Zahnle, R.L. McNutt Jr., Escape processes at Pluto and Charon, in *Pluto and Charon*, ed. by S.A. Stern, D.J. Tholen (Univ. Arizona Press, Tucson, 1997), pp. 475–522

L.M. Trafton, D.L. Matson, J.A. Stansberry, Surface/atmosphere interactions and volatile transport (Triton, Pluto and Io), in *Solar System Ices*, ed. by B. Schmitt, C. de Bergh, M. Festou (Kluwer, Dordrecht, 1998), p. 773

K.A. Tryka, R.H. Brown, V. Anicich, Near-infrared absorption coefficients of solid nitrogen as a function of temperature. Icarus **116**, 409–414 (1995)

L. Tyler et al., The Radio EXperiment (REX) on New Horizons. Space Sci. Rev. (2008, this issue). doi:10.1007/s11214-007-9302-3

H.A. Weaver, S.A. Stern, M.J. Mutchler, A.J. Steffl, M.W. Buie, W.J. Merline, J.R. Spencer, E.F. Young, L.A. Young, The discovery of two new satellites of Pluto. Nature **439**, 943 (2006)

P.R. Weissman, H.F. Levison, The population of the trans-neptunian region: the Pluto–Charon environment, in *Pluto and Charon*, ed. by S.A. Stern, D.J. Tholen (Univ. of Arizona Press, Tucson, 1997), pp. 559–604

R.V. Yelle, J.L. Elliot, Atmospheric structure and composition: Pluto and Charon, in *Pluto and Charon*, ed. by S.A. Stern, D.J. Tholen (Univ. of Arizona Press, Tucson, 1997), pp. 347–390

L.A. Young, J.L. Elliot, A. Tokunaga, C. de Bergh, T. Owen, Detection of gaseous methane on Pluto. Icarus **127**, 258 (1997)

E.F. Young, K. Galdamez, M.W. Buie, R.P. Binzel, D.J. Tholen, Mapping the variegated surface of Pluto. Astron. J. **117**, 1063–1076 (1999)

E.F. Young, R.P. Binzel, K. Crane, A two-color map of Pluto's sub-Charon hemisphere. Astron. J. **121**, 552–561 (2001a)

L.A. Young, J.C. Cook, R.V. Yelle, E.F. Young, Upper limits on gaseous CO at Pluto and Triton from high-resolution near-IR spectroscopy. Icarus **153**, 148–156 (2001b)

L.A. Young, C.B. Olkin, E.F. Young, R.G. French, International Astronomical Union Circular 8570 (2005)

E.F. Young et al., Vertical structure in Pluto's atmosphere from the 12 June 2006 stellar occultation. Astron. J. **136**, 1757–1769 (2008)

Ralph: A Visible/Infrared Imager for the New Horizons Pluto/Kuiper Belt Mission

Dennis C. Reuter · S. Alan Stern · John Scherrer · Donald E. Jennings · James W. Baer · John Hanley · Lisa Hardaway · Allen Lunsford · Stuart McMuldroch · Jeffrey Moore · Cathy Olkin · Robert Parizek · Harold Reitsma · Derek Sabatke · John Spencer · John Stone · Henry Throop · Jeffrey Van Cleve · Gerald E. Weigle · Leslie A. Young

Originally published in the journal Space Science Reviews, Volume 140, Nos 1–4, 129–154.
DOI: 10.1007/s11214-008-9375-7

Abstract The New Horizons instrument named Ralph is a visible/near infrared multi-spectral imager and a short wavelength infrared spectral imager. It is one of the core instruments on New Horizons, NASA's first mission to the Pluto/Charon system and the Kuiper Belt. Ralph combines panchromatic and color imaging capabilities with SWIR imaging spectroscopy. Its primary purpose is to map the surface geology and composition of these objects, but it will also be used for atmospheric studies and to map the surface temperature. It is a compact, low-mass (10.5 kg) power efficient (7.1 W peak), and robust instrument with good sensitivity and excellent imaging characteristics. Other than a door opened once in flight, it has no moving parts. These characteristics and its high degree of redundancy make Ralph ideally suited to this long-duration flyby reconnaissance mission.

D.C. Reuter (✉) · D.E. Jennings · A. Lunsford
Code 693, NASA/GSFC, Greenbelt, MD 20771, USA
e-mail: dennis.c.reuter@nasa.gov

S.A. Stern
Space Sciences and Engineering Division, Southwest Research Institute (SwRI), 1050 Walnut St., Suite 400, Boulder, CO 80302, USA

J. Scherrer · J. Hanley · J. Stone · G.E. Weigle
SwRI, 6220 Culebra Rd., San Antonio, TX 78228, USA

J.W. Baer · L. Hardaway · R. Parizek · H. Reitsma · D. Sabatke · J. Van Cleve
Ball Aerospace and Technology Corporation (BATC), 1600 Commerce St., Boulder, CO 80301, USA

S. McMuldroch
SSG Precision Optronics, 65 Jonspin Rd., Wilmington, MA 01887, USA

J. Moore
MS 245-3, NASA/Ames Research Center, Moffett Field, CA 94035-1000, USA

C. Olkin · J. Spencer · H. Throop · L.A. Young
Dept. of Space Studies, Southwest Research Institute (SwRI), 1050 Walnut St., Suite 400, Boulder, CO 80302, USA

Keywords New Horizons · Pluto/Charon · Kuiper Belt · Ralph · Multi-spectral imager · Infrared spectral imager · Hyperspectral imager · Wedged filter · TDI

1 Introduction

New Horizons, a flyby mission to the Pluto/Charon system and the Kuiper Belt, is the first in NASA's New Frontiers line of moderate-scale planetary missions and is the first mission to explore Pluto and its moons Charon, Nix and Hydra. Launched on January 19, 2006, it is scheduled for a closest approach of about 10,000 km on July 14, 2015. The scientific rationale for New Horizons and the overall mission planning are described in detail in several papers in this issue (Stern 2008; Young et al. 2008, and Fountain et al. 2008). The New Horizons mission is led by Principal Investigator Alan Stern of the Southwest Research Institute of Boulder, CO and is managed by SwRI and the Johns Hopkins Applied Physics Laboratory in Laurel, MD. A core remote-sensing instrument on New Horizons is Ralph, a visible/NIR camera and shortwave infrared spectral mapper. The instrument's primary purpose is to measure surface characteristics, including geological processes, geomorphology, photometric properties, and surface composition. In addition, surface temperature will be inferred from the shapes and positions of well-established, thermally diagnostic reflectance spectral features in H_2O, CH_4, and N_2 ices. Ralph will also be used to measure haze optical depths (if present) and to search for rings and small satellites. This paper describes Ralph and specifies its characteristics (see also Reuter et al. 2005).

The Ralph instrument is mounted to the exterior of the New Horizons spacecraft (Fountain et al. 2008). Ralph consists of a single telescope that feeds two sets of focal planes: (1) the Multi-spectral Visible Imaging Camera (MVIC), a visible, near-IR imager and (2) the Linear Etalon Imaging Spectral Array (LEISA), a short-wavelength, IR, spectral imager. The telescope uses an unobscurred, off-axis, three-mirror anastigmat design. The entire telescope assembly, including the three diamond turned mirrors, is constructed from grain aligned 6061-T6 aluminum. The housing, excluding the attached covers but including the integral optical mounts, is diamond turned from a single Al block. This combination of an all Al structure and optics is lightweight, athermal and thermally conductive. It ensures that the optical performance of the system is insensitive to temperature and that thermal gradients are minimized. The highly baffled 75-mm aperture, VIS/IR telescope provides ample sensitivity at Pluto/Charon, while minimizing size and mass. The $f/8.7$ system's approximately 658 mm Effective Focal Length offers a good compromise between photometric throughput and alignment stability. The mirrors are direct diamond turned off axis aspheres, each with surface roughness below 6.0 nm RMS, fabricated by Corning Diamond Turning Division of Keene, NH. Stray light control is improved by using a field baffle at an intermediate focus between the secondary and tertiary mirrors, and a Lyot stop at the exit pupil after the tertiary mirror. A dichroic beamsplitter transmits IR wavelengths longer than 1.1 μm to LEISA and reflects shorter wavelengths to MVIC.

MVIC is composed of 7 independent CCD arrays on a single substrate. It uses two of its large format (5024×32 pixel) CCD arrays, operated in time delay integration (TDI) mode, to provide panchromatic (400 to 975 nm) images. Four additional 5024×32 CCDs, combined with the appropriate filters and also operated in TDI mode, provide the capability of mapping in blue (400–550 nm), red (540–700 nm), near IR (780–975 nm) and a narrow methane absorption band (860–910 nm). TDI operates by synchronizing the parallel transfer rate of each of the CCD's thirty-two 5024 pixel wide rows to the relative motion of the image across the detector's surface. In this way, very large format images are obtained as the spacecraft scans the FOV rapidly across the surface. The presence of 32 rows effectively increases the integration time by that same factor, allowing high signal-to-noise measurements. The panchromatic (pan) channels of MVIC will be used to produce hemispheric maps of Pluto and Charon at a double sampled spatial resolution of 1 km^2 or better. The static FOV of each of the TDI arrays is $5.7^\circ \times 0.037^\circ$. In addition to the TDI arrays, MVIC has a 5024×128 element panchromatic frame transfer array operated in staring mode, with an FOV of $5.7^\circ \times 0.15^\circ$. The primary purpose of the framing array is to provide data for optical navigation (OpNav) of the spacecraft. The IFOV of a single MVIC pixel is 20×20 $\mu radian^2$.

LEISA is a wedged filter infra-red spectral imager that creates spectral maps in the compositionally important 1.25–2.5 micron short wave infrared (SWIR) spectral region. It images a scene through a wedged filter (linear variable filter, LVF, Rosenberg et al. 1994) placed about 100 μm above a 256×256 pixel Mercury Cadmium Telluride (HgCdTe) detector array. An image is formed on both the wedged filter and the array simultaneously (there is less than 5% spectral broadening by the $f/8.7$ beam). The LVF is fabricated such that the wavelength varies along one dimension, the scan direction. LEISA forms a spectral map by scanning the FOV across the surface in a push broom fashion, similar to that of the MVIC TDI channels. The frame rate is synchronized to the rate of the scan, so that a frame is read out each time the image moves by the single pixel IFOV. The difference between a LEISA scan and a TDI scan is that in LEISA the row-to-row image motion builds up a spectrum while in TDI the motion increases the signal in a single spectral interval. The filter is made in two segments. The first covers from 1.25 to 2.5 microns at an average spectral resolving power ($\lambda/\Delta\lambda$) of 240. This section of LEISA will be used to obtain composition maps. The second segment covers 2.1 to 2.25 microns with an average spectral resolving power of 560. It will be used to obtain both compositional information and surface temperature maps by measuring the spectral shape of solid N_2.

The MVIC and LEISA components of Ralph were originally developed in 1993 for what was then called the "Pluto Fast Flyby" mission using grants from NASA's "Advanced Technology Insertion" (ATI) project. At the time, they were combined with a UV mapping spectrometer (also developed under the ATI grant) into a fully integrated remote-sensing package called HIPPS (Highly Integrated Pluto Payload System, Stern et al. 1995). For the New Horizons mission, HIPPS evolved into the Pluto Exploration Remote Sensing Instrument (PERSI), in which the UV spectrometer, now named Alice, was decoupled from the MVIC and LEISA components. This allowed the UV and Vis/IR optics to be separately optimized and reduced the chances of contamination of the sensitive UV optics. Versions of Alice are flying on the Rosetta comet orbiter mission and have been chosen for the Lunar Reconnaissance Orbiter (see Stern et al. 2008 and references therein). Instruments based on the ATI LEISA concept have flown on the Lewis mission (Reuter et al. 1996) and the EO-1 mission (Reuter et al. 2001; Ungar et al. 2003). The Lewis spacecraft prematurely re-entered the atmosphere before any instrument aboard could take data, but the version of LEISA aboard EO-1 provided numerous images. Now that the UV Alice spectrometer is a separate entity from the MVIC/LEISA sub-assembly, the latter is named Ralph in honor of Ralph and Alice Kramden of "Honeymooner's" fame. New Horizons is the first mission on which the Ralph instrument has flown.

Ralph is a joint effort of the Southwest Research Institute (SwRI, San Antonio, TX and Boulder, CO which is the home institution of Alan Stern, the Ralph Principal Investigator), Ball Aerospace and Technologies Corp. (BATC, Boulder, CO) and NASA's Goddard Space Flight Center (GSFC, Greenbelt, MD).

2 Ralph Science Overview

The full scientific rationale for the New Horizons mission to the Pluto system (and beyond into the deeper Kuiper Belt) is given in detail in another paper in this issue (Young et al. 2008) and will not be repeated at length here. In brief, the Kuiper Belt is an extended disk containing numerous primordial bodies whose planetary evolution was arrested early in Solar System formation. Kuiper Belt Objects (KBOs) are the "fossils" of planetary evolution and the Kuiper Belt is the prime "archeological site" in the Solar System. Pluto is among the largest known KBOs and is a full-fledged dwarf planet. Because Charon is approximately half of Pluto's size, the center of mass of the Pluto-Charon system lies between the two objects, making it a true binary system. The first exploration of the Pluto-Charon system and the Kuiper Belt is both scientifically and publicly exciting. It will provide invaluable insights into the origin of the outer solar system and the ancient outer solar nebula. It will explore the origin and evolution of planet–satellite systems and the comparative geology, geochemistry, tidal evolution, atmospheres, and volatile transport mechanics of icy worlds.

The Ralph instrument will play a leading role in this exploration. It directly addresses two of the three Group 1, or primary, mission requirements (see Young et al. 2008): (1) to characterize the global geology and morphology of Pluto and Charon and (2) to map the surface composition of Pluto and Charon. It also contributes to the third Group 1 requirement of searching for atmospheric haze. High spatial resolution ($\leq$ 1 km/line pair) panchromatic maps generated by the MVIC component of Ralph will be used to address the first Group 1 requirement. These maps will be obtained for the hemisphere visible at closest approach, and will address many of the outstanding questions about these bodies. What is their cratering history? What types of structures are found on their surfaces? What is the spatial variability and scale size of surface features? What is the effect of seasonal volatile transport on the "smoothness" of surface features? Answers to these questions will revolutionize our understanding of the formation and evolution of the Pluto/Charon system.

The second Group 1 goal will be addressed both by MVIC and by LEISA. LEISA will obtain hemispheric maps in the 1.25 to 2.5 μm spectral region with an average resolving power ($\lambda/\Delta\lambda$) of 240 and a spatial resolution of less than 10 km. Similarly, MVIC will provide hemispheric surface color maps and maps of surface methane (CH_4) at even higher spatial resolutions. Pluto's surface is known to contain the species CH_4, N_2, and CO, while Charon's is primarily H_2O but is also likely to contain ammonium hydrates. LEISA's hemispheric maps will allow us to address questions pertaining to composition. What is the surface distribution of the main species? Are there areas of pure frost and mixed areas? What is the effect of seasonal transport? Are there more complex species in selected regions of the surface? Is there a connection between geology and composition? Answers to these questions will significantly advance our understanding of the chemical and physical processes that occur on icy objects and of the processes that occurred in the cold outer regions of our solar system during its formation. Figure 1 shows simulated spectra for Pluto and Charon. As is evident from this figure, there is a wealth of information to be gleaned from this spectral region even from globally averaged spectra. LEISA's spectral maps will permit the correlation of composition with both geology and atmospheric transport of volatiles. They will also enable the study of Pluto's crustal composition where craters or other windows into the interior so permit.

In addition to the Group 1 objectives, Ralph will address numerous Group 2 and Group 3 measurement goals. These include: obtaining stereo images of Pluto and Charon (MVIC), mapping the terminators (MVIC), obtaining high resolution maps in selected regions (MVIC and LEISA), refining the bulk parameters and orbits of the Pluto system and searching for

Fig. 1 Reflectance spectra of Pluto (*top*) and Charon (*bottom*) in the LEISA spectral range at the LEISA spectral resolution (Cruikshank and Dalle Ore 2000). Pluto's known spectrum contains CH_4, N_2 and CO bands, while Charon's is dominated by H_2O

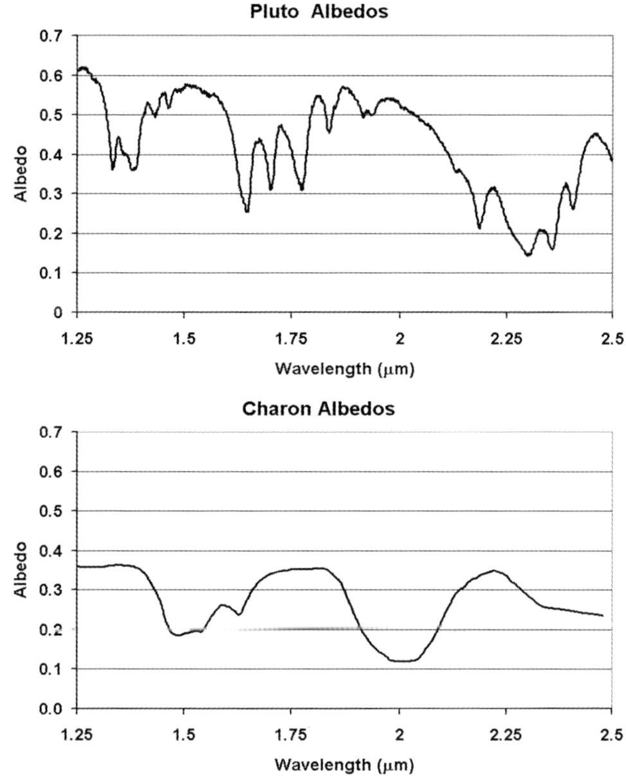

rings and additional satellites (MVIC). LEISA will use its high resolution ($\langle\lambda/\Delta\lambda\rangle = 560$) 2.1 to 2.25 μm segment to obtain surface temperature maps of Pluto employing a technique that relates temperature to the spectral shape of the N_2 transition near 2.15 μm. This technique is particularly sensitive for temperatures near 35 K, the temperature at which N_2 undergoes a transition from α phase to β phase (see Grundy et al. 1993 for a discussion of the change of band-shape as a function of temperature). Thirty-five K is close to the predicted surface temperature of Pluto at the time of the flyby. The accuracy of the retrieved surface temperature is relatively insensitive to absolute radiometric calibration, but is affected by wavelength calibration errors. It is expected that the effective surface temperature will be determined to within $\pm 2°$ K at a spatial resolution of 60 km. For Pluto, additional temperature information will be obtained from the band shape of the CH_4 features measured using the lower resolution LEISA filter. For Charon, which does not have a prominent N_2 band, reasonably accurate temperature maps can be deduced from the shape of the water bands observed with the lower resolution LEISA filter. These secondary and tertiary objectives, while not mission critical like the Group 1 goals, will add substantially to our understanding of Pluto and Charon. MVIC panchromatic and color maps and LEISA spectral maps will also be obtained of Nix and Hydra, two recently discovered satellites in the Pluto system.

Table 1 summarizes the science objectives that determined the Ralph design, the measurement strategies that address these objectives and the derived instrument performance requirements. Except for the spectral resolution and coverage of the high-resolution segment of LEISA, the performance requirements were determined by the need to address the Group 1 goals. The high-resolution LEISA segment was added specifically to address the

Table 1 Science objectives and derived instrument requirements

Science objective	Measurement strategy	Derived instrument requirements			
		Spectral coverage	Resolution	Image quality	Signal-to-noise
Global geology and morphology of Pluto/Charon	Panchromatic images: spatial resolution of 1 km/line pair	400–975 nm	N/A	MTF ≥ 0.15@20 cycles/milliradian	50 (33 AU, 0.35 I/F)
Map the surface composition of Pluto/Charon	Color images: spatial resolution <10 km/line pair	400–550 nm (blue) 540–700 nm (red) 780–975 nm (NIR) 860–910 nm (CH_4)	N/A	No additional requirement	50 (33 AU, 0.35 I/F) 50 (33 AU, 0.35 I/F) 50 (33 AU, 0.35 I/F) 15 (33 AU, 0.35 I/F)
Map the surface composition of Pluto/Charon	SWIR spectral images: spatial resolution <10 km	1.25–2.5 μm	$\lambda/\Delta\lambda \geq 250$	No additional requirement	32 (1.25 μm; Pluto) 27 (2.00 μm; Pluto) 18 (2.15 μm; Pluto)
Map Pluto's surface temperature	High spectral resolution images in the 2.15 μm N_2 band	2.10–2.25 μm	$\lambda/\Delta\lambda \geq 550$	No additional requirement	No additional requirement

Group 2 goal of mapping Pluto's surface temperature, but it will prove useful in the surface composition mapping as well. The MVIC framing camera will be used to perform optical navigation. This gives rise to the additional requirement that it be capable of measuring a 10th magnitude star with a signal-to-noise ratio of 7 in a 0.25 second exposure.

3 Opto-Mechanical Design

Figure 2 shows a model of Ralph and a picture of the assembled instrument before addition of the final MLI. The major elements are labeled in the model. The mass of the instrument is 10.5 kg and the maximum peak power load is 7.1 Watts. The low power and mass are especially important for the New Horizons mission where both of these resources are at a premium. As shown in Fig. 2, Ralph has two assemblies, the telescope detector assembly (TDA) and the electronics assembly. The TDA consists of the telescope optical elements, the baffling, the MVIC and LEISA focal planes, the two-stage passive radiator that cools the focal planes and the flat fielding Solar Illumination Assembly (SIA). The aperture of the TDA was closed by a one-time use door with a partially transmitting window (about 20% throughput). The door protected the optics from contamination prior to and during launch and protected the focal planes from accidental solar exposure during the early flight stage. The door was opened when the spacecraft was 2.3 AU from the sun and can not be closed. The TDA is mounted to the spacecraft by thermally isolating titanium flexures. The in-flight temperature of the TDA is about 220 K, The temperature of the electronics box, which is mounted directly to the spacecraft, is about 290 K. The low temperature of the TDA reduces the conductive and radiative thermal load on the focal planes. It also limits the background signal at the long wavelength end of LEISA. The inner stage of the externally mounted passive radiator cools the LEISA detector to < 130 K. The outer annulus maintains the MVIC CCDs at temperatures below 175 K and lowers the temperature of an $f/2.4$ cold shield for LEISA to below 190 K. The 75 mm aperture, 657.5 mm focal length, $f/8.7$ telescope provides diffraction limited image quality over the $5.7° \times 1.0°$ field of view spanned by the MVIC and LEISA arrays. The instrument parameters for Ralph are summarized in Table 2. Figure 3 shows a model of the TDA interior with a raytrace diagram.

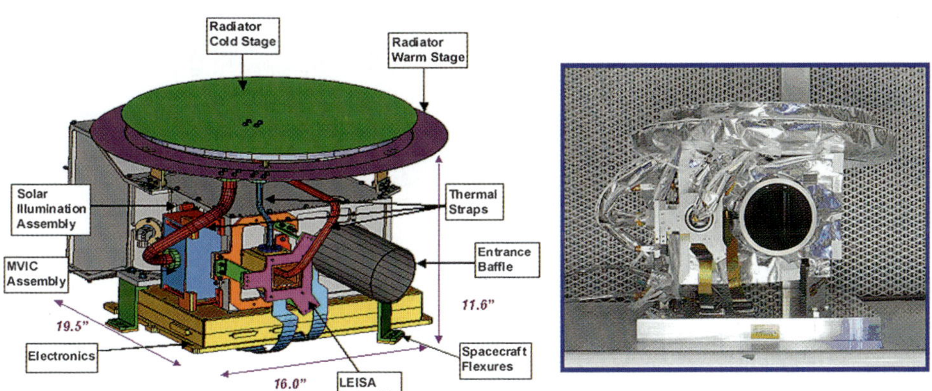

Fig. 2 (*Left*) Model of the Ralph instrument with principle structures labeled. (*Right*) Picture of Ralph, looking down the aperture, before the addition of most of the multi-layer insulation (MLI)

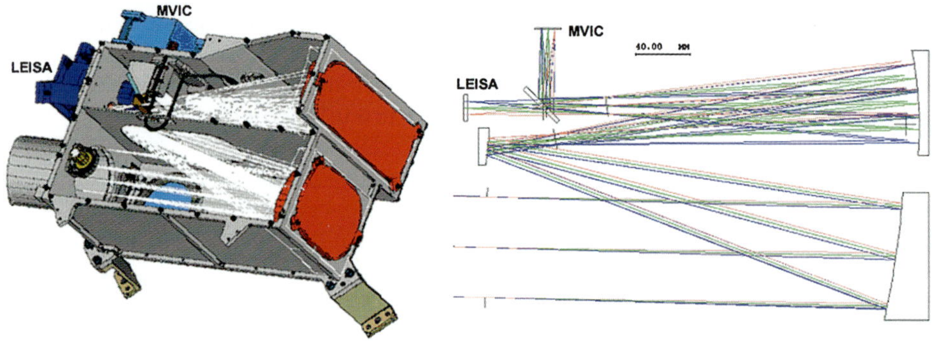

Fig. 3 (*Left*) Interior of the Ralph TDA Showing the light path. Note the stainless steel tube loop containing the SIA fibers. (*Right*) Raytrace diagram showing the path to the LEISA and MVIC focal planes

Fig. 4 (*Top*) Schematic showing the arrangement of the CCDs on the MVIC focal Plane. The *shaded areas* are photo-active regions. The *white strips* are the serial readouts. (*Bottom*) Photograph of the flight MVIC focal plane in its mounting holder

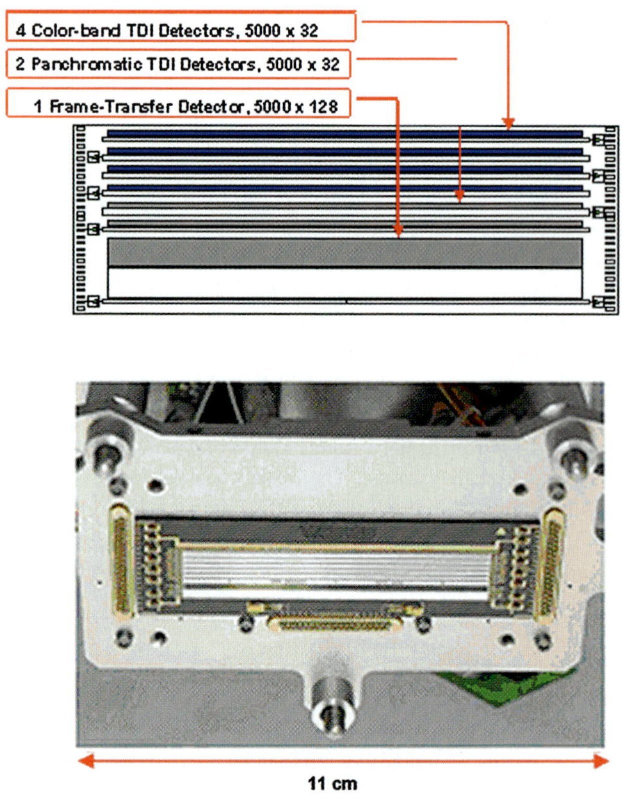

3.1 The MVIC Focal Plane

The MVIC focal plane assembly consists of a customized front side illuminated CCD array provided by E2V Corp. of Chelmsford, England. It is mounted to a heat sink plate and placed directly behind a "butcher block" filter assembly. The array has six identical 5024 × 32 pixel TDI CCDs (5000 × 32 pixel photoactive area) and one 5024 × 264 pixel frame transfer CCD (5000 × 128 pixel photoactive area) on a single substrate. Figure 4 shows a schematic

Table 2 Ralph instrument parameters

Mass: 10.5 kg
Power: 7.1 Watt (maximum)
Telescope Aperture: 75 mm
Focal Length: 657.5 mm
f/#: 8.7
MVIC: Time Delay and Integrate (TDI) and framing arrays
2 Redundant 5024 × 32 Pixel Panchromatic TDI CCDs (400–975 nm)
Four 5024 × 32 Pixel Color TDI CCDs
Blue (400–550 nm)
Red (540–700 nm)
NIR (780–975 nm)
Methane (860–910 nm)
5024 × 128 Frame Transfer Pan CCD
13 μm × 13 μm pixels
Single pixel Field of View: 19.77 μrad × 19.77 μrad
TDI array FOV: 5.7° × 0.037°
Framing camera FOV: 5.7° × 0.146°
Focal plane temperature: < 175 K
Pan TDI rate: 4–84 Hz
Color TDI rate: 4–54 Hz
Frame transfer integration time: 0.25–10 s
LEISA: 256 × 256 element HgCdTe array operated in pushbroom mode
40 μm × 40 μm pixels
Single pixel Field of View: 60.83 μrad × 60.83 μrad
FOV: 0.9° × 0.9°
Focal plane temperature: < 130 K
Filter segment 1 (1.25–2.5 μm) average resolving power ($\lambda/\Delta\lambda$): 240
Filter segment 2 (2.1–2.25 μm) average resolving power ($\lambda/\Delta\lambda$): 560
Frame rate: 0.25–8 Hz

of the array indicating the positions of the CCDs and a picture of the actual flight array. The long dimension of 5000 elements was chosen to obtain a 1 km/line pair image across Pluto's 2300 km diameter in a single scan which requires at least 4600 pixels. The remaining 400 pixels allow for pointing inaccuracies and drift during the scan. All MVIC pixels are $13 \times 13\ \mu m^2$.

The frame transfer array consists of two regions; the 5024 × 128 pixel image gathering area, and a 5024 × 136 pixel image storage area. The extra eight rows in the image storage area contain injection charge that reduces charge traps. For both the TDI and framing arrays, the extra 24 dark pixels (12 on each side of the 5000 pixel active region) are used as reference pixels and for injected charge. The filter, which was provided by Barr Associates of Chelmsford, MA, is mounted about 700 microns above the surface of the array. It has five segments, four with the passbands described in Table 2 placed directly over the four CCDs forming the color segment of MVIC. The remaining two TDI CCDs and the frame transfer array are overlain by a clear filter so that the focus position is the same for all seven arrays. In TDI mode, the spacecraft is rotated to scan the image of each segment of the sur-

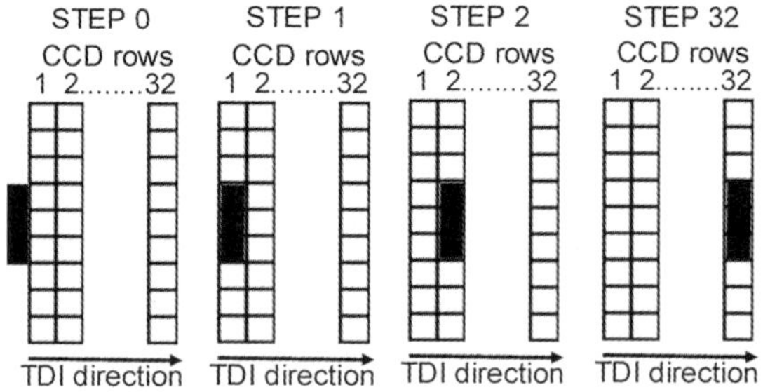

Fig. 5 Illustration of TDI in operation. In step 0, a target is about to enter the field of view of a TDI array. In step 1, the image of the target has moved a single pixel width into the first row of the array. In step 2, the charge is transferred to the second row of the array and integration continues. A new target has moved into row 1. The next charge transfer occurs when the image of the target has moved another row. In step 32, the target has moved 32 rows, with the charge being transferred each time the image moves another row. At the next charge transfer, the charge in row 32 is transferred to a serial column for pixel readout

face across the focal plane in a pushbroom fashion. The entire surface may be imaged in this way. As Fig. 5 illustrates, the clocking of each 5024 element pixel row is synchronized with the spacecraft motion using attitude control knowledge obtained from the spacecraft so that a row charge transfer occurs each time the image has moved a pixel width. The effective integration time is thus a factor of 32 greater than the row transfer period. In this way, the signal-to-noise ratio of the observations is increased while the time required to obtain a full image is reduced. TDI mode takes advantage of the ability of the spacecraft to scan smoothly in attitude and does not need the multiple pointing operations that a mosaic of framing images would require. The nominal rotation rates are about 1600 μrad/s for the pan band and about 1000 μrad/s for the color bands. These correspond to integration times of about 0.4 and 0.6 seconds respectively. In flight, the CCD clocking rate errors in the TDI (along scan) direction have been shown to cause less than 1/4 of a pixel of excess image "smear" for integration times of 0.7 seconds or less.

3.2 The LEISA Focal Plane

The LEISA SWIR spectral imager also works in pushbroom mode, except in this case, the image motion is used to scan a surface element over all spectral channels. The wedged filter effectively makes each column of the array responsive to only a narrow band of wavelengths, so that, conceptually, the filter may be considered as consisting of 256 adjacent narrow band filters. As with MVIC, the image is scanned over the LEISA focal plane by rotating the spacecraft. The nominal rotation rate is about 120 μrad/s for a frame rate of 2 Hz. Here again, the frame rate is synchronized to the spacecraft-measured rotation rate, so that, in principle, the image moves one column per frame. The synchronization accuracy is limited by the error in the measured rotation rate, which is 2.5 μrad/s (1σ). For the nominal rotation rate, this corresponds to an error of 2% of a pixel per frame. Post acquisition processing allows the actual scan trajectory to be determined very accurately, and the maximum misregistration of the spectrum for any scan rate is limited to 50% of a pixel over the entire spectrum.

The LEISA detector is a 1.25- to 2.5-μm HgCdTe PICNIC array, supplied by Rockwell Scientific Corporation (now Teledyne) of Camarillo CA. The architecture of the PICNIC array is based on the NICMOS array developed for the Near Infrared Camera for Multi-Object Spectrometry on HST. The array is a 256×256 pixel array and each pixel is $40 \times$

Fig. 6 (*Left*) Picture of the LEISA focal plane showing the fanout board/cable assembly mounted to the same molybdenum plate as the PICNIC array. (*Right*) The full focal plane assembly including the wedged filter. The two segments of the filter appear as *different colors*

40 μm^2 in area. However, several modifications were made to the standard PICNIC array. The HgCdTe was grown on a CdTe substrate using Molecular Beam Epitaxy (MBE) to provide good lattice matching and low dark currents. The detector was bump bonded to a standard PICNIC multiplexer and the resulting hybrid was mounted to molybdenum pad. This process reduces mechanical stress induced during cooling to operational temperature. It is estimated that the assembly can safely undergo at least 1000 thermal cycles. The electrical interface to the array is provided by two ribbon cables and a multilayer fan-out board that were fabricated into a single element. The LEISA array is back illuminated, but the substrate has been thinned from the standard 800 μm to 200 μm so that the active area of the array is significantly closer to the surface than is usual. This puts both the filter and the array within the depth of focus. The filter, supplied by JDSU Uniphase/Optical Coating Laboratories Inc. of Santa Rosa CA, was made in two segments. The first, covering from 1.25 to 2.5 μm at a constant resolving power (constant $\lambda/\Delta\lambda$) of about 240, provides information primarily for surface composition mapping. The second, covering from 2.1 to 2.25 μm at a constant resolving power of about 560, uses temperature dependent changes in the spectral structure of solid N_2 near the α to β phase transition at 35 K to provide surface temperature maps. In both segments, a constant resolving power is achieved by making the transmitted wavelength depend logarithmically on position. The two segments were bonded together to form a single filter element. This filter was, in turn, bonded into a holder and mounted such that the filter surface is about 100 μm above the surface of the array. The refractive index of the array is approximately 2.7 so that the total optical path between the filter and photo-active area of the array is less than 200 μm. In this distance, the $f/8.7$ beam spreads about 0.5 pixel. When the focus position is optimized between the array and filter surface, the convolved image smear is about 0.04 pixel. A picture of the array and the complete focal plane assembly is shown in Fig. 6.

3.3 The Solar Illumination Assembly (SIA)

The SIA is a second input port whose FOV is along the spacecraft antenna pointing direction at an angle of 90 degrees with respect to the main aperture. It is designed to provide diffuse solar illumination to both the MVIC and LEISA focal planes. In practice, it illuminates all of the LEISA array and about 3000 pixels of each MVIC array with a reproducible pattern that can be used for determining the stability of the pixel-to-pixel response (flat-fielding) during the mission. The SIA consists of a small fused silica lens (4 mm aperture, 10 mm focal length) that images the sun onto the input end of a 125 μm core fiber. The output end of the fiber illuminates a pair of lenses, which are directly under the Lyot stop (the exit pupil) and about 10 cm from both focal planes. To obtain an SIA measurement, the spacecraft is oriented so that the SIA lens images the sun onto the fiber. At Pluto, the diameter of the image of the Sun on the fiber would be about 50 μm. This is significantly larger than the diffraction limited image size because of chromatic aberration in the single element lens. Nevertheless, the solar image underfills the fiber, so the intensity level is relatively insensitive to pointing errors. The fiber is about 10 cm long and is contained in a stainless steel tube. It is more than 50% transmitting over the full spectral band from 0.4 to 2.5 μm. There is a second fiber in the SIA with an attenuation factor of about 40 that can be used for flat fielding nearer the Sun (e.g. at Jupiter).

A second possible use of the SIA is as a solar limb viewing port. In this mode, an atmospheric spectrum can be measured as a function of tangent height as a planet's atmosphere occludes the Sun. Vertical spectral profiles would be obtained using this capability. To increase sensitivity in sparse atmospheres, such as Pluto's, the spectra from all rows may be summed into a single spectrum. The SIA is co-aligned with the Alice solar occultation (SOC) port (Stern et al. 2008, this issue).

4 Electronics

The Ralph control electronics consist of three boards; detector electronics (DE), command and data handling (C&DH) and a low voltage power supply (LVPS). These are contained within an electronics box (EB) mounted directly to the spacecraft, below the TDA (see Fig. 2), and operate essentially at the spacecraft surface temperature, which is near room temperature. The DE board provides biases and clocks to both focal planes, amplifies the signals from the arrays and performs the A/D conversion of the imaging data. The science data are converted using 12 bits per pixel. The C&DH board interprets the commands, does the A/D conversion of the low speed engineering data and provides both the high-speed imaging data interface and the low-speed housekeeping data interface. The LVPS converts the 30 V spacecraft power to the various voltages required by Ralph.

In a long duration mission such as New Horizons, reliability of the electronics is of paramount importance, particularly for a core instrument that addresses all three Group 1 objectives. To ensure that Ralph is robust, almost all of the electronics are redundant. As illustrated in Fig. 7, Ralph can operate on two separate sides (side A or B) which have very few components in common. The only common elements are: (1) the relays that choose whether side A or side B is to be powered, (2) The arrays themselves and (3) the interface to the spacecraft. However, the spacecraft interface has two identical circuits and is inherently redundant. For MVIC, the potential single point failure mode of the array is mitigated by dividing the six TDI arrays into two groupings, each containing two color CCDs and one panchromatic CCD. The first grouping is composed of a pan band and the red and CH_4 channels. The second grouping is composed of the other pan band and the blue and NIR channels. If either group should fail, the other would still be able to meet the science requirement of observations in 2 color bands and 1 panchromatic band. LEISA has 4 outputs corresponding to the four 128×128 quadrants of the array. If any one quadrant should fail, all science can still be completed. The same is true for the four out of six possible two-quadrant failures that still have active pixels at all wavelengths.

MVIC always produces image data in correlated double sampling (CDS) mode in which the reset level is subtracted from the integrated level. LEISA can send either CDS data, which is its standard operating mode, or "raw" data that contains both the reset levels and the integrated levels. The "raw" mode produces twice the data volume of the CDS mode and is used to set the LEISA A/D converter offset level for each quadrant of the array. The same offset is used in both CDS and "raw" modes. The selectable offset compensates for

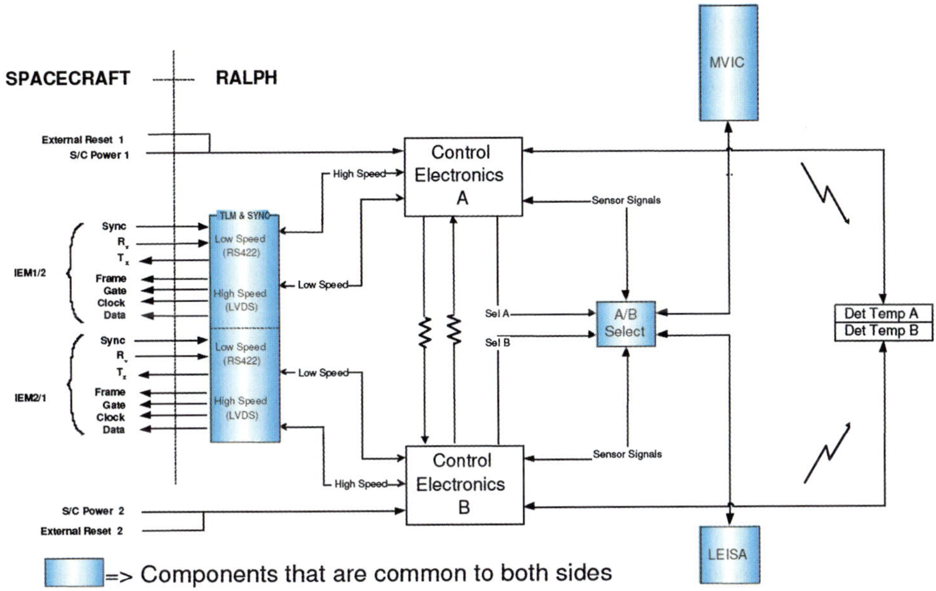

Fig. 7 Schematic diagram of Ralph electronics showing the high degree of redundancy in the system

voltage drifts in the analog signal train over the life of the mission, maintaining dynamic range without sacrificing signal resolution (increasing quantization noise). For both focal planes the measured spacecraft rotation rate is fed back to the instrument to optimize the TDI or frame rate. That is, after a scan has been initiated the spacecraft determines the actual rotation rate and sends that information to Ralph, which uses it to calculate a TDI or frame rate such that the image moves a single row during a clock period. This reduces smear in the along-track direction.

For both MVIC and LEISA, the dominant noise source at low light levels is the system electronics noise, including array read noise. For MVIC this is about 30 e- ($\sim$ 200 μV) and for LEISA it is about 50 e- ($\sim$ 100 μV). The overall average gain for MVIC is about 58.6 e-/DN (Digital Number, or least significant bit), while for LEISA it is approximately 11 e-/DN.

5 Pre-Launch Instrument Characterization

An extensive pre-launch program of performance verification measurements was carried out for Ralph at both the component level and the full instrument level. The component level characterization included measurements of the wavelength dependent quantum efficiency for the MVIC and LEISA array/filter assemblies and measurements of the wavelength dependence of the other optical elements (i.e. reflectance of the mirrors, transmission of the filters and throughput of the beamsplitter). Full instrument level testing was carried out under spaceflight-like conditions in a thermal vacuum chamber at BATC. The primary performance characteristics verified in these tests were relative system throughput (relative radiometric sensitivity) and image quality. The directional characteristics of the SIA were also measured.

5.1 Component Level Measurements

LEISA Spectral Lineshape The instrument line shape was determined for the LEISA filter/array focal plane assembly over the entire 1.25 to 2.5 μm band by using a combination of multi-order grating and narrow band filter measurements. In this way a pixel-by-pixel table of the central wavelength, resolving power and out-of-band transmission was generated. Figure 8a shows an example of the readout along a single LEISA row when the focal plane assembly was illuminated using the output of a grating monochromator. The first five peaks correspond to orders 6 through 10 of the grating. The intensity decrease is primarily caused by the spectral shape of the source and by the decreasing efficiency of the grating at higher order. The line at 2.2 μm occurs in both segments of the filter. Figure 8b shows the intensity measured at a single pixel as the grating is scanned in small wavelength increments (0.33 nm). This instrument lineshape is representative of all pixels and is approximately Gaussian. At this wavelength (2165 nm), the full width at half maximum (FWHM) is 8.3 nm, giving a resolving power of 260. Figure 9 shows the average resolving power for both filter segments of LEISA generated using the measurement technique described above. As may be seen from this figure, the average resolving power for the lower resolution segment is 240, with variations of 10 to 15 percent lower and higher. This is slightly below the requirement of 250. The resolution determines the reliability with which pure materials may be

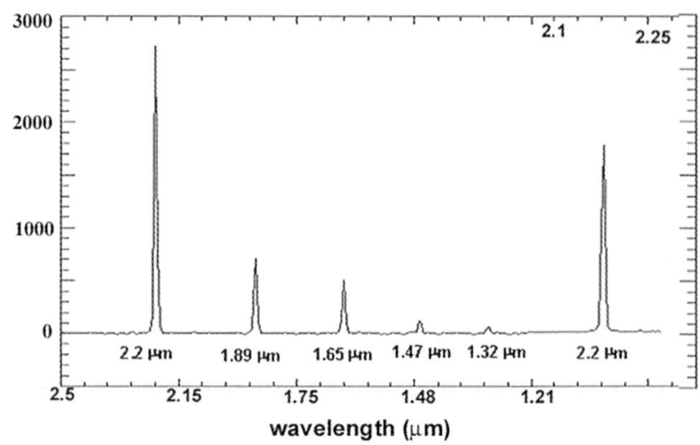

(a)

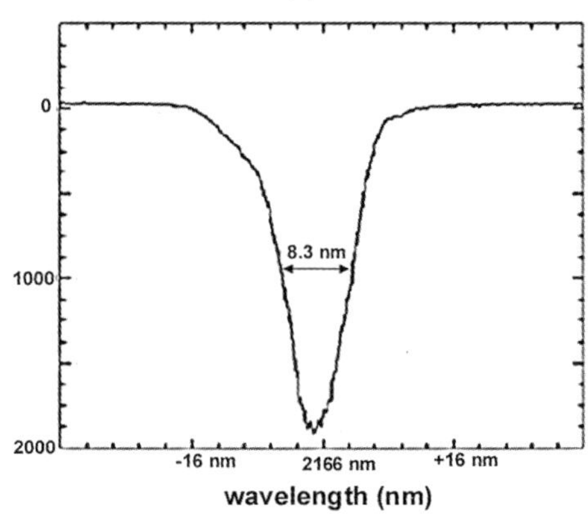

(b)

Fig. 8 (**a**) A single row of LEISA showing multiple orders of a grating. Note the non-linear (logarithmic) wavelength scale and the presence of the 2.2 μm order in both the low resolution (1.25 to 2.5 μm, $\lambda/\Delta\lambda \sim 240$) and the high resolution (2.1 to 2.25 μm, $\lambda/\Delta\lambda \sim 560$) filter segments. (**b**) Wavelength dependence of transmitted intensity (instrument lineshape function) at a single pixel as the wavelength is varied from 2133 to 2200 nm in 0.3 nm steps

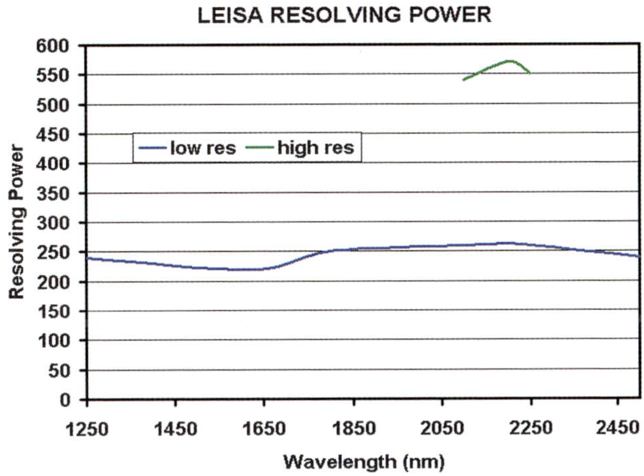

Fig. 9 Measured resolving power ($\lambda/\Delta\lambda$) of the LEISA array/filter assembly for the lower-resolution (1.25–2.5 µm) and higher-resolution (2.1–2.25 µm) segments

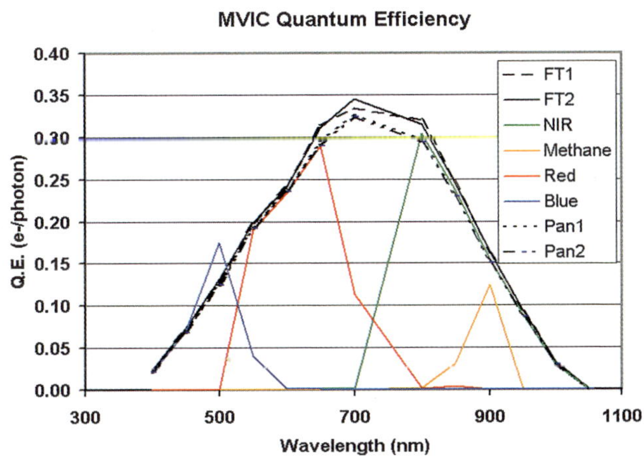

Fig. 10 Spectral dependence of the sensitivity of the MVIC filter/CCD assembly in each channel in terms of quantum efficiency (e-/photon). Measured in 50 nm steps

differentiated from mixtures and the accuracy of temperatures determined from line shapes. The slight decrease in accuracy in the lower resolution regions will not have significant scientific impact. Temperatures will be determined most accurately using the high-resolution segment. As required, the resolving power of the higher-resolution segment is greater than 550 in the region of the 2.15-µm N_2 band.

MVIC Spectral Response The spectral resolution of the MVIC channels is much lower than for LEISA and the CCD response varies relatively slowly, so a coarser measurement of the wavelength dependence of the filter/CCD assembly is acceptable. Figure 10 shows the measured sensitivity of the combined array/filter focal plane assembly. These measurements were obtained at 50 nm intervals using a broad spectral source filtered with 50 nm wide spectral filters as the input. In order to account for observational sources with greater spectral variation, the spectral responses of the filters were determined at higher spectral resolution and finer point spacing. Figure 11 shows the spectral response curves for the MVIC color filters. In addition to allowing for the retrieval of more accurate spectral models, these data will be used to calculate the sensitivity of MVIC to the source spectral distribution (e.g. a solar reflectance spectrum vs. the blackbody spectrum of a volcano on Io). This is necessary

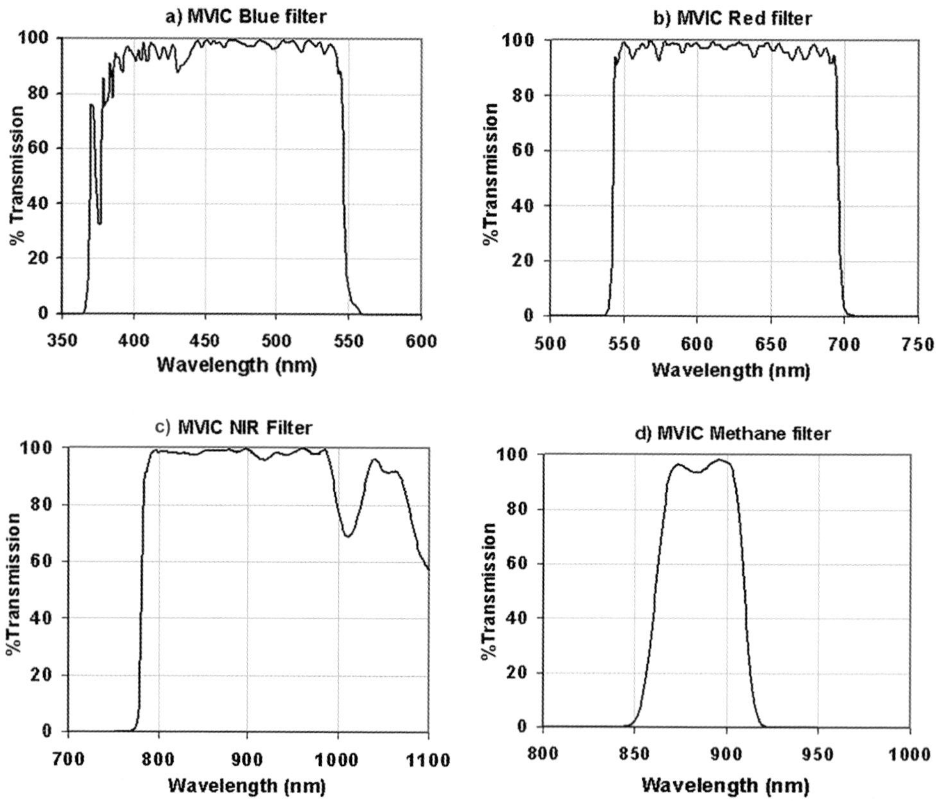

Fig. 11 Wavelength dependent transmissions for the four MVIC color filters. At wavelengths above 950 nm, the responsivity of the CCD falls rapidly, limiting the long wavelength NIR system response cutoff to 975 nm. Similarly, the short wavelength system response cutoff for the blue filter is 400 nm because of the decreased responsivity at shorter λ

because the number of counts generated in a channel for a given photon flux (photon/cm^2/s) depends upon the details of the spectral distribution.

5.2 Full Instrument Level Measurements

Absolute Radiometry A spectrally calibrated integrating sphere filling about a 1° field of view was used as the radiometric source. Radiometric response over the full MVIC image plane was measured by rotating the sphere illumination using a cryomirror assembly in the chamber. This technique was used to determine accurate relative radiometry (flatfields). The MVIC absolute radiometry was determined with an accuracy of approximately ±30%. The accuracy of the LEISA absolute radiometry was lower because water contamination in the integrating sphere caused significant absorption at some wavelengths. Radiometric calibration is being performed in-flight and the combined results of the pre-flight and in-flight calibration are shown in the next section. It is expected that the MVIC absolute radiometric response will be determined to an accuracy of 1% by in-flight calibration, while LEISA will be calibrated to the 3 % to 5% level.

Image Quality The image quality, defined in terms of system MTF (Modulation Tranfer Function), was determined using a collimated point source to simulate a distant object. Collimation was verified interferometrically, as were corrections for pressure and thermally induced optical power in the chamber window. By defocussing the point source in a controlled fashion, this system could also be used to determine the best focus position for the focal planes. At the best focus position, a point source produced a spot in the focal plane whose FWHM was 1.2 ± 0.1 pixel. After the tests were completed, it was found that there was some optical power in the cryomirror that was not accounted for in determining the focus. However it was determined that for the Ralph system with its 650-mm focal length, the focus error caused by the cryomirror curvature was negligible. The pre-flight image quality measurements were used to determine an expected budget for the full system MTF that is shown in the next section.

SIA Pointing The pointing direction and spatial distribution of focal plane illumination for the SIA were also determined in the instrument level tests. When illuminated by a source simulating the angular size of the sun at Pluto, the SIA produced a stable pattern in both the MVIC and LEISA focal planes that was insensitive to the exact source position over a range of 0.5° in each dimension.

6 Combined Pre-Launch and In-Flight Instrument Calibration Results

Calibration of Ralph is being carried out in flight. To date, standard stars and Jupiter observations are being used to determine image quality, to measure over-all system radiometric sensitivity and, for LEISA, to verify the spectral calibration. Additionally, dense star-fields are being used to determine overall optical distortion. A Jupiter gravity assist occurred in 2007 with closest approach on February 28. It provided additional opportunities for measuring the flat field response, and will permit radiometric and spectral calibration. Because of motional broadening, MVIC and LEISA response is expected to be the same for point- and extended sources. Afterwards, every year during the flight, there will be a 50-day checkout period during which calibration will be checked. Extensive calibration will also be carried out during the 6-month period prior to and following the planned Pluto encounter on 14 July, 2015. Because the initial in-flight calibration analyses are not complete, the following results should be considered to be preliminary estimates.

So far in flight, the MVIC and LEISA system random noises have remained at their pre-launch values: ≤ 1 count (≤ 60 e-) for all the MVIC CCDs and 4.5 counts (50 e-) for LEISA. In addition to the random noise, the MVIC CCDs have some periodic noise (as much as 2 counts) that may be removed by post-processing. This noise was present before launch. The MVIC dark current is completely negligible; it is not measurable for the integration times allowed by the Ralph electronics. The LEISA dark current is ~ 40 counts/second (~ 440 e-/s), and does not contribute significantly to the noise even for the longest allowed integration time (4 s). The Ralph decontamination heaters have been left on except for 24-hour periods around data collection events in order to minimize the condensation of contaminants produced by spacecraft outgassing and thruster operation. This means that, as of this writing, the LEISA focal plane has not reached its quiescent operating temperature when acquiring data. A cool-down period of greater than 24 hours is common in low temperature, passively cooled systems. When the heaters are left off for longer periods, it is expected that the focal plane temperature will drop a few more degrees. The dark current will decrease by about a factor of two for every five degrees drop in the focal plane temperature.

6.1 Image Quality and MTF

The point spread function of MVIC's panchromatic TDI channels determined in-flight is well represented by a 2-D Gaussian function. Fittings of the point source intensity distribution to this PSF for about 30 relatively bright stars has yielded FWHMs of 1.48 ± 0.13 pixels in the in-track direction and 1.40 ± 0.1 pixels in the cross-track direction. These observations are all for integration times of 0.5 seconds or less, for which the contribution of the uncorrected spacecraft motion is less than half a pixel. Taking the Fourier transform of the fitted PSF yields the expression:

$$\mathrm{MTF}(d) = \mathrm{e}^{-(d\pi 0.01977\mathrm{FWHM}/(2(\ln 2)^{1/2}))^2} \tag{1}$$

for the value of the MTF at spatial frequency, d (cycles/milliradian). Using this expression, $\mathrm{MTF}(20) = 0.30 \pm 0.07$ (in-track) and $\mathrm{MTF}(20) = 0.34 \pm 0.06$ (cross-track). Both the in-track and cross-track MTF(20) values surpass the requirement that they must be greater than 0.15 (see Table 1). These results are summarized in Fig. 12, which shows the pre-launch MTF curves, determined solely by the optical characteristics, and the pre-flight model of the full system. In this figure the top area in each panel ("measured spots") represents the measured contribution of the instrumental characteristics to the MTF. Off array integration is a term that accounts for the effect of insufficient masking of the bulk silicon from the light. The other factors are based on predicted spacecraft behavior at the 3σ level. The temporal aperture and the TDI error have to do with uncompensated spacecraft motion. The point labeled "requirement" is the 15% minimum overall MTF at 20 cycles/milliradian (0.4 cycles/pixel) that satisfies the science requirements. The measured in-flight image characteristics clearly exceed requirements and agree remarkably well with the predicted behavior based on the pre-flight instrument level measurements.

The MVIC color channels and the LEISA spectral channels have no additional requirements on image quality beyond the requirement that they be in focus to within 0.5 pixel. This is because the required spatial resolutions for the observations using the MVIC color and LEISA channels are significantly lower than for the MVIC panchromatic channel. The MVIC color channels are at the same focal distance as the panchromatic channel by virtue of being on the same substrate. The LEISA focal plane was focused separately. Table 3 summarizes the preliminary results of the analysis of the PSF for all the Ralph channels. The effect of diffraction is apparent in the FWHM measurements for the MVIC color channels.

Table 3 Summary of full width at half maximum for Ralph channels[1]

Channel	In-track FWHM	Cross-track FWHM
MVIC pan	1.48 ± 0.13	1.40 ± 0.10
MVIC blue	1.48 ± 0.10	1.29 ± 0.07
MVIC red	1.55 ± 0.12	1.38 ± 0.08
MVIC NIR	1.95 ± 0.15	1.97 ± 0.15
MVIC CH_4	1.78 ± 0.13	1.81 ± 0.20
[2]LEISA	1.40 ± 0.13	1.40 ± 0.10

[1]Initial results for FWHM in units of pixels

[2]This is the result of a single stellar observation and is the average FWHM for wavelengths from 1.25 to 1.8 μm. There is no apparent wavelength dependence over this range

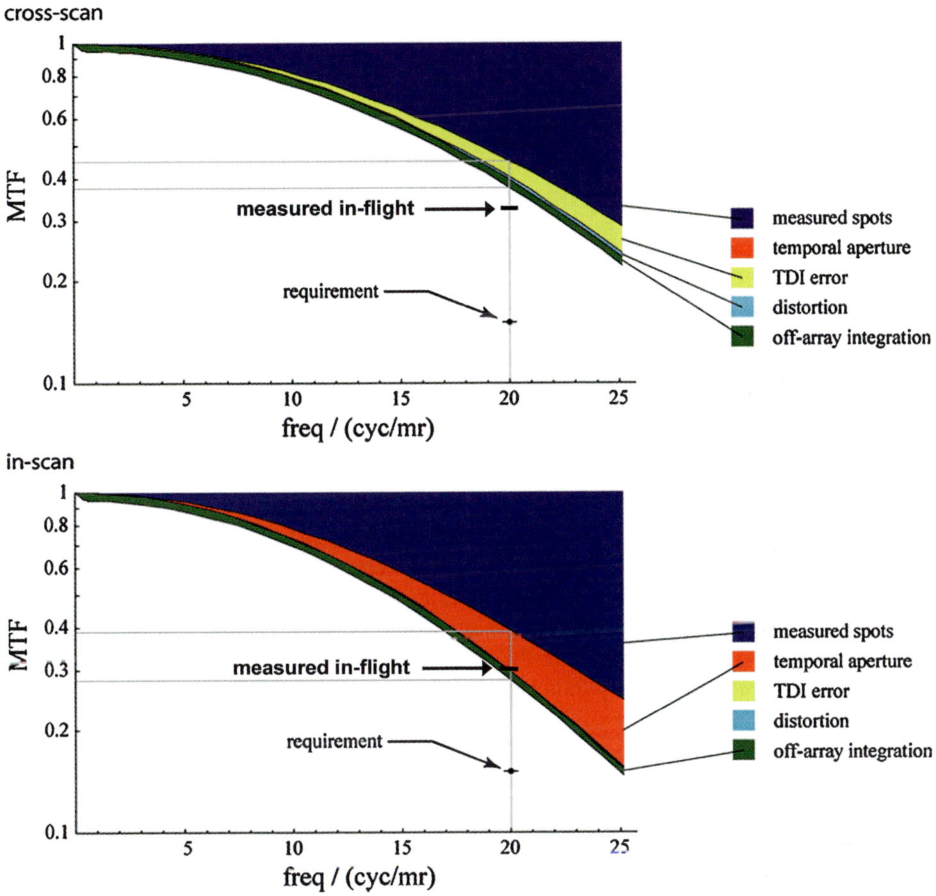

Fig. 12 Modeled and measured MTF for the MVIC panchromatic TDI channel. "Measured spots" represents the measured contribution of the optical characteristics to the MTF, while the other factors are based on predicted behavior at the 3σ level. The temporal aperture and the TDI error have to do with uncompensated spacecraft motion. Off array integration is a term that comes about because of insufficient masking of the bulk silicon from the light. The points labeled "measured in-flight" are derived from stellar observations and represent the full system MTF. These clearly exceed the requirement

The LEISA results are based on a single stellar observation that only covered from 1.25 to 1.8 μm. Diffraction effects are not apparent in these LEISA measurements because the LEISA pixels are three times larger than the MVIC pixels.

6.2 Optical Distortion

The MVIC focal plane covers 5.7° in the cross track direction, which means that it is expected that there will be optical distortion on the order of a few pixels at the ends of the FOV. In addition, the three-mirror anastigmat design is known to be anamorphic. That is, the spatial scale in the cross track direction is slightly different from that in the along track direction. These effects both make the apparent position of objects different from their true position. In order to account for this effect, both to maintain spatial fidelity in the science data and to allow accurate observations for optical navigation, a distortion correction must be found that

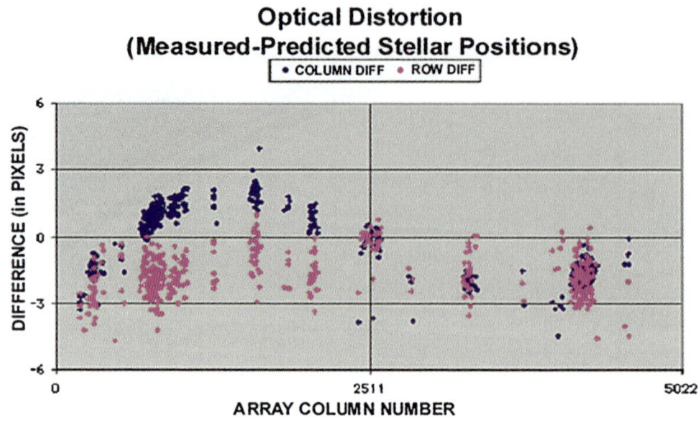

Fig. 13 Effect of optical distortion for MVIC framing camera. The *plotted points* are the differences between predicted row and column stellar positions and true row and column stellar positions as determined using dense star fields. The differences are plotted as a function of array column number

maps the apparent position to the true position. The distortion effect is illustrated in Fig. 13 for a set of observations of a dense star field using MVIC's panchromatic framing camera. This plot shows the difference between the true position of a star and the position obtained assuming that each pixel is 19.77 μm square. The stellar positions are typically known to better than 5 μrad. As may be seen from this figure, there is about a three-pixel distortion at the ends of the focal plane. Note that the column distortion is sinusoidal about the center of the FOV, while the row distortion is the same sign on opposite sides of the array. The anamorphism is the principle driver for the row distortion. These results are preliminary and are being refined further, but they indicate the effects are easily modeled.

6.3 Radiometric Calibration

The absolute radiometric calibration of the MVIC and LEISA components of Ralph is primarily being carried out in-flight, using point sources (standard stars) and Jupiter. The flat-fields determined during the pre-launch testing are used to interpolate the stellar and Jupiter results over the entire focal planes. This work is still progressing. The results below are preliminary, and are expected to be accurate to about 10%. Figure 14 shows an example of the stellar source calibration measurements for the MVIC framing array. This figure shows a plot of the integrated signal at the focal plane *vs.* the visual magnitude of a star, for a number of A-, B- and K-type stars. A fit was done using data for B9-type stars that is linear to a very high degree of accuracy (it accounts for 99.5% of the standard deviation of the data). The other A- and B- type stars are well described by this fit as well. The colder, K-type star is a bit of an outlier, but no correction has been made for color temperatures of the stars. These data indicate that, for the framing array, a 14th magnitude B9 star will produce an integrated signal of one count per second. Given the MVIC frame noise characteristics (~ 0.6 counts of noise at low illumination levels), a tenth magnitude star will produce a measurement with a signal-to-noise ratio of 14 in 0.25 seconds if it is focused onto a single pixel. If the same signal is spread out over 4 adjacent pixels it produces an average SNR of 8, which still meets the OpNav requirement.

Table 4 shows the sensitivities and the predicted signal-to-noise ratios for the MVIC channels and for the LEISA wavelengths at which the LEISA SNR performance requirements were defined. The predictions are made for the conditions specified in Table 1 which are representative of the flux levels expected at Pluto. The table also lists the noise performance requirements set forth in the Announcement of Opportunity (AO) for the mission. As can be seen from this table, all MVIC channels easily meet their sensitivity requirements.

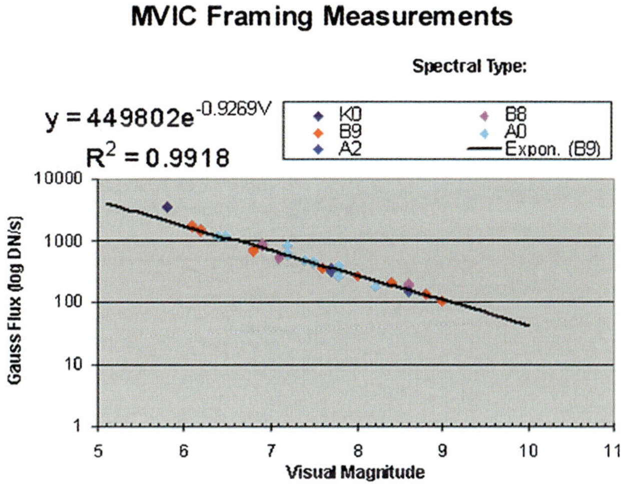

Fig. 14 MVIC frame camera. Log of integrated signal *vs.* visual magnitude for a number of stars. The B9-type stars are fitted to a line that explains 99.5% of the standard deviation. The A- and B-type stars not included in the fit align very well with the fitted line. The fit indicates a 14th magnitude star will produce one count per second above the background

Table 4 Ralph radiometric sensitivity, predicted signal-to-noise ratios and mission requirements

Channel	Sensitivity (DN/photon)[1]	Predicted SNR[2]	Required SNR
MVIC pan	1.89×10^{-3}	150	50
MVIC blue	1.38×10^{-3}	68	50
MVIC red	2.25×10^{-3}	122	50
MVIC NIR	1.59×10^{-3}	106	50
MVIC CH_4	2.00×10^{-3}	48	15[3]
[4]LEISA 1.25 μm	2.45×10^{-3}	32	31
[4]LEISA 2.00 μm	5.82×10^{-3}	35	27
[4]LEISA 2.15 μm	8.32×10^{-3}	24	18

[1]For MVIC, there are on the average, 58.6 e- per DN, where DN (digital number) is the least significant bit of the A/D. For LEISA there are about 11 e- per DN

[2]Predicted SNR at Pluto. MVIC: Assumes 35% albedo; LEISA: Assumes Pluto model albedo in Fig. 1. MVIC Pan is for 0.4 s integration, MVIC color is for 0.6 s integration, LEISA is for 0.5 s integration

[3]The methane specification is an internal goal, not a requirement

[4]The LEISA SNRs are for the average of two planned scans

Similarly, LEISA meets the AO radiometric performance specifications, but with significantly smaller margin, particularly at shorter wavelengths. The decreased performance in this spectral region is caused by a known drop in quantum efficiency of the array at wavelengths shorter than 1.6 μm, and by lower transmittance of the filter at shorter wavelengths. However, solar flux increases at the shorter wavelengths so the decreased efficiency is compensated for by the larger flux. Figure 15 shows the LEISA radiometric sensitivity as a function of wavelength. Figure 16 shows the predicted signal-to-noise ratio of LEISA observations (average of two scans, 0.5 s/pixel integration time) of Pluto and Charon for the nominal albedos plotted in Fig. 1. At the Pluto-system flux levels the noise is dominated by the system noise (read noise), so that the signal-to-noise ratio grows nearly linearly with increased integration time.

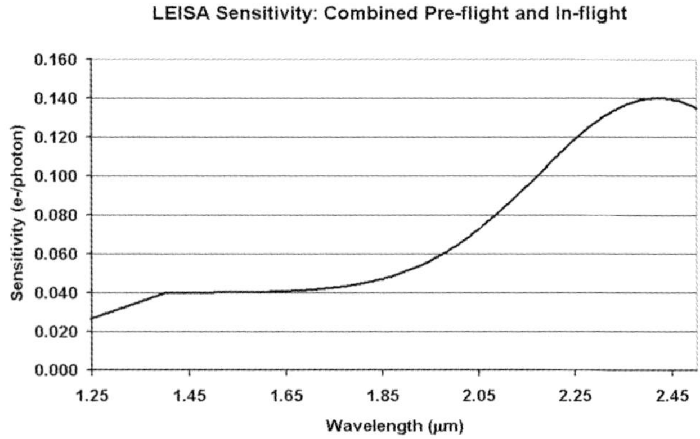

Fig. 15 LEISA wavelength dependent sensitivity in terms of e-/photon. This is for the flux contained in a LEISA spectral resolution element and within the LEISA single pixel AΩ of 1.66×10^{-7} cm^2 sr

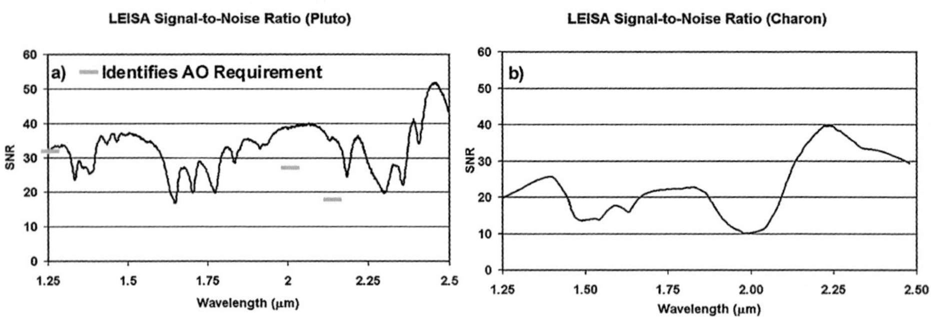

Fig. 16 The predicted SNRs for the LEISA observations of Pluto (**a**) and Charon (**b**), assuming the albedo spectrum shown in Fig. 1. The specified requirements for the Pluto scan are shown in *green*. The predicted SNRs are for the sum of two 0.5 second integration spectral maps with 10 km spatial resolution

6.4 Anomalous Solar Light Leak

In-flight testing of Ralph has shown the presence of an anomalous background signal in the LEISA imager that appears to be caused by the transmission of a very small fraction of the ambient solar flux into the area behind the focal plane. These photons pass between the filter and the array and give rise to a "solar light leak" signal that, in the worst case, is less than 1 part in 10^7 of the ambient solar flux. The background may be eliminated by using structures on the spacecraft to shield Ralph from the sun. This behavior, and the R^2_{sun} dependence of the magnitude of the background signal are both evidence of the sun as the source of this anomaly. Ralph can not always be shielded. When Ralph is not shielded, the magnitude of the effect is a slowly varying function of the position of the sun relative to Ralph and it may be removed to a high degree of accuracy. This means that the primary result of the light leak is to increase the system noise because of photon counting statistics. At Pluto's heliocentric distance, and for the integration times that are possible with LEISA, the excess background introduced by the effect is at least a factor of two less than the read-noise equivalent input flux. Therefore, after the background is removed, there will be little or no measurable effect of this excess flux on the LEISA observational results. As of yet, the root manufacturing cause of the light leak has not been determined. It is possible that light is more or less uniformly penetrating the multi-layer insulation (MLI) thermal shield that encloses the instrument, and propagating along LEISA's interface cables to the focal plane.

However, definitive answers, if they can be determined, await further laboratory and in-flight measurements. Nevertheless, for practical purposes, the phenomenology is well understood. MVIC shows no evidence of this effect.

7 In-Flight Instrument Operation

Ralph data collection operations fall into one of four categories: (1) panchromatic MVIC TDI, (2) color MVIC TDI, (3) LEISA and (4) panchromatic MVIC framing. Ralph may operate in only one of these categories at any time meaning, for example, that LEISA data may not be taken simultaneously with MVIC data. The panchromatic MVIC TDI category covers operation in either of the panchromatic TDI CCD arrays, but only one pan array can be operated at any time. All four color TDI CCDs operate simultaneously. As a result, a color MVIC scan is slightly longer than the equivalent Pan TDI scan, because the target must be scanned over all 4 color CCDs. The category is chosen by command and implemented by a set of relays.

For the MVIC TDI and LEISA data categories operation may be either in calculated rate mode or forced rate mode. In calculated rate mode, the array read-out rate is set using the measured rotation rate about the scan axis (the Z-axis) as determined by the spacecraft's star trackers and gyroscopes. This scan rate is provided to Ralph at the beginning of a scan by the spacecraft's guidance and control (G&C) system. The array readout rate is set such that the scan moves a spot on the image a single row between reads. In calculated rate mode, the MVIC TDI arrays may scan only in one direction, but LEISA may scan in either direction. In forced rate mode, the array readout rate is set by a command and is not coupled to the G&C scan rate. For all categories, data collection is initiated with a start command and continues until a stop command is received. MVIC framing data is always obtained in forced rate mode.

There are two types of scans. The normal scan type is used when the target is sufficiently distant that the effective scan rate of the boresight caused by the relative motion of the target and the spacecraft remains constant during the scan. For a normal scan, once the rotation rate has stabilized, the along track thrusters are disabled and the spacecraft is allowed to rotate at a constant rate. In calculated rate mode the rotation rate, measured to a 3σ accuracy of ± 7 μrad/s is passed to Ralph and the calculated frame rate is matched to the boresight motion. A normal type scan using calculated rate mode is the most common form of either MVIC TDI or LEISA operation. Typically, in a normal LEISA scan, the cross track thrusters are enabled so that the target does not drift off the focal plane in the direction perpendicular to the scan during the data collection event. For a given target, the MVIC scans are usually almost an order of magnitude faster than the LEISA scans, so the cross track thrusters are typically disabled in MVIC TDI scans. Correct operation of the normal scan type has been verified in flight for both MVIC and LEISA using stellar sources.

When the target is sufficiently close that the effective rotation rate induced by the relative motion of the target and the spacecraft changes during the scan, the scan becomes more complex. In this case, the rotation rate of the spacecraft is changed during the scan by thruster firings. This type of scan, called a pseudo CB3 scan, is implemented by making the boresight track an artificial object whose ephemeris is defined in such a fashion to keep the combined boresight rotation rate constant. The frame rate is set by the commanded scan rate, and not by the changing spacecraft rotation rate as measured by the G&C system. The scan is controlled to within ± 34 μrad/s. Pseudo CB3 scans are only used when the target is close to the spacecraft, such as the LEISA and MVIC Pluto scans near closest approach. For this type of operation, both the in-track and cross-track thrusters are enabled during LEISA scans, while only the in-track thrusters are enabled during MVIC scans. Correct operation of the pseudo CB3 scan type has been verified in flight for both MVIC and LEISA using an asteroid near encounter (Olkin et al. 2008).

8 Conclusion

This paper describes the design, operation and performance of Ralph, a highly capable remote sensing imager/SWIR spectral imager flying on the New Horizons mission to the Pluto/Charon system and the Kuiper Belt beyond. Ralph consists of a telescope feeding two focal planes, the visible/NIR MVIC imager and the LEISA SWIR spectral imager. MVIC will provide very sensitive, high fidelity, full hemispheric panchromatic maps of Pluto and Charon at a spatial resolution of 1 km/linepair and VIS/NIR color maps at a spatial resolution of better than 4 km/linepair. LEISA will provide full hemispheric, SWIR, spectral maps with spatial resolutions of 10 km/pixel or better. These will be used to accurately characterize the surface composition. At closest approach, MVIC will obtain images with spatial resolutions on the order of a few hundred meters in selected areas, while LEISA will measure spectra at the 1–2 km spatial scale. Ralph has been extensively tested at the component and full instrument level and its in-flight operation has verified that it meets all its performance requirements with margin. Ralph will provide a wealth of information on the geology, composition, morphology and thermal characteristics of the Pluto/Charon system. The data it produces during its flyby will revolutionize our understanding of Pluto and its neighbors and will shed new light on the evolution of our solar system and the nature of the objects in the Pluto system.

Eight years prior to arriving at Pluto, Ralph observed Jupiter for a period of about one-month during its close approach for a gravity assist. The closest approach of $\sim 33R_J$ occurred on 28 February, 2007. During this period, numerous Ralph observations of Jupiter and its moons were made to provide encounter practice and to obtain calibration measurements on the last fully resolved object prior to the Pluto encounter 8 years hence. The LEISA observations obtained at this time included some of the highest spectral/spatial data ever obtained of Jupiter in the SWIR spectral range. Thus New Horizons has provided exciting new science, even before its rendezvous with Pluto.

9 Glossary

A/D	Analog to Digital
AO	Announcement of Opportunity
ATI	Advanced Technology Insertion
AU	Astronomical Units
BATC	Ball Aerospace and Technologies Corp
C&DH	Command and Data Handling
CCD	Charge Coupled Device
CDS	Correlated Double Sampling
DE	Detector Electronics
DN	Digital Number
EB	Electronics Box
EO-1	Earth Observing-One

FOV	Field Of View
FWHM	Full Width at Half Maximum
G&C	Guidance and Control
GSFC	Goddard Space Flight Center
HIPPS	Highly Integrated Pluto Payload System
IFOV	Instantaneous Field Of View
KBOs	Kuiper Belt Objects
LEISA	Linear Etalon Imaging Spectral Array
LVF	Linear Variable Filter
LVPS	Low Voltage Power Supply
MBE	Molecular Beam Epitaxy
MLI	Multi-Layer Insulation
MTF	Modulation Transfer Function
MVIC	Multi-spectral Visible Imaging Camera
NICMOS	Near Infrared Camera for Multi-Object Spectrometry
NIR	Near Infrared
OpNav	Optical Navigation
PICNIC	Pluto Infrared Camera using NICMOS
SIA	Solar Illumination Assembly
SNR	Signal-to-Noise Ratio
SWIR	Shortwave Infrared
SwRI	Southwest Research Institute
TDA	Telescope Detector Assembly
TDI	Time Delay and Integrate
UV	Ultra Violet
VIS/IR	Visible/Infrared

Acknowledgements The authors would like to thank the entire Ralph support teams at BATC and SwRI and the LEISA support team at GSFC for their untiring efforts in making Ralph a reality. The contributions of JDSU/Uniphase, E2V, Barr Associates, Corning Diamond Turning Division and Teledyne are also gratefully acknowledged.

References

D. Cruikshank, C.M. Dalle Ore, Private communication, 2000

G.H. Fountain, D.Y. Kusnierkiewicz, C.B. Hersman, T.S. Herder, T.B. Coughlin, W.C. Gibson et al., The New Horizons spacecraft. Space Sci. Rev. (2008), this issue

W.M. Grundy, B. Schmitt, E. Quirico, The temperature dependent spectra of alpha and beta nitrogen ice with application to triton. Icarus **105**, 254 (1993)

C.B. Olkin, D.C. Reuter, A. Lunsford, R.P. Binzel, S.A. Stern, The New Horizons distant flyby of asteroid 2002 JF56 (2008, in preparation)

D.C. Reuter, D.E. Jennings, G.H. McCabe, J.W. Travis, V.T. Bly, A.T. La et al., Hyperspectral sensing using the linear etalon imaging spectral array, in *SPIE Proceedings of the European Symposium on Satellite Remote Sensing III*, Conference on Sensors, Systems, and Next Generation Satellites II, September 23–26, 1996, Taorima, Sicily, Italy, vol. 2957 (Int. Soc. Opt. Eng., Bellingham, 1996), pp. 154–161

D.C. Reuter, G.H. McCabe, R. Dimitrov, S.M. Graham, D.E. Jennings, M.M. Matsumura et al., The LEISA/Atmospheric Corrector (LAC) on EO-1, in *IGARS Proceedings*, IEEE 2001 International Geoscience and Remote Sensing Symposium, July 9–13, 2001, Sydney, Australia, vol. 1, pp. 46–48

D.C. Reuter, A. Stern, J. Baer, L. Hardaway, D. Jennings, S. McMuldroch et al., Ralph: A visible/infrared imager for the New Horizons Pluto/Kuiper Belt mission, in *SPIE Proceedings of the Optics and Photonics Conference, Astrobiology and Planetary Missions*, July 31–August 4, 2005, San Diego CA, vol. 5906, 59061F-1 to 59061F-11 (Int. Soc. Opt. Eng., Bellingham, 2005), paper 5906-51

K.P. Rosenberg, K.D. Hendrix, D.E. Jennings, D.C. Reuter, M.D. Jhabvala, A.T. La, Logarithmically variable infrared etalon filters, in *SPIE Proc. Opt. Thin Films IV N. Dev.*, San Diego, CA, vol. 2262 (Int. Soc. Opt. Eng., Bellingham, 1994), pp. 25–27

S.A. Stern, D.C. Slater, W. Gibson, H.J. Reitsema, A. Delamere, D.E. Jennings, D.C. Reuter, J.T. Clarke, C.C. Porco, E.M. Shoemaker, J.R. Spencer, The highly integrated Pluto payload system (HIPPS): A sciencecraft instrument for the Pluto mission, in *SPIE Proceedings, EUV, X-RAY and Gamma-Ray Instrumentation for Astronomy VI*, San Diego, CA, vol. 2518 (Int. Soc. Opt. Eng., Bellingham, 1995), pp. 39–58

S.A. Stern, The New Horizons Pluto Kuiper Belt mission: An overview with historical context. Space Sci. Rev. (2008), this issue

S.A Stern, D.C Slater, J. Scherrer, J. Stone, G. Dirks, M. Versteeg, M. Davis, G.R. Gladstone, J.W. Parker, L.A. Young, O.H.W. Siegmund, ALICE: The ultraviolet imaging spectrograph aboard the New Horizons Pluto-Kuiper Belt mission. Space Sci. Rev. (2008), this issue

S.G. Ungar, J.S. Pearlman, J.A. Mendenhall, D. Reuter, Overview of the Earth Observing One (EO-1) mission. IEEE Trans. Geosci. Remote Sens. **41**(6), 1149–1159 (2003). doi:10.1109/TGRS.2003.815999

L.A. Young, S.A. Stern, H.A. Weaver, F. Bagenal, R.P. Binzel, B. Buratti et al., New Horizons: Anticipated scientific investigations at the Pluto system. Space Sci. Rev. (2008), this issue

ALICE: The Ultraviolet Imaging Spectrograph Aboard the New Horizons Pluto–Kuiper Belt Mission

S. Alan Stern · David C. Slater · John Scherrer · John Stone · Greg Dirks · Maarten Versteeg · Michael Davis · G. Randall Gladstone · Joel W. Parker · Leslie A. Young · Oswald H.W. Siegmund

Originally published in the journal Space Science Reviews, Volume 140, Nos 1–4, 155–187.
DOI: 10.1007/s11214-008-9407-3

Abstract The ALICE instrument is a lightweight (4.4 kg), low-power (4.4 watt) imaging spectrograph aboard the New Horizons mission to the Pluto system and the Kuiper Belt. Its primary job is to determine the relative abundances of various species in Pluto's atmosphere. ALICE will also be used to search for an atmosphere around Pluto's moon, Charon, as well as the Kuiper Belt Objects (KBOs) that New Horizons is expected to fly by after Pluto-Charon, and it will make UV surface reflectivity measurements of all of these bodies, as well as of Pluto's smaller moons Nix and Hydra. The instrument incorporates an off-axis telescope feeding a Rowland-circle spectrograph with a 520–1870 Å spectral passband, a spectral point spread function of 3–6 Å FWHM, and an instantaneous spatial field-of-view that is 6 degrees long. Two different input apertures that feed the telescope allow for both airglow and solar occultation observations during the mission. The focal plane detector is an imaging microchannel plate (MCP) double delay-line detector with dual solar-blind opaque photocathodes (KBr and CsI) and a focal surface that matches the instrument's 15-cm diameter Rowland-circle. In this paper, we describe the instrument in greater detail, including descriptions of its ground calibration and initial in flight performance. New Horizons launched on 19 January 2006.

Keywords NASA missions · Instrumentation · Calibration · Ultraviolet · Spectroscopy · Pluto · Kuiper belt

S.A. Stern · J.W. Parker (✉) · L.A. Young
Southwest Research Institute, 1050 Walnut St., Suite 400, Boulder, CO 80302, USA
e-mail: joel@boulder.swri.edu

D.C. Slater · J. Scherrer · J. Stone · G. Dirks · M. Versteeg · M. Davis · G.R. Gladstone
Southwest Research Institute, 6220 Culebra Rd., San Antonio, TX 78238, USA

O.H.W. Siegmund
Sensor Sciences, 3333 Vincent Road, Pleasant Hill, CA 94523, USA

1 Background

New Horizons (NH) is the first mission in NASA's New Frontiers line of mid-scale planetary exploration missions. More specifically, New Horizons is a scientific reconnaissance mission to reconnoiter the Pluto system in 2015 and then possibly one or more Kuiper-Belt Objects (KBOs) thereafter. En route to Pluto, the mission conducted a Jupiter Gravity Assist flyby that occurred in 2007; that flyby involved extensive scientific observations.

New Horizons ALICE (also referred to as P-ALICE, for Pluto-ALICE) is a compact, low-cost instrument designed to perform spectroscopic investigations of planetary atmospheres and surfaces at extreme (EUV) and far-ultraviolet (FUV) wavelengths between 520 and 1870 Å. It is a direct derivative of the Pluto mission "HIPPS" UV spectrograph (HIPPS/UVSC), developed at Southwest Research Institute (SwRI) in the mid-1990s with funds from NASA, JPL, and SwRI (Stern et al. 1995). Later, the HIPPS/UVSC instrument was re-optimized for the European Space Agency (ESA) Rosetta comet orbiter mission by increasing its sensitivity, instantaneous field-of-view (FOV), and wavelength coverage, and by adding a lightweight microprocessor. This instrument, called Rosetta-ALICE (R-ALICE; Slater et al. 2001), was successfully launched on the Rosetta spacecraft in March 2004, and is operating successfully on a 12-year flight to rendezvous with and orbit comet 67P/Churyumov-Gerasimenko. For NH, notable ALICE modifications include a solar occultation channel (SOCC) for atmospheric solar occultation observations, various reliability enhancements, and a slightly different passband than R-ALICE. Extensive information on the New Horizons mission and its scientific objectives can be found in the articles by Stern (2008), Young et al. (2008), Fountain et al. (2008), and others in this issue.

2 Scientific Objectives

As described in Young et al. (2008) in this issue, one of the three primary (i.e., Group 1) NH mission objectives is to "Characterize the neutral atmosphere of Pluto and its escape rate." P-ALICE is specifically designed to address this measurement objective. Moreover, P-ALICE addresses both Group 1 and Group 2 mission science objectives at Pluto, including important questions about Pluto's atmosphere, such as determining or constraining:

- The mixing ratios of N_2, CO, CH_4 and noble gases;
- The vertical density and temperature structure (e.g., gradient) of the upper atmosphere;
- The atmospheric haze optical depth;
- The atmospheric escape rate and escape regime;
- The brightness of airglow emissions from Pluto's atmosphere.

P-ALICE will also address New Horizons objectives concerning the search for an atmosphere around Charon and it will study the UV reflectance properties of the surfaces of Pluto, Charon, Pluto's small moons, and KBO targets. Searches for atmospheres around KBOs will also be performed.

As an illustration of expected scientific performance of this instrument, Fig. 1 shows results of modeling of solar occultation and airglow observations by P-ALICE. N_2 continuum and band absorption structure at wavelengths <100 nm dominates Pluto's predicted EUV opacity, allowing sampling of the uppermost atmosphere (Krasnopolsky and Cruikshank 1999). From 100–150 nm, CH_4 dominates the opacity, providing a window on the middle atmosphere from roughly 300 to 1200 km. At wavelengths >150 nm, longer hydrocarbons with strong FUV absorption bands, such as C_2H_2 and C_4H_2, along with hazes, are expected

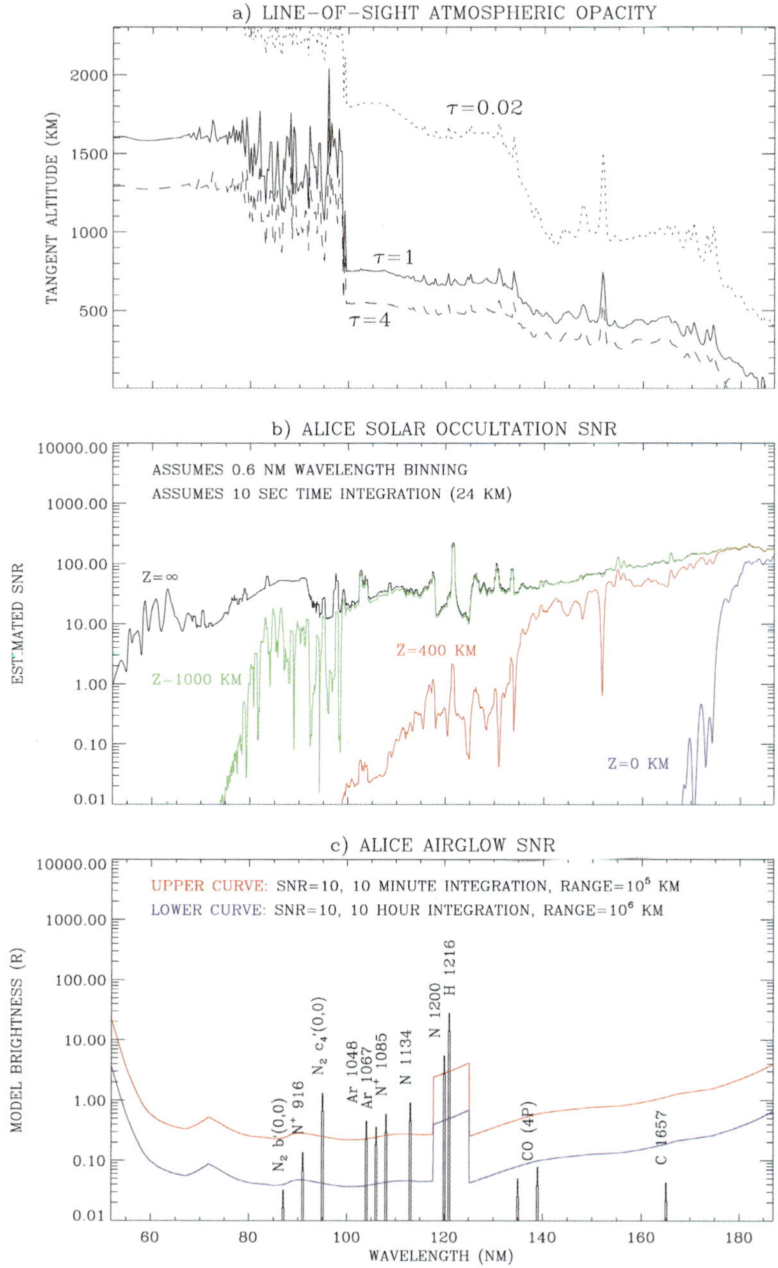

Fig. 1 **a**) Limb-viewing, line-of-sight opacity for model Pluto as a function of wavelength and tangent altitude, for several values of optical depth τ; **b**) Solar occultation SNR estimates based on the line-of-sight transmitted solar flux through Pluto's atmosphere, for several tangent altitudes, as a function of wavelength for a representative integration time of 10 sec (the observations will be made in time-tagged "pixel list" mode, so the data will be combined on the ground after the fact with various time windows and effective exposure times); **c**) Model dayglow brightnesses for emissions resulting from photoelectron impact excitation, dissociative excitation, and resonant scattering of sunlight. Here the upper and lower curves show expected P-ALICE SNR ~ 10 for 10-min and 2-hr integrations at a range of 10^5 km

to provide information about the lowest ~100 km of Pluto's atmosphere (Elliot et al. 1989). Figure 1 also shows predicted airglow spectra at Pluto indicating the potential feasibility of the detection of atomic species like H, N, and Ar, the N^+ ion, as well as molecular CO and N_2.

Of course, the actual ability of P-ALICE to detect and study Pluto's atmosphere depends sensitively on the degree of atmospheric collapse that might occur on Pluto (e.g., Spencer et al. 1997; Trafton et al. 1997) by the time of NH's 2015.5 encounter.

In addition to the planned solar occultation during the flyby, P-ALICE could also observe serendipitous stellar occultations of either Pluto or Charon or both though the airglow channel. Based on models of stars in the Yale Bright Star Catalog, we find 1534 stars with UV fluxes that are large enough (count rates higher than 200 Hz in the P-ALICE passband) to yield potentially useful occultations. Of those, we find that there are seven occultation candidates during the two days surrounding the flyby closest approach, and one other candidate that could be used to search for a ring system. These observation opportunities will be examined in more detail as the Pluto system flyby sequence is developed in coming years.

More details of the mission scientific objectives and how they are met by P-ALICE and the other instruments are given in Young et al. (2008) in this issue.

3 Instrument Description

3.1 Overview

As shown in Fig. 2, the P-ALICE UV spectrograph is comprised of a telescope, a Rowland-circle spectrograph, a detector at the focal plane, and associated electronics and mechanisms. Figure 3 shows a photograph of the exterior of the P-ALICE flight model.

P-ALICE has two separate entrance apertures that feed light to the telescope section of the instrument: the airglow channel (AGC) aperture, and the SOCC aperture. The AGC's $40 \times 40\ \text{mm}^2$ entrance aperture is the front end of the instrument; the SOCC aperture is a small 1-mm diameter opening located perpendicular to the side of the telescope section of the instrument (see Fig. 2). The SOCC stops down the entrance area relative to the AGC by a factor of ~3000 to allow for solar occultation studies without detector saturation. The SOCC is orthogonal to the AGC to boresight it with the REX Radio Science Experiment on NH for the nearly simultaneous solar and Earth occultations that occur in the Pluto system. A small relay mirror reflects the SOCC beam into the telescope optical path. The SOCC and AGC may also be used for stellar occultations by solar system targets.

Light entering either aperture is collected and focused by an $f/3$ off-axis paraboloidal (OAP) primary mirror at the back end of the telescope section onto the instrument's entrance slit. After passing through the entrance slit, the light falls onto a toroidal holographic diffraction grating, which disperses the light onto a double-delay line (DDL) microchannel plate (MCP) detector (Siegmund et al. 1992). The 2-D format MCP detector uses back-to-back, solar-blind photocathodes—potassium bromide (KBr) and cesium iodide (CsI)—and has a cylindrical, curved MCP-stack that matches the Rowland-circle. P-ALICE is controlled by an Intel 8052 compatible microcontroller, and utilizes lightweight, compact, surface-mount electronics to support the science detector, as well as redundant power supplies, command and data handling, instrument support, and other spacecraft interface electronics.

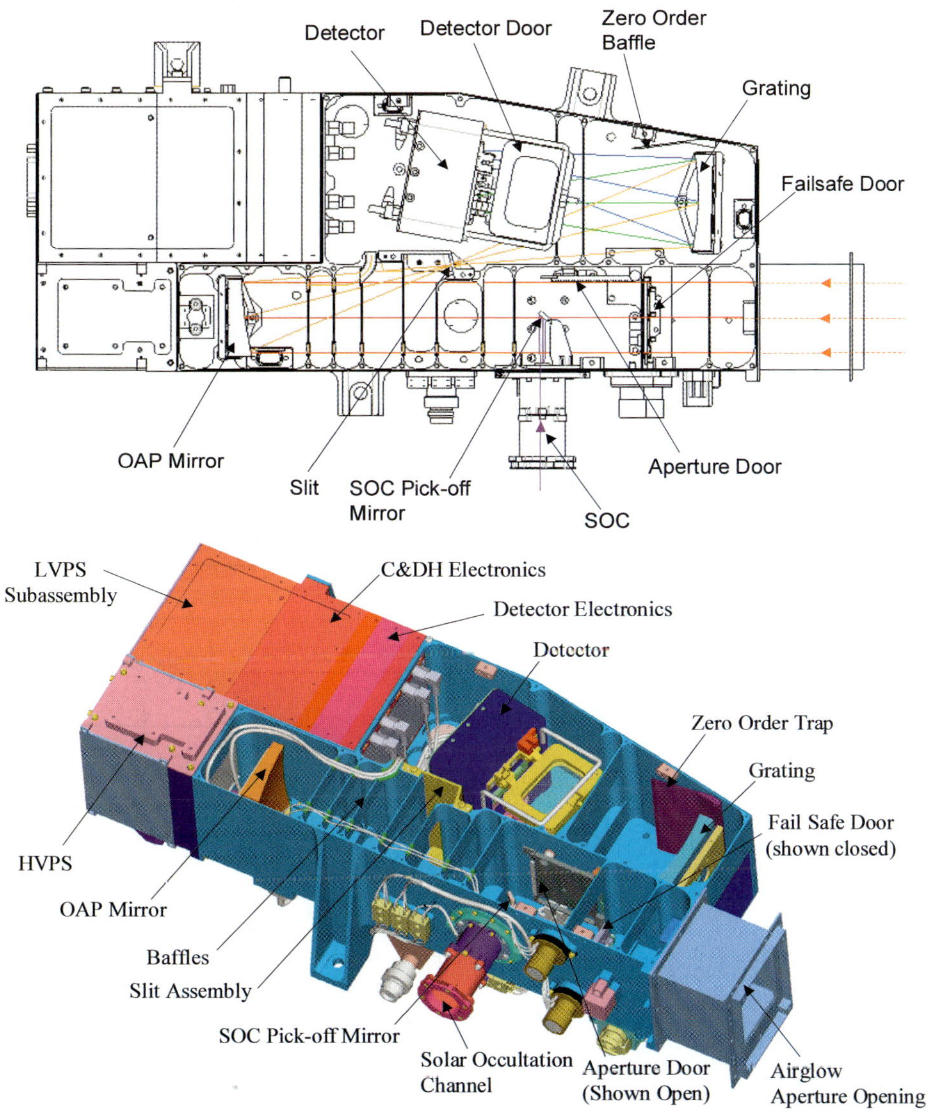

Fig. 2 (*Top*) An opto-mechanical drawing of P-ALICE with central-axis light rays shown. (*Bottom*) 3-D opto-mechanical view. P-ALICE is approximately 45-cm long

3.2 Opto-Mechanical Design Overview

P-ALICE's telescope mirror and diffraction grating are both constructed from monolithic pieces of aluminum, coated with electroless nickel and polished using low-scatter polishing techniques. These aluminum optics, in conjunction with the aluminum housing, form an athermal optical design.

The P-ALICE optics are over-coated with sputtered SiC for optimum reflectivity within the instrument's spectral passband (Osantowski et al. 1991). Additionally, P-ALICE's zero-order baffle light trap is treated with a nickel-phosphorus (Ni-P) black coating with very

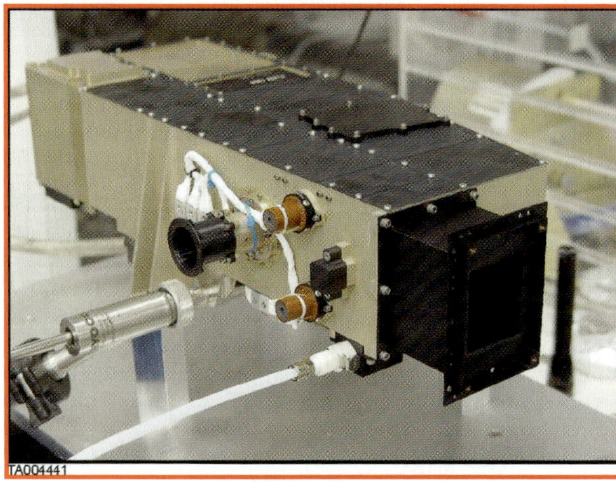

Fig. 3 A photograph of P-ALICE

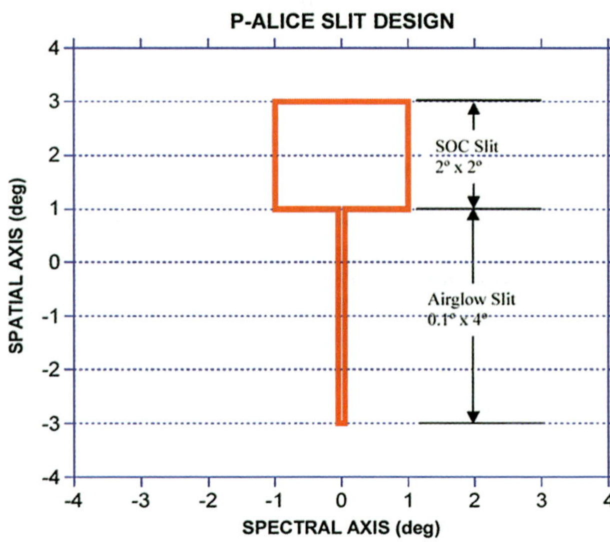

Fig. 4 The P-ALICE entrance slit design. The narrow 0.1° × 4° section is called the "slot" or the "airglow slit"; the larger 2° × 2° opening above the airglow slit is called the "box" or the "SOCC slit". The airglow optical boresight is centered at (0°, 0°)

low surface reflectance at EUV/FUV wavelengths (Brown et al. 2002). Additional control of internal stray light is achieved via (i) UV-absorbing internal baffle vanes within the telescope and spectrograph sections of the housing, (ii) a holographic diffraction grating that has low scatter and near-zero line ghost problems, and (iii) alodyned internal surfaces (Jelinsky and Jelinsky 1987; Moldosanov et al. 1998).

The spectrograph entrance slit design is shown in Fig. 4. It has two contiguous sections: a narrow, rectangular slit segment with a field-of-view of 0.1° × 4.0°; and a fat, square slit segment with a 2.0° × 2.0° field-of-view. The large, 2.0° × 2.0° slit opening is designed to ensure that the Sun is captured within the instrument's field-of-view during the solar occultation observations of Pluto and Charon that must occur nearly simultaneously with the radio science occultation observations performed with the REX instrument (Tyler et al. 2008) that uses the spacecraft's high gain antenna (HGA). During solar occultation observations, the HGA would be pointed back at the Earth, with P-ALICE pointed back at the Sun. P-ALICE is aligned on the spacecraft with a 2° tilt along the spectrograph's spatial axis so that the Sun is centered within the SOCC square FOV when the HGA is pointed at the Earth (to improve the quality of the radio occultation data, the flyby is planned to occur when Pluto is near opposition as seen from Earth, i.e., around mid-July). A misalignment up to ±0.9° between the SOCC FOV and the HGA boresight is accommodated by the large SOCC FOV. The spectral PSF of the SOCC is ~5 Å across the passband (Slater et al. 2005).

Heaters are mounted to the back surfaces of the OAP mirror, the SOCC relay mirror bracket, and the grating, to prevent condensation of contaminants onto the optics during flight. To protect the sensitive photocathodes and MCP surfaces from exposure to moisture and other harmful contaminants during ground operations, instrument integration, and the early stages of the mission, the detector tube body assembly is enclosed in a vacuum chamber with a front hermetic door that includes a magnesium fluoride (MgF_2) UV-transparent window that was permanently opened during the commissioning phase of the flight. The front aperture door can be commanded closed for additional protection of the optics and detector from direct solar exposure during propulsive maneuvers.

The only multiple-use actuator on P-ALICE is used to move the aperture door, which can be opened and closed via command; it is qualified for 10,000 cycles, but is expected to be used <1,000 over the course of the New Horizons mission. A limited angle torque (LAT) motor with a direct tie to the aperture door hinge pin provides the torque required to open and close the door. The aperture door remained closed until the spacecraft had sufficient time to fully outgas (i.e., ~50 days according to models) after launch. The aperture door was then opened to allow residual outgassing of the interior of the instrument and for heater decontamination sessions of the optics.

Following these activities, the detector door, which uses a Wax Pellet Actuator (WPA) was also opened; this was followed by a slow, controlled, high voltage ramp up of the detector MCP stack to begin detector check out. The detector door design is an improved version of that flying aboard the R-ALICE instrument (Slater et al. 2001). During ground operations the door was opened numerous times and manually reset; during flight, once open, the door need not and cannot be re-closed. A MgF_2 window provides a method to get light to the detector during ground operations when the detector door is closed; this window also provided the capability to meet science objectives at wavelengths longer than 120 nm in the event the detector door had not opened in flight.

A one-time opening failsafe door is available on the aperture door in the event that it ever becomes stuck in the closed state. With the failsafe door open, the AGC sensitivity is ~10% that of the fully opened AGC with the aperture door in the open state. A larger failsafe aperture was not considered prudent owing to the long cruise exposure (up to 9+ years) that could risk UV photolysis of hydrocarbons on the optics. The SOCC pinhole aperture cover was opened in July 2007 when the risk of accidental SOCC aperture Sun pointing that can cause UV photolysis of the optics has been sufficiently reduced owing to the spacecraft heliocentric distance.

3.3 Detector and Detector Electronics Overview

P-ALICE's detector utilizes an MCP Z-stack that feeds its DDL readout array (Siegmund et al. 1992). To cover the 520–1870 Å passband and 6° spatial FOV, the size of the detector's active area is 35 mm (in the dispersion direction) by 20 mm (in the spatial dimension), with a pixel format of (1024 × 32)-pixels. The 6° slit-height is imaged onto the central 22 of the detector's 32 spatial channels; the remaining spatial channels are used for dark count monitoring. The pixel format allows Nyquist sampling with a spectral resolution of

3.6 Å, and a spatial resolution of ~0.6°. The input surface of the Z-stack is coated with opaque photocathodes of KBr (520–1180 Å) and CsI (1250–1870 Å) (Siegmund 2000), in two different wavelength regions.

The MCP Z-stack is composed of three 80:1 pore length-to-diameter (L/D) MCPs that are cylindrically curved with a radius-of-curvature of 75 mm to match the Rowland-circle for optimum focus; the total Z-Stack resistance at room temperature is ~300 MΩ. The MCPs are rectangular in format (46×30 mm^2), with 12-μm diameter pores. Above the MCP Z-Stack is a repeller grid that is biased ~900 V more negative than the top of the MCP Z-Stack. This repeller grid reflects electrons liberated in the interstitial regions of the MCP back down to the MCP input surface to enhance the detector quantum efficiency. The MCP Z-stack requires a high negative voltage bias of approximately −3 kV; an additional −600 V is required between the MCP Z-stack output and the anode array (the anode array is referenced to ground). The intrinsic dark count rate of the flight MCP stack is quite low, less than 3 counts sec^{-1} over its entire active area (Slater et al. 2005); however, in flight radiation (both natural and from the NH RTG) has raised the background count rate to almost 150 Hz.

To prevent saturation effects on the detector electronics during observations, it is necessary to attenuate the solar Hydrogen Lyman-alpha (Ly α) emission brightness to an acceptable count rate level well below the maximum count rate capability of the electronics (i.e., below 3×10^4 c sec^{-1}). An attenuation factor of at least an order of magnitude is required to achieve this lower count rate. This is achieved by physically masking the MCP active area around Ly α during photocathode coating so it remains bare. The bare MCP glass has a quantum efficiency about 10 times less than that of KBr at 1216 Å. This masking technique has been successfully demonstrated in the past with the DDL detector aboard the Rosetta-ALICE instrument (Slater et al. 2001) and the SUMER instrument on SOHO (Siegmund et al. 1994).

Surrounding the detector tube body is a vacuum chamber housing made of aluminum and stainless steel (Fig. 5). As described above, this vacuum chamber is designed to protect the MCP stack and the KBr and CsI photocathodes against damage from moisture exposure during ground handling and from outgassing constituents during the early stages of flight. It also allows the detector to remain under vacuum during ground operations, testing, handling, and transportation. The MCP/anode tube body assembly mounts to the rear stainless steel vacuum flange. This flange mates to the aluminum vacuum housing with a vacuum tight O-ring seal, and contains two high-voltage (HV) feedthrough connectors that are welded to the flange (HV input and return), as well as four welded microdot feedthrough connectors for the four analog signal outputs from the DDL anode.

The DDL detector's electronics include preamplifier circuitry, time to digital converter circuitry (TDC), and pulse-position analyzer circuitry (PPA) cards. These three boards are mounted inside a separate enclosed aluminum housing that mounts to the rear of the spectrograph section (just behind the detector itself). The detector electronics require ±5 VDC, and draw ~1.1 W.

The detector electronics amplify and convert the detected output pulses from the MCP Z-stack to pixel address locations. Only those analog pulses output from the MCP that have amplitudes above a set threshold level are processed and converted to pixel address locations. For each detected and processed event, a 10-bit x-position (spectral) address and a 5-bit y-position (spatial) address are generated by the detector electronics and sent to the P-ALICE C&DH electronics for data storage and manipulation. In addition to the pixel address words, the detector electronics also digitizes the analog amplitude of each detected event output by the preamplifiers and sends these data to the C&DH electronics. Histogramming of these "pulse-height" data creates a pulse-height distribution (PHD) used to monitor

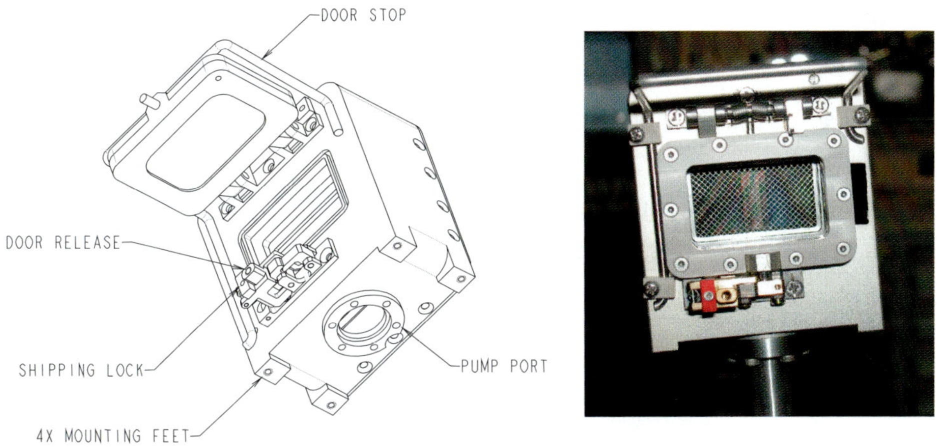

Fig. 5 (*Left*) 3-D schematic of DDL detector vacuum housing with the detector door shown in the open position. (*Right*) Photograph of the flight DDL detector/vacuum housing. The repeller mesh and MCP Z-stack are visible through the MgF_2 window

the health and status of the detector during operation, e.g., the output gain of the MCP stack as a function of HV applied, and the shape of the PHD can help to identify what sources (ion, UV or gamma-ray photons) are stimulating the detector.

An analog count rate signal is also output from the detector electronics to the C&DH to allow monitoring and recording of the detector total array count rate. These count rate data are updated once per second and are reported in the instrument housekeeping (HK) data.

A built-in electronics "stim-pulser" (hereafter referred to as a "stim") is included in the electronics that simulates photon events at two pixel locations on the array (located in the upper right and lower left corners of the active array). This stim can be turned on and off by command and allows testing of the entire detector and C&DH electronics signal path without having to power on the detector high-voltage power supplies or put light on the detector.

3.4 Instrument Electrical Design Overview

The P-ALICE instrument support electronics include the redundant low-voltage power supplies (LVPS), actuator electronics, C&DH electronics, optics decontamination heaters, and redundant detector high-voltage power supplies (HVPSs).

A simplified block diagram of these electronic subsystems is shown in Fig. 6; Figure 7 contains a more detailed block diagram. These sub-systems are controlled by a radiation-hardened version of the Intel 8052 microprocessor with 32 kB of fuse programmable PROM, 128 kB of EEPROM, 32 kB of SRAM, and 128 kB of acquisition memory. The C&DH electronics are contained on four circuit boards located just behind the detector electronics (see Fig. 2). We now describe each major block.

Low-Voltage Power Supply Electronics LVPS electronics and associated drivers are redundant, and are each composed of DC/DC converters designed to convert the +30 V supplied spacecraft power to ±5 VDC (power ∼4 W) and +2.5 VDC (power ∼0.4 W) to the P-ALICE electronics. Five boards make up the LVPS: two LVPS boards (both primary and

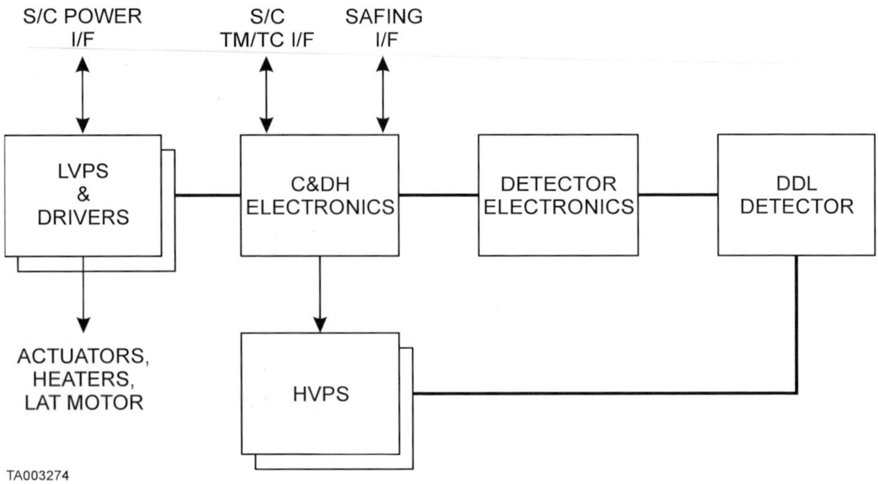

Fig. 6 Simplified P-ALICE electronics block diagram

redundant), an EMI filter board, a heater/actuator board, and a motherboard. The LVPS electronics interfaces to the spacecraft via three links: two links to the +30 V spacecraft power bus for (i) instrument power and (ii) actuator/heater power; the third link is a safing connector interface that prevents inadvertent HVPS and/or actuator (i.e., launch latch, detector door, fail-safe door, SOCC door, and/or detector vent valve) operations. The switching circuit for the decontamination heaters and the motor controller that operates the front aperture door is in the LVPS electronics.

Command-and-Data-Handling Electronics The P-ALICE C&DH electronics are controlled by the 8052 microprocessor, and provide the command interface as well as the science and telemetry interfaces with the NH spacecraft. They handle the following instrument functions: (i) interpretation and execution of commands to the instrument, (ii) collection of raw event data from the detector to the dual-port acquisition memory (see Fig. 7) in either the pixel list or image histogram data collection modes (described in the following section), (iii) telemetry formatting of science and housekeeping data, (iv) control of the redundant detector HVPSs, (v) operation of all actuator and heater functions, (vi) control of the front aperture door LAT motor, (vii) control of the housekeeping ADCs used to convert analog housekeeping data to digital data for inclusion into the TM data stream, and (viii) monitoring of detector health and status via the detector analog count rate, and the HVPSs' high voltage and MCP strip current values. For P-ALICE flight operations, the Intel 8052 microprocessor operates with a clock frequency of 4.0 MHz. P-ALICE low-speed HK telemetry and telecommand interface is accomplished via an RS-422 asynchronous interface to the NH spacecraft. A second FPGA within the C&DH electronics implements a UART that sends the low-speed data to the spacecraft.

Decontamination Heaters As described earlier, a redundant decontamination heater ($\sim$1 W resistive) is bonded to the backside surface of both the OAP mirror and the grating substrates. Along with each heater, two redundant thermistors are also mounted to the back of each substrate to monitor and provide control feedback to the heaters. The C&DH electronics can separately control each heater.

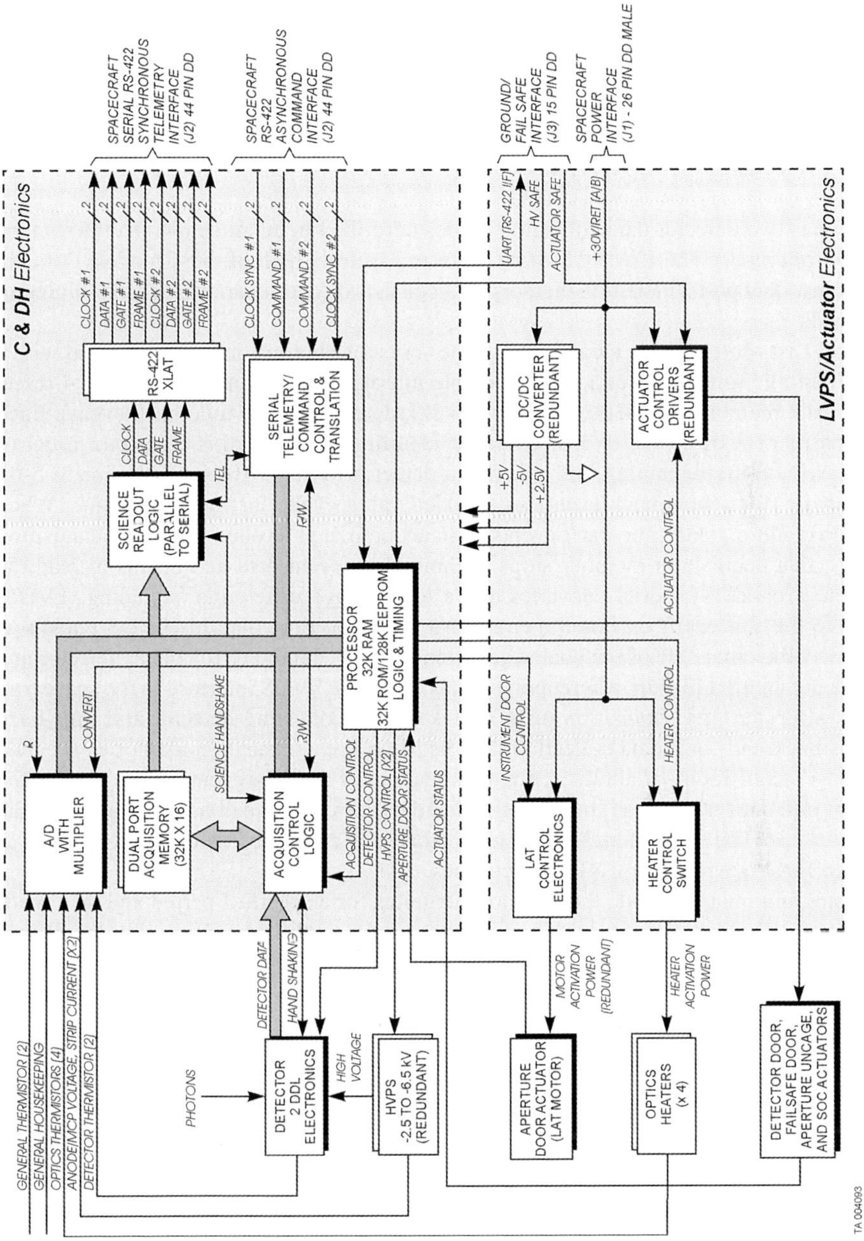

Fig. 7 Detailed ALICE instrument block diagram

High Voltage Power Supplies Redundant HVPSs are located in a separate enclosed bay behind the OAP primary mirror (see Fig. 2). The HV output of these two supplies are diode-or'ed together to the single HV terminal on the MCP/anode assembly. Each supply can output a maximum of −6 kV to the top of the MCP/repeller grid assembly—well above the nominal operational HV level of −4.5 kV. The voltage between the MCP output and the anode array is fixed at −600 V using Zener diodes between the output and ground. The voltage to the MCP Z-stack is fully programmable by command between 0 and −6 kV; they require ±5 VDC and consume ~0.3 W.

3.5 Data Collection Modes Overview

P-ALICE has two detector data collection modes described in detail below: *pixel list* mode and *histogram* mode. The P-ALICE flight software controls both of these modes. Data are collected in a dual port acquisition memory that consists of two separate 32k × 16-bit memory channels.

In Pixel List Mode (PLM) a continuous time series of photon counts is generated with a series of periodic time hacks at a programmable interval, which can be as short as 4 msec. In PLM, each memory channel can hold up to 32k detector and/or time-hack events. Each event consists of a 16-bit word, where one bit identifies if the word is a detector event or a time hack, and the remaining bits are: (for a detector event) a 10-bit x-position, a 5-bit y-position; or (for a time hack event) an encoded value of the instrument on-time. When 32k detector address and time-hack events have accumulated in one of the two acquisition memories, that acquisition memory stops accumulating event data and begins to read the data out to a parallel-to-serial converter and a low-voltage differential signaling (LVDS) interface to the spacecraft on-board memory. At the same time that this LVDS readout is taking place, the other side of the dual acquisition memory continues to collect detector and time-hack data until it fills up, whereupon it reads out to the LVDS interface to the spacecraft memory, while the first acquisition memory takes over collecting detector and time-hack data. This back-and-forth data collection flow between both acquisition memories is called "ping-pong" acquisition—it allows contiguous readout of detector event data as long as the data event rate does not exceed the rate at which the data can be read out of memory to the LVDS interface. The "ping-pong" acquisition process is controlled using logic encoded in one of two FPGAs within the C&DH electronics.

In Histogram imaging mode the detector integrates for a specified period and then reads out its entire 2-D array as an image. In histogram mode, each memory location thus represents a pixel location in the detector 2-D 1024 × 32 array. When a detector event address is sent from the detector to the C&DH, the acquisition control logic updates the proper pixel address location in memory with a single additional count. Each pixel location in the detector array corresponds to a 16-bit word in the acquisition memory and can hold up to $2^{16} - 1$ events before the counter saturates at this maximum value. When a histogram exposure is complete, the memory is read out to the LVDS interface to the spacecraft. A second exposure can then begin using the other acquisition memory bank while the first bank is reading out to the spacecraft. Besides the collection and binning of detector events, histogram mode allows the collection of MCP pulse-height distribution (PHD) data from the detector electronics. These PHD data are collected and binned into a 64-bin histogram that is stored within the first two rows of the detector histogram, in a location where no physical pixel within the detector active area exists (therefore, the PHD data do not interfere with the collected detector data). The same C&DH FPGA that controls the pixel-list "ping-pong" acquisitions also controls the histogram data collection mode.

The advantage of having these two separate modes is that the exposures can be tailored to optimize the scientific goal of each observation and the data volume produced. Pixel list mode provides time resolution, which is essential for some observations, such as occultations. However, since in pixel list mode each event is saved as a word in memory, the data volume can be quite large. *A priori* estimates of data volumes are important for reserving spacecraft memory for an observation and meeting constraints for data downlink, and such estimates can be difficult in pixel list mode because the data volume is a function of the source flux. In histogram mode, the data volume is deterministic; each exposure requires a fixed amount of memory and downlink capability. So, when time resolution is not needed (e.g., observations of calibration stars or fixed points with no temporal variation), the histogram mode is optimal.

P-ALICE also includes a mechanism to filter out events from selected areas of the detector: up to 8 rectangular data exclusion regions of 32×4 pixels can be defined using configuration parameters. This can be used to suppress 'hot pixels' that could develop in the detector, or regions receiving high count rates that are not of interest for a particular observation (e.g., to "ignore" Ly α counts when lines at other wavelengths are being studied). This filtering is performed before events are processed in either histogram or pixel list mode, and is especially useful in the pixel list mode, where high count rates from undesirable sources might otherwise consume a large fraction of the available acquisition memory.

4 Ground Calibration Radiometric Performance

Ground radiometric characterization and absolute calibration of the instrument was performed in 2005 at SwRI's UV space instrument calibration facility located in SwRI's Space Science and Engineering Division in San Antonio, Texas.

The radiometric vacuum chamber consists of a 4-inch diameter off-axis parabolic collimator mirror that is fed by a differentially pumped hollow-cathode UV light source (Danzmann et al. 1988) and an Acton Research Corporation VM-502 vacuum monochromator. A variable slit and pinhole assembly at the output of the monochromator (and situated at the focus of the collimator mirror) allowed for point source illumination of the P-ALICE airglow and SOCC input apertures.

For these calibrations, P-ALICE was mounted to motorized translation and tip/tilt rotation stages within the vacuum chamber that allowed for instrument motion with respect to the collimated input beam. A set of NIST-calibrated photodiodes (one windowless and one silicon PN photodiode) and an AmpTek channeltron were used to measure the relative and absolute fluxes of the collimated beam for absolute effective area measurements. The chamber pressure during the radiometric calibration data runs were in the 5×10^{-7} to 5×10^{-6} Torr range. The higher-pressure periods were during operation of the hollow-cathode light source when gas was flowing through the light source.

Radiometric characterization tests included the detector dark count rate, wavelength calibration, spectral and spatial point source function (PSF) vs. wavelength, filled slit spectral resolution, off-axis stray light attenuation, and absolute effective area measurements as a function of wavelength.

4.1 Ground Dark Count Rate

The native (i.e., ground) detector dark count rate was measured via a set of pixel-list exposures that totaled 9.1 hours of accumulated exposure time. These datasets allowed us to

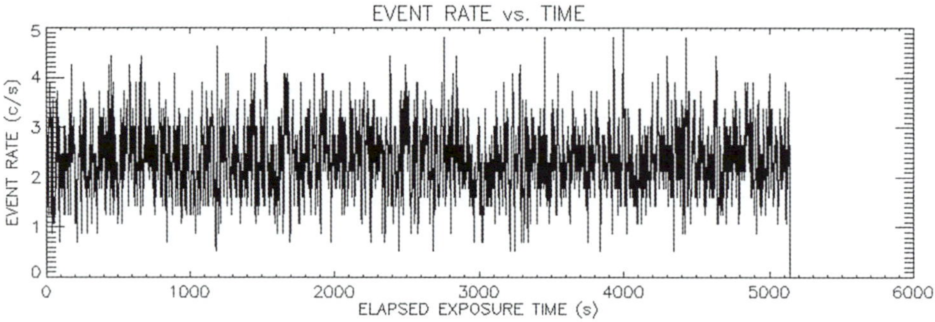

Fig. 8 Detector dark count rate as a function of time (smoothed over a 5.12 sec period) during one set of dark calibration pixel list exposures totaling ∼1.4 hours. The average rate in this exposure was 2.37 ± 0.67 counts sec^{-1}

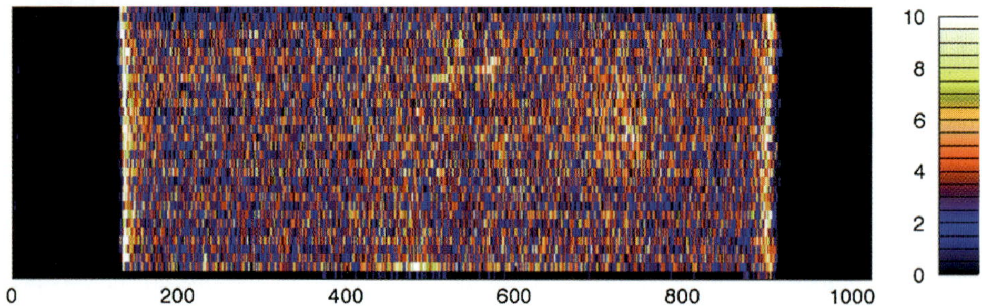

Fig. 9 Image histogram of the total, 9.1 hour, accumulated dark count exposure. The *horizontal axis* is the spectral axis (780 active pixels); the *vertical axis* is the spatial axis (30 active pixels). The *vertical color bar scale at right* is in units of accumulated counts per pixel over the entire 9.1-hour exposure

examine both the temporal and spatial distribution of the dark noise and measure its rate. The time hack repetition rate was set to 0.512 sec. Figure 8 shows the event rate during one 1.4-hour exposure. The average dark count rate over the 9.1-hour total exposure interval varied between 2.35 and 2.46 counts sec^{-1} (0.29–0.31 count cm^{-2} sec^{-1}) with Poisson counting statistics ($\sigma \sim \pm 0.67$ counts sec^{-1} averaged over 5.12 second bins). This is well below the dark rate specification of <1 count cm^{-2} sec^{-1}. Figure 9 shows the accumulated 9.1-hour dark exposure in 2-D histogram format; Figure 10 shows the row and column sums of the total dark exposure. The dark events are uniformly distributed over the entire array with three slightly more active dark regions at (i) center right, (ii) top center, and (iii) bottom center (max rate $\sim 3 \times 10^{-4}$ counts sec^{-1} pixel^{-1}). Event pileup at the left and right edges of the array is a characteristic of the DDL detector/electronics, and is evident in Figs. 9 and 10.

4.2 Spatial/Spectral Resolution

4.2.1 Airglow Channel Point Spread Function

The spectral and spatial PSFs were measured before launch with the airglow aperture at 0.5° intervals along the length of the slit and at specific wavelengths across the P-ALICE passband using argon and neon gasses in the hollow-cathode UV light source. Point source images were acquired in histogram format at each slit location beginning at the center of

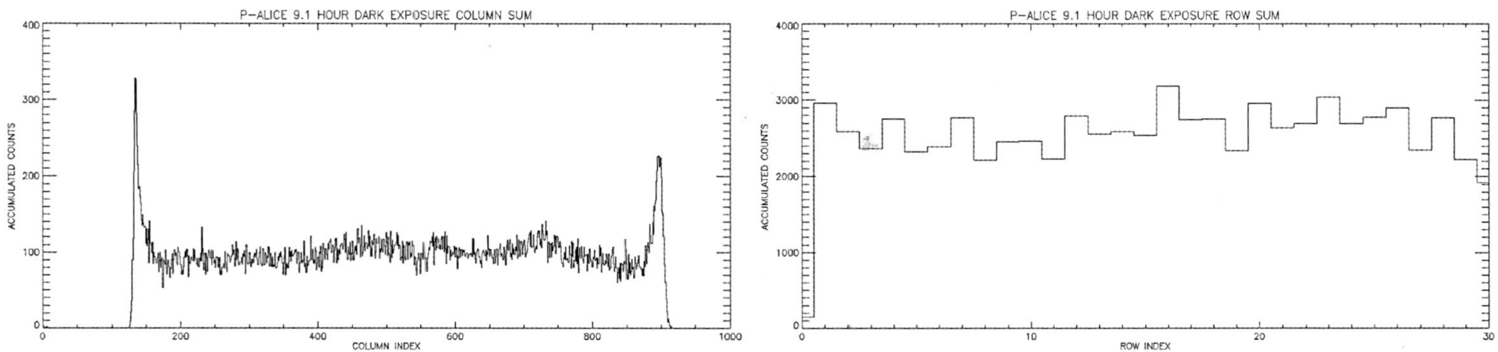

Fig. 10 (*Left*) Column sum of the totaled 9.1-hours of dark exposure calibration. Note the uniform distribution of events across the width of the detector active area. (*Right*) Row sum of the total 9.1-hour dark exposure. Note that a few rows (*every third row to left and right of center*) show slightly more output than expected with just pure spatially uniform noise characteristics (indicative of a pure Poisson process). This periodic noise is a characteristic of the flat field performance of the detector and may be attributed to differential nonlinearity (DNL) noise in the analog-to-digital converters in the detector electronics

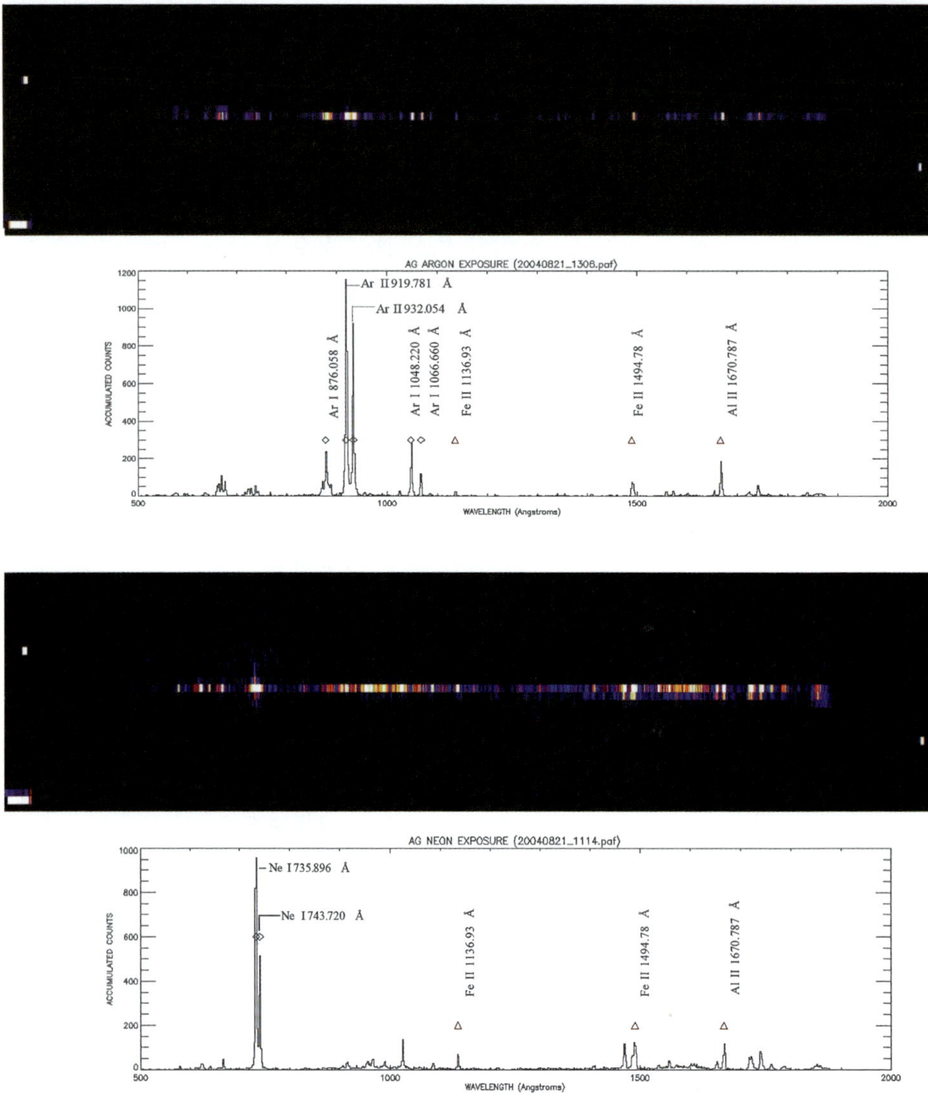

Fig. 11 Histogram images and spectral plots of Ar (*top*) and Ne (*bottom*) emission spectra using a collimated point source input to the airglow channel at slit center. The stim pixels were enabled during the exposures. PHD data is also recorded and embedded in the histogram images. The *blue diamonds* show the emission lines from each noble gas; the *red triangles* are emission lines of iron and aluminum from the lamp cathode

the slit (spatial offset of 0°) and at offset angles of ±0.5°, ±1.0°, ±1.5°, ±2.0°, ±2.5°, and ±3.0° with respect to slit center. Figure 11 shows two histogram image exposures showing the recorded emission lines of Ar and Ne at slit center across the P-ALICE passband with the Acton monochromator set at zero order. The two stim pixels and the embedded pulse-height distribution data are also visible in the images.

For each exposure, the image row with the maximum number of counts was used to fit a series of Gaussians for each identified emission line in the spectral axis. In the spatial

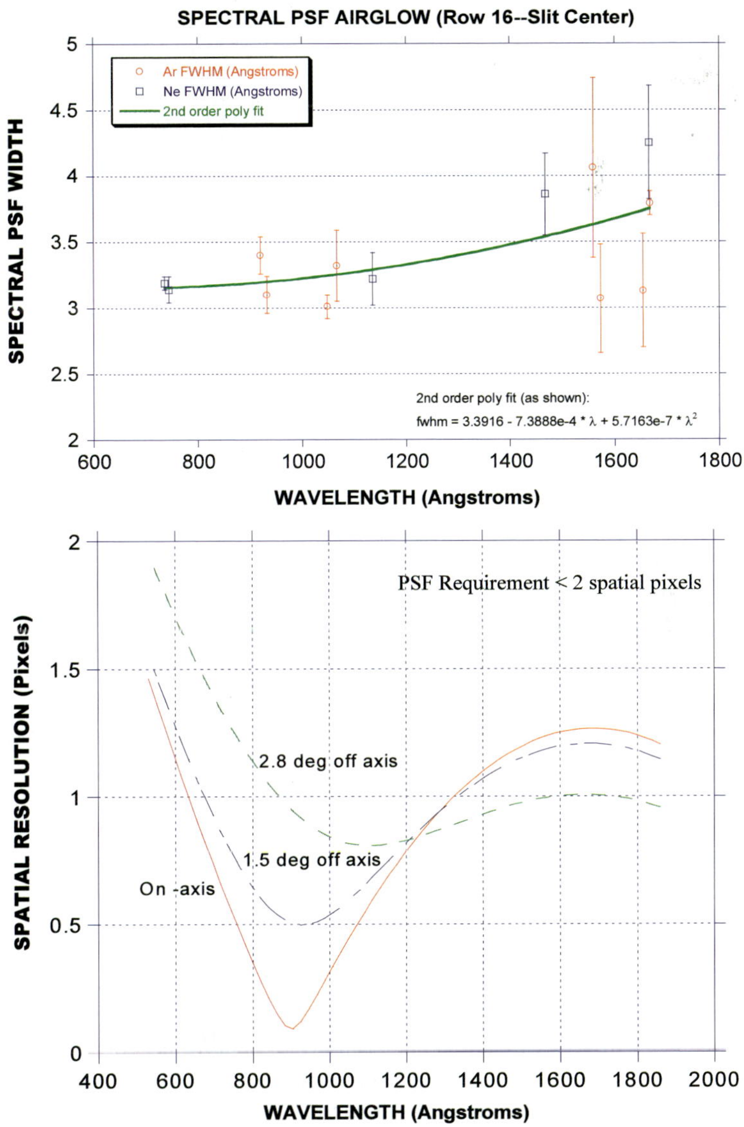

Fig. 12 (*Top*) A plot of the measured airglow aperture spectral PSF as a function of wavelength at the center of the P-ALICE slit; *error bars* are 1σ errors based on Gaussian fits of the spectral profiles. The *solid line* is a 2nd order polynomial fit to the data. Lower count rates at longer wavelength emission lines produce more PSF scatter. (*Bottom*) A plot of the spatial PSF resolution as a function of wavelength

axis each column was fit with a Gaussian to determine the spatial PSF. Figure 12 shows the spectral PSF as a function of wavelength at slit center using the Ar and Ne exposures. The measured values varied between 3 and 4.5 Å, which satisfied the P-ALICE specification at slit center of <6 Å FWHM. The measured spatial PSF across the P-ALICE passband was <1.6 spatial pixels FWHM, satisfying the requirement of a spatial PSF of <2 spatial pixels FWHM. The spectral and spatial PSFs were very similar in size and shape (even though $-2°$

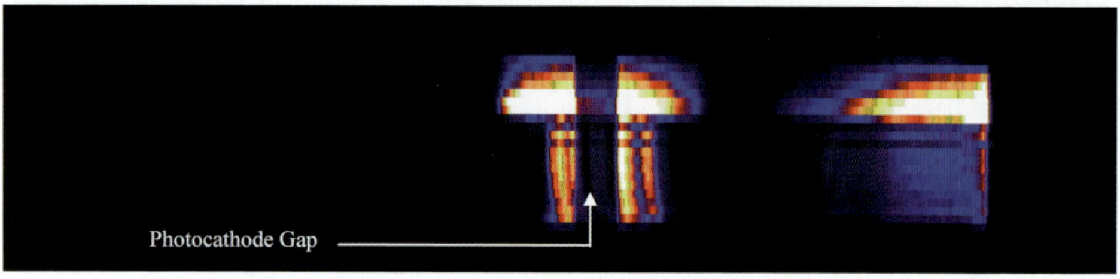

Fig. 13 A histogram image acquired using a D_2 light source flooding P-ALICE's input airglow aperture

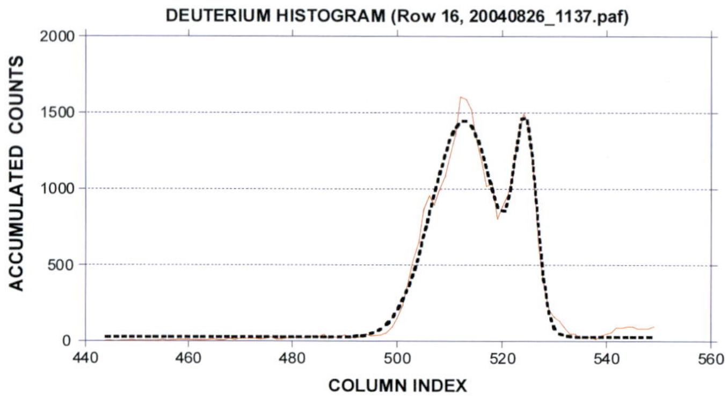

Fig. 14 Plot of row 16 in the D_2 spectral image shown in Fig. 13 (*solid curve*). A two Gaussian sum plus DC offset was fit to this profile (*dashed line*). The narrow emission on the right was used to calculate the filled slit spectral resolution

is in the narrow slit segment and +2° is in the wide slit segment), and varied between 4.4 and 6.5 Å FWHM, with an average of ~5 Å FWHM across the passband. The spatial PSF remained <1.6 spatial pixels FWHM, but grew an average of ~ 0.2 spatial pixels as the off axis angle was increased.

4.2.2 Airglow Channel Filled Slit Resolution

Filled slit images as in Fig. 13 using a deuterium (D_2) lamp to flood the airglow entrance aperture were used to determine the filled slit spectral resolution and the spatial plate scale. A few discrete spectral emission lines exist about the Ly α emission at 1216 Å (i.e., emission lines are evident on either side of the photocathode gap), and a continuum of emission appears at the right side of the active area. The histogram clearly shows the image of the slit with the wide 2° × 2° SOCC FOV at the top (the large square pattern) followed by the narrower 0.1° × 4° airglow FOV just underneath the SOCC FOV (refer to the slit design in Fig. 4). In addition, the design "slit curvature" aberration is clearly evident. This aberration was well characterized and implies that the wavelength scale offset is a function of the spatial location along the length of the slit.

One isolated emission line just to the left of the gap was used to compute an upper bound to the filled slit resolution. Figure 14 shows this emission at the center of the slit along with a Gaussian fit to it and its companion. The narrow emission has a FWHM spectral width of 9.0 ± 1.4 Å. This width is within the filled-slit resolution requirement of <18 Å FWHM for the airglow channel. The spatial plate scale was also derived from these data and was found

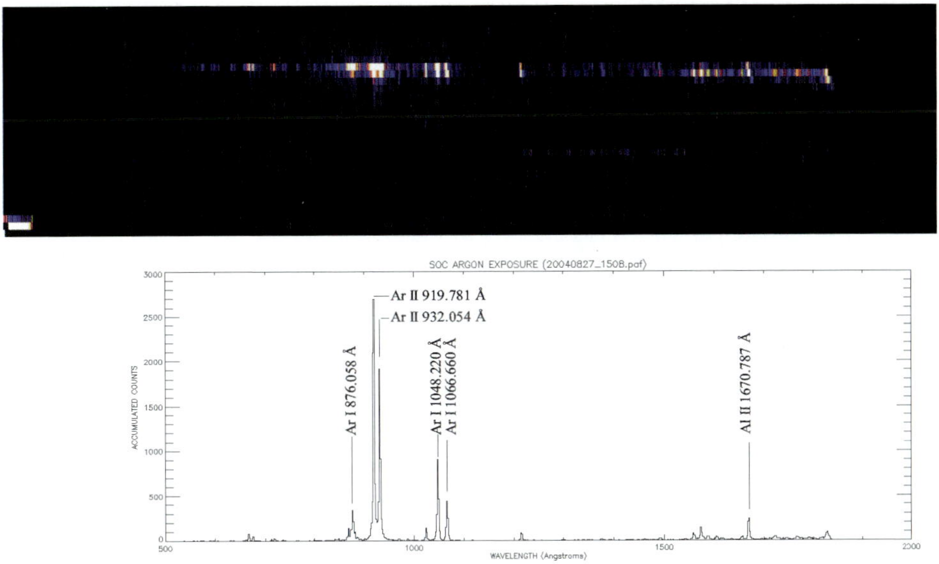

Fig. 15 (*Top*) Histogram image of the argon emission spectrum through the SOCC. (*Bottom*) Spectrum of the above histogram image summed across rows 21–23

to be $0.27° \pm 0.01°$ per spatial pixel, which meets our $<0.3°$ per spatial pixel requirement. The filled slit spectral resolution in the wide portion of the slit was measured to be 172 Å FWHM.

4.2.3 Solar Occultation Channel Point Spread Function

As shown in Fig. 15, the SOCC spectral and spatial PSFs were measured at the center of the $2° \times 2°$ slit SOCC opening using argon and neon gasses in the hollow-cathode UV light source. The image row with the maximum number of counts was used to fit a series of Gaussians for each identified emission line in these images. The measured spectral PSF as a function of wavelength varied between 2.7 and 3.5 Å FWHM (see Fig. 16), easily meeting the instrument's <6 Å FWHM specification.

4.3 Ground Wavelength Calibration

Image histograms taken to determine the airglow PSF values were also used to determine the wavelength calibration at room temperature (+22°C). Figure 17 shows the pixel locations of the four principal Ar emission lines at 919.781, 932.054, 1048.22, and 1066.66 Å and two bright Ne emission lines at 735.896 and 743.72 Å at slit center. The spectral plate scale and linear offset were determined with a simple linear fit to these data: $d\lambda/dx = 1.815 \pm 0.004$ Å/pixel; offset (slit center) $= 229.5 \pm 1.5$ Å (at pixel = 0). The linear fit is good—the χ^2 statistic for the fit is 1.93 with a linear correlation coefficient of 0.99999. The standard deviation of the wavelength residual between the linear fit and the six absolute emission line wavelengths is ±0.62 Å.

The wavelength passband was computed with the above measured plate scale and offset values. The active area starts at spectral pixel 130 and ends at pixel 910, which corresponds

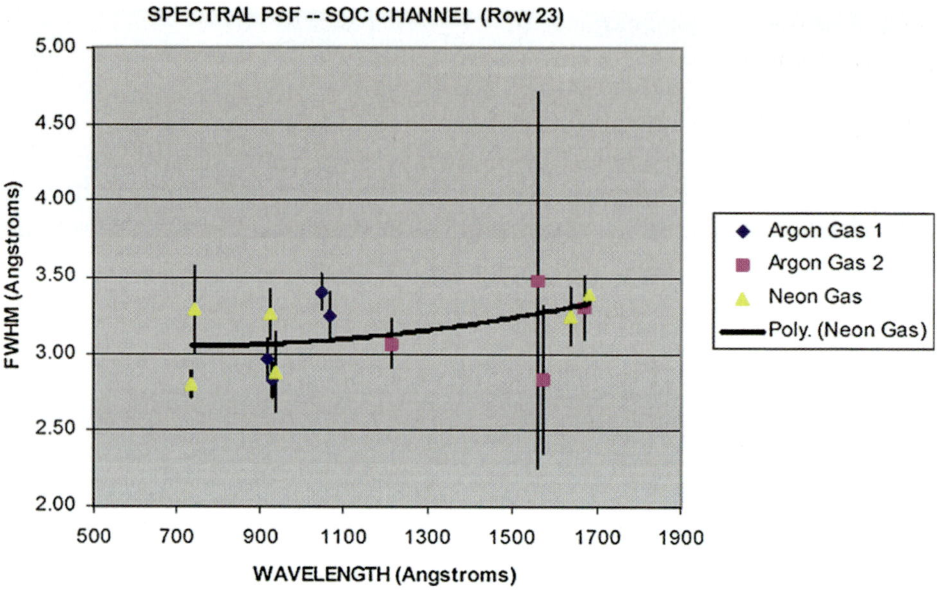

Fig. 16 The measured spectral PSF at the center of the SOCC field of view opening in the slit; *error bars* are 1σ errors based on Gaussian fits of the spectral profiles. The *solid line* is a 2nd order polynomial fit to the PSF data

Fig. 17 Measured positions of discrete Ar and Ne emissions versus wavelength at slit center. The *solid line* shows a linear fit to the data to determine the spectral plate scale and offset

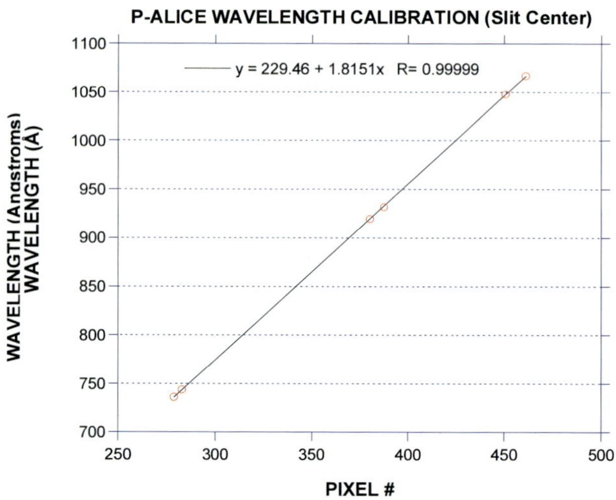

to a total wavelength passband of 465–1881 Å, which more than satisfies our passband requirement of 520–1870 Å.

The wavelength offset varies slightly with the temperature of the detector electronics. Therefore, care must be taken to apply the proper offset according to the detector electronics temperature. This dependence of roughly 0.1 pixel (deg C)$^{-1}$ towards longer wavelengths[1] was measured during detector subsystem and instrument thermal vacuum testing. The wavelength offset also varies with the (x, y)-location of the point source image within the slit due to both instrument pointing and to slit curvature aberrations.

4.4 Scattered Light Characteristics

4.4.1 Off-axis Light Scatter

The off-axis light scattering characteristics of the airglow channel were measured prior to launch in both the horizontal and vertical axes of P-ALICE. Histogram images using a collimated input source and a vacuum UV hollow-cathode lamp gas mixture of H/He were acquired at angular input angles that varied between $-8.9°$ and $+9.2°$ with respect to the airglow boresight axis in the horizontal plane (perpendicular to the slit length with the vertical axis fixed at the center of the slit), and at input angles between $-2.5°$ to $+5.0°$ in the vertical plane (parallel to the slit length with the horizontal axis fixed at the center of the slit).[2]

The acquired histogram images were analyzed to determine the point source transmittance (PST) as a function of incident input angle to the P-ALICE boresight. The PST is defined as follows: $PST = E_{FP}/E_{\text{input}}$ where E_{FP} is the irradiance at the focal plane; and E_{input} is the input irradiance at the entrance aperture of the instrument. This expression can be written in terms of the measured off-axis angle count rate to the on-axis count rate ratio (R_{off}/R_0); the effective area averaged over the P-ALICE passband $(\overline{A_{\text{eff}}})$; the focal plane active area (A_{FP}); the average in-band QE of the detector photocathodes $(\overline{QE})$; and the ratio of the input beam area (A_{beam}) to the geometric area (A_g) of the airglow entrance aperture (since the input beam under fills the entrance aperture). Notice that the PST is not normalized to unity at the zero degree offset position (on axis). Using these parameters, the PST equation becomes:

$$PST = \left(\frac{\overline{A_{\text{eff}}}}{A_{FP}\overline{QE}}\right)\left(\frac{A_{\text{beam}}}{A_g}\right)\left(\frac{R_{\text{off}}}{R_0}\right). \tag{1}$$

The average effective area and QE values used in (1) to compute the PST were values measured during instrument radiometric tests and detector QE tests at the detector vendor (Sensor Sciences, LLC), respectively: $\overline{A_{\text{eff}}} = 0.13\ \text{cm}^2$; $\overline{QE} = 0.27$; $A_{\text{beam}}/A_g = 0.05$; $A_{FP} = 7.5\ \text{cm}^2$. With these input values, (1) gives a PST of 0.003 at an off-axis angle of zero degrees (i.e., on axis along the boresight with $R_{\text{off}}/R_0 = 1$).

Figure 18 shows the measured PST as a function of off-axis angle in the spectral and spatial axes. Our requirement of a PST $<10^{-6}$ at off-axis angles $>7°$ was met in both axes. This requirement improves the viability of high-phase (e.g., night-side) observations of Pluto and Charon during the flyby.

4.4.2 Ly α Scatter

The total integrated scatter (TIS) at Ly α was measured prior to launch using histogram exposures and a monochromatic collimated input light source at 1216 Å. This scatter is due

[1] The wavelength offset value, therefore, shifts towards the blue at a rate of -0.2 Å (deg C)$^{-1}$. The plate scale variation with temperature is negligible and can be ignored.

[2] Volume constraints within the radiometric vacuum chamber limited the input offset angles to these values.

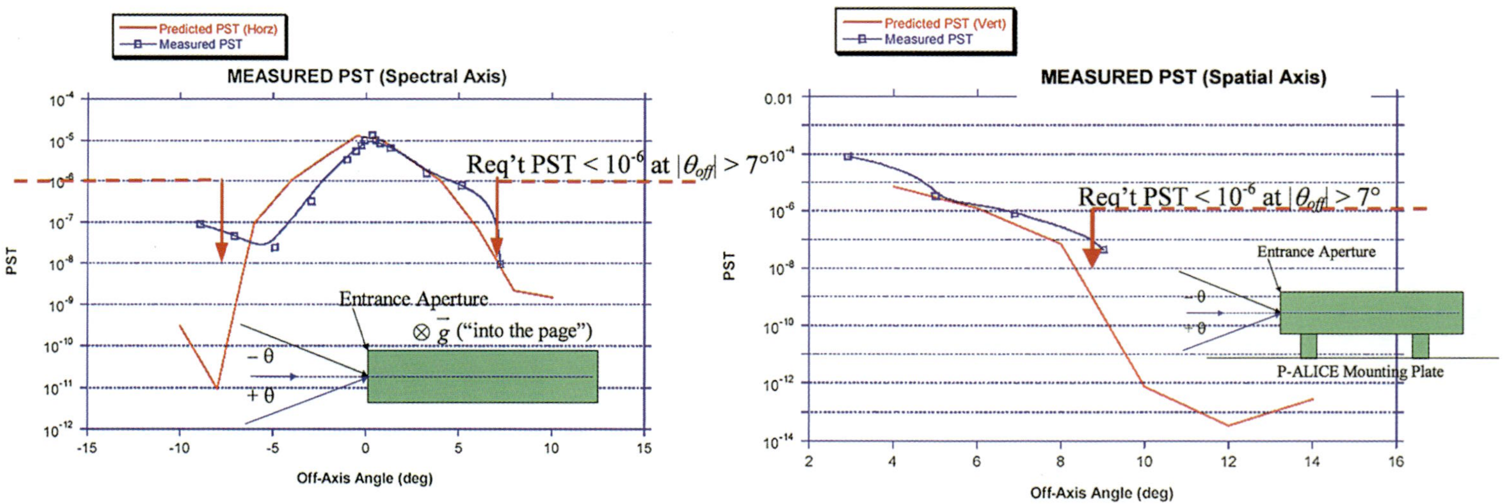

Fig. 18 The measured PST as a function of the input off-axis angle with respect to the airglow boresight in the spectral axis (*left*) and spatial axis (*right*). Both the measured data values (*blue squares*) and the predicted PST (*red solid curve* based on the P-ALICE stray light analysis) are shown along with the specification requirement. The PST values near the 0 degree off-axis angle in the spectral axis are all outside the FOV of the slit in the *horizontal axis* (closest off-axis angle shown is at an off-axis angle of $-0.07°$ where $PST = 1.0 \times 10^{-5}$). The "inset" diagrams define the sign of the input offset angle

primarily to imperfections in the grating surface. The TIS is defined as the ratio of detected counts outside the photocathode (PC) gap to the total counts within the emission line inside the gap (after normalizing the inside rate with the ratio of the QE outside to inside the gap). This scatter can be expressed as follows:

$$TIS = \frac{R_o}{R_t}, \tag{2}$$

where R_o is the rate outside the PC gap; and R_t is the total normalized detector rate (i.e., corrected for electronic deadtime effects). This ratio can be expressed in terms of the two measurable values r_t (the total detector rate) and r_i (the measured rate inside the gap). Equation (2) becomes

$$TIS = \frac{r_t - r_i}{(k-1)r_i + r_t}, \tag{3}$$

where k is the ratio of the photocathode QE to that of the bare MCP glass inside the gap. Using the measured ratio of effective area just outside the PC gap to that inside the gap (this ratio is nearly equivalent to the ratio of QE values), we find that $k = 9.2$. Inserting the above values into (3) gives a TIS for Ly α of 0.024 (2.4%). This result is well within the TIS goal of <10%.

Figures 19 and 20 show, respectively, the 300 sec histogram exposure and spectral profile taken to make this measurement. The Ly α emission line within the PC gap is evident as a peak in the center that drops off strongly to either side within the gap, along with bright

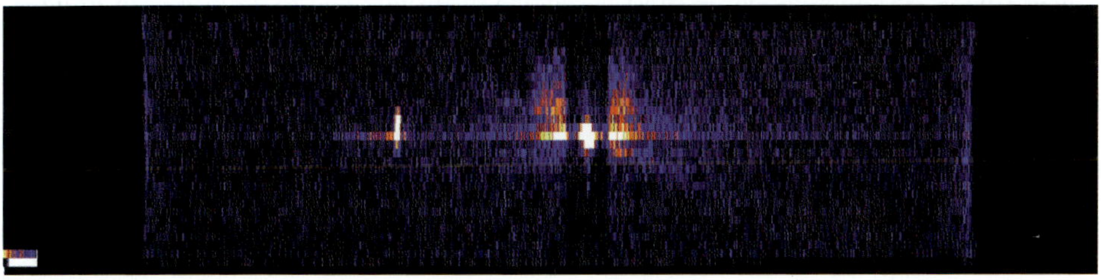

Fig. 19 Histogram exposure of the H Ly α emission. The primary emission is centered in the photocathode gap near the center of the active area; but the wings of the emission and scattered light from the emission are evident outside the gap. The fainter emission to the left of center is the Ly α grating ghost discussed in the text

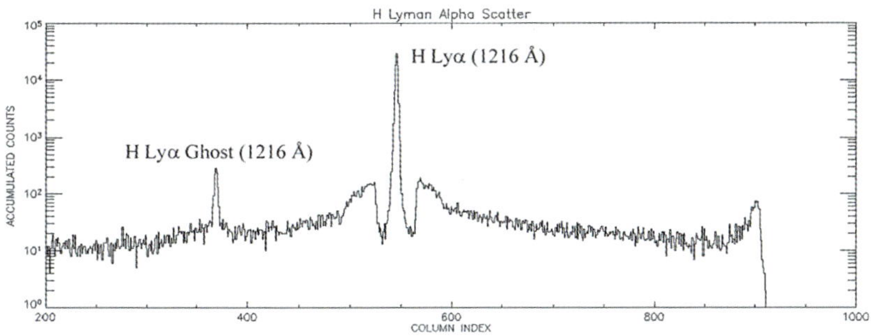

Fig. 20 Spectral profile of the above histogram image showing the H Ly α emission line centered in the PC gap and the Ly α ghost at 893 Å

wings of the emission line protruding on either side of the PC gap, as well as light scatter across the active area. In addition, a ghost image of the Ly α line shows up left of center in the image. This ghost was also noted in the R-ALICE instrument. Ray trace studies of P-ALICE show that this ghost is caused by the reflection of the Ly α emission line off the front surface of the detector MCP back towards the grating, which then re-diffracts the light back onto the detector surface (see Slater et al. 2005). The intensity of this ghost is nearly identical to that measured with R-ALICE. This "narcissistic ghost" (McCandliss et al. 1998) is of different origin than the more commonly known Rowland spectral ghost that arises from periodic errors in the spacing of the grating rulings.

4.5 Absolute Effective Area

4.5.1 Airglow Channel

The effective area of the airglow channel was measured in the laboratory across the P-ALICE passband at discrete wavelengths using emission lines from neon, argon, and a hydrogen/helium gas mix covering a wavelength range of 584–1603 Å. Histogram exposures were made with the chamber's monochromator set to the specific measurement wavelength. The monochromator output slit was set to provide a detector output count rate that did not show any saturation effects in the pulse-height distribution (i.e., too high a local flux on the MCP stack will cause a local gain drop that is evident in the PHD data). Background exposures were also taken for each illuminated exposure to allow for dark subtraction.

The beam flux was measured for each effective area measurement using a NIST calibrated silicon (Si) photodiode. In most cases, the beam flux had to be increased in intensity to allow enough flux for a measurement with the photodiode. However, this increased level flux was too high for a measurement with P-ALICE causing local MCP saturation. An Amptektron channeltron was used to measure the ratio of the beam fluxes at low and high intensity levels to provide a correction factor to the flux measured by the NIST photodiode.

Figure 21 shows the measured effective area as a function of wavelength λ, $A_{\mathrm{eff}}(\lambda)$. The equation used to compute A_{eff} is

$$A_{\mathrm{eff}}(\lambda) = \frac{A_g R(\lambda)}{\Phi(\lambda)}, \tag{4}$$

where A_g is the geometric area of the P-ALICE airglow entrance aperture (16 cm^2); $R(\lambda)$ is the detector count rate (corrected for the detector deadtime and the detector background), and $\Phi(\lambda)$ is the input flux to the instrument based on measurements with the NIST Si photodiode and the channeltron. The red circles in the Figure show the measured effective area values; the green squares show the predicted values based on reflectivity values of the OAP mirror and grating (including the grating diffraction efficiency), and the photocathode QE response measured during subsystem level optics and detector tests. The blue solid line (no data points) is the minimum effective area requirement set for the instrument. As is evident in Fig. 21, the effective area requirement was met over the full bandpass and with significant margin at wavelengths shorter than 1600 Å.

4.5.2 Solar Occultation Channel

The effective area of the SOCC was measured prior to launch, and the relative sensitivity compared with the airglow channel was computed. The airglow channel is 3000–7300 times

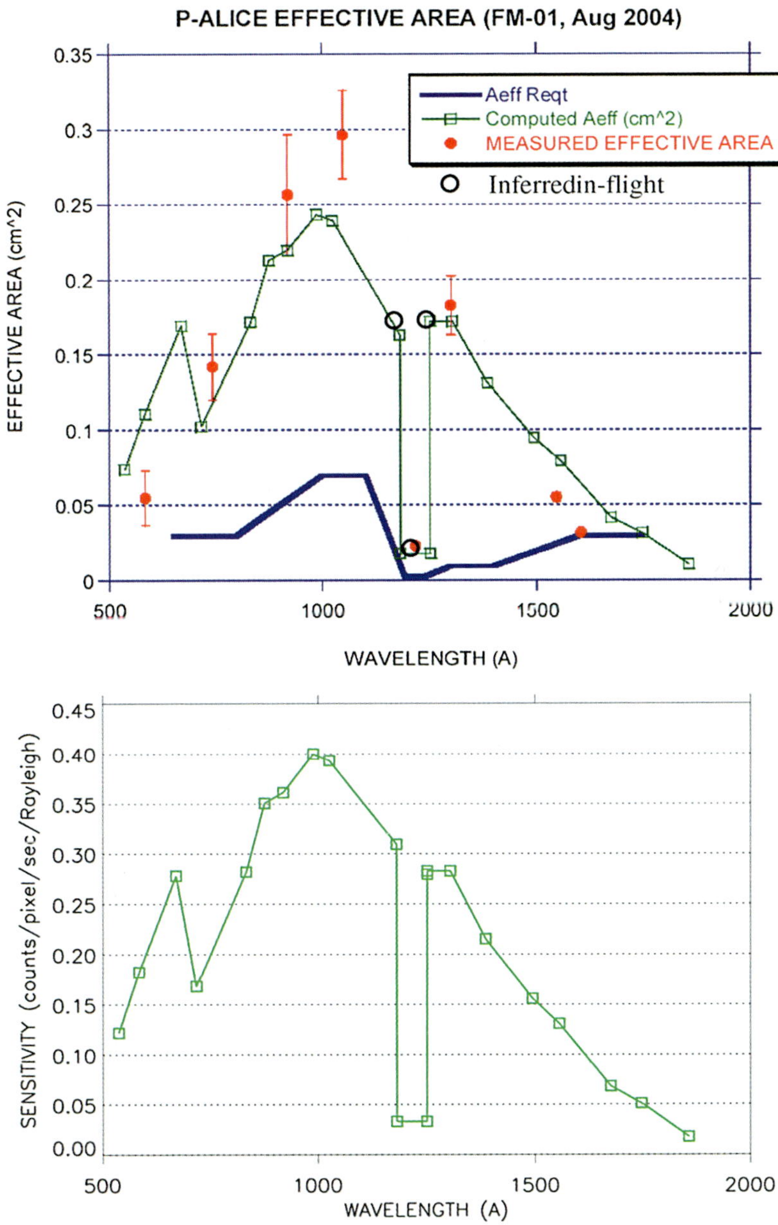

Fig. 21 (*Top*) The ground-measured effective area for the airglow channel (*red solid circles*) plotted versus wavelength. Also shown are the computed effective area estimates (*green open squares*) based on previously measured SiC reflectivity values of the OAP mirror and grating and detector QE values. The effective area specification is shown as the solid blue line in the lower portion of the graph. Superimposed on this plot are the initially inferred in-flight effective area values within and just outside the PC gap (*black open circles*) assuming a model-calculated emission brightness H Ly α skyglow of 200 Rayleigh. (*Bottom*) The physical sensitivity for an extended source based on the computed effective area

Table 1 Summary of P-ALICE ground radiometric test performance

Radiometric test	Test requirement/goal	Ground measured values
Dark count rate	$\leq$1 cm^{-2} sec^{-1} (total array)	0.3 cm^{-2} sec^{-1} (total array)
Wavelength calibration	<1 Å calibration precision	±0.62 Å (1σ using a linear fit)
PSF vs. λ (airglow & SOCC)	<6 Å (SOCC)	3.0–3.5 Å (SOCC)
	<18 Å (airglow)	3.0–4.5 Å (airglow)
	<2 spatial pixels (spatial)	0.4–1.5 spatial pixels (spatial)
Wavelength passband	520–1870 Å (minimum coverage)	465–1881 Å
Filled slit spectral resolution	<18 Å FWHM (airglow)	9.0 ± 1.4 Å (FWHM)
Spectral plate scale	1.7 ± 0.2 Å/pixel	1.832 ± 0.003 Å/pixel
Spatial plate scale	<0.30 per spatial pixel	0.27° ± 0.01°
Off-axis light scatter test (airglow)	PST* $<10^{-6}$ at $\theta_{\mathrm{off}} > 7°$	PST $< 4.6 \times 10^{-8}$ at $\theta_{\mathrm{off}} > 7°$ (spectral axis)
		PST $< 8.3 \times 10^{-7}$ at $\theta_{\mathrm{off}} > 7°$ (spatial axis)
Ly α attenuation/scatter	<10% (total integrated scatter)	TIS = 2.4% (outside PC gap)
Ly α gap λ boundaries	$\Delta\lambda \sim$ 70–75 Å centered at 1216 Å	Gap: 1178.6 Å to 1251.9 Å ($\Delta\lambda$ = 73.3 Å)†
Absolute effective area	650–800 Å: >0.03 cm^2	650–800 Å: 0.07–0.17 cm^2
(airglow aperture)	1000–1100 Å: >0.07 cm^2	1000–1100 Å: 0.20–0.30 cm^2
	1181–1251 Å: 0.003–0.03 cm^2	1181–1251 Å: 0.02 cm^2
	1300–1400 Å: >0.01 cm^2	1300–1400 Å: 0.15–0.18 cm^2
	1600–1750 Å: >0.03 cm^2	1600–1750 Å: 0.03–0.05 cm^2
SOCC relative effective area	<1/2500 of AGC effective area**	1/7300 to 1/3000 of AGC effective area

*PST = Point Source Transmittance = $E_{\mathrm{det}}/E_{\mathrm{in}}$, where E_{det} is the irradiance at the detector; E_{in} is the input irradiance

**Assumes a nominal AGC effective area

†With point source centered in AGC slit; gap pixel boundaries: cols 525.5 to 565.4 at room temperature, +22°C

more sensitive than the SOCC channel (depending on the wavelength), with an average ratio of ~5000. The ratio of airglow to SOCC effective area requirement is set to limit the detector count rate to <30 kHz during the solar occultation observations at Pluto/Charon. This ratio must be >2400 to meet this requirement; this requirement was easily met.

5 In-flight Performance

5.1 Commissioning Test Overview

In-flight commissioning of P-ALICE began in February 2006, about one month after launch of NH. The first set of commissioning tests consisted of the first in-flight power on of the instrument followed by a set of sequences that tested the program memory and command and telemetry interfaces with the spacecraft. In addition, the aperture door launch latch was fired open and the door exercised to verify its function. These early tests while the spacecraft was still potentially outgassing and therefore when the detector risk to arcing was non-trivial did not involve detector high voltage.

A second set of commissioning tests was run in May 2006 that included sequences to test the optics heaters for decontamination control, to open the detector door, and a ramp up of the detector HVPS to full operational voltage. These activities were followed by sequences to obtain the first-light exposures of the Ly α sky glow and exposures of the dark background.

Performance testing continued in March 2006 with a set of commissioning tests that were entirely nominal. These tests occurred with P-ALICE at room temperature ($\sim$20°C); the light-time distance of NH from Earth was between 112 and 115 seconds. The measured power draw of P-ALICE was nominal at $\sim$3.6 W, and all the initial mode commanding and memory/C&DH tests showed proper performance. The heaters on the OAP mirror, grating and SOCC mirrors were activated and showed nominal heating profiles; in addition, the detector stim pixels were activated and exposures taken that verified proper stim pixel locations and stim widths. Finally, the aperture door launch latch was successfully fired open, and the door was exercised to verify operation of the door LAT motor. At the conclusion of these first commissioning tests, P-ALICE was powered off until early May, when another set of tests were run.

The May 2006 set of commissioning tests began with initial power on of the instrument followed by mode checks, stim pixel exposures, and two 24-hour long optical heater decontamination sessions. Performance in all of these tests was nominal. The detector door was successfully opened to allow the interior of the detector housing to outgas to space. Twenty-four hours after opening the detector door, the HVPSs were activated, and tests to verify the on-board safety triggers were successfully run. A noise immunity test was also performed that showed no anomalous behavior of the instrument electronics with HV on. After all HV tests were completed successfully, the HV was powered off, and the heaters were activated for 12 hours for optical decontamination. After completion of these activities, P-ALICE was turned back on and a set of dark and sky glow images were obtained in both pixel list and histogram modes. The detailed results of dark and sky glow exposures are discussed next.

5.2 Dark Count Rate

Two 600-second histogram exposures were taken with the aperture door closed and the detector HV set to -4.5 kV. The detector electronics temperature was 21.9°C and the detector housing temperature was 19.6°C. Figure 22 shows the image histogram of both exposures co-added together; Figure 23 shows the row and column sums of this dark exposure.

The stim pixels were activated during these exposures and are indicated in Figs. 22 and 23. The average detector count rate of the co-added exposure was 98.5 counts per second after subtraction of the stim pixel rate of 18 c/sec. This background rate includes contributions from 1) the intrinsic MCP dark noise (typically 1–5 c/sec as measured during ground vacuum tests); 2) atomic particles in the local environment (protons, electrons, other ions); and 3) radiation from the spacecraft's RTG (neutrons and gamma-rays). Assuming a similar combined MCP intrinsic dark rate and ambient space background rate to that observed with Rosetta-ALICE of $\sim$ 20 c/sec, the contribution from the RTG is $\sim$80 c/sec, due primarily to gamma-rays. This RTG background rate is within predicted values based on previous RTG exposure calculations and tests with the P-ALICE detector during the NH integration and test program.

Figure 24 shows the recorded PHD of the background exposures obtained in May 2006. The background events contribute to the negative exponential shape of the curve—this is

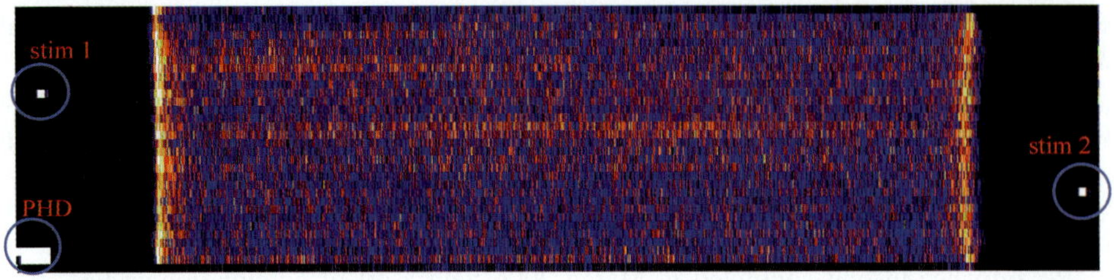

Fig. 22 Detector background (*dark*) histogram image made with the P-ALICE aperture door shut after the detector door was opened. The two stim pixel locations and the PHD data are highlighted in the image. The higher row amplitudes are due primarily to DNL electronic noise caused by the high gain pulse contribution in the dark background pulse-height distribution

the expected shape caused by gamma-ray events being converted to electrons throughout the entire thickness of the MCP stack.

5.3 First Light Images

At the conclusion of the background exposures, P-ALICE was commanded to take histogram and pixel-list exposures of the Ly α skyglow. P-ALICE's airglow boresight was centered at coordinates (J2000) RA = 107.150° (7h 8m 36s) and DEC = +22.205° (+22° 12′ 18″), which is near the center of the constellation Gemini. The ecliptic coordinates of this nominal boresight RA, DEC position is longitude = 105.84°; latitude = −0.3°. Three 600-second histogram exposures and a 60-second pixel list exposure were taken with the detector HV set at −4.5 kV, and the aperture door in the open state.

5.3.1 Image Histogram Exposures

Figure 25 shows an image of the three initial sky histogram exposures co-added together. The Ly α emission line is the strongest in the spectrum, and the exposure shows a nice image of the slit aperture. Also note that Ly β at 1026 Å, Ly γ at 972 Å, and He I at 584 Å are seen, along with the known Ly α "ghost" at 893 Å. Counts at wavelengths >1216 Å are due primarily to scattered Ly α light off the grating (see Slater et al. 2005), and possibly of a faint UV star within the SOCC field-of-view (FOV) (see Fig. 25). Figure 26 shows the emission line profiles from the three co-added exposures with the background removed within the narrow airglow slit region only (co-addition of rows 5–17). The best filled-slit spectral resolution occurs in rows 15–17 (where the P-ALICE boresight falls), with FWHM values in the range of 10.0–10.5 Å. This closely matches with the measured value of 9.0 ± 1.4 Å measured during ground radiometric test (see Slater et al. 2005). Note also the scattered light plateau of ∼100 counts in Fig. 26 from ∼900–1800 Å. This translates to a scattered light contribution of ∼0.006 c/s per pixel in this region of the focal plane (or ∼1% of the total flux from the skyglow flux illuminating the focal plane).

The count rates in both the narrow and wide slit segments can be used to estimate the effective area provided we know the brightness of the emission. For the Ly α emission, the estimated Ly α sky brightness in the direction of the P-ALICE boresight was very roughly 200 Rayleighs (with an estimated error of ±30%). This emission line fills both the 2° × 2° SOCC FOV (top rectangular region in Fig. 25), and the 0.1° × 4° narrow airglow FOV. At Ly α, the narrow slit illuminates only the photocathode gap region which has a detective quantum efficiency one-tenth that of each cathode on either side of the gap. The SOCC

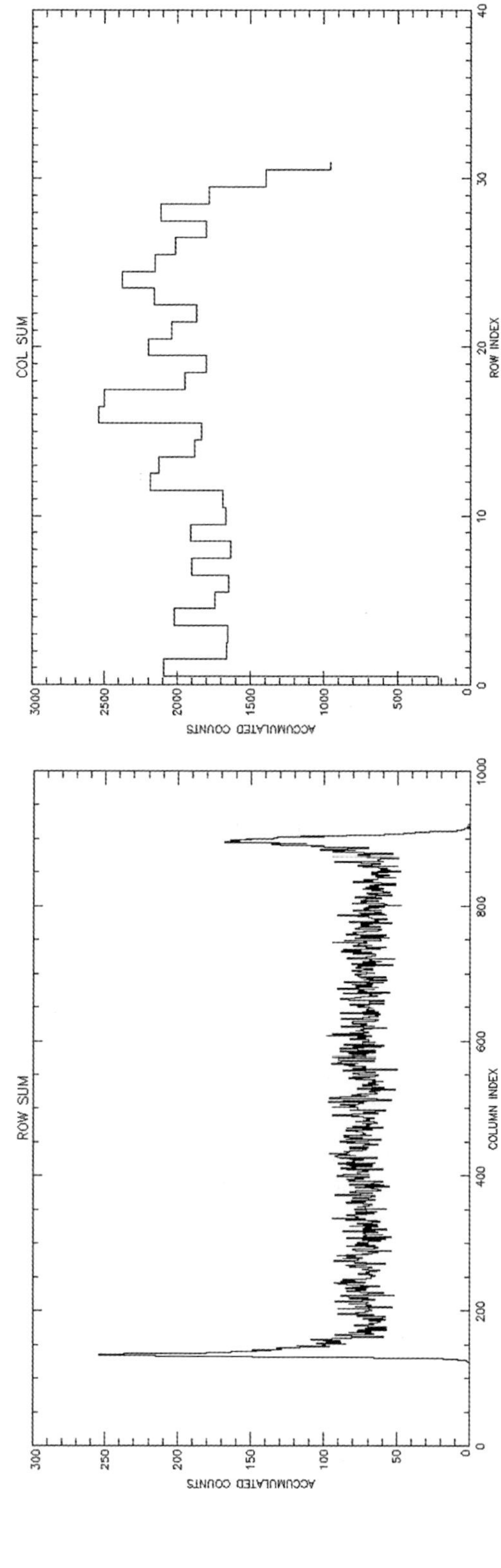

Fig. 23 The row (*left*) and column (*right*) sum of events of the first background exposure

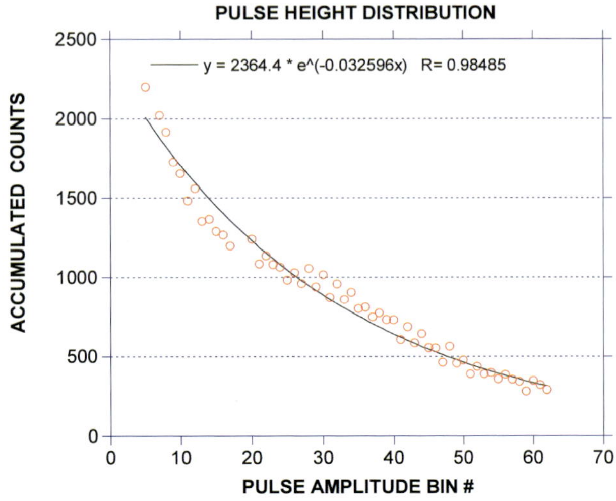

Fig. 24 The PHD of the dark background histogram exposures showing its negative exponential shape. The *black solid line* through the data points is a negative exponential fit to the data

FOV, on the other hand, fills both the 70-Å wide gap and ~110 Å of photocathode split nearly evenly on either side of the gap. This is clearly evident in the histogram image shown in Fig. 25 (top), which shows the two bright emission features on either side of the gap in the top portion of the slit image, and the fainter filled gap region along the stem of the slit image. Note that the Ly α emission is slightly "blueward" of the gap center by approximately 7 Å. This is within the measured preflight value of 3 ± 5 Å blueward of the gap center (the uncertainty here is based on the exact position of the point source within the 0.1° wide airglow gap).

For Ly α, the count rate in the combined background-subtracted image histogram within the SOCC 2° × 2° slit was 2220 c/sec, and within the narrow 0.1° × 4.0° slit it was 45 c/sec. This translates to an effective area in the gap region of ~0.02 cm^2, and an effective area on either side of the gap of 0.17 cm^2. Figure 21 shows the effective area as a function of wavelength measured during ground test; note that the new in-flight measured values are close to the ground measurements. Stellar calibrations taken later provided more exact in-flight effective area values at a variety of wavelengths, but these early measurements provided strong encouragement at the time that the instrument was operating to specifications in space.

Table 2 lists the inferred brightness of the Ly β and He I 584 Å emissions assuming the ground-measured effective area values shown in Fig. 21. Note that both emissions are in the 1.0–1.5 Rayleigh brightness level (for He I, this emission strength is comparable to measurements by EUVE—e.g., see Flynn et al. 1998). The brightness of the Ly γ emission was difficult to estimate because of its low SNR of its profile.

5.3.2 Pixel List Mode Test Exposures

A 61-second pixel list integration was taken of the skyglow after completion of the three histogram skyglow exposures discussed above. The time hack interval was 0.004 seconds; the discriminator voltage was 0.5 V. The average digital count rate was 2608 c/sec (including stims, which were on) with a standard deviation of 51 c/sec.

Figure 27 shows a histogram of the number of events per time hack interval, along with a Gaussian fit to the data. The excellent fit to the data (reduced $\chi^2 = 4.7$) indicates Poisson

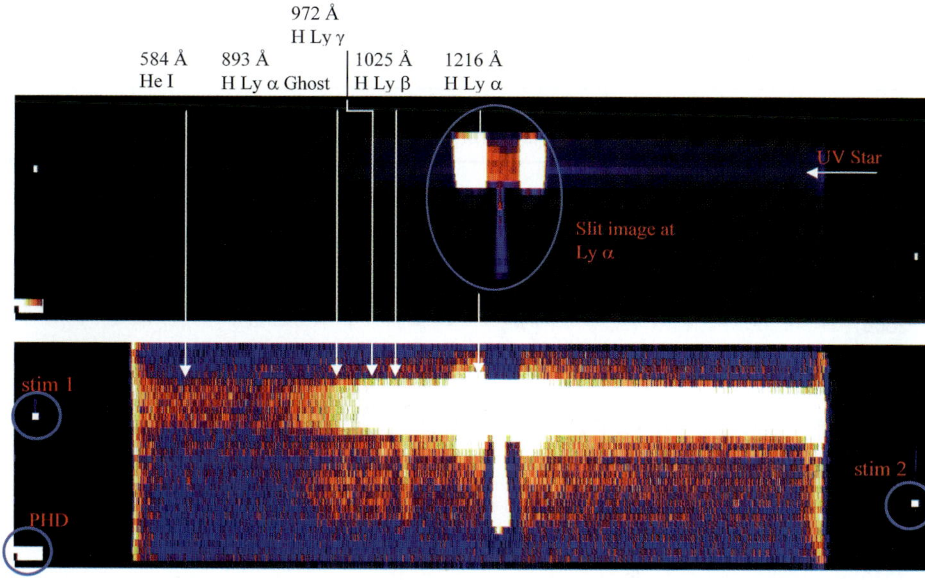

Fig. 25 The Ly α skyglow "first-light" image histogram made up by the co-addition of the three separate 600-second exposures. (*Top*) Image lookup table set to show off the Ly α emission at 1216 Å. A possible UV star's continuum spectrum also appears in the lower portion of the SOCC FOV. (*Bottom*) A stretched lookup table that shows the fainter emissions at 1026 Å (Ly β); 972 Å (Ly γ); 584 Å (He I); and the Ly α ghost at 893 Å. The two stim pixels and PHD data are also shown. Note that image is distinctly wider at the lower end of the slit; this is due to off-axis coma from the OAP mirror (just below the SOCC opening the coma is near zero since it is nearly on-axis light)

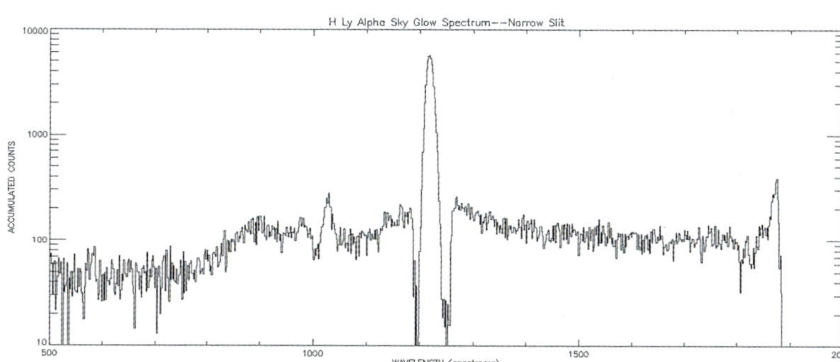

Fig. 26 Semi-log plot of the background-subtracted accumulated counts as a function of wavelength accumulated within the narrow airglow slit showing the emissions from Ly α (1216 Å); Ly β (1026 Å); Ly γ (972 Å); and He I (584 Å). Note that the Ly α emission is within the photocathode gap; hence, the much lower background events immediately surrounding the emission line

counting statistics dominate (as expected for photon signals), and that there are no major contributions in the dataset that are non-Poisson in nature, as can sometimes result from hot emission points from the MCP stack.

Table 2 Emission brightnesses from the skyglow exposures assuming the ground-based effective area values. The count rate in column 4 is the net rate after subtracting the dark background over the entire slit FOV (including both the narrow and wide slit segments)

Emission line		Effective area (cm^2)	Count rate (c/s)	Brightness (Rayleighs)
H Ly-beta	1026 A	0.25	41	1.5
He I	584 A	0.06	6	1.0

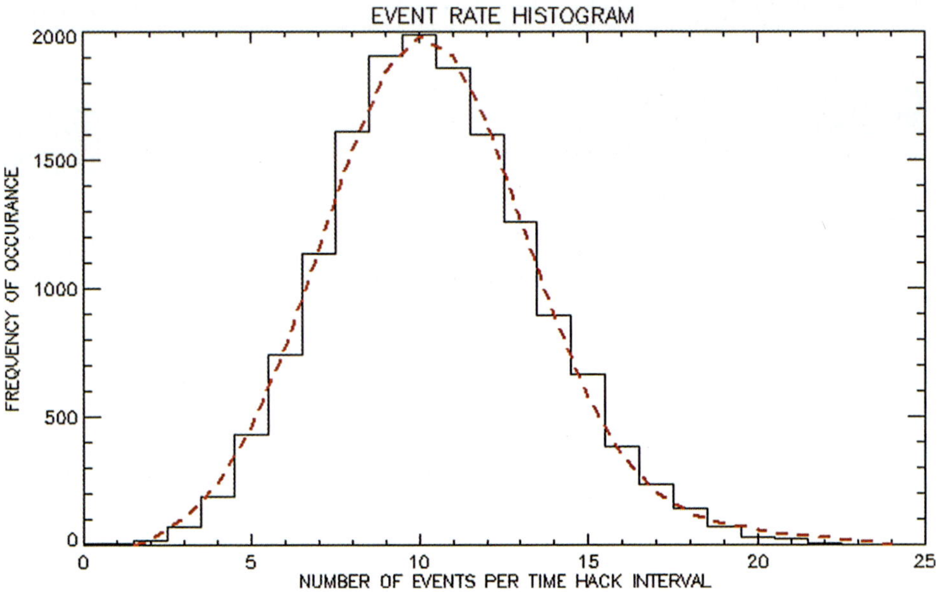

Fig. 27 The event rate histogram data (*solid line*). The *red-dashed curve* is a Poisson fit to the histogram data

6 Conclusion

The New Horizons P-ALICE UV spectrometer has been described above. It was successfully launched on 19 January 2006 and is operating normally in space. All in of the early flight performance tests indicate performance within specification; more detailed testing and calibration observations performed in late 2007 and annual checkouts will be discussed in a formal calibration paper to be published in the future.

Acknowledgements We thank the entire P-ALICE engineering and support team at SwRI, as well as John Vallerga and Rick Raffanti of our detector supplier at Sensor Sciences, and our French collaborator Jean-Loup Bertaux for providing optics. We also thank the New Horiozns spacecraft and mission teams. This work was supported by NASA contract #NASW02008.

References

R.J.C. Brown, P.J. Brewer, M.J.T. Milton, The physical and chemical properties of electroless nickel-phosphorus black surfaces. J. Mater. Chem. **12**, 2749–2754 (2002)

K. Danzmann, M. Günther, J. Fischer, M. Kock, M. Kühne, High current hollow cathode as a radiometric transfer standard source for the extreme vacuum ultraviolet. Appl. Opt. **27**, 4947–4951 (1988)

J.L. Elliot, E.W. Dunham, A.S. Bosh, S.M. Slivan, L.A. Young, L.H. Wasserman, Pluto's atmosphere. Icarus **77**, 148 (1989)

Fountain et al., Space Sci. Rev. (2008, this issue). doi:10.1007/s11214-008-9374-8

B. Flynn, J. Vallerga, F. Dalaudier, G.R. Gladstone, EUVE measurement of the local interstellar wind and geocorona via resonance scattering of solar He I 584-Å line emission. J. Geophys. Res. **103**, A4 (1998)

P. Jelinsky, S. Jelinsky, Low reflectance EUV materials: a comparative study. Appl. Opt. **26**(4), 613–615 (1987)

V.A. Krasnopolsky, D.P. Cruikshank, Photochemistry of Pluto's atmosphere and ionosphere near perihelion. J. Geophys. Res. **104**(E9), 21,979–21,996 (1999)

S.R. McCandliss, J.B. McPhate, P.D. Feldman, Narcissistic ghosts in Rowland-mounted, concave gratings with $\upsilon = 0°$: a cautionary note. Appl. Opt. **37**, 5070 (1998)

K.A. Moldosanov, M.A. Samsonov, L.S. Kim, R. Henneck, O.H.W. Siegmund, J. Warren, S. Cully, D. Marsh, Highly absorptive coating for the vacuum ultraviolet range. Appl. Opt. **37**(1), 93–97 (1998)

J.F. Osantowski, R.A.M. Keski-Kuha, H. Herzig, A.R. Toft, J.S. Gum, C.M. Fleetwood, Optical coating technology for the EUV. Adv. Space Res. **11**(11), 185–201 (1991)

O.H.W. Siegmund, Microchannel plate imaging detector technologies for UV instruments, in Conference Proceedings *From X-rays to X-band—Space Astrophysics Detectors and Detector Technologies*, Space Telescope Science Institute, 2000

O.H.W. Siegmund, J. Stock, R. Raffanti, D. Marsh, M. Lampton, Planar delay line readouts for high resolution astronomical EUV/UV spectroscopy, in *UV and X-Ray Spectroscopy of Astrophysical and Laboratory Plasmas*, Proceedings from the 10th International Colloquium, Berkeley, CA, 3–5 February 1992, pp. 383–386

O.H. Siegmund, J.M. Stock, D.R. Marsh, M.A. Gummin, R. Raffanti, J. Hull, G.A. Gaines, B.Y. Welsh, B. Donakowski, P.N. Jelinsky, T. Sasseen, J.L. Tom, B. Higgins, T. Magoncelli, J.W. Hamilton, S.J. Battel, A.I. Poland, M.D. Jhabvala, K. Sizemore, J. Shannon, Delay-line detectors for the UVCS and SUMER instruments on the SOHO Satellite, in *EUV, X-Ray, and Gamma-Ray Instrumentation for Astronomy V*, ed. by O.H.W. Siegmund, J.V. Vallerga, Proceedings of SPIE, vol. 2280 (1994), pp. 89–100

D.C. Slater, S.A. Stern, T. Booker, J. Scherrer, M.F. A'Hearn, J.L. Bertaux, P.D. Feldman, M.C. Festou, O.H.W. Siegmund, Radiometric and calibration performance results for the Rosetta UV imaging spectrometer *ALICE*, in *UV/EUV and Visible Space Instrumentation for Astronomy and Solar Physics*, ed. by O.H.W. Siegmund, S. Fineschi, M.A. Gummin, Proceedings of SPIE, vol. 4498 (2001), pp. 239–247

D.C. Slater, M.W. Davis, C.B. Olkin, J. Scherrer, S. Alan Stern, Radiometric performance results of the *New Horizons' ALICE* UV imaging spectrograph, in *X-Ray, UV, Visible, and IR Instrumentation for Planetary Missions*, ed. by O.H.W. Siegmund, G. Randall Gladstone, Proceedings of SPIE, vol. 5906B, 2005

J.R. Spencer, J.A. Stansberry, L.M. Trafton, E.F. Young, R.P. Binzel, S.K. Croft, Volatile transport, seasonal cycles, and atmospheric dynamics on Pluto, in *Pluto and Charon*, ed. by S. Alan Stern, David J. Tholen (University of Arizona Press, Tucson, 1997), p. 435

S.A. Stern, D.C. Slater, W. Gibson, H.J. Reitsema, A. Delamere, D.E. Jennings, D.C. Reuter, J.T. Clarke, C.C. Porco, E.M. Shoemaker, J.R. Spencer, The highly integrated Pluto payload system (HIPPS): A sciencecraft instrument for the Pluto mission, in *EUV, X-Ray, and Gamma-Ray Instrumentation for Astronomy VI*, ed. by O.H.W. Siegmund, J. Vallerga, Proceedings of SPIE, vol. 2518 (1995), pp. 39–58

S.A. Stern, Space Sci. Rev. (2008, this issue). doi:10.1007/s11214-007-9295-y

L.M. Trafton, D.M. Hunten, K.J. Zahnle, R.L. McNutt Jr., Escape processes at Pluto and Charon, in *Pluto and Charon*, ed. by S. Alan Stern, David J. Tholen (University of Arizona Press, Tucson, 1997), p. 475

G.L. Tyler, I.R. Linscott, M.K. Bird, D.P. Hinson, D.F. Strobel, M. Pätzold, M.E. Summers, K. Sivaramakrishnan, The New Horizons radio science experiment (REX). Space Sci. Rev. (2008, in press)

Young et al., Space Sci. Rev. (2008, this issue)

Long-Range Reconnaissance Imager on New Horizons

A.F. Cheng · H.A. Weaver · S.J. Conard · M.F. Morgan · O. Barnouin-Jha · J.D. Boldt · K.A. Cooper · E.H. Darlington · M.P. Grey · J.R. Hayes · K.E. Kosakowski · T. Magee · E. Rossano · D. Sampath · C. Schlemm · H.W. Taylor

Originally published in the journal Space Science Reviews, Volume 140, Nos 1–4, 189–215.
DOI: 10.1007/s11214-007-9271-6

Abstract The LOng-Range Reconnaissance Imager (LORRI) is the high-resolution imaging instrument for the New Horizons mission to Pluto, its giant satellite Charon, its small moons Nix and Hydra, and the Kuiper Belt, which is the vast region of icy bodies extending roughly from Neptune's orbit out to 50 astronomical units (AU). New Horizons launched on January 19, 2006, as the inaugural mission in NASA's New Frontiers program. LORRI is a narrow-angle (field of view= 0.29°), high-resolution (4.95 μrad pixels), Ritchey-Chrétien telescope with a 20.8-cm diameter primary mirror, a focal length of 263 cm, and a three-lens, field-flattening assembly. A 1,024 × 1,024 pixel (optically active region), thinned, backside-illuminated charge-coupled device (CCD) detector is used in the focal plane unit and is operated in frame-transfer mode. LORRI provides panchromatic imaging over a bandpass that extends approximately from 350 nm to 850 nm. LORRI operates in an extreme thermal environment, situated inside the warm spacecraft with a large, open aperture viewing cold space. LORRI has a silicon carbide optical system, designed to maintain focus over the operating temperature range without a focus adjustment mechanism. Moreover, the spacecraft is thruster-stabilized without reaction wheels, placing stringent limits on the available exposure time and the optical throughput needed to satisfy the measurement requirements.

Keywords Pluto · Jupiter · Satellites · Telescopes · Calibration · Imaging science

1 Introduction

The New Horizons mission launched on January 19, 2006, on its way to perform the first reconnaissance of the Pluto–Charon system and the Kuiper Belt. First, however, New Hori-

A.F. Cheng (✉) · H.A. Weaver · S.J. Conard · M.F. Morgan · O. Barnouin-Jha · J.D. Boldt · K.A. Cooper · E.H. Darlington · M.P. Grey · J.R. Hayes · T. Magee · E. Rossano · C. Schlemm · H.W. Taylor
The Johns Hopkins University Applied Physics Laboratory, 11100 Johns Hopkins Road, Laurel, MD 20723, USA
e-mail: andrew.cheng@jhuapl.edu

K.E. Kosakowski · D. Sampath
SSG Precision Optronics, Wilmington, MA, USA

Table 1 Pluto and Charon prioritized measurement objectives

Group	Goal	LORRI Contribution
1.1	Characterize the global geology and morphology of Pluto and Charon	Hemispheric panchromatic maps of Pluto and Charon at best resolution exceeding 0.5 km/pixel
1.2	Map surface composition of Pluto and Charon	
1.3	Characterize the neutral atmosphere of Pluto and its escape rate	Search for atmospheric haze at a vertical resolution <5 km
2.1	Characterize the time variability of Pluto's surface and atmosphere	Long time base of observations, extending over 10 to 12 Pluto rotations; panchromatic maps of the far-side hemisphere
2.2	Image Pluto and Charon in stereo	Panchromatic stereo images of Pluto and Charon
2.3	Map the terminators of Pluto and Charon with high resolution	High resolution panchromatic maps of the terminator region
2.4	Map the surface composition of selected areas of Pluto and Charon with high resolution	
2.5	Characterize Pluto's ionosphere and solar wind interaction	
2.6	Search for neutral species including H, H_2, HCN, and C_xH_y, and other hydrocarbons and nitriles in Pluto's upper atmosphere, and obtain isotopic discrimination where possible	
2.7	Search for an atmosphere around Charon	
2.8	Determine bolometric Bond albedos for Pluto and Charon	Panchromatic, wide phase angle coverage of Pluto and Charon
2.9	Map the surface temperatures of Pluto and Charon	
3.1	Characterize the energetic particle environment of Pluto and Charon	
3.2	Refine bulk parameters (radii, masses, densities) and orbits of Pluto and Charon	Orbital parameters, bulk parameters of Pluto and Charon
3.3	Search for magnetic fields of Pluto and Charon	
3.4	Search for additional satellites and rings	Search for satellites and rings; refine orbits, sizes, shapes of Nix and Hydra

zons made a Jupiter swingby with closest approach on February 28, 2007. Extensive observations of the Jovian atmosphere, rings, and satellites were acquired. New Horizons' closest approach to Pluto will occur on July 14, 2015. The prioritized measurement objectives of the New Horizons mission, and the contributions from LOng-Range Reconnaissance Imager (LORRI) observations, are summarized in Table 1.

Pluto is an icy dwarf planet (Cheng and Young 2002) with a significant atmosphere consisting mainly of nitrogen, CO, and methane. High-resolution images from LORRI will yield important information on Pluto's geology and surface morphology, collisional history, and atmosphere–surface interactions. Will Pluto have a young surface, with evidence of endogenic activity like plumes or cryovolcanism? Will there be evidence for tectonism in the form of faulting or ridge and groove formations? Will there be layered terrains? Will there be evidence for atmospheric hazes or for surface winds forming dunes (dunes on Pluto may be mostly grains of nitrogen ice)? Pluto is known to have an active surface, with changes in surface colors and reflectances observed by Earth-based telescopic monitoring. LORRI's high-resolution images will reveal features as small as 100 m on Pluto (260 m on Charon).

Charon is Pluto's giant satellite: at about half the size of Pluto, it is larger than any other planetary satellite relative to its primary. Unlike Pluto, Charon has no detectable atmosphere, and it probably has an old surface that may preserve a cratering record from collisional evolution within the Kuiper Belt. LORRI data will play a critical role in determining the crater size distribution and morphologies on Charon. Equally important, LORRI images will provide precise measurements of the shapes and sizes of both bodies. In addition, LORRI will obtain high-resolution images of the newly discovered moons Nix and Hydra (Weaver et al. 2006), comparable in terms of resolution elements across the illuminated disk to the Galileo images of Gaspra.

After the Pluto–Charon encounter, New Horizons will make the first visit to one or more Kuiper Belt objects. Owing to the likely small size of these targets, LORRI's high resolution is especially important to capture as much surface detail as possible. Will these Kuiper Belt objects look like the asteroid Eros, or will there be bizarre surface features like the flat-floored, steep-walled depressions (craters?) found on the nucleus of comet 81P/Wild 2?

The New Horizons mission has a long focal length, narrow-angle imager for several reasons. Pluto is a dwarf planet, and New Horizons flies by quickly, so the encounter science observations occur within one Earth day. With LORRI, however, New Horizons will be able to image the Pluto system at higher resolution than any Earth based telescope can (even the Hubble Space Telescope, or its successor in 2015) for 90 days prior to encounter. These images will provide an extended time base of observations, for studies of the shapes, rotations, and mutual orbits of both Pluto and Charon, and for characterizing surface changes.

Moreover, Pluto and Charon both rotate at the same rate as for their 6.38 day mutual orbit, always keeping the same faces towards each other. Hence, during the near encounter, which lasts less than an Earth day, only one hemisphere of each body—that which faces New Horizons—can be studied at the highest resolution. The opposite faces of both Pluto and Charon are last seen some three days earlier, when the spacecraft is still ∼4 million km away. Despite this distance, LORRI will obtain images with 40-km resolution. These will be the best images of the portions of Pluto and Charon which are not visible during the near encounter period.

Finally, we have not yet discovered the Kuiper Belt object(s) to which New Horizons will be targeted after the Pluto–Charon encounter, and extensive Earth-based observing campaigns are searching for potential targets. However, even after discovery, the heliocentric orbits of the targets cannot be measured with sufficient accuracy from Earth to enable the New Horizons spacecraft to fly to them, unless the targets are also observed directly from the spacecraft. The direction in which the target is seen from the spacecraft is then used to steer the spacecraft to the target by optical navigation. LORRI is expected to play a key role, by making the first and highest resolution detections of the Kuiper Belt target object from New Horizons, more than 40 days before closest approach.

The New Horizons instrument payload includes three imaging investigations: LORRI, the focus of this manuscript; Alice, an ultraviolet imaging spectrometer; and Ralph, a visible imager and infrared imaging spectrometer. These optical instruments are all approximately coaligned to view a common boresight direction, and the spacecraft will maneuver as required to provide pointing during the various planetary encounters. The Ralph instrument includes a multispectral imaging channel (MVIC), which is a four-color, medium-angle imager with a field-of-view (FOV) 5.7° wide, scanned by time-delay integration. LORRI provides complementary imaging data, as a panchromatic, narrow angle (field-of-view 0.29°) framing camera.

2 LORRI Requirements

LORRI is a long focal length imager (see Fig. 1) designed for high resolution and responsivity at visible wavelengths. LORRI will perform its primary measurements while New Horizons approaches Pluto and its satellite Charon in July, 2015, obtaining images from a closest approach distance of 11,100 kilometers to Pluto, at a flyby speed of 13.77 kilometers per second. New Horizons performed a Jupiter flyby, with closest approach on February 28, 2007. During the Jupiter system flyby, LORRI imaged the atmosphere of Jupiter, its ring system, and several of its satellites. Finally, after the Pluto and Charon encounter, the New Horizons spacecraft will be targeted to encounter a Kuiper Belt object (KBO), at which LORRI will again obtain high-resolution images.

LORRI is required to obtain high-resolution, monochromatic images under low light conditions. Other instruments on New Horizons have multispectral and hyperspectral capabilities. At Pluto encounter, 33 AU from the Sun, the illumination level is $\sim 1/1{,}000$ that at Earth, but Pluto is an unusually bright object with a visible albedo of ~ 0.55. At the Kuiper Belt object, likely to be encountered outside 40 AU from the Sun, the illumination will be still lower, and the object will be darker, with an albedo of typically ~ 0.1.

The main objectives of LORRI are: (1) to obtain high-resolution images of Pluto and Charon during the approach phase, including the hemisphere that will not be observed during closest approach; (2) to take images at closest approach with instantaneous field of view (IFOV) of approximately 50 m per detector element; and (3) to obtain optical navigation images required to support trajectory corrections. LORRI's reflective telescope is a Ritchey-Chrétien design, with a field-flattening lens group near the focal plane. The digital image is captured by a frame transfer CCD detector. The FOV is $0.29^\circ \times 0.29^\circ$; the $1{,}024 \times 1{,}024$ square detector elements have a pixel resolution (or IFOV) of 4.94 μrad. A summary of the characteristics of the LORRI instrument appears in Table 2.

The LORRI instrument is mounted inside the New Horizons spacecraft, the interior of which is designed to remain near room temperature. As the telescope views cold space, and the CCD is designed to operate below -70°C, the thermal implementation of the system was a challenge. In order to maintain optical performance, a material with high thermal conductivity and low coefficient of thermal expansion (CTE) was required. As a result, the LORRI optical telescope assembly (OTA) has both primary and secondary mirrors and a metering structure fabricated from silicon-impregnated silicon carbide (see Fig. 1).

Fig. 1 (*Left*) LORRI telescope assembly, showing SiC mirrors and metering structure; (*right*) LORRI composite baffle and flexure mount on test stand

Table 2 Summary of LORRI characteristics

Visible Panchromatic Imager
Telescope Aperture 208 mm
Focal Length 2630 mm
Passband 0.35–0.85 μm
Field-of-view 0.29° × 0.29°
Instantaneous field-of-view 4.95 μrad
Back-thinned, frame transfer CCD
Nominal exposure times 50–200 ms
On-chip 4 × 4 pixel binning available
Autoexposure

Fig. 2 LORRI mounted within New Horizons spacecraft, with outer panels of spacecraft removed. LORRI is wrapped in thermal blanketing, cover door closed

During the approach to Pluto, which occurs under high sun conditions (small phase angle or Sun–Pluto–spacecraft angle), LORRI is required to image the surface of Pluto at signal-to-noise ratio (SNR) > 100 in single frames. The encounter geometry is such that near closest approach to Pluto, where the highest resolution images would be obtained, LORRI views regions near the terminator under low sun conditions and still less illumination; here LORRI has the goal of imaging at SNR > 20 in single frames. Likewise, at the Kuiper Belt object LORRI has an SNR goal > 20.

The resolution requirement near Pluto closest approach is for LORRI to resolve 100 meters per line pair at a distance of 10,000 km from the surface. An IFOV required to be <5 μrad (Table 3) was derived, with a modulation transfer function (MTF) goal of 0.05 for spatial modulation at 1 cycle per 10 μrad. LORRI is designed to meet imaging requirements not only at nominal operating temperature (as low as approximately −100°C for the telescope), but also at room temperature.

For optical navigation, LORRI is required to be able to image a star of visual magnitude $V = 11.5$ at SNR > 7 in a single 100 ms exposure, with full width at half maximum (FWHM) > 1 pixel. It is not desirable for too great a fraction of the energy from a point source to be imaged onto a single pixel, because stellar images become too undersampled. LORRI has a 4 × 4 pixel binning mode, for which its limiting magnitude requirement is

Table 3 LORRI measurement requirements at Pluto

Resolution	Resolve 100 m per line pair at 10,000 km
Derived requirement	Map full illuminated disk of Pluto at better than 1 km per pixel resolution with a 3×3 mosaic
Derived requirement	IFOV < 5 μrad with FOV 0.29°; image frame pixel format $1,024 \times 1,024$
Signal-to-noise ratio	Achieve SNR > 100, single pixel, single exposure at Pluto during approach, for albedo 0.55 at low phase angle
Derived requirement	Achieve SNR > 20 (goal), single pixel, single exposure at Pluto, at 110° phase angle near terminator
Optical navigation	Achieve SNR > 7 for star of visual magnitude 11.5 with a single exposure of 100 ms
Derived requirement	Achieve stellar limiting magnitude > 17 for 9.9 second exposure in 4×4 binned mode for KBO search at SNR > 7

Table 4 LORRI planned observations of the Pluto system

Approach imaging starts c/a-90 days, Pluto and Charon already resolved, Nix and Hydra detectable (refine orbits)
Through at least ten full orbits ending c/a-14 days, both Pluto and Charon can be acquired in single frame or 2×1 mosaic together with background stars
At c/a-14 days, Pluto subtends 28 pixels
By c/a-7 days Pluto subtends 57 pixels, both Pluto and Charon can be acquired with 3×1 image strip
At c/a-3.2 days, Pluto subtends 123 pixels, imaging far side of Pluto (unseen side during c/a)
At c/a-10 hours, last full frame image of Pluto, 2.5 km/pixel at nadir
Near encounter Pluto observations—3×3 mosaics of full disc better than 1 km/px; image strips at progressively finer resolutions to 110 m/px; terminator imaging sequence with 50 m/px resolution (goal)
Near encounter Charon observations—full frame illuminated disk images; 3×3 mosaics of full disc better than 0.5 km/px; terminator observation 130 m/px
Near encounter observations of Nix and Hydra, better than 200 m/px resolution (goal)

$V > 17$ in a single exposure of 9.9 s. This 4×4 pixel-binning mode will be used to search for the target KBO and to perform optical navigation on approach. A special spacecraft guidance mode is available for the KBO search in which the spacecraft will hold the target within the 4×4 pixel pointing tolerance for ten-second exposures. At 40 AU from the Sun, LORRI is predicted to be able to detect a 50 km diameter object, of albedo 0.04 and at phase angle 25°, from a distance of 0.35 AU, more than 40 days before the object would be encountered. This is ample time for targeting of the spacecraft.

Observations of the Pluto system begin at least 90 days prior to the July 2015 encounter, with both Pluto and Charon already resolved (see Table 4). Initial observations are planned to refine the orbits of Pluto, Charon, and the two newly discovered moons Nix and Hydra. Pluto and Charon can be imaged in single frames or in 2×1 mosaics through at least ten full orbits (orbit period 6.38 days) ending about 14 days before Pluto closest approach (c/a), to refine orbital ephemeredes and especially the eccentricity. In the last week prior to c/a, searches for librations of Pluto and Charon are performed, where Pluto subtends 123 pixels one half rotation before c/a. Nix and Hydra are expected to be detectable in

Table 5 LORRI observations of the Jovian system

Global imaging of Jupiter atmosphere, full rotation sequences, during approach
Near encounter imaging of Jovian cloud and storm dynamics near c/a, best resolution 12 km/pixel
Global imaging of Io, illuminated portion, best resolution 12 km/pixel
Io plume inventory (plumes higher than 60 km)
Io nightside imaging from eclipse; Io hot spots and auroral emissions
Global imaging of Europa; map broad, regional-scale arctuate troughs, best resolution 15 km/pixel
Map Europa nightside auroral emissions in eclipse
Map Ganymede nightside auroral emissions in eclipse
Global imaging of Callisto, best resolution 23 km/pixel
Jovian ring plane crossing; map vertical structure of ring systems
Map longitudinal structure of Jovian rings
Resolved images and phase curves of irregular satellites Himalia and Elara

4×4 binned images about 90 days before c/a, and in unbinned single frames within the last 14 days.

The last full frame image of Pluto is acquired about 10 hours before c/a, and two 3×3 global mosaics of the full illuminated disk are acquired during the near encounter. The full illuminated disk observations will be useful to construct global base maps, to determine the global shape of Pluto, and to search for oblateness and tidal bulges. Additional images near c/a are obtained at successively higher resolutions but covering smaller portions of the illuminated disk. The Pluto near encounter image data set will be used to study surface morphology, geologic processes, and atmosphere–surface interactions. Likewise, the full illuminated disk of Charon will be imaged with 3×3 LORRI mosaics at resolution of better than 0.5 km per pixel, meeting the Group I panchromatic imaging requirement for Charon (the corresponding 0.5 km per pixel imaging requirement for Pluto will be met by MVIC).

The LORRI data sets acquired during the Jupiter encounter in January through March of 2007 are summarized in Table 5. Images of the Jovian atmosphere and its clouds and storms include full disk rotation sequences, acquired up to two months before Jupiter c/a on February 28, as well as high-resolution 2×2 mosaics of specific features acquired near c/a. Imaging sequences were executed for each of the Galilean satellites, including observations of the nightsides of Io, Europa, and Ganymede acquired while the respective satellites were in eclipse (i.e., in Jupiter's shadow). The Jupiter ring system was imaged at both low and high solar phase angles, as well as during the ring plane crossing. Images to study the shapes and photometric properties of Elara and Himalia, two of Jupiter's irregular satellites, were obtained.

3 LORRI Instrument Description

3.1 LORRI Overview

The LORRI instrument was designed and fabricated by a combined effort of The Johns Hopkins University Applied Physics Laboratory (JHU/APL) and SSG Precision Optronics

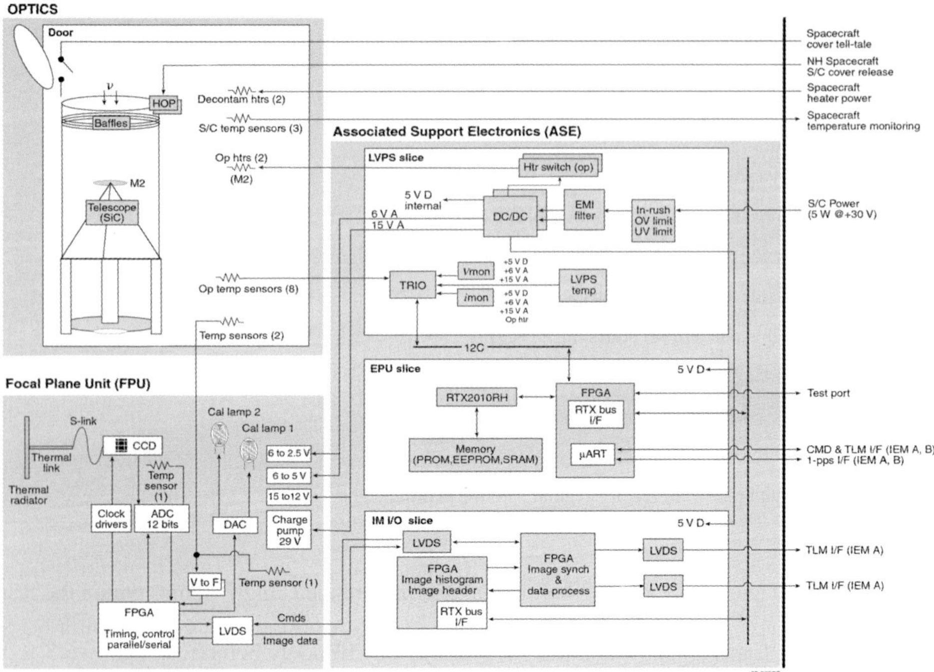

Fig. 3 LORRI block diagram, showing subassemblies: optics (OTA) and aperture door; focal plane unit (FPU); and associated support electronics (ASE) with three slices, which are the low-voltage power supply, the event processing unit, and the imager input/output board

Incorporated (SSG), of Wilmington, MA. LORRI has four subassemblies in close proximity connected by electrical harnesses. These are the OTA, the aperture cover door, the associated support electronics (ASE), and the focal plane unit (FPU). Except for the door, all are mounted inside the spacecraft on its central deck; the door is mounted to an external spacecraft panel. LORRI is electronically shuttered, with no moving parts aside from the cover door. The ASE implements all electrical interfaces between LORRI and the spacecraft except for the door control, several spacecraft thermistors, and two decontamination heaters. Figure 3 is a block diagram of the instrument. Conard et al. (2005) gave a detailed description of LORRI design, manufacture and test.

A summary of LORRI instrument specifications is given in Table 6. The LORRI boresight is required to be aligned within 0.1° of the boresight of the Ralph imager (MVIC). For a combined summary of the fields-of-view for the three imaging instruments on New Horizons, see the payload overview companion paper (Weaver et al. 2007).

The OTA, aside from the focal plane unit and thermal blanketing, was designed and built by SSG Precision Optronics, Inc. The primary and secondary reflecting optical elements are constructed of SiC. The telescope is a 2,630 mm focal length, $f/12.6$ Ritchey-Chrétien design. Three field-flattening fused silica lenses, located in front of the focal plane unit, are the only refractive elements in the system.

The SiC metering structure of the telescope holds the mirrors and field-flattening lens cell. It is a monolithic structure consisting of a primary mirror (M1) bulkhead, short cylindrical section, and three-blade spider with secondary mirror (M2) mounting. The field flattener assembly mounts to the M1 mounting plate and protrudes through the M1 mir-

Table 6 LORRI instrument specifications

Optical telescope assembly mass 5.6 kg
Total mass 8.6 kg
Electrical power 5 W
Heater power 10 W
Focal plane calibration lamps (two)
Data interfaces Low-Voltage Differential Signaling (LVDS) and RS-422 (both dual redundant)
ADC 12 bit
Image format 0: 1024 × 1028 (including four dark columns)
Image format 1: 256 × 257 (4 × 4 binned, including one dark column)
Embedded image header (first 408 bits of image data per frame, either format)
32-bin image histogram provided for every image

ror. The metering structure is mounted to the graphite composite baffle using three titanium, vibration-isolating feet. The baffle assembly is mounted to the spacecraft using six glass-epoxy legs, which provide thermal isolation. The entire OTA is covered with multilayer insulation (vented away from the OTA), except for the entrance aperture.

LORRI is protected from contamination and solar illumination using a one-time-open door mechanism. The door is mounted to the exterior of the spacecraft, and the LORRI baffle tube extends into the door to form a contamination seal. The door is aluminum, with thermal blankets for temperature control prior to deployment. The mechanism uses redundant springs and redundant paraffin actuators for deployment. A port allowing for installation of a witness mirror or small window is also part of the door. The door was opened successfully in flight on August 29, 2006.

The ASE provides the data and control interfaces to the spacecraft and to the focal plane unit. It consists of three 10 cm × 10 cm printed circuit cards electrically interfaced to one another via stackable connectors. They are in a magnesium housing, mounted directly to the spacecraft deck, a short distance from the OTA.

The LORRI FPU has a back-illuminated, thinned, high-quantum-efficiency CCD (an E2V Technologies Model 47-20). The FPU consists of a magnesium box mounted to the spacecraft deck, housing a 15 cm × 10 cm circuit card that controls the frame transfer CCD and provides interfaces to the imager board of the ASE. This circuit card is connected by a flex circuit to a small electronics board mounted at the focal plane. The small focal plane board holds the CCD itself and is mounted on thermal stand-offs, allowing the CCD and the small board to operate at $\leq -70°C$ while the magnesium box operates near room temperature. The CCD is in a windowless mount to avoid scatter and multiple reflections, with a black, anodized aluminum plate installed over the CCD storage readout area.

3.1.1 Design Requirements and Trades

The stringent optical, thermal, and structural requirements for the LORRI OTA presented many design challenges. The primary design driver for LORRI instrument was the resolution requirement. The resolution is limited by a number of factors, including the stability of the spacecraft while an image is being exposed. The pointing stability of the spacecraft is characterized as a typical drift rate of 25 μrad per second. The minimum exposure time is limited by the frame transfer time of the CCD. In order to remove completely the image smear which occurs during transfer, the exposure time should preferably exceed the frame transfer time (approximately 13 milliseconds), although acceptable image quality has been achieved in flight at exposures as short as 1 millisecond. LORRI was designed for an exposure time range between 50 and 200 milliseconds, with 100 milliseconds the nominal design value. Over this exposure range, the spacecraft orientation would drift ∼2 to ∼7 μrad.

After the range of exposure times was determined, IFOV had to be traded. A smaller IFOV yielded higher resolution, though it would ultimately be limited by spacecraft stability. Diffraction limited system resolution when the entrance pupil became appreciably smaller than 200 mm diameter. Strict mass limitations combined with cost limitations prevented increasing the aperture much beyond 200 mm diameter. An aperture of 208 mm was selected.

With reasonable assumptions about the type of telescope and detector and their associated efficiencies, and with the range of nominal exposure times, a nominal IFOV of 5 μrad was chosen, which defined an effective focal length of 2,630 mm. The FOV of the final design was 0.29° square.

The aperture requirement drove the telescope to reflecting optics. Mass and cost limitations, combined with the FOV and imaging requirements, drove the design to a Ritchey-Chrétien design. Refractive elements were used as field-flattening lenses, as the Ritchey-Chrétien focal plane curvature over the flat CCD would have limited imaging performance without them. The field-flattening lenses allow LORRI to meet the requirement of <0.1% distortion at all points over the full field of view. There was no requirement for color imaging. An optical ray trace layout for LORRI is shown in Fig. 4.

During the New Horizons mission, LORRI is exposed to the following radiation environments: a total dose of ionizing radiation 15 krad and a neutron fluence (worst case) of 1.4×10^{10} n/cm^2. The neutron fluence is from the radioisotope thermal generator over the course of the mission. In the OTA, the refractive optics, antireflective coatings, and reflective coatings were required to be tolerant to the specified radiation environments. The LORRI CCD is operated below −70°C to mitigate degradation of charge transfer efficiency from exposure to the neutron fluence.

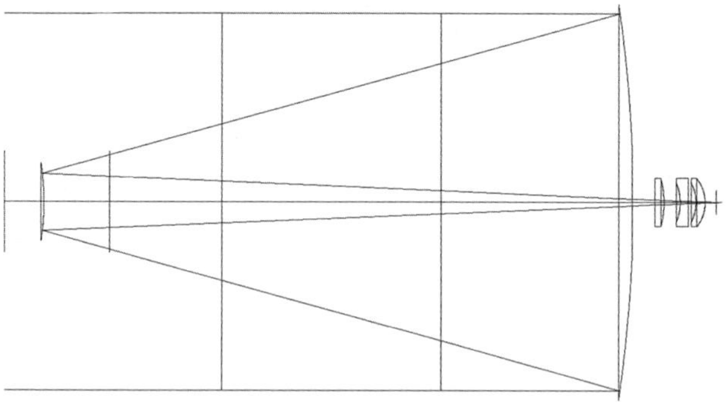

Fig. 4 Optical ray trace layout, showing primary and secondary mirror surfaces, plus fused silica lenses

3.1.2 Optical Design

The LORRI OTA is a Ritchey-Chrétien design, with high system throughput required because of the short allowed exposure time and low light level at Pluto. The complete LORRI OTA design was evaluated with a computer-aided design model including stray light analysis. Specular reflections and bidirectional reflectance distribution functions of the Aeroglaze Z-306 black paint, primary and secondary mirrors, field group optics and focal plane were included in the model. The primary and secondary baffle tubes were sized to minimize obscuration and suppress direct paths to the FPA's active area. The telescope magnification and obscuration were balanced, affecting the optical sensitivity and MTF, respectively. Out-of-field stray light was evaluated by generating point source transmittance curves with angular scans across the boresight in two orthogonal directions (-70° to $+70^\circ$ for each scan) to search for any obscured paths with unacceptable amplitude.

The OTA design required multiple baffle vanes, fabricated from graphite composite material, surrounding the metering structure to suppress stray light. In addition, an inner baffle is used extending from the hole in the primary mirror. This inner baffle has both inner and outer vanes plus threading in its interior. All baffle design features were optimized through TracePro ray tracing analysis. This analysis shows that out-of-field stray light is adequately suppressed and that ghosting is acceptable. The derived OTA system root-mean-squared (rms) wavefront error requirement based on the MTF requirement was <0.10 waves @ 632.8 nm, over an operating environmental temperature range of -125°C to 40°C and after g-release. The system throughput requirement drove a surface roughness requirement on all optics to a goal of <20 Å.

3.1.3 Thermal Requirements and Design

The extreme thermal environment posed a design challenge. LORRI is mounted within the interior of a spacecraft with deck temperature in the range 0 to $+40^\circ$C temperature, while it has an open, 208-mm aperture radiating into cold space. The challenge was to minimize defocus due to gradients developed in the OTA metering structure. The derived requirement was to limit the heat loss from the spacecraft to the OTA to less than 12 W, requiring conductive isolation at the interface and near zero radiative coupling to the spacecraft. The gradients in the OTA metering structure were limited to 2.5°C axial and 1.0°C lateral, and 0.5 W heater power was available for mirror gradient control without any in-flight focus adjustment mechanism.

The thermal requirements were met by an athermal, single-material solution, which is self-compensating under soak conditions. SSG's SiC 55A formulation was chosen due to its inherent high conductivity, which acts to minimize gradients, and its low coefficient of thermal expansion, which minimizes the thermal strain impact of such gradients. Invar 36, a good match to SiC 55A over the temperature range of interest, was chosen for the metallic inserts that allow bolting together of the OTA assembly. All invar inserts, as well as the secondary mirror foot, were epoxy bonded to the SiC. In addition, the OTA is mounted inside of the telescope baffle tube, made of highly conductive K13C graphite composite. The baffle tube provides a relatively uniform cold sink along the length of the telescope which helps to reduce longitudinal thermal gradients.

Additional thermal control features on the LORRI telescope help to reduce system gradients. The telescope is mounted to the spacecraft via G-10 isolators which isolate the OTA conductively from the spacecraft deck. Covering the appropriate OTA and spacecraft surfaces with multilayer insulation minimizes radiative coupling between the OTA and the spacecraft deck. The LORRI multilayer insulation, which represents 15% of the instrument mass, consists of 23 separate pieces.

Areas where material mismatches occur are flexured or otherwise configured to minimize induced strain and resulting optical degradation. The secondary mirror is mounted to a flexured Invar 36 mount plate, which in turn mounts to the end of the SiC metering structure. The secondary and primary mirror magnesium baffles are flexure mounted to the OTA structure (SiC) and the lens cell (Invar 36), respectively. The system aperture stop is made of aluminum 6061-T6 and is flexure-mounted to the middle ring of the OTA structure. The CCD mount plate is attached to the OTA structure by titanium flexures which have the dual purpose of mitigating thermal strain and thermally isolating the CCD from the structure. The G-10 mounts have titanium post flexures on either end. Additionally, the OTA itself is mounted to the K13C baffle tube via titanium isolators, which act to mitigate thermal strain and vibrations and to provide conductive isolation.

The baseline mission requires decontamination of the CCD at temperatures $>-18°C$ before door opening, with decontamination heaters. Decontamination can be achieved at lower temperatures, but long time periods may be necessary to desorb water.

An additional thermal requirement is that the CCD must be maintained at a temperature of $<-70°C$ while acquiring science data. This requirement is met by mounting the CCD to a bracket which is mounted to the OTA via conductively isolating titanium flexures; the CCD bracket is in turn attached to a gold-coated beryllium S200F conduction bar that is bolted to a gold-coated beryllium S200F thermal radiator whose outside surface is painted white with Aeroglaze A276. Since the radiator is mounted to a separate spacecraft panel from the OTA itself, there must be some compliance to allow motion between the two, provided by a highly conductive 1100 series aluminum alloy S-link.

The LORRI in-flight temperatures were predicted via the creation of a finite difference thermal model that included all conductive and radiative heat transfer. A thermal balance test of the instrument was performed that validated the thermal model in five separate test cases. The results of thermal balance testing predict that the 0.5-watt gradient control heater will not be required in flight. Flight data shows a gradient of 1.5°C, well within the 2.5°C requirement for gradient.

3.1.4 Telescope Mechanical Requirements and Design

The structural design objectives of the OTA were to obtain minimum mass while maintaining performance over operational temperatures and allowing for the stiffest design that will survive the launch environment. The requirements were a maximum mass of 5.64 kg, a minimum resonant frequency of 60 Hz, and survival under launch-induced vibration and stresses.

The inherently high stiffness-to-weight ratio of SiC, ($\sim$4.5 times that of aluminum) allowed the fabrication of a low-mass structure with a light-weighted primary mirror of open back, hub-mounted design to minimize weight. The main baffle tube was fabricated from graphite composite (K13C2U, M55J, and T300), another very high stiffness-to-weight material ($\sim$2.5 to 4 times that of aluminum). The smaller internal baffles were fabricated from a light-weight magnesium alloy (ZK60A).

Vibration isolation was required to survive the launch environment. Titanium isolators were incorporated to mount the structure to the main baffle tube at three points approximately at its center of gravity location. Another important structural design consideration was to minimize any potential for mount-induced distortion of the optics. Intimately connected to this design consideration is the requirement for the OTA to mount to a surface with only moderate mounting coplanarity, namely, a spacecraft aluminum honeycomb panel. To avoid degradation of optical quality, a three-point mount was adopted, with the bases of each of the three mount locations on the OTA outfitted with a ball joint that can be loosened and retightened if necessary.

At the interface between the main baffle tube and the OTA inner assembly, the vibration isolators also act to mitigate any mount-induced strains. The flexurized mount plate at the secondary mirror serves the dual purpose of controlling thermally induced distortions, as well as mount-induced distortions; the CCD flexure mounts also serve this dual purpose.

Due to its mass, the primary mirror was not flexure-mounted, as the low resultant frequency and dynamic responses would have increased the risk to the OTA under vibration. Instead the primary mirror is hub-mounted, with a post and a foot bolted via three invar inserts to a mount plate. The mount plate is in turn bolted to the structure. Because the lens cell is made of Invar and mounts to the structure, close to the primary mirror, the mount plate helps to separate any induced thermal strain in the structure from being transferred to the primary mirror.

3.1.5 Instrument Integration, Focus, and Alignment

When the OTA was assembled at SSG, a convex spherical reflector was centered at the focus using interferometry. Metrology, combined with knowledge of the shim size used to connect the reflector to the carrier plate, determined the location of the focus relative to the interface location on the carrier plate. SSG also provided a reference mirror on the back of the secondary mirror mount, such that the telescope line of-sight was parallel to the reference mirror's normal. Two optical reference flats at right angles to the line-of-sight were mounted to the LORRI metering structure for use in alignment monitoring after mounting to the spacecraft.

The depth of focus for LORRI at the detector, with mechanical tolerances on the CCD, allowed initial shim sizes to be selected for system focusing at the SSG-provided focus location. A 300-mm aperture, $f/5$ off-axis parabolic collimator was used to project a point-like image into LORRI. This image was produced by a laser unequal path interferometer (LUPI) at the collimator focus. A series of exposures were made at nine points in the LORRI field by moving within the field using a fold mirror. At each location, the spot was centered on a detector element by viewing the live image from the CCD through ground support electronics, and balancing the wings of the image symmetrically about the center detector element. This was repeated for seven focus adjustments of the collimator, with slight measured changes from nominal to allow for deterministic shimming of LORRI. Data from these exposures were examined to find the best LORRI focus versus the collimator adjustment. New shims were then installed to move the plane of the CCD onto the plane of best focus.

Once at nominal focus, the CCD was centered on the optical axis by use of a theodolite viewing both the reference mirror on the back of the secondary mount and the primary mirror. The theodolite was autocollimated on the reference mirror, and the azimuth and elevation recorded. The theodolite then viewed the four corners of the CCD off the primary mirror, and these azimuth and elevation values were recorded and averaged. The CCD location was shifted in the plane of best focus such that the average of the four CCD corners was within tolerance of the normal to the reference mirror.

After the best focus was found in ambient conditions, LORRI was installed in the NASA Goddard Space Flight Center's Diffraction Grating Evaluation Facility for a focus check at flight-like thermal and vacuum conditions. This was done by viewing a collimated beam, which projected a small point-like image into LORRI, and stepping the spot over the field by tilting LORRI on a gimbal platform.

Prior to delivery to the spacecraft, LORRI's line-of-sight was measured relative to the optical reference flat mounted to the back of the secondary mirror support. Additionally, the roll angle was measured by viewing the CCD corners, and referencing to the orthogonal alignment mirrors. These data, combined with measurements performed referencing the flats to the spacecraft coordinate system, showed that LORRI's line-of-sight was within mission requirements. Tracking of the LORRI alignment references through the spacecraft environmental test program did not show any significant movement relative to the spacecraft coordinate system.

3.1.6 Contamination Control

The LORRI telescope assembly remained under a nitrogen purge during all phases of integration and test until launch, except for limited times when images were taken or when put under vacuum. The internal cleanliness requirements for the LORRI OTA per the LORRI performance specification are a beginning-of-life specification of 250 A/2 and an end-of-life specification of 300 A per MIL STD 1246C. The resulting predicted reduction in optical efficiency (Conard et al. 2005) is $<4\%$.

3.2 Electronics

As shown in Fig. 3, LORRI electronics consist of the ASE and FPU. The ASE contains three printed circuit cards. These are the low-voltage power supply (LVPS), the event processor unit (EPU), and the imager input/output (IM I/O). The ASE is the primary interface between the spacecraft and the FPU, which mounts and controls the CCD. Additional information can be found in Conard et al. (2005).

3.2.1 Focal Plane Unit

The LORRI FPU is required to read out a complete image in one second, with the charge level in each pixel represented by a 12-bit binary word. A highly sensitive CCD with antiblooming was required, leading to the choice of the E2V Technologies CCD47-20. This is a $1{,}024 \times 1{,}024$ pixel frame transfer CCD with 13-micron square pixels. This device is back-illuminated for high quantum efficiency and has a frame transfer time of 13 milliseconds. The FPU noise is required to be <40 electrons per pixel, well above the CCD read noise which is calculated to be about 10 electrons at the readout time of about 0.7 microsecond per pixel. In general, exposures of 50 to 200 milliseconds are typical for LORRI, although the maximum exposure time is 29.9 seconds. The FPU design includes switchable 4×4 pixel on-chip binning. The FPU also includes two small incandescent bulbs that can illuminate the CCD through multiply scattered light, so that testing can be performed even when imaging is not possible through the optics (e.g., when the cover door is closed).

All CCDs have a number of clocks that must be driven to specific levels for satisfactory operation. The E2V CCD uses three phase clocks for image zone, memory zone, and line transfer. These clocks are highly capacitive, particularly for the image and memory areas of the chip, and they have capacitive coupling between different phases. LORRI uses Micrel MIC4427 drivers which are designed to drive high capacitance loads at the required voltage levels from logic level inputs. They are switching, not linear, devices so that low- and high-voltage levels are obtained by suitable choice of supply voltages, and transition rates must be adjusted at the output. This is done with series resistance which adds to the internal switch resistance of the drivers, forming a simple time constant with the capacitance of the CCD phase. The CCD requires 29 volts bias for the output field effect transistor. A charge pump with pre- and post-regulation was used to generate this voltage.

The LORRI FPU uses an Analog Devices AD9807 integrated circuit that performs correlated double sampling, signal amplification, and analog to digital conversion to 12 bits at maximum rates of 6 MHz, comfortably above the pixel readout rates which are close to 1.5 MHz. The AD9807 is susceptible to latch-up from ionizing radiation in space, and the LORRI FPU incorporates latch-up protection circuitry. The CCD output is low enough for the AD9807 amplifier to contribute significant noise, so a low-noise, wide-band operational amplifier is added between the CCD and the correlated double sampler, and the AD9807 is run at low gain.

3.2.2 Associated Support Electronics

The LORRI EPU controls the instrument via interfaces to the LVPS and IM I/O boards. The EPU communicates to the spacecraft using an RS-422 link, which receives commands and transmits engineering data. The EPU uses a RTX2010RH processor and runs FORTH code.

The main function of the IM I/O board is to receive serial image data from the FPU and transmit that data to either of two Integrated Electronics Modules (IEMs) in the required format. There are dual, redundant IEMs on the New Horizons spacecraft, which provide command, data handling, and telemetry functions. One IEM is active and one is a back-up; LORRI provides both interfaces (LVDS and RS-422) to both IEMs. Secondary functions of the LORRI IM I/O board include the ability to: store and transmit the image header, receive commands from the RTX processor, calculate a 32-bin histogram, generate test patterns without an FPU present, and command the FPU mode and exposure times based on input from the RTX.

The Imager I/O board contains two field-programmable gate array (FPGA) designs. The first is called the imager-interface FPGA, and the second is called the RTX-bus FPGA. The main function of the imager-interface FPGA is to read images from the FPU and send them to the IEM. As an added feature, the IM I/O board can generate test pattern images across the IEM interface without an FPU present. The first pattern consists of a horizontal ramp and the second pattern consists of a vertical ramp. The imager interface FPGA can also receive data from the RTX. These data are sent across the ASE backplane and through the RTX-bus FPGA. The data are used to set the FPU mode and exposure time, to set the active IEM low-voltage differential signaling port and to write the 408-bit header. The FPU mode data is transmitted to the FPU across the pixel data signal at the beginning of each second. The header data replaces the first 34 pixels when data are sent to the IEM, to provide redundancy in associating instrument engineering data with each image. The instrument engineering data are transmitted by the spacecraft within a separate data stream from the science data, and the engineering data must be associated with individual images in ground processing. For LORRI, the critical header information is also encoded into the images themselves, at the cost of 34 pixels in the first row of each image.

The RTX-bus FPGA also calculates a 32-bin histogram of the FPU image data currently being transmitted. This histogram is then made available to the RTX for future exposure time calculations. The RTX FPGA also collects the FPU status and temperature data, making it available to the RTX.

The LVPS provides 2.5 V, 6 V, and 15 V power as required by the other boards within the ASE and by the FPU. The input voltage from the spacecraft is 30 ± 1 V. The LVPS also provides for current, voltage, and temperature monitoring via an I^2C serial interface to the EPU board. The LVPS board provides switching to control power on/off to the FPU and the telescope trim heaters.

3.2.3 *Flight Software*

The LORRI RTX-2010 processor shares a common design with that of the PEPSSI instrument on New Horizons (McNutt et al. 2007). This common design extends into the software. The common flight software provides packet telemetry and command handling services. Besides handling LORRI-specific packets, the common software automatically generates a variety of standard packets, including housekeeping/status, command echo, memory dump, and alarm packets. Similarly, besides handling LORRI-specific commands, the common software also handles standard commands for memory loads and memory dump requests. The command handling software also provides storage and execution of command sequences. The common software has timekeeping, voltage and current monitoring, and memory management services and a standard boot program.

The LORRI-specific flight software controls heaters, collects voltages, currents, and temperatures from the LVPS board, and manages the FPU. In the FPU, the software controls the exposure time, either by manual command or automatically based on the hardware-provided image histogram, generates an image header, and enables routing of the image to the spacecraft. The software also controls the FPU's test patterns and calibration lamps. Whenever the software has nothing to do, it reduces the processor's clock rate to save power.

3.3 Laboratory Test and Calibration

The LORRI instrument was subjected to an environmental qualification program. Performance and environmental testing was performed at both the subassembly and instrument levels. Typically, performance tests or calibrations were done before and after environmental testing.

3.3.1 *Subassembly Test and Calibration*

The OTA was tested by SSG. Wavefront testing was performed using a LUPI operated in double-pass mode. Wavefronts were measured before and after vibration test and during thermal vacuum test. No change was detected due to the vibration test. Some change was noted at cold temperature during thermal vacuum testing, but it was determined that the level of change was acceptable within the LORRI performance requirements. As a secondary verification of performance, modulation transfer function testing was also performed by projecting small features into the OTA, and recording highly magnified images on a nonflight detector. This testing was performed only in ambient conditions and showed excellent correlation to wavefront data.

The FPU and CCD were also subjected at JHU/APL to testing at the subassembly level. They were calibrated at predicted operating temperatures and then subjected to environmental testing. The calibration of the FPU and CCD was not repeated after the environmental series due to time limitations.

3.3.2 *Instrument-Level Calibration*

Results of the LORRI instrument-level laboratory calibrations were discussed in detail by Morgan et al. (2005). The instrument-level optical calibrations under ambient conditions were performed at JHU/APL. Due to schedule conflicts with another instrument development program at JHU/APL, LORRI instrument-level optical calibrations in thermal vacuum were performed at the NASA Goddard Space Flight Center, in the Diffraction Grating Evaluation Facility (DGEF).

The objective of radiometric calibration is to determine the conversion from the observed signal in instrumental units to the scene radiance in physical units. For LORRI, the radiometric calibration equation relates the digital output S expressed as DN (12-bit "data numbers") to the scene radiance (or surface brightness) I

$$I_{x,y,\lambda} = \frac{S_{x,y,\lambda,T,t} - Bias_{x,y,T} - Dark_{x,y,\lambda,T<t} - Smear_{x,y,\lambda,T,t,\phi} - Stray_{x,y,\lambda,T,t,\Phi}}{EF_{x,y,\lambda,T}\, R_{\lambda,T}\, t}, \tag{1}$$

where subscripts x, y denote dependence on detector position (i.e., line and pixel numbers), λ on wavelength, T on temperature, t on exposure time, ϕ on optical power at other points within the FOV, and Φ on total optical power within the telescope. The quantities and their units are: I the scene radiance in $\mathrm{W\,m^{-2}\,sr^{-1}\,nm^{-1}}$; S the observed signal in DN; *Bias* the electronic offset of the CCD signal in DN; *Dark* the detector dark current in DN; *Smear* the signal in DN acquired by a pixel as it is shifted through the scene during frame transfer; *Stray* the signal in DN due to stray light; *FF* the (dimensionless) flat field response which normalizes responses of individual pixels using a uniform diffuse source; R the absolute responsivity in $[\mathrm{DN\,s^{-1}\,pixel^{-1}}]/[\mathrm{W\,m^{-2}\,sr^{-1}\,nm^{-1}}]$; and t the exposure time in seconds.

Of the terms in the radiometric calibration equation, bias and smear were characterized over the relevant parameters; dark is insignificant at operational temperatures; stray light was tested at discrete off-axis source angles; flat field response was measured, but there were difficulties obtaining the desired 0.5% accuracy at all pixels from ground-based data, and in-flight measurements are planned to replace or correct the current flat field; and absolute response was measured with sufficient accuracy for LORRI science goals. In addition to radiometric calibration, instrument characteristics such as detector read noise, the imaging point spread function (PSF), effective focal length and field-of-view were determined during calibration. The PSF was measured at several locations across the FOV, and the field-of-view and read noise were determined.

3.3.3 Calibration Setup

To confirm stability of LORRI's focus and response, key measurements were conducted at the nominal operating temperature as well as temperatures at the upper and lower ends of the operational range. The telescope assembly temperature ranged from −97°C to −60°C during calibration, while the CCD temperature ranged from −96°C to −78°C. At these low temperatures it was necessary to operate LORRI in vacuum to avoid condensation. The calibration was conducted at the NASA/GSFC DGEF which operates as a large vacuum chamber for optical calibrations. Calibrations at DGEF were conducted in July 2004 prior to instrument-level environmental (vibration and thermal vacuum) testing of LORRI, and again in September 2004 after environmental testing, to establish stability of the calibration.

Figure 5 is a block diagram of the DGEF calibration setup for LORRI. A light source outside the vacuum chamber was fiber optically coupled into the chamber, feeding an integrating sphere. The integrating sphere backlit a rotating target wheel that contained various apertures such as pinholes and resolution targets. The target wheel was at the focal plane of a 38-cm diameter collimator, the output of which illuminated LORRI, which was mounted on a two-axis gimbal approximately 2 m from the collimator aperture. Between LORRI and the collimator, a 30.5-cm diameter off-axis paraboloid (OAP) reflector could be rotated into the beam to focus a pinhole image onto a calibrated photodiode, providing a measurement of the integrating sphere port radiance for radiometric calibration.

A 150-watt, high-pressure, ozone-free xenon arc lamp was used for the LORRI calibration in order to provide a high radiance, solar-like spectrum. Since there are significant

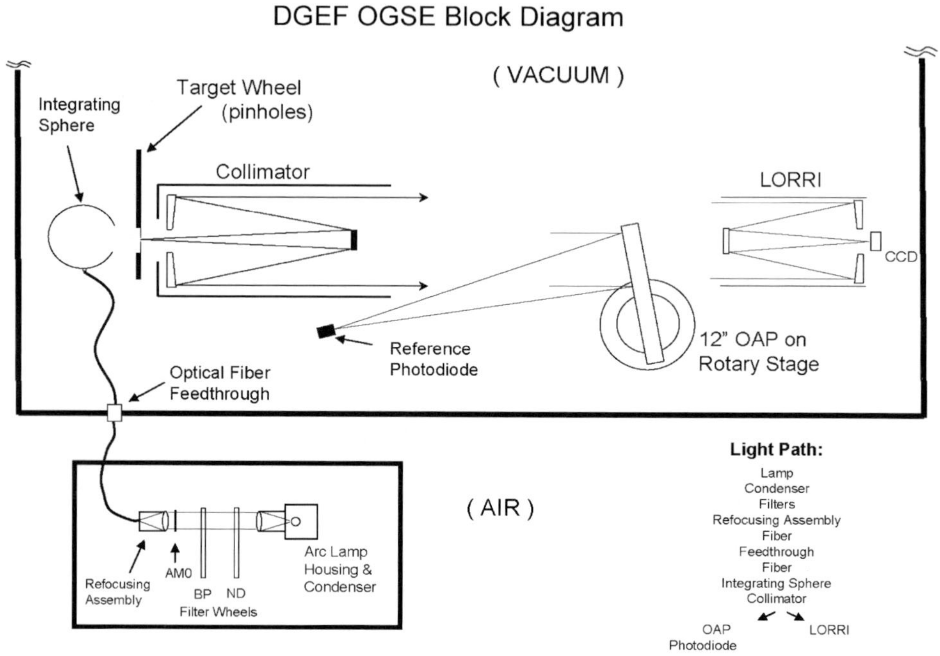

Fig. 5 Calibration facility at the NASA Goddard Space Flight Center

spectral variations in response over the LORRI bandpass, it is important to approximate the expected spectral radiance distribution in order to infer broadband radiometric response from calibration measurements with reasonable accuracy. Pluto, Charon, Jupiter, and Jovian satellites have spectra ranging from generally neutral to moderately red relative to a solar spectrum.

The arc lamp output was collimated using a fused silica condenser and transmitted through two six-position filter wheels, one with bandpass (BP) filters and the other with neutral density (ND) filters. The output then passed through an airmass zero (AM0) filter and was refocused into a fiber optic providing input to an integrating sphere inside the vacuum chamber. The AM0 filter attenuates the near-infrared portion of the spectrum to make the overall spectral energy distribution more solar-like. The BP filters were centered at 400 nm, 500 nm, 600 nm, 700 nm, and 850 nm. Bandpasses at full width half maximum (FWHM) for the filters were 65 nm for the 400 nm filter, 119 nm for the 850 nm filter, and approximately 90 nm for the others.

The radiance at the integrating sphere output port, spectrally integrated over the LORRI bandpass, was 46 $\mu W\,cm^{-2}\,sr^{-1}$. Figure 6 shows the source spectrum. Although the lines appear to dominate the spectrum in Fig. 6, they contribute a modest fraction of the total energy in the spectrum because their widths are relatively narrow. The calibration source spectrum is compared to the spectra of Pluto and Jupiter, as well as LORRI's calculated spectral response in Fig. 6 (Morgan et al. 2005).

The target wheel (Fig. 5) contained image masks, including pinholes ranging in size from 5 μm to 1,000 μm, an NBS 1963A resolution target, and an open position. The target wheel was located at the focal plane of a Cassegrain collimator with a 38-cm aperture, 460-cm focal length, and 0.5° unvignetted field of view. The collimator aperture was large enough to overfill the LORRI aperture without vignetting over the LORRI FOV and gimbal motion range,

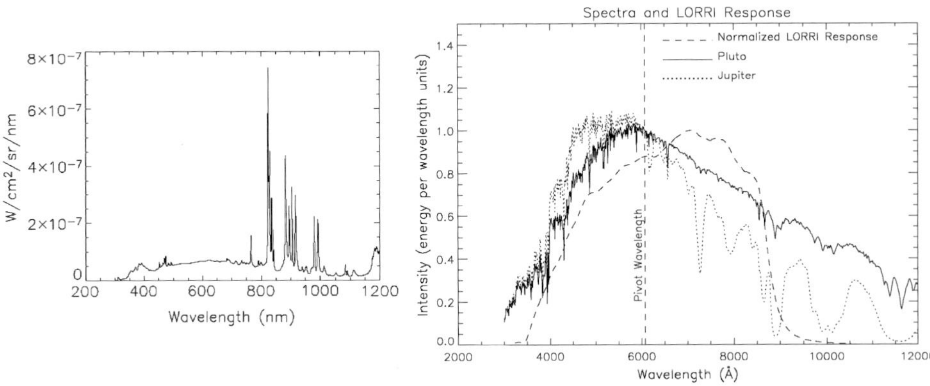

Fig. 6 (*Left*) Arc lamp calibration source spectrum and (*right*) spectra of Pluto, Jupiter normalized to unity at 607.6 nm and LORRI response normalized to unity at maximum (Morgan et al. 2005)

Table 7 LORRI calibration temperatures

	Nominal (°C)	Cold (°C)	Hot (°C)
CCD	−80	−93	−75
FPU board	26	10	35
Primary mirror	−72.5	−98.9	−62.1
Secondary mirror	−73.3	−99.3	−63.5

with tolerance for easy coalignment. The two-axis gimbal was stepper motor-driven with a Unidex controller at 63 steps per arcsecond to scan LORRI across targets. A LORRI pixel subtends roughly one arcsecond. The reference photodiode was a Hamamatsu S1336-8BQ, whose responsivity was calibrated by the vendor at 10 nm intervals from 200 to 400 nm, and at 20 nm intervals from 400 to 1,180 nm.

For thermal control during calibration, a shroud cooled by liquid nitrogen was constructed surrounding the telescope. A small annular shroud encircling the telescope inside the cold shroud was heated to control the telescope temperature to the desired set point. The thermal conductivity of the telescope structure was sufficient to guarantee a uniform temperature distribution despite the uneven heating. The shrouds were fixed, with LORRI gimbaled inside them. LORRI viewed the collimator through a ten-inch aperture in the main cold shroud, which kept the shroud edge well out of the LORRI FOV over the maximum gimbal motion range used. LORRI calibration temperatures are listed in Table 7.

3.3.4 Laboratory Calibration Results

Results of preflight calibrations are summarized here, and preliminary analyses of in-flight calibrations from instrument commissioning are summarized in the next section. Preflight and in-flight calibrations are fully consistent, considering limitations of ground test equipment.

The detector read noise was estimated from zero exposure time exposures taken in the DGEF at nominal temperature with no illumination. One hundred of these dark images were acquired, and the standard deviation of the signal of each pixel was evaluated. The distribution of the standard deviations is plotted in Fig. 7. The most probable standard deviation is 1.2 DN, which is adopted as a read noise estimate.

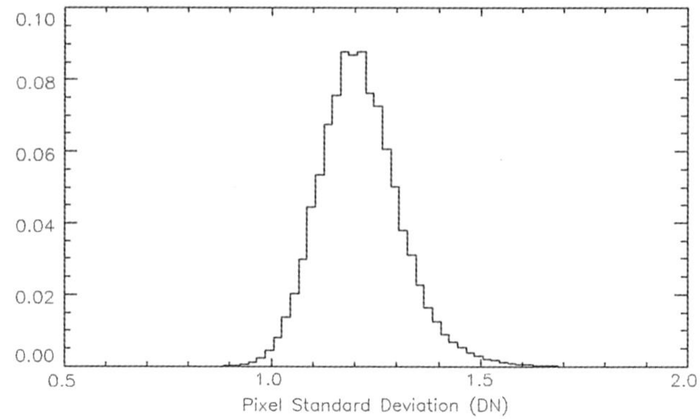

Fig. 7 Read noise distribution, from 100 dark images measured preflight. The estimated read noise is 1.2 DN

The detector bias (signal at zero illumination due to electronic offset) depends on the focal plane electronics board temperature. From subsystem level tests, the bias in unbinned 1×1 mode is

$$Bias\ [\mathrm{DN}] = 509.941\ (\pm 0.151) + 1.073\ (\pm 0.005) \times T \tag{2}$$

in units of DN as a function of board temperature T in °C. The observed in-flight board temperature range is $+34$ (± 0.7)°C. The bias ranges from 546 to 547 DN over that range. The bias in 4×4 binned mode is 2 DN higher. Detector dark current is negligible over the flight CCD temperature range, for exposure times up to at least one second. Planned exposure times are on the order of 100 ms. Also from subsystem level testing, the FPU gain is 22 ($+0$, -0.5) electron/DN for CCD temperatures of -50°C and -70°C and for board temperatures of 0°C and 40°C. This was determined from analyses of photon counting noise in flat field exposures using constant illumination with a green light emitting diode. A gain estimate of 22 ± 0.4 electron/DN was also obtained with analyses of noise in flat field exposures versus net signal with eight different exposure times. The linearity of the FPU was found to be within $\pm 1\%$ over DN values from 900 to 3,900, where the saturation level is 4,095.

During calibration at the DGEF, point source imaging performance was evaluated by observing a 3×3 grid of 5 μm pinholes. The grid target was arranged with four pinholes near the corners of the FOV, four along the edges, and one near FOV center. LORRI was gimbaled through a 6×6 subpixel grid of positions spanning approximately 1.5 pixels, acquiring 10 images at each step. A fitting process was applied to the images to estimate the subpixel irradiance distribution. Results indicated a PSF with FWHM on the order of two pixels, with no discernible dependence on position within the FOV. No significant variation of the FWHM with temperature was observed, verifying focus stability from ambient temperature to -100°C. These tests were adversely affected by diffraction by the collimator secondary obscuration, which was larger than the LORRI secondary.

More accurate PSF measurements were obtained in bench tests at ambient temperature at APL. A source was projected through a Laser Unequal Pathlength Interferometer (LUPI) and then an off-axis parabola collimator with an unobstructed aperture to generate a collimated beam that completely filled LORRI's aperture. For LORRI tests, the beam was reflected from an optical flat mounted with tip/tilt controls, which allowed focus and PSF measurements at five positions in the LORRI FOV (near the corners and near the center). Image analysis indicated a PSF with FWHM of 1.5 pixels with little variation across the

Fig. 8 Provisional flat-field measured at APL: logarithmic stretch from 0.95 (*black*) to 1.05 (*white*). Dust particles form "donuts" and dark spots. There is 4% vignetting in the *corners*, but in the *upper left corner* and in the *center of the field*, there is excess brightness from image ghosting which is not removed in preliminary processing

FOV, when fit with a two dimensional Gaussian function. With the spot near pixel center, ensquared energy in a pixel exceeded 0.3 at all four corners and the center of the CCD.

Although LORRI flat-field images were obtained during the post-environmental calibration at the DGEF, the CCD was subsequently cleaned by blowing ionized nitrogen across it. New flat-field measurements conducted at APL with a quartz halogen lamp (see Fig. 8) showed that the number of pixels affected by particle shadowing was reduced to 62 from more than 150. These new flat-field measurements were conducted in air at room temperature. There is vignetting of approximately 4% in the corners, consistent with expectations based on the design of the baffles and field stop. Low amplitude, dark "doughnut" rings (up to 1% amplitude) are seen from particles probably on the field-flattening lenses, and dark spots (typical 20% signal loss in single pixels) are seen from particles on the CCD. Comparison with in-flight images of Jupiter and with calibration lamp images shows that the particles have not moved since prelaunch calibration, and no new particles have been introduced since launch.

A central region of approximately 200 pixel radius is brightened by a maximum of 1.5% by ghost images that result from multiple reflections between surfaces in the field flattening lens group. There is also excess diffuse brightness from image ghosting in the upper left corner of Fig. 8 such that the known vignetting is masked, and arc-like ghost features have been removed for preliminary processing. The ghosts are dominated by out-of-field illumination at the red extreme of the LORRI passband, depending on the radiance distribution over field angles just outside the FOV up to approximately 0.37° off-axis. The ghosts are strongly dependent on source spectrum and will be characterized extensively with Jupiter observations. Also seen in Fig. 8 is fringing of 0.5% amplitude from constructive and destructive interference in the CCD.

The DGEF flat fields showed no significant temperature dependence. The maximum differences between the flat field at nominal temperature and those at the hot and cold extremes were on the order of 1%, with fewer than 250 pixels exceeding 0.5% variation in either case. Bandpass-filtered flat fields showed somewhat more variation with temperature, but for most of the filters the variations are not statistically significant. Only ten images were used for the filtered flats, compared to 100 for the panchromatic flats. There appeared to be significant temperature dependence in the 850 nm flat, possibly due to interference in the CCD with the emission lines in the xenon arc lamp spectrum. These strong lines do not occur in natural targets, and the panchromatic flat fields are practically independent of temperature for flight conditions.

However, laboratory flat field calibrations apply only imperfectly to the flight system, because of limitations such as use of nonsolar source spectra. The provisional flat field of Fig. 8 will be replaced with in-flight flat field observations, using Jupiter as an extended source during the Jupiter flyby, either to use directly or to be combined with the laboratory flat field data. Additional in-flight flat field data are planned using solar stray light.

The absolute radiometric response of LORRI was also determined at the DGEF. When system responsivity varies significantly over the spectral bandpass, as is the case for LORRI, the measured signal depends on the scene spectrum, which will depend on the target. As can be seen in Table 3, LORRI has requirements for relative, not absolute, radiometry. However, the source spectra that LORRI will view in flight will be measured by other instruments on New Horizons (Reuter et al. 2007). With the aid of independent spectral observations, LORRI can provide useful radiometric data.

LORRI's absolute responsivity as a function of wavelength was determined (Morgan et al. 2005) using panchromatic absolute measurements together with a calculated relative response curve. First, the LORRI responsivity spectrum was calculated from geometrical throughput of the optical design and from spectral characteristics of optical components and the CCD. It was assumed that an unknown constant factor would scale this response spectrum to the correct absolute level. The calculated spectral responsivity accounted for the measured mirror reflectance curves and lens transmissions, and it used E2Vs typical quantum efficiency curve for the model 47-20 CCD which was not individually measured for the flight CCD.

Second, the absolute LORRI response was estimated by scaling the calculated spectral responsivity to match the DGEF flat-field observations. The flat-field source spectrum was corrected to an absolute scale using reference photodiode measurements taken immediately before each set of flat field images. The 1 mm pinhole was imaged onto the calibrated photodiode with the OAP mirror to obtain the spectral radiance of the integrating sphere port (see Fig. 6). Correction was made for signal reduction from the partial blockage of LORRI's aperture by the collimator secondary, which is larger than LORRI's secondary. The LORRI obscuration by spider and secondary is 11%. The LORRI signal was taken from the median flat-field signal, which was 15,908 DN/s/pixel viewing this diffuse source spectrum. From this signal, the absolute LORRI response curve shown in Fig. 9 was derived (correcting the figure shown in Morgan et al. 2005).

Finally, the response to realistic scene spectra was calculated numerically by integrating the product of the assumed scene spectra and the LORRI response spectrum derived from calibration. From the calibrated response curve, LORRI's responsivity to specified scene spectra can be calculated by integrating the product of the LORRI absolute response curve and the normalized scene spectrum. For a source with the average Pluto spectrum, the absolute response of LORRI is estimated from the DGEF calibrations as

$$R_{\lambda,T}\ (607.6\ \text{nm}) = 2.2 \times 10^{11}\ \frac{\text{DN/s/pixel}}{\text{W/cm}^2\text{/sr/nm}} \tag{3}$$

giving the conversion from spectral radiance at the LORRI pivot wavelength (Horne 2004), which is 607.6 nm, to signal expressed in units of DN/pixel/s. The spectral radiance in the

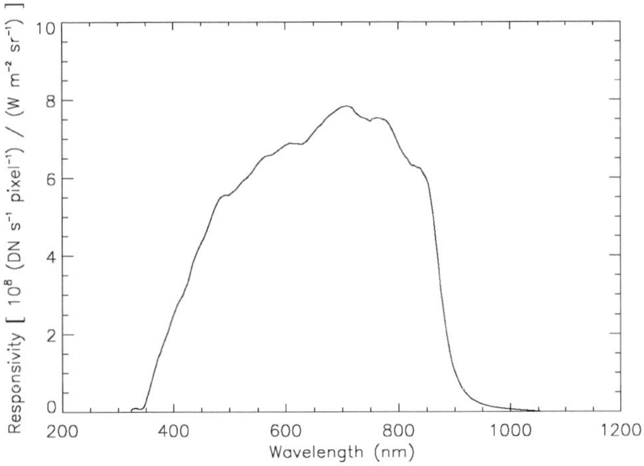

Fig. 9 Absolute monochromatic response from calculated spectral responsivity, correcting Morgan et al. (2005)

Table 8 LORRI in-flight commissioning and calibration testsummary

Dates and description	Sequence name
Feb 24, 2006: all currents, voltages, power, and temperatures nominal	LORRI-005
April 23–24, and May 2–3, 2006: in-flight noise test	LORRI-025
Jul 30, 2006: Dark series bias images, cover door closed	LORRI-006
29 Aug 2006: Door opening verification images (Messier 7)	LORRI-007
Aug 31, 2006: Radiometric/PSF images (Messier 7)	LORRI-010
Aug 31, 2006: MVIC co-alignment images	LORRI-018
Sep 04, 2006: Jupiter exposure/auto exposure/scattered light tests	LORRI-027
Sep 10,19, 2006: Uranus	LORRI-029
Sep 10,19, 2006: Neptune	LORRI-030
Sep 21, 24, 2006: Pluto	LORRI-023a

denominator is understood to correspond to the pivot wavelength. The absolute response depends on the shape of the adopted scene spectrum; for additional details see Morgan et al. (2005) and the following discussion of in-flight observations.

3.3.5 In-Flight Calibration

LORRI has successfully completed its in-flight commissioning tests. LORRI's cover door opened on August 29, 2006, and successful observations have been acquired of the open star cluster Messier 7 and the prime mission target Pluto as well as the planets Uranus, Neptune, and Jupiter. Instrument commissioning tests through September, 2006 are summarized in Table 8.

The commissioning test sequences LORRI-005, LORRI-006, and LORRI-025, with the door still closed, were used to verify the FPU noise and bias characteristics in-flight versus the ground calibrations. Results of the in-flight tests have been consistent with the ground calibrations of the FPU. For instance, the difference of two consecutive zero-exposure time images, obtained from LORRI-005 on February 24, 2006, yielded a read noise estimate of 1.11 ± 0.12 DN, where DN ~ 22 e^-. The cosmic ray "hit rate" was also measured from LORRI-005, with the rate of hits (mean $\pm$ standard deviation) determined as 15.7 ± 13.3

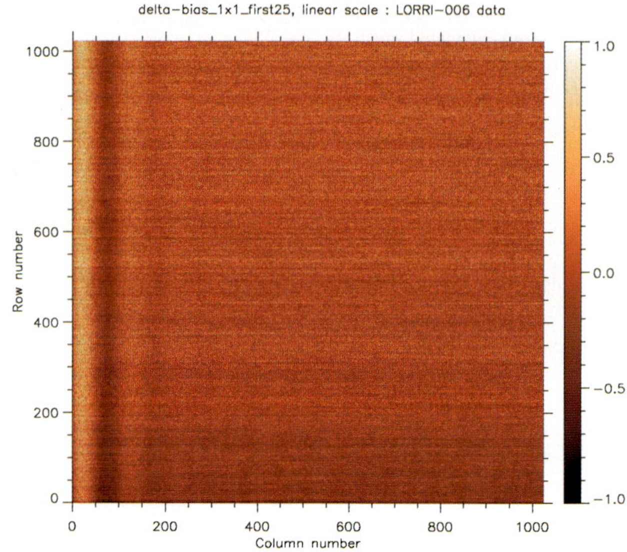

Fig. 10 *Delta_bias* frame: in-flight measurement of FPU bias correction, based on analysis of 25 frames acquired July 30, 2006. This image is linearly stretched between −1 DN and +1 DN

per frame for nominal exposures <0.5 s, where a "hit" was defined to be a pixel with DN at least five standard deviations above the mean.

The result of preliminary in-flight measured bias correction, from analysis of 25 frames acquired during the LORRI-006 test performed on July 30, 2006, is shown in Fig. 10. A total of 100 frames were acquired in that test and are under analysis. The preliminary analysis used 25 of the available dark frames, all of which are exposures of zero seconds. For each of these dark frames, the median DN is found of all the pixels in the four dark columns (which are the masked columns at the right side of the image), and the resulting scalar is subtracted from each pixel value in the image zone to obtain a corrected dark frame. After removal of outlier pixel values (more than three standard deviations from the mean of the 25 values at the same row and column numbers), the 25 corrected dark frames are then averaged to obtain the so-called *delta_bias* frame which is shown in Fig. 10. This frame shows the pixel-level variations of the bias and demonstrates the excellent uniformity of the LORRI flight CCD. The low-level vertical banding in Fig. 10 is the result of amplifier oscillation in the FPU; the maximum peak-to-trough amplitude of this wave-like feature is ~0.8 DN.

The *delta_bias* frame is used in the LORRI pipeline data processing to remove the FPU bias from a LORRI image as follows. First, the median of the pixels in the four dark columns of the raw image is determined and subtracted from each pixel in the image zone of the raw image. Then, the *delta_bias* frame is subtracted to obtain the final bias-corrected result. The statistics of *delta_bias* frame are: the mean value in DN is −0.0049; the standard deviation is 0.26; the maximum value is 1.64; the minimum value is −1.56. Since the pipeline processing uses the absolute bias level estimated from the dark columns of every image, the bias estimate from (2) is not used, although the latter provides an independent check.

The next step in LORRI pipeline processing is to apply the flat-field correction. Acquisition of in-flight flat-field observations is planned during and after Jupiter encounter in 2007. The in-flight observations to date have been reduced with flat fields determined in pre-flight laboratory testing; see Morgan et al. (2005).

Also required to complete the data reduction is removal of an effect called readout smear. This arises from the operation of the frame transfer CCD used in LORRI, where first the image zone is flushed, then an exposure is taken, and finally the image is transferred into

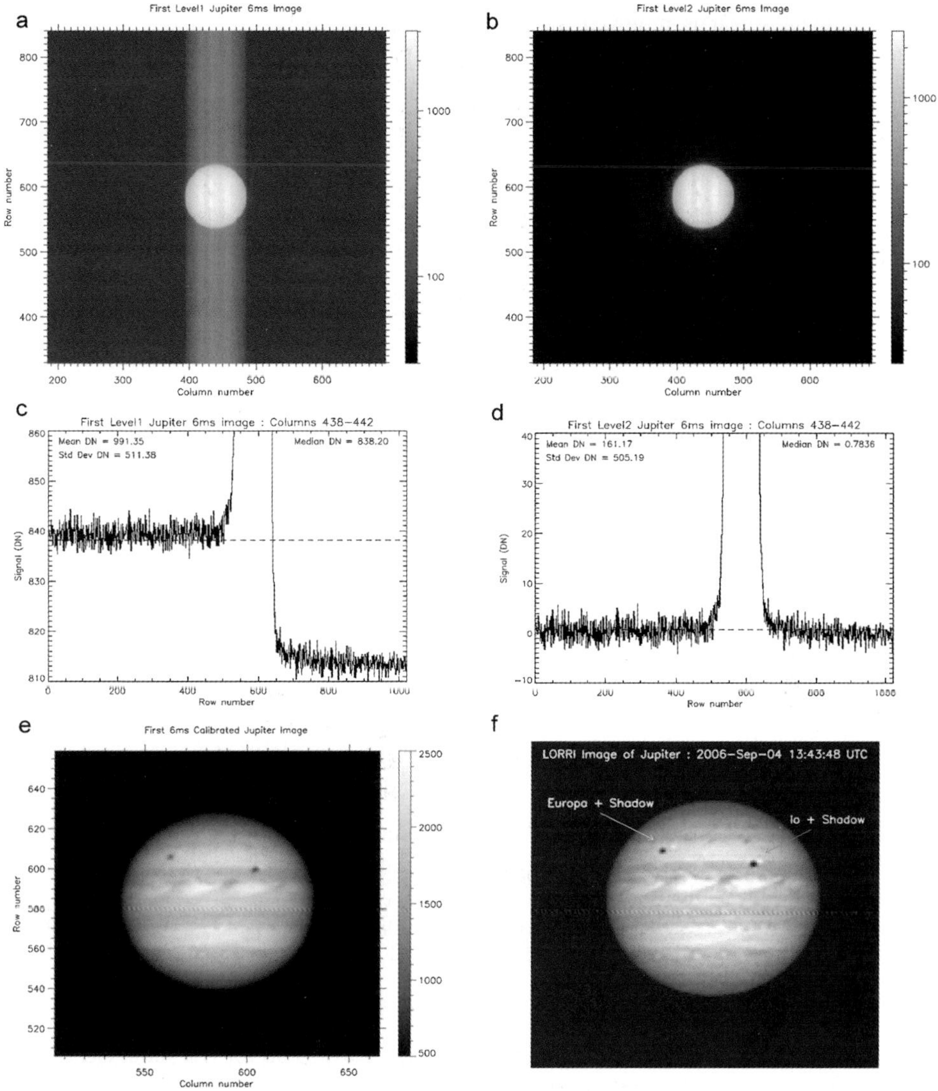

Fig. 11 Jupiter observed September 4, 2006. Panel **a**: Raw image MET 19683344 on logarithmic scale, showing readout smear. Panel **b**: Same image after calibration on same logarithmic scale, showing bias and smear removal. Panel **c**: Average of indicated columns showing different readout smear on either side of Jupiter. Panel **d**: Same as **c** for calibrated image, showing removal of readout smear. Panel **e**: Image of panel **a** after calibration, on linear scale. Panel **f**: Five LORRI images including that of panel **a**, all with 6 ms exposures, after calibration and co-addition. Io and Europa (both barely resolved) and their shadows are indicated

the storage zone. Hence a pixel of the raw image is exposed to the scene radiance from the corresponding geometrical element of the scene, but it is also exposed to the radiances of all the scene elements in the same image column during the image transfers. Thus the raw image is the superposition of the scene radiance and the signal acquired during frame transfers, which is called readout smear.

The readout smear is removed as follows. Let $P_{i,j}^{\text{meas}}$ = measured image array in DN where i, j are the column and row indices, respectively. Let the exposure time be written T_{exp}, with the transfer times for the frame scrub T_{f1} and the frame storage T_{f2} and with N the number of rows (which is 1,024 for 1×1 images). Let $T_{f\text{avg}}$ be the average of T_{f1} and T_{f2} to define the constant

$$A = \frac{T_{\text{exp}}}{T_{\text{exp}} - T_{f\text{avg}}/N}.$$

Finally we define the $N \times N$ constant matrix

$$\varepsilon_{k,j} = \begin{cases} T_{f1}/T_{f\text{avg}} & \text{for } k < j, \\ 1 & \text{for } k = j, \\ T_{f2}/T_{f\text{avg}} & \text{for } k > j \end{cases}$$

with $k, j = 1, \ldots, N$, and we calculate the $N \times N$ matrix

$$\lambda_{i,j}^{(1)} \cdot T_{\text{exp}} = A\left[P_{i,j}^{\text{meas}} - \frac{AT_{f\text{avg}} \sum_k P_{i,k}^{\text{meas}} \varepsilon_{k,j}}{N(T_{\text{exp}} + AT_{f\text{avg}})}\right].$$

The desmeared image is then

$$P_{i,j}^{\text{desmear}} = A\left[P_{i,j}^{\text{meas}} - \frac{AT_{f\text{avg}}[\sum_k P_{i,k}^{\text{meas}} \varepsilon_{k,j} + E_{i,j}/A]}{N(T_{\text{exp}} + AT_{f\text{avg}})}\right]$$

with

$$\frac{E_{i,j}}{A} = T_{f\text{avg}}\left[\sum_k \lambda_{i,k}^{(1)} \varepsilon_{k,j} - \frac{1}{N} \sum_l \sum_k \lambda_{i,k}^{(1)} \varepsilon_{k,l} \varepsilon_{l,j}\right].$$

In-flight tests have verified desmear by this technique using observations of Jupiter obtained at exposure times as short as 1 ms. For the image in Fig. 11, $T_{f\text{avg}} = 10.5$ ms and $T_{f1}/T_{f\text{avg}} = 1.044$, while $T_{f2}/T_{f\text{avg}} = 0.956$.

In-flight photometric calibration was obtained using observations of the open star cluster Messier 7 obtained from LORRI-007 on August 29, 2006. The visual magnitude is given by

$$V = -2.5 \log S + 18.94 + CC$$

with S in [DN/second]. We find the color correction $CC = -0.06$ for OB stars, $CC = 0$ for FG stars, and $CC = 0.4$ for K stars.

The Messier 7 observations (see Fig. 12) also confirm from stellar images that the point source function of the LORRI system, including the effects of spacecraft pointing jitter, is 1.8 pixels FWHM. Stars to at least 12th magnitude are detected. The plate scale, based on comparisons of pixel separations of bright stars in the field versus cataloged positions, is 4.96 μrad/pixel. Any geometric distortion in this image is less than 0.5 pixel.

4 Summary

The LORRI instrument was successfully developed, qualified, and calibrated, and it was delivered to the New Horizons spacecraft on time and within budget. It has operated successfully in flight, and its performance is completely nominal. It will obtain the first high-resolution imaging observations in the Pluto system and at one or more additional Kuiper Belt objects.

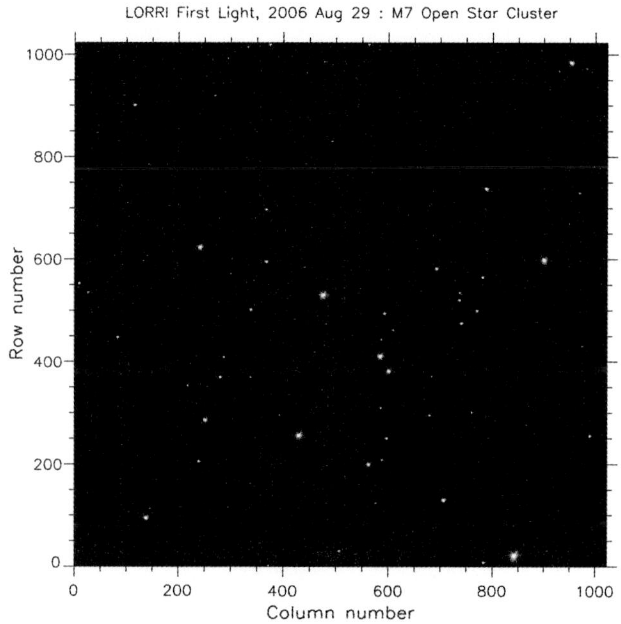

Fig. 12 LORRI image of open cluster Messier 7 obtained in flight. This image has been logarithmically stretched. North is up, east is to the left

Acknowledgements We thank NASA and the New Horizons Project teams for support. We also thank many more members of the LORRI team whose contributions have helped to make LORRI a success.

References

A.F. Cheng, L.A. Young, EOS Trans. AGU, **83**, 384, 389 (2002).

S. Conard, F. Azad, J. Boldt, A. Cheng, K. Cooper, E. Darlington, M. Grey, J. Hayes, P. Hogue, K. Kosakowski, T. Magee, M. Morgan, F. Rossano, D. Sampath, C. Schlemm, H. Weaver, in *Astrobiology and Planetary Missions*, ed. by G.R. Gladstone. Proc. SPIE 5906, 2005.

K. Horne, star-www.stand.ac.uk/kdh1/pub0/ada/woe/woe.ps, University of St. Andrews, 2004.

R.L. McNutt et al., Space Sci. Rev. (2007, this issue)

F. Morgan, S.J. Conard, H.A. Weaver, O. Barnouin-Jha, A.F. Cheng, H.W. Taylor, K.A. Cooper, R.H. Barkhouser, R. Boucarut, E.H. Darlington, M.P. Grey, I. Kuznetsov, T.J. Madison, M.A. Quijada, D.J. Sahnow, J.M. Stock, in *Astrobiology and Planetary Missions*, ed. by G.R. Gladstone. Proc. SPIE 5906, 2005.

D. Reuter et al., Space Sci. Rev. (2007, this issue)

H.A. Weaver, S.A. Stern, M. Mutchler, A. Steffl, M. Buie, W. Merline, J. Spencer, E. Young, L. Young, Discovery of two new satellites of Pluto. Nature **439**, 943–946 (2006)

H.A. Weaver et al., Space Sci. Rev. (2007, this issue)

The New Horizons Radio Science Experiment (REX)

G.L. Tyler · I.R. Linscott · M.K. Bird · D.P. Hinson · D.F. Strobel · M. Pätzold · M.E. Summers · K. Sivaramakrishnan

Originally published in the journal Space Science Reviews, Volume 140, Nos 1–4, 217–259.
DOI: 10.1007/s11214-007-9302-3

Abstract The New Horizons (NH) Radio Science Experiment, REX, is designed to determine the atmospheric state at the surface of Pluto and in the lowest few scale heights. Expected absolute accuracies in n, p, and T at the surface are $4 \cdot 10^{19}$ m^{-3}, 0.1 Pa, and 3 K, respectively, obtained by radio occultation of a 4.2 cm-λ signal transmitted from Earth at 10–30 kW and received at the NH spacecraft. The threshold for ionospheric observations is roughly $2 \cdot 10^9$ e$^-$ m^{-3}. Radio occultation experiments are planned for both Pluto and Charon, but the level of accuracy for the neutral gas is expected to be useful at Pluto only. REX will also measure the nightside 4.2 cm-λ thermal emission from Pluto and Charon during the time NH is occulted. At Pluto, the thermal scan provides about five half-beams across the disk; at Charon, only disk integrated values can be obtained. A combination of two-way tracking and occultation signals will determine the Pluto system mass to about 0.01 percent, and improve the Pluto–Charon mass ratio. REX flight equipment augments the NH radio transceiver used for spacecraft communications and tracking. Implementation of REX required realization of a new CIC-SCIC signal processing algorithm; the REX hardware implementation requires 1.6 W, and has mass of 160 g in 520 cm^3. Commissioning tests conducted after NH launch demonstrate that the REX system is operating as expected.

G.L. Tyler (✉) · I.R. Linscott · D.P. Hinson · K. Sivaramakrishnan
Dept. of Electrical Engineering, 350 Serra Mall, Stanford, CA 94305-4020, USA
e-mail: len.tyler@stanford.edu

M.K. Bird
Argelander-Institut für Astronomie, Universität Bonn, Auf dem Hügel 71, 53121 Bonn, Germany

D.F. Strobel
Johns Hopkins University, 1121 Olin Hall, 34th and North Charles Streets, Baltimore, MD 21218, USA

M. Pätzold
Abt. Planetenforschung, Rheinisches Institut für Umweltforschung an der Universität zu Köln, Aachener Strasse 209, 50931 Köln, Germany

M.E. Summers
Dept. of Physics and Astronomy, George Mason University, 4400 University Drive, MS-3F3, Fairfax, VA 22030-4444, USA

Keywords Pluto · Charon · New Horizons · Pluto atmosphere · Pluto mass · Pluto thermal emission · Radio science · Radio occultation · SCIC-CIC filter

1 Introduction

1.1 Overview

The New Horizons (NH) Radio Science Investigation, or Radio Experiment, 'REX,' addresses several specific questions related to the Pluto–Charon system. Primary among these is determination of surface conditions on Pluto manifest in atmospheric temperature and pressure, with the expectation that the structure of the neutral atmosphere in the first few scale heights will be determined to useful accuracies. Secondary objectives include a search for and, if possible, measurement of the ionosphere of Pluto and measurement of the 4 cm-λ thermal emission temperatures on the nightsides of Pluto and Charon. Tertiary objectives include determination of the occultation chords of Pluto and Charon to very high accuracy as a constraint on the sizes, and an independent determination of the Pluto–Charon system mass and possible separation of the individual masses of Pluto and Charon. These goals are summarized in Table 1.

The Pluto Science Definition Team (SDT) also identified the neutral atmosphere of Charon as a goal for a first mission. This is omitted from the current list as the presence of such an atmosphere appears to be ruled out at the level of REX sensitivity by ground-based stellar occultation measurements (Person et al. 2006). Similarly, the SDT objective of observations of solar wind effects on the ionospheres of Pluto and Charon can be inferred if sensible ionospheres are extant at the time of the NH flyby (Person et al. 2006).

The experiment employs occultation geometry for sensing of Pluto's atmosphere. This observation requires the use of a highly stable source to produce a signal of precisely known characteristics and a receiver of comparable stability to capture the refracted signal after it emerges from the atmosphere. The REX instrument must capture the physical properties of the emergent signal. For this purpose, NH makes use of a new radio system design, flown once previously on CONTOUR (Bokulic et al. 2003), which implements an onboard receiving system that captures signals transmitted from the ground. The REX experiment is enabled by adding a small amount of signal processing hardware to the radio system planned for communication and tracking of the NH spacecraft. This differs from most other U.S. missions conducting occultation experiments in which transmissions from the spacecraft are received on the ground. As explained below, this change is necessary to meet the NH objectives, owing primarily to the great distance from Earth to the Pluto system.

In order to understand the importance of this change it is necessary to relate the approach taken in the design of the NH system to earlier methods. Previous practice for U.S. planetary spacecraft has been to use radio transponders in which 'uplink' signals derived from

Table 1 Objectives of NH REX investigations

Atmosphere (I)	$n(h)$, $T[p(h)]$	Pluto
Ionospheres (II)	$n_{\bar{e}}(h)$	Pluto & Charon
Surface Emission Temperature (II)	T_{sur}	Pluto & Charon
Occultation Chord (III)	L entry–exit	Pluto & Charon
Masses (III)	GM_{sys}, GM_{P}, GM_{C}	System & P, C

ground-based H_2 masers are received, filtered, amplified, and then used to derive a coherently related downlink signal which is transmitted immediately to a receiving Earth station (for Voyager and Cassini examples *v.* Tyler (1987) and Asmar et al. (2005). Under normal operating conditions this method guarantees a known relationship between the frequencies and phases of the 'uplink' and 'downlink' signals received and transmitted by the spacecraft, respectively; the ratio between the uplink and downlink frequencies, and hence the phases, is by design always of the form M/N, where M and N are integers, and where the phase may be offset by an unknown but constant amount. This approach results in a round trip stability that approaches that of the master atomic clock used to derive the original uplink signal.

The use of coherent transponders is extremely effective for precision tracking of deep space vehicles, and was also used for the conduct of radio occultation experiments using the NASA Deep Space Network prior to Voyager (see below). For radio occultation this method works well in the case of 'thin' atmospheres, such as that of Mars, where it was first used, but is susceptible to loss of coherence as a result of rapid changes in signal frequency and strength (i.e., scintillations) and the occurrence of multiple atmospheric paths when applied to 'dense' atmospheres, such as those of Venus and the gas giants.

The problem of loss of coherence was addressed at the time of Voyager by the inclusion of a precision quartz oscillator frequency reference onboard the spacecraft for use as the downlink reference during occultation events. Such oscillators, referred to as 'USO's (for ultra-stable oscillators) achieve levels of coherence over time intervals of 1–100 s that approach that of H_2 masers. (USOs are 1 to 1-1/2 orders of magnitude less stable than masers on time intervals of roughly 1,000–10,000 s, however.) Onboard USO frequency references have also been used effectively on several other missions to the outer planets, Mars, and Venus. Over time, USOs have proven to be extremely effective in a number of other areas related to mission operations design, as well as for science. In the Voyager-style implementation, though, the USO is considered to be an auxiliary device external to the standard transponder.

The REX investigation team will also NH cruise and far encounter data for studies of the solar wind, the solar corona, and for investigation of the 'Pioneer Anomaly' (Anderson et al. 1998, 2002) to the extent that this can be supported by the in-flight performance of the radio tracking system. These possibilities are not discussed further here.

1.2 Implementation

The REX investigation is based on the use of radio signals traveling between large ground transmitters, operated by the NASA Deep Space Network at three locations around the world, and a receiving system onboard the New Horizons spacecraft. For the primary investigation of the atmosphere of Pluto and the ionospheres of Pluto and Charon, the method of sensing is the detection of small perturbations in these signals. The direction of transmission is exclusively from the ground to the spacecraft. Ancillary investigations of gravitational parameters make use of round-trip Earth–spacecraft–Earth measurements with the NH tracking system.

The use of an uplink signal for these investigation is driven, directly and indirectly, by the great distance to Pluto. First, large signal-to-noise ratios (SNRs) at Pluto are required to support accurate measurements of Pluto's tenuous atmosphere, leading directly to the use of large transmitter powers. Second, the large flyby velocities limit the total available observation time for occultation measurements of Pluto's atmosphere to minutes, and the observations of the lower atmosphere to several seconds, increasing the required SNR further.

While occultation measurements based on the use of spacecraft transmitters are routine for thick atmospheres from orbiting and flyby spacecraft at bodies closer to Earth, this is not possible for NH at Pluto. The use of large Earth-based transmitters providing increases in transmitter power by 3–4 orders of magnitude over that of spacecraft systems, together with signal reception at the spacecraft, makes radio occultation measurements of Pluto's atmosphere practical.

As REX requires the coordinated use of Earth-based transmitters and a NH spacecraft receiver, REX comprises two elements. The first is the 'Ground Element,' made up of the DSN hardware and operations facilities supporting the NH mission, while the second, 'Flight Element,' consists of a small amount of specialized signal processing hardware onboard the NH spacecraft. As the DSN support capabilities for radio science activities are well documented (e.g., Asmar et al. 2005), we will not discuss the REX Ground Element further.

The New Horizons radio system incorporates a USO as an inherent component in its design. In the approach adopted by NH, the spacecraft transmitter is always referenced to an onboard USO which is entirely independent in its functioning from the received uplink signals. This is unusual in precision deep space Doppler tracking systems. Typically such systems make use of the uplink signal transmitted from the ground to establish the downlink signal frequency, so that there is a fixed relationship between the frequency of the uplink received onboard to that of the downlink radiated by the spacecraft. Knowledge of this relationship allows an accurate calculation of the spacecraft radial velocity from a comparison of the transmitted uplink and received downlink frequencies. In the NH case the necessary link between the uplink signals received by the spacecraft and the downlink signals received by the ground tracking station is established by an onboard measurement of the received uplink frequency relative to an onboard USO. This is accomplished by counting the difference between the number of radio frequency cycles arriving at the spacecraft on the uplink relative to the number of cycles of oscillation of the USO in the same period, with the frequency difference returned to the ground by telemetry. In this system of tracking it is necessary to solve for the USO frequency as well as the round trip Doppler shift. Observation on the ground of the received one-way signal derived from the USO together with the onboard measurements of the frequency difference—taking into account the inherent difference in frequency between the ground transmitter and the USO—is sufficient to solve for the two unknowns.

The advantages of this approach are *i*) simplification of the radio system design, *ii*) increased stability of the downlink in the absence of an uplink signal from the ground, and *iii*) increased operational flexibility in conducting scientific investigations based on use of the radio system. Because the downlink in this type of system is independent of the uplink, the NH radio system is referred to as a 'transceiver' rather than a 'transponder.' This system is discussed further by Deboy et al. (2005), Fountain et al. (2006), Jensen and Bokulic (2000), and others.

Given the a priori requirement for a highly-stable onboard frequency source, we note the functional and operational simplicity of the NH transceiver as compared with a conventional phase-locked transponder. The standard approach would require a phase-locked loop around the receiver amplifier chain (see discussion of transceiver, below), as well as an offset frequency conversion chain to drive the downlink transmitter, and considerable additional switching circuitry. (We note that the use of a closed-loop transponder alone, without a stable, on-board, downlink frequency reference, would be inadequate for NH given the need to establish and maintain an uplink frequency reference for a maneuvering spacecraft with critical events at intervals that are shorter than the round-trip light time.)

Figure 1 illustrates the receiver portion of the NH transceiver and the integration of the REX hardware within the receiver. The amplifier chain is a conventional heterodyne design.

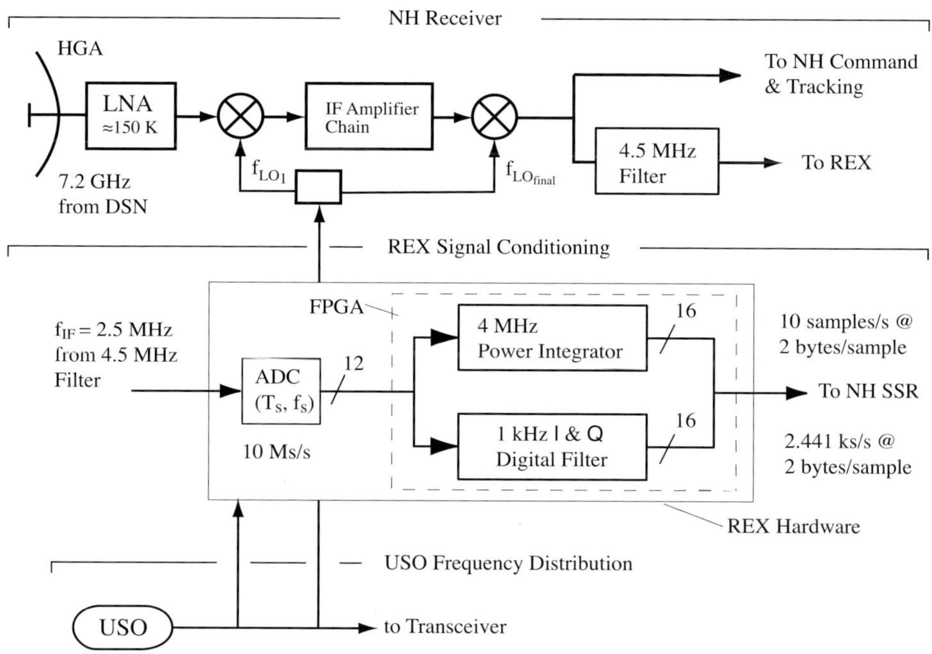

Fig. 1 Overview of REX onboard Implementation. Top section of figure shows functions of receiver portion of the NH transceiver. Following the 'IF Amplifier Chain,' the lower signal path with 4.5 MHz Filter drives the input to the REX hardware imbedded in the transceiver; the upper path drives the uplink data processing functions of the transceiver. REX signal conditioning hardware, located with the labeled rectangle, functionally comprises an ADC and a gate array programmed as a digital down-converter to 0 Hz, followed by a filtering and decimation step to reduce data volume, see Sect. 1.2, 5.3, and the Appendix. The output data are transferred to the spacecraft memory for storage and later return to Earth by telemetry. The transmitter side of the transceiver shares the HGA. USO, bottom, serves as frequency reference for essential functions in transceiver and REX

The noise performance of the receiver has been improved over previous implementations by locating the leading low-noise amplifier (LNA) close to the antenna to reduce the physical temperature of the wave guide connecting the LNA to the high-gain antenna (HGA). The various mixing frequencies, f_{LO}, for the intermediate frequency (IF) amplifier stages are derived from the USO, as are the clock reference frequencies used to drive the analog-to-digital converter. The REX portion of the system, which follows the 4.5 MHz buffer and anti-aliasing filter, is made up of an analog-to-digital converter (ADC) which feeds a triply-redundant programmed gate array (FPGA). This gate array implements the two data processing functions required by the REX experiment. These are *i*) calculation of the total power in the 4.5 MHz bandwidth containing the uplink signal that enters the antenna, and *ii*) processing of the 4.5 MHz data stream to isolate the $\approx$1 kHz portion of the frequency spectrum containing the occultation signals in order that these can be returned to the ground efficiently. The output of both processes is passed to the NH spacecraft data memory for later transmission to Earth.

Table 2 provides the basic physical and power attributes of REX. The values in the table refer to the ADC and FPGA shown in Fig. 1 plus the electronic structure supporting these elements.

Table 2 REX onboard resource usage. In addition to the dedicated ADC and FPGA, Fig. 1, REX receives stable clock signals from USO used for navigation and amplified input signals from the NH transceiver

Mass (g)	Power (W)	Volume (cm^3)
160	1.6	520

Table 3 Summary of REX link parameters

λ	P_{rad}	G_{gnd}	$G_{s/c}$	T_{sys}
4.18 cm	10 kW (min)	$\approx$66 dBi (min)	$\approx$42 dBi	$\approx$150 K

λ = operating wavelength, P_{rad} = power radiated by DSN ground station, G_{gnd} = gain of 34 m dia. ground station antenna, $G_{s/c}$ = gain of NH 2.1 m spacecraft antenna, T_{sys} = effective system noise temperature of NH transceiver

Table 4 NH/REX expected SNR. Signal-to-noise ratio achieved will depend on ground configuration extant at the time of Pluto system encounter. Capabilities depend on the combination of transmitter power and antenna available, or their equivalent. The 10 kW, 34 m dia. combination satisfies basic REX requirements

Uplink power	Ground antenna dia.	
(kW)	34 m	70 m
10	49.8 dB	55.8 dB
20	52.8	58.8

Table 5 NH/REX radiometric performance

$\delta f/f$	$\delta P/P$	$\Delta P/P$	δT
$\approx 2 \cdot 10^{-13}$	$\sqrt{\text{SNR}}/s$	$\approx$ 0.05 dB	$\approx$ 0.1K

$\Delta f/f$ = expected limiting uncertainty in uplink frequency measurement (see text), $\delta P/P$ = statistical error due to noise, $\Delta P/P$ = typ. error from HGA pointing, δT = expected 1-σ uncertainty in thermal emission temperature, for conditions at Pluto

Tables 3 and 4 together summarize the system aspects of REX. In Table 3 note that the ground transmitter power and antenna gain are minimum values. Table 4 lists the expected signal-to-noise ratios (SNR) for four combinations of transmitter power and antenna size in routine use at DSN ground stations. As the DSN is currently studying options for future configurations of ground stations, the levels of transmitter support available during the Pluto system encounter in 2015 are not known with certainty, but are expected to be within the range reflected by the options in Table 4.

Table 5 summarizes the radiometric performance expected of the REX instrument. Values in the table represent the limiting statistical uncertainties in measurement of frequency (f) and power (P) in coherent signals received by REX ($\delta f/f$ and $\delta P/P$) and the statistical uncertainty in the measurement of emission temperature in the absence of a coherent signal (δT). The variation in power represented by $\Delta P/P$ results primarily from uncertainties in spacecraft antenna pointing. Although pointing errors vary over periods of tens of seconds or longer they are highly correlated on much shorter periods of time, in contrast with noise-driven uncertainties that essentially change with each data sample. For occultation observations over intervals of 1 s the system SNR is the limiting error in determination of received power, while $\delta f/f$ is determinant for retrieval of the atmospheric structure. For other REX observations in the Pluto system $\delta f/f$, $\Delta P/P$, and δT represent the primary experimental limitations.

The value of $\delta f/f$ in Table 5 requires comment. the value given is the design value anticipated for encounter with the Pluto system. Typically the performance of precision oscillators improves with uninterrupted operation. Laboratory values seen prior to launch were in the high parts of 10^{-13}; values observed during the Jupiter encounter thirteen months after launch were $\approx 5 \cdot 10^{-13}$. Values as of this writing are $\approx 3 \cdot 10^{-13}$ (Nov. 2007).

1.3 Pluto Encounter

The REX encounter with the Pluto system is focused on occultations by Pluto and Charon with the goals listed above. These occultations provide opportunities for characterization of Pluto's atmosphere and possibly a sensible ionosphere, and to search for a sensible atmosphere and ionosphere of Charon. For each body the time difference between the entry and exit events will provide the length of the occultation chords. Occultation events also provide opportunities to observe thermal emission from Pluto and Charon. By maintaining the pointing direction of the HGA in the Earth direction while NH is behind each body, the motion of the spacecraft results in a sweep of the antenna beam across the nightsides of Pluto and Charon. At Pluto this strategy will yield approximately five half-beam resolution elements arranged diametrically in a line across the disk. At Charon, however, the distance behind the body is such that the antenna beam is larger than the obscured disk. Results for gravity will be derived from tracking data obtained during approach to and recession from the system, and from occasional opportunities around the times of occultation observations. Baseline occultation data analyzed to obtain the uplink Doppler shift referenced to the on-board USO, though of lesser quality than two-way Doppler observations, also may prove useful in this regard. Details of these measurements are discussed individually in subsequent sections of this paper.

1.4 Encounter Geometry

Figure 2 shows the nominal New Horizons encounter geometry. Not shown are the relative positions of Earth and Sun. Some details of timing and other conditions at closest approach can be found in Table 6. In the figure the path of NH through the system passes between the orbits of Pluto and Charon, but at a time when Charon is located opposite Pluto from the spacecraft. After closest approach to Pluto, the spacecraft proceeds to fly behind Pluto's disk at a distance of about 37 R_{P}, and then behind Charon at a much greater distance of roughly 177 R_{C}. The occultation at Pluto is targeted to be diametric with respect to the Earth for the radio event, which also works well for the UV solar occultation experiment owing to the closeness of NH behind Pluto and the closeness of Earth and Sun in the sky. At Charon the UV observations are optimized by making this event close to diametric with respect to the sun. As a result the radio occultation chord is expected to lie a large fraction of the Charon's radius from the center of the disk, depending on trajectory delivery errors.

Not shown in Fig. 2 are the relative positions of Earth and Sun. These have been optimized for the radio and UV occultation events by the choice of the NH trajectory and an

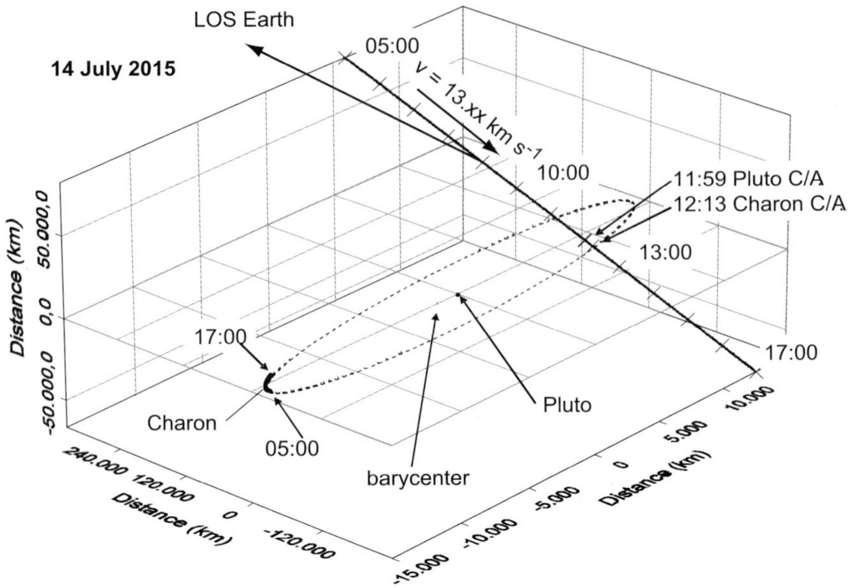

Fig. 2 Geometry of NH encounter with the Pluto–Charon system. Nominal NH flight through the Pluto–Charon system follows a nearly straight line $\approx 165°$ from the line-of-sight to Earth and relative to Pluto; the bending angle of the trajectory is $\approx 0.04°$. Trajectory angle relative to the Sun is $\approx 171°$

Table 6 Pluto system encounter geometry. Nominal geometric parameters for NH closest approach (C/A) to Pluto and Charon, July 14, 2015. Included angle is that between the direction of motion of the spacecraft relative to the sun and the LOS direction from spacecraft to Earth. See Tables 9 and 10 for details

NH spacecraft	Distance to Earth	31.90 AU
	Included angle, NH velociy and LOS	171.4°
	Flyby velocity	13,785 km s^{-1}
Pluto C/A	Time	11:59:00 UTC
	Distance	11,094 km
Charon C/A	Time	12:12:51 UTC
	Distance	26,925 km

arrival date in the Pluto system with Earth at inferior solar conjunction. In this regard, two factors played into the trajectory selection. First, in planetary occultations there is often a conflict between UV solar and radio experiments owing to the differing directions from the spacecraft to the sun and to Earth, respectively. Second, the anticipated radio occultation effects, expressed as an apparent change in radio path length, will be a small fraction of the radio wave length.

In order to understand the significance of the last point, consider that centimeter-λ radio signals traveling between Earth and a vehicle somewhere in Earth's orbit about the sun typically encounters phase fluctuations of many meters to kilometers in passing through the solar wind, depending on path geometry. Much larger effects can be anticipated for various $\approx$33 AU paths between Earth and Pluto. Solar wind fluctuations arise primarily from the radial advection of irregularities in solar wind density across the propagation path, thereby introducing a time-variation on scales of seconds to hours in the radio (optical) path length. This effect is reduced significantly for radio paths that lie along radials from the Sun since the total electron content along such a path can vary only slowly as solar wind material enters and departs the path at Earth and NH, respectively, and the path position with respect to sources on the Sun changes slowly with respect to the speed of the radial flow. Nevertheless, changes in path structure can occur to varying degrees along the radial direction on time scales of minutes to hours.

NH mission planning addressed the two issues of *i*) how best to obtain both solar UV and radio occultation observations and *ii*) the effects of solar wind induced noise on radio occultation data. For the chosen encounter time *i*) the angular separation of Sun and Earth as viewed from NH is only 0.23°, which allows the UV sensor and the NH high-gain antenna to be effectively co-aligned while *ii*) placing the radio propagation path as close as possible to a solar radial through Pluto at the Earth end. This choice eliminates the potentially conflicting requirements for the UV-solar and radio-Earth occultation observations; solar wind scintillation are reduced by more than an order-of-magnitude relative to an Earth–NH quadrature geometry, and by about three orders of magnitude relative to geometries in which the Earth is located within a few degrees of and behind the sun as seen from NH. We note that the approximate integrated effect of the solar wind along the $\approx$32 AU radial path to Pluto from Earth during the encounter in 2015 will be to add an apparent 0.5 m length to the path. The radial geometry also provides nighttime conditions at the DSN tracking sites, leading to reduced ionospheric effects as well.

2 Radio Occultation Sounding of Pluto and Charon

2.1 Introduction

The reconnaissance of the Neptune system by Voyager 2 included a radio occultation experiment that successfully sounded the tenuous neutral atmosphere of Triton (Tyler et al. 1989). This experiment was conducted in the conventional downlink configuration, using radio signals transmitted by Voyager 2 and received by terrestrial tracking stations equipped with large-diameter antennas. The data at 3.6 cm-λ had a fractional frequency stability $\delta f/f$ of $\sigma_{\text{Allan}} \approx 2 \cdot 10^{-12}$, where σ_{Allan} is the Allan deviation set by an ultra-stable oscillator (USO) onboard Voyager 2, and a concomitant signal-to-noise ratio (SNR) of 45 dB in a 1-Hz bandwidth. Analysis of these data initially yielded a surface pressure of 1.6 ± 0.3 Pa (16 ± 3 μbar), later refined to 1.4 ± 0.1 Pa (Gurrola 1996), and a constraint on the average scale height of the lower atmosphere, but the observational data were too noisy to permit reliable atmospheric profiles to be obtained through data inversion.

Pluto's atmosphere is believed to be similar to Triton's in that both atmospheres comprise predominantly N_2 plus other gases such as CO and CH_4 in uncertain abundances, and with surface pressure determined by vapor-pressure equilibrium. In designing a radio occultation experiment to sound Pluto's atmosphere we drew heavily on lessons learned with Voyager 2 at Triton.

The New Horizons radio system includes two key enhancements relative to Voyager 2. First, New Horizons carries two USOs, each with a fractional frequency stability anticipated to be $\sigma_{\text{Allan}} \approx 2 \cdot 10^{-13}$ at Pluto. (See Table 5 and discussion, above.) Additionally, the New Horizons experiment will be conducted in an uplink configuration that exploits high-power transmitters available on the ground at tracking stations of the NASA Deep Space Network

(DSN), thereby making it possible to obtain precision measurements in the short interval of time available during the Pluto occultation. The resulting SNR of the data recorded onboard New Horizons will be about 55 dB in a 1-Hz bandwidth, commensurate with the enhanced frequency stability of the onboard USOs. For New Horizons the two critical parameters of frequency stability and SNR are improved substantially with respect to the performance of Voyager 2 at Titan.

2.2 Background and Motivation

Before further discussion of the radio occultation experiment, we review briefly the fragmentary understanding of Pluto's atmosphere that has emerged from Earth-based observations during the past few decades. Our goals here are to provide a context for describing the capabilities of REX and to identify major questions that the instrument is uniquely qualified to address.

Observations of sunlight reflected by Pluto in the near-infrared spectral region provide strong constraints on the identity and relative abundances of surface ices. High-albedo surface units are dominated by solid N_2, with frozen CH_4 and CO as trace constituents (Owen et al. 1993; Cruikshank et al. 1997; Grundy and Buie 2001). Among the ices identified on the surface, N_2 is not only the most abundant but also the most volatile (Brown and Ziegler 1979), so that it is widely believed to be the dominant component of the atmosphere (e.g., Cruikshank et al. 1997; Yelle and Elliot 1997). The atmosphere should contain traces of both CH_4 and CO, but their mixing ratios are unknown.

Photometric observations of Pluto's occultation of the star P8 in 1988 provided decisive evidence for the presence of an atmosphere and important constraints on its thermal structure (Hubbard et al. 1988; Elliot et al. 1989). Yelle and Elliot (1997) reviewed the implications of these measurements. In high-quality data recorded by the Kuiper Airborne Observatory (KAO), the slope of the light curve changes abruptly near the half-light level at both ingress and egress, marking a significant transition in atmospheric structure at a radius of about 1,215 km. Above this level the atmosphere is nearly isothermal with a pressure scale height of approximately 56 km, corresponding to a temperature of about 104 K for a pure N_2 atmosphere (Elliot and Young 1992; Millis et al. 1993). The stellar occultation results are ambiguous with respect to structure of the lower atmosphere, as the distinctive slope change in the light curve could arise either through extinction by atmospheric aerosols (Elliot et al. 1989; Elliot and Young 1992; Millis et al. 1993) or from the refractive effects of a strong temperature inversion (Eshleman 1989; Hubbard et al. 1990). By enhancing the refractive bending of starlight, such an inversion could also conceal the presence of an underlying troposphere (Stansberry et al. 1994).

Neither Pluto's radius nor the pressure at the surface can be determined with confidence from observations of the 1988 stellar occultation (Stansberry et al. 1994; Yelle and Elliot 1997). The data preclude a radius larger than about 1,200 km and a surface pressure less than roughly 0.3 Pa, but the measurements cannot discriminate among diverse models for the lower atmosphere, including one with a 40-km deep troposphere, a radius of 1,160 km, and a surface pressure of about 2.4 Pa (Stansberry et al. 1994).

An occultation of the star P131.1 by Pluto, as viewed by an array of ground-based telescopes in 2002, revealed a significant increase in the total mass of Pluto's atmosphere (Elliot et al. 2003; Sicardy et al. 2003). In the region above roughly 1,215 km radius, the atmosphere remains isothermal and the temperature appears essentially unchanged since 1988, but the pressure at a fixed radius of 1,215 km has doubled, increasing from about 0.23 Pa in 1988 to about 0.50 Pa in 2002. It is expected that the surface pressure has probably increased by a similar factor, although the structure of the lower atmosphere and the surface pressure remain highly uncertain.

Hansen and Paige (1996) predicted such an increase in surface pressure, and modeled the seasonal evolution of an N_2 atmosphere in vapor pressure equilibrium with surface frost. In this model, sublimation and condensation of N_2 are controlled by the heat balance of surface frost and, consequently, vary with latitude and season. The net result is a surface pressure that depends not only on the distance of Pluto from the sun but also on the subsolar latitude. The vapor-pressure equilibrium model predicts that the peak surface pressure on Pluto would occur not at perihelion in 1989 but would be delayed by two to three decades to the time when the subsolar latitude moves from the equator into the southern hemisphere, exposing a polar cap of solid N_2 to increasing solar heating. In this senario, a precipitous decrease in surface pressure is expected to begin in roughly the period around 2025 as the south polar N_2 cap is depleted, the north polar cap grows, and Pluto continues to recede from perihelion (Fig. 11, Hansen and Paige 1996). The accuracy of this model is limited by large uncertainties in the albedo and emissivity of the frost, the albedo and thermal inertia of the substrate, the total nitrogen inventory, and the behavior of multicomponent ices, particularly N_2 contaminated with traces of CH_4 (Hansen and Paige 1996; Spencer et al. 1997).

No observations of Pluto's extended upper atmosphere are currently available, but the atmospheric chemistry and the characteristics of the ionosphere have been explored with photochemical models (Summers et al. 1997; Krasnopolsky and Cruikshank 1999). In the ionosphere, the models predict a strong dependence of both ion composition and peak electron density on the currently unknown abundances of CO and CH_4. The vertical distribution of CH_4 depends in turn on its abundance near the surface and the rate of vertical transport. If the Pluto CH_4 mixing ratio near the surface is relatively small, as on Triton, it is expected that CH_4 would be destroyed photochemically in the lower atmosphere, leaving the ionosphere dominated by relatively long-lived atomic ions such as N^+ and C^+. In this instance the peak electron density can be as large as about 2,000 cm^{-3} (Summers et al. 1997, Fig. 8). On the other hand, if the CH_4 mixing ratio near the surface is sufficiently large, then the resulting enhancement in CH_4 abundance at high altitudes would ensure that the ionosphere is dominated by molecular ions. Consequently, the relatively fast recombination then limits the peak electron density to roughly 800 cm^{-3} (Summers et al. 1997, Fig. 10; Krasnopolsky and Cruikshank 1999).

Pluto's size has been the subject of numerous investigations using several diverse techniques (Tholen and Buie 1997). The best results to date have been obtained from mutual event photometry (e.g., Young and Binzel 1994) and direct imaging with the Hubble Space Telescope (Albrecht et al. 1994; also see Sect. 3.1, below). According to these observations, Pluto's radius lies between about 1,150 and 1,180 km, with a typical uncertainty of 15 km (Tholen and Buie 1997, Table 3). Pluto's volume is therefore known to an accuracy of approximately 4 percent.

Pluto's mass has been determined from HST images of Pluto and its three known satellites (Buie et al. 2006). The combined mass of Pluto and Charon is $(1.4570 \pm 0.0009) \times 10^{22}$ kg, while the Charon-to-Pluto mass ratio is 0.1165 ± 0.0055, yielding a mass for Pluto of $(1.3050 \pm 0.0064) \cdot 10^{22}$ kg. As the mass uncertainty is only about 0.5 percent, the radius is the limiting factor in current estimates of Pluto's density.

2.3 Experiment Objectives, Methods of Analysis, and Predicted Performance

2.3.1 Pluto's Radius

The New Horizons radio occultation experiment will determine Pluto's radius to an uncertainty of much less than 1 km. The occultation is nearly diametric by design, minimizing the sensitivity of this measurement to errors in the spacecraft trajectory. In addition, because extinction by aerosols is negligible at centimeter wavelengths the radio occultation radius is insensitive to the presence of an optically thick limb haze, e.g., as in the radio occultation of Voyager 1 by Titan (Lindal et al. 1983). When this new radius measurement is combined with previous determinations of Pluto's mass, the mean density of Pluto will be known to less than 1 percent, comparable to the current uncertainty for Triton (Tyler et al. 1989; McKinnon et al. 1995).

The data recorded by REX during the Pluto occultation will contain not only the effects of Pluto's atmosphere, as discussed below, but also an extended diffraction pattern caused by the surface. We will remove these diffraction effects from the data through use of an established technique, alternately called "inverse Fresnel filtering" or "back propagation." First developed and applied in connection with radio occultation studies of planetary rings (e.g., Marouf et al. 1986; Tyler 1987; Gresh et al. 1989) and later adapted for use in atmospheric occultations (e.g., Tyler et al. 1989; Karayel and Hinson 1997; Hinson et al. 1997, 1998), this technique for excising limb effects is well understood. When applied to Pluto, this method of analysis will allow atmospheric sounding without interference from diffraction effects while yielding an accurate solution for Pluto's radius.

As an example of this technique, Fig. 3 shows results obtained by applying inverse Fresnel filtering to radio occultation measurements obtained with Mars Global Surveyor. This experiment sounded the atmosphere of Mars at 293.3E, 67.2N on 31 December 1998; the data quality is comparable to what will be obtained with New Horizons at Pluto (e.g., Hinson et al. 2001). The unfiltered intensity data (Fig. 3a) are strongly modulated by diffraction from the limb of Mars. Information about the location of the limb is distributed across a

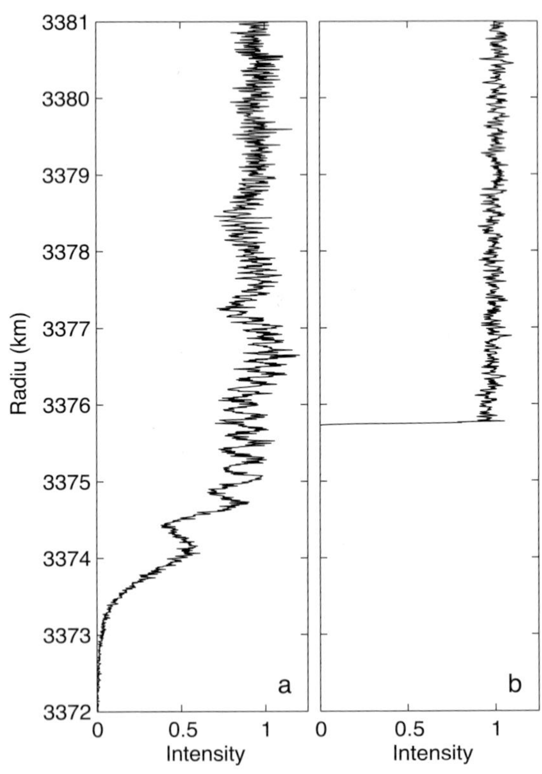

Fig. 3 Measurements of signal intensity during a radio occultation experiment conducted with Mars Global Surveyor on 31 December 1998. (**a**) Original data with a sample spacing of approximately 5 m. The diffraction pattern produced by the limb of Mars contains >40 diffraction fringes and extends more than 5 km below the location of the limb. Refractive bending in the lower atmosphere deflects the diffraction pattern into the geometric shadow of Mars by ≈2 km. (**b**) Intensity measurements after inverse Fresnel filtering. Abrupt drop in intensity marks the location of the surface

pattern of more than 40 diffraction fringes extending to more than 5 km in radius below the geometric limb. Inverse Fresnel filtering of these data removes not only the limb diffraction pattern but also the amplitude effects due to refractive bending in the atmosphere, as shown in Fig. 3b. The filtered data resemble a step function, with the signal intensity dropping by nearly 30 dB, or a ratio of 1,000 : 1, in a radial span of 30 m, yielding a clearly defined local radius of $3,375,742 \pm 10$ m. This result is consistent with independent measurements of the surface radius at this location by the Mars Orbiter Laser Altimeter (Smith et al. 2001). The uncertainty in radius, about 10 m, is smaller by a factor of 50 than the characteristic Fresnel scale of the diffraction pattern, which is 490 m for the geometry and wavelength of this MGS experiment. The Fresnel scale for New Horizons will be about 1,350 m, suggesting that Pluto's radius at the occultation point can be determined with a precision of about 30 m.

2.3.2 Pluto's Neutral Atmosphere

The New Horizons radio occultation experiment will sound the near-surface neutral atmosphere of Pluto at both ingress and egress. The sensitivity of these measurements is sufficient to allow retrieval of accurate vertical profiles of number density (n), pressure (p), and temperature (T) that extend from the surface to about the 0.2 Pa pressure level. Key results will include the first unambiguous measurements of atmospheric structure below approximately 1,215 km radius, as well as the values of n, p, and T, at the surface. The experiment will reveal the detailed structure of any low altitude temperature inversion and will determine whether or not a deep troposphere is present and, if so, its characteristics.

In analyzing occultation data we will remove diffraction effects through inverse Fresnel filtering, as described above. Atmospheric profiles can then be obtained via the conventional "Abel-transform" retrieval algorithm (e.g., Fjeldbo et al. 1971; Tyler 1987; Karayel and Hinson 1997; Hinson et al. 1999), which has been applied extensively in occultation experiments elsewhere throughout the solar system. By this procedure we will obtain vertical profiles of n, p, and T at both ingress and egress. The profiles will be registered accurately in both planetocentric radius and altitude above the surface.

We conducted numerical simulations to characterize the expected performance of the radio occultation experiment at Pluto. For these calculations we adopted an uplink SNR of 55 dB in a 1-Hz bandwidth and a USO stability $\sigma_{\text{Allan}} = 2 \cdot 10^{-13}$, consistent with prelaunch laboratory measurements. As discussed here, the NH trajectory will produce a diametric occultation. The plane-of-sky velocity of the spacecraft relative to Pluto will be 3.52 km s^{-1}, with a nominal spacecraft-to-Pluto distance of 43,400 km at the midpoint of the occultation.

The atmospheric model used in the simulations is based loosely on the predictions of pure N_2, radiative-conductive models for Pluto's atmosphere (Strobel et al. 1996). The model pressure and temperature at the surface are 1.0 Pa (10 μbar) and 37 K, respectively, consistent with vapor pressure equilibrium (β-N_2). We based the model surface conditions on the results of the 2002 stellar occultation extrapolated downward to a conservative, i.e., relatively large, estimate of the surface radius of 1,190 km. The model includes a strong temperature inversion in the lowest scale height above the surface, with the temperature increasing to 100 K at 0.5 Pa; at higher altitudes the profile is nearly isothermal. Figure 4 shows the vertical structure of the model in the lowest 50 km above the surface.

We used geometrical optics to calculate the signal variations that would be observed by REX during a radio occultation of the model atmosphere, above. A radio signal that grazes the surface of Pluto as it traverses this atmosphere will experience a phase shift of roughly 1 rad, about half as large as the effect produced by the atmosphere of Triton (Tyler et al.

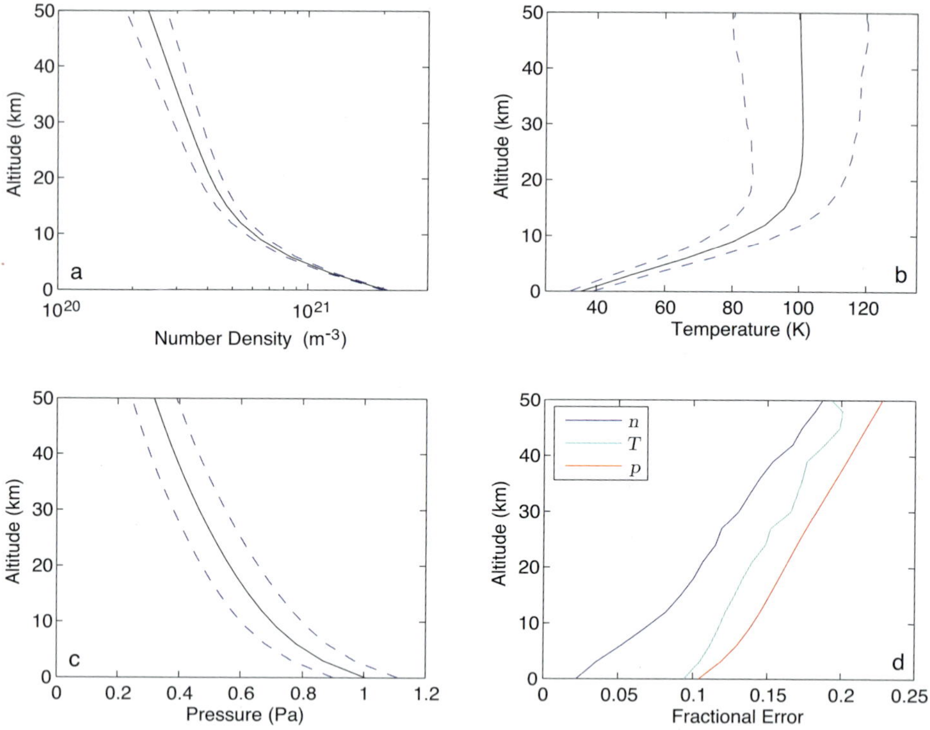

Fig. 4 Simulated performance of radio occultation sounding of Pluto's atmosphere with REX. *Solid lines* in the first three panels show profiles of (**a**) number density, (**b**) temperature, and (**c**) pressure of the model atmosphere used in the simulations; *dashed lines* in these three panels show 1-σ deviations of the radio occultation retrievals, as characterized by a 900-case Monte Carlo simulation. (**d**) Fractional errors in n, p, and T

1989); the equivalent increase in the propagation path length as a result of the atmosphere is about 6.6 mm. The peak angular deflection caused by refractive bending is 0.8 μrad, and the corresponding Doppler shift observed by REX is roughly 70 mHz, which exceeds the frequency-noise fluctuations associated with the USO by a factor of 50 : 1 in a one second observation time.

We used the Monte Carlo method to characterize the accuracy of the retrievals in this simulated experiment. For this we applied the Abel-transform retrieval algorithm to 900 sets of simulated observations, each constructed by adding computer-generated random noise to the results of the forward calculations. The noise is uncorrelated from one sample to the next with a standard deviation determined by the simulated experimental SNR and the USO stability. Panels (a) through (c), Fig. 4, show the standard deviations in the retrieved profiles of n, p, and T. The radio occultation retrievals are inherently most accurate near the surface, as shown in Fig. 4d. At the surface, the uncertainties in n, p, and T are $4 \cdot 10^{19}$ m^{-3}, 3 K, and 0.1 Pa, respectively, corresponding to fractional uncertainties of about 2 percent in n and 10 percent in both p and T.

The uncertainty in surface pressure can be reduced in several ways, such as by averaging results from ingress and egress or by imposing additional physical constraints to reduce the number of degrees for freedom in the result. The Abel-transform retrieval algorithm derives a pressure profile from the number density profile by assuming hydrostatic balance.

An alternate estimate of surface pressure can be obtained by assuming vapor pressure equilibrium. Use of the equation of state with the vapor pressure formula for N_2 then allows the surface pressure to be derived directly from the number density at the surface. Owing to the very strong dependence of vapor pressure on temperature, the uncertainty in surface pressure is then commensurate with the uncertainty in number density, reducing its value from 10 percent (0.1 Pa) to about 2 percent (0.02 Pa). The actual method(s) of analysis chosen will depend on the nature of the data obtained.

2.3.3 Pluto's Ionosphere

We will also examine the radio occultation measurements for evidence of an ionosphere. On the basis of predictions by photochemical models (Summers et al. 1997; Krasnopolsky and Cruikshank 1999), we expect the effect of Pluto's ionosphere on the radio signal received by REX to be comparable to the noise threshold. Detection should be possible if the peak electron density exceeds about $2 \cdot 10^3$ cm^{-3}, at the high end of current predictions, particularly if the lower ionosphere contains a sharp vertical gradient of electron density (e.g., Fig. 8c, Summers et al. 1997). Such an ionospheric feature would produce an appreciable phase shift in a relatively short time interval. If an ionosphere is detected, we will determine or constrain basic characteristics such as the peak density, the height of the peak, and the topside scale height.

2.3.4 Charon's Atmosphere and Ionosphere

The neutral atmosphere of Charon is likely to be undetectable by the REX occultation observations, given current upper bounds on the surface pressure (e.g., Sicardy et al. 2006). REX will be able to detect an ionosphere of Charon should it exist with number densities comparable to or exceeding the threshhold given in Sect. 2.3.3 above.

3 Gravity Investigations

3.1 Approach

Radio-Doppler tracking of spacecraft falling past a planetary body is a powerful and precise tool for the determination of the acceleration constant product, GM, the gravitational constant G times the body mass M, and hence M. Given M and the radii as determined by a combination of calibrated imaging and radio occultation, the bulk density is found readily. The importance of a flyby determination of GM in the Pluto–Charon system lies in the opportunity to obtain values that are essentially independent of astrometry.

Radiometric observations by the New Horizons spacecraft during the Pluto system encounter will provide an opportunity to separate the masses of Pluto and Charon by first-order observation of their fields. To the extent that radiometric observations are available during the periods which include the approximate closest approaches to Pluto and Charon, it will be possible to determine the masses of Pluto and Charon independently at comparable or better accuracies than current Earth-based observations. We note the Voyager fllyby determination of the GM for Neptune (Standish 1993), which obviated the need for "Planet X."

Using conservative values from past missions (e.g., Voyager, Tyler 1987; Mars Global Surveyor, Tyler et al. 2001; Rosetta, Pätzold et al. 2001), and extrapolating the near-Earth performance of New Horizons to Pluto, typical accuracies in Doppler determination of the

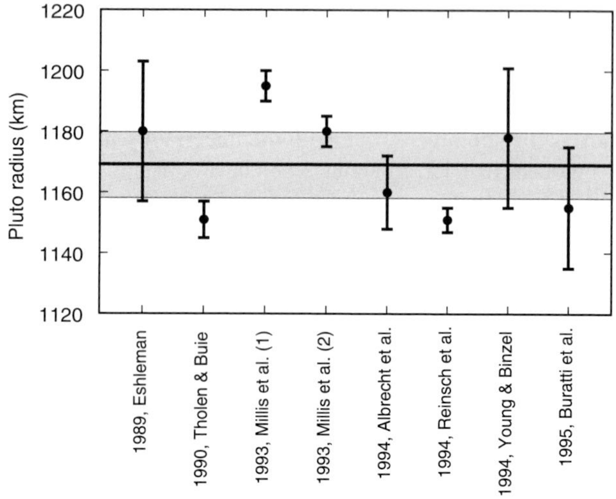

Fig. 5 Observed radii of Pluto from mutual events. Occultation results for Millis et al., (1) and (2), are upper bounds subject to resolution of the atmospheric opacity near the surface; value from Eshleman is an upper bound derived from analysis of stellar occultation radii

spacecraft line-of-sight velocity averaged over 10–60 s of approximately 20 μm s^{-1} can be expected, and values as low as 10 μm s^{-1} might be achieved, subject to conditions at the DSN tracking stations and in Earth's ionosphere and the solar wind. This level of performance may lead to an achievable accuracy in GM of, e.g., $5.6 \cdot 10^{-5}$ for the adopted values of Pluto's bulk parameters (cf. Table 7) if the accuracy is driven by the measurement of $(\Delta v)_\infty$ alone (Anderson et al. 1992).

We consider the approach outlined in the remainder of this section to the estimation of the Pluto and Charon masses. For this purpose we made use of the currently planned flyby geometry, Table 6 and Fig. 2, as well as the bulk parameters discussed below and listed in Table 7.

Recent derivation of the system mass including Pluto and its three moons (Buie et al. 2006) yielded,

$$GM_{\text{sys}} = (972.44 \pm 0.61) \cdot 10^9 \text{ m}^3 \text{ s}^{-2} \tag{1}$$

using $G = (6.67428 \pm 0.00067) \cdot 10^{-11}$ m^3 kg^{-1} s^{-2} and a mass ratio of

$$k = \frac{M_{\text{C}}}{M_{\text{P}}} = 0.1165 \pm 0.0055. \tag{2}$$

The actual mass estimates M_{P} and M_{C} for Pluto and Charon, respectively, are computed from a combination of the system mass, (1) and the mass ratio (2), and are shown in Table 7.

Figure 5 summarizes the various determinations of radii for Pluto together with their stated 1-σ uncertainties. Seven determinations of the radius of Pluto made after 1990 during the mutual events are roughly consistent with each other. All but one have uncertainties that either fall within or touch the range of their average behavior. The individual values with the smaller reported uncertainties, however, do not overlap within their error bars (Fig. 5). From Fig. 5 it is clear that the variation in recent determinations of Pluto's radius is in the range of 20 km. These determinations are also subject to an unknown, systematic bias since the limb of Pluto may be obscured by haze and there is no clear evidence that observed stellar occultation rays have reached the surface. Our adopted mean value and its one sigma error is listed in Table 7.

Table 7 Adopted bulk parameters for Pluto and Charon. Pluto radius is weighted mean of values in Fig. 5. Charon radius is from Person et al.; GM_P and GM_C are from (1) and (2). Masses of P and C are from the respective GMs plus value of G. Errors in radii, GM_{sys}, k, and G have been propagated to tabulated values

Pluto		
Radius	$1{,}169 \pm 12$ km	1%
GM_P	$8.709 \pm 0.043 \cdot 10^{11}$ m^3 s^{-2}	0.5%
Mass	$1.305 \pm 0.0065 \cdot 10^{22}$ kg	0.5%
Density	$1{,}950 \pm 61$ kg m^{-3}	3.1%
Charon		
Radius	606 ± 1.5 km	0.2%
GM_C	$1.01 \pm 0.05 \cdot 10^{11}$ m^3 s^{-2}	4.2%
Mass	$0.1520 \pm 0.007 \cdot 10^{22}$ kg	4.2%
Density	$1{,}630 \pm 78$ kg m^{-3}	4.3%

Until recently, the radius of Charon was even less well determined, although the absence of an obscuring atmosphere eliminates this as a source of bias. Person et al. (2006) determined the radius of Charon from stellar observations to a high degree of precision (see Table 7).

The densities shown in Table 7 are those obtained using the estimates of the radii, the system mass, and the mass ratio just discussed.

The quoted uncertainties of the values in Table 7 are a consequence of statistical uncertainties in the system mass, the mass ratio, and the radii. Bulk density values from the literature often assume a known fixed-body radius without error and result in bulk density values—particularly for Charon—that vary sufficiently as to lead to different models of the interior (cf. Gulbis et al. 2006; Buie et al. 2006). As noted above, the radius of Pluto is subject to the possibility of a systematic overestimation error as a result of atmospheric obscuration near the surface.

From Table 7, it is clear that the limiting factors in our knowledge of the densities of Pluto and Charon are reversed for the two bodies. For Pluto, the 1 percent uncertainty in radius is limiting as it contributes 3 percent uncertainty to the density. For Charon it is the 4.7 percent uncertainty in mass which is limiting. The near-diametric radio occultation of Pluto will be definitive as to the radius whether or not the limb is obscured optically. By its flyby through the system, REX can also provide an independent means of separating the masses.

3.2 REX Contributions

For the estimated values in Table 7 we integrated the equation of motion for the planned 14 July 2015 New Horizon trajectory within the Pluto system based on an initial state 24 h prior to closest approach. We consider both the dominant gravitational forces and a model of the stochastic non-gravitational forces acting on the spacecraft. The primary observable is the line-of-sight (LOS) velocity, v_{LOS}, between the spacecraft and the Earth-based receiving station, obtained by projecting the total velocity onto the direction to Earth. The angle between the direction to Earth and the direction of motion of the New Horizons spacecraft is approximately 171° (Fig. 2). As we are interested in the accelerations of the spacecraft as a result of the gravity of Pluto and Charon, we calculated the expected change in velocity

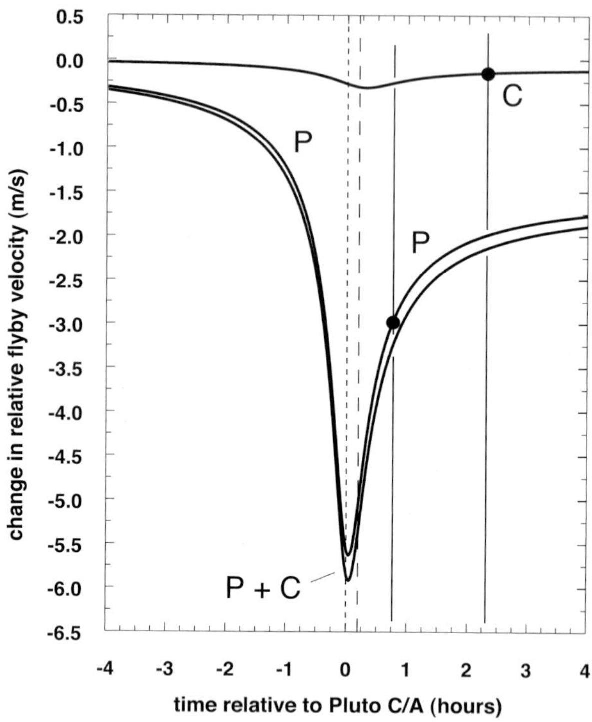

Fig. 6 Line-of-sight velocities for Pluto and Charon. The three curves represent the individual contributions of Pluto (P) and Charon (C) to the total change in velocity, 'P + C.' The time scale is ±4 hrs from the time of closest approach of NH to Pluto. *Stipled line* is closest approach to Pluto; *dashed line* marks the closest approach to Charon, which occurs about 14 minutes later. *Solid lines* crossing marked points in P and C curves show values at times of occultation measurements

component along the line-of-sight, Δv_{LOS}, computed for the time interval within ±4 hrs of the closest approach to Pluto.

Figure 6 shows the nominal variations in Δv_{LOS} for Pluto and Charon calculated for the trajectory with detailed parameters given in Tables 9 and 10. It should be clear that increasing/decreasing the masses of Pluto and Charon to first order increases/decreases the amplitude of the corresponding curve.

In Fig. 6, the units on the vertical scale, $0.5\ \mathrm{m\,s^{-1}}$, represent $5{,}000\ \sigma$, if the measurement error were to be as large $\sigma = 100\ \mu\mathrm{m\,s^{-1}}$. The units of 1 h on the horizontal scale represent 60 samples of velocity obtained with a conservative 1-min integration time. These measurement accuracies, together with the offsets between the points of maximum curvature, indicate that radiometric tracking data can be effective in determining the individual masses of Pluto and Charon.

Figure 6 illustrates the separate contributions of Pluto and Charon to the change in spacecraft velocity, again based on integration of the motion. Curve 'P+C' shows the variation in the total line-of-sight spacecraft velocity from Pluto and Charon. The maximum absolute rates of change in the individual velocity curves for Pluto and Charon occur some minutes before and after the times of closest approach to the two bodies. This is due to the projection of the Δv components across and along the trajectory into the LOS to Earth.

After correcting the data for other forces, e.g., third body gravity plus non-gravitational forces, the values of GM for Pluto and Charon can be extracted from a series of observations during the fly-by interval by constraining the integrated equations of motion to the observed points. The degree of accuracy to which this can be accomplished is not known as of this writing owing to uncertainties in the availability of tracking data during the encounter period, which is limited by the need to employ attitude maneuvers for instrument pointing, thereby taking the spacecraft high-gain antenna beam off the Earth direction. Two periods of planned Earth pointing are the radio occultations of Pluto and Charon, indicated by the two vertical lines in Fig. 6.

There is a simple relationship illustrative of a flyby in which there is only slight bending of the spacecraft trajectory. Let α be the angle between the line-of-sight of observation and the velocity and the change in $\sin\alpha \ll 1$, and let v_{LOS} be the line-of-sight velocity. Then the magnitude of the total change in the spacecraft velocity as a result of the close encounter with the Pluto system is

$$(\Delta v)_{\infty} = |\lim_{t=-\infty}(v_{\mathrm{LOS}}) - \lim_{t=+\infty}(v_{\mathrm{LOS}})| \tag{3}$$

and

$$(\Delta v)_{\infty} = -2\frac{GM_{\mathrm{sys}}}{v_0 d}\sin\alpha^{+}, \tag{4}$$

where v_0 is the relative flyby velocity at the point of closest approach of the spacecraft to the barycenter, d is the distance of closest approach, and α^{+} is the angle between the final direction of asymptotic motion and the LOS to Earth (Pätzold et al. 2001). As a practical matter, given the relatively small mass in the Pluto system and the high velocity of New Horizons approach to the Pluto system, $(\Delta v)_{\infty}$ can be based on observations within a few days before and after leaving the system, subject mainly to modeling of non-gravitational accelerations during the interval bracketing $(\Delta v)_{\infty}$.

The post-encounter change in $(\Delta v)_{\infty}$ for Pluto and Charon together, i.e., the change associated with the system, is approximately $-1.8\ \mathrm{m\,s^{-1}}$, Fig. 6. Again, this is to be compared with the conservative estimate of expected thermal and system noise corresponding to LOS velocity errors of the order of $20\ \mu\mathrm{m\,s^{-1}}$ in 60 s. The post-encounter change in velocity $(\Delta v)_{\infty}$ for Pluto and Charon separately are estimated as $1.78\ \mathrm{m\,s^{-1}}$ and $0.086\ \mathrm{m\,s^{-1}}$, respectively, also in Fig. 6.

The utility of (4) is that it shows the direct connection between the change in velocity of a spacecraft 'test particle' and the planetary mass. More precisely, the solution for the mass M_{sys} is proportional to $(\Delta v)_{\infty}$ since $(\Delta v)_{\infty}$ depend on M for a given flyby velocity and closest approach distance. Typical uncertainties in line-of-sight velocities from modern solutions for spacecraft trajectories are in the range of $20\ \mu\mathrm{m\,s^{-1}}$, while distances of closest approach are usually accurate to within about 1 km; uncertainties in values for $\sin\alpha$ are negligible.

We may use the analytic solution, (4), for a first-order error analysis. The relative error contributions to the determined GM value are,

$$\left(\frac{\sigma_{GM}}{GM}\right)^2 = \left(\frac{\sigma_{v0}}{v_0}\right)^2 + \left(\frac{\sigma_d}{d}\right)^2 + \left(\frac{\sigma_{\Delta v}}{(\Delta v)_{\infty}}\right)^2, \tag{5}$$

where σ_{GM}, σ_{v_0}, σ_d, and $\sigma_{\Delta v}$ are the 1-σ errors contributing to the error in the value of GM. The relative flyby velocity v_0 and the closest approach distance d will be determined from orbit determination after the encounter and are expected to be accurate to much less than $0.0001\ \mathrm{m\,s^{-1}}$ and about 1 km, respectively. This implies that the fractional error associated with uncertainty in the velocity observations is,

$$\frac{\sigma_{v_0}}{v_0} = \frac{10^{-4}\ \mathrm{m\,s^{-1}}}{13{,}785\ \mathrm{m\,s^{-1}}} = 7.2\cdot 10^{-9} \tag{6}$$

which is a very small contribution to the total error.

The error contribution from the closest approach distance is

$$\frac{\sigma_d}{d} = \frac{10^3}{11.1 \cdot 10^6} \frac{\mathrm{m}}{\mathrm{m}} = 9 \cdot 10^{-5} \tag{7}$$

for Pluto, and $\sigma_d/d = 3.7 \cdot 10^{-5}$ for Charon.

Again the error in the observed $(\Delta v)_\infty$ is of the order,

$$\frac{\sigma_{\Delta v}}{(\Delta v)_\infty} = \frac{0.02 \cdot 10^{-3}}{1.8} \frac{\mathrm{m\,s^{-1}}}{\mathrm{m\,s^{-1}}} = 1.1 \cdot 10^{-5} \tag{8}$$

for Pluto and $\sigma_{\Delta v}/(\Delta v)_\infty = 2.3 \cdot 10^{-4}$ for Charon.

For the values in (6)–(8), the individual first order relative accuracy of GM is of the order of

$$\frac{\sigma_{GM}}{GM} \approx 1 \cdot 10^{-4} \tag{9}$$

for Pluto and Charon. We note that the relative accuracy of the gravitational constant G is similar, $1.0 \cdot 10^{-4}$.

As comparisons of the values found in (6) and (8), with those of (7) show, the major determinant of the accuracy of GM is the uncertainty in the distance of closest approach, d. A potentially better accuracy in the determination of $(\Delta v)_\infty$ would make the closest approach distance the sole driver of the GM accuracy. The reverse is also true, namely, any poorer knowledge of the closest approach distance than 1 km would worsen the accuracy of the derived GM values significantly.

Referring again to Fig. 6, it is clear that since the total velocity signature is the sum of the individual velocity changes, that the key to separation of the velocity curves, and hence to separating the masses of Pluto and Charon, is the availability and quality of data that can be obtained in within an hour or two of the center of the encounter. The NH encounter addresses a prioritized list of objectives which in turn lead to specific observations at particular times in the encounter sequence. Since the HGA is body mounted, during the encounter the pointing of the antenna is controlled by the overall mission objectives since the specialized instruments are also body mounted. The amount of tracking that can be obtained practically depends mostly on the occurrence of favorable opportunities to acquire tracking data. This has not be analyzed in detail to date.

The radio occultation observations of Pluto and Charon provide two known opportunities to obtain tracking data since the HGA will be Earth-pointed for fractions of an hour in each instance. These opportunities occur about 51 min and 2 h 18 min following closest approach to Pluto, *v.* Fig. 6 and Tables 9 and 10. The number and timing of other opportunities to obtain additional tracking data during the encounter period cannot be predicted at this time.

In principle, it should be possible to extract the individual masses of both bodies with an accuracy of at least one to two orders of magnitude better than the current best values (see Table 7), but this will be difficult. A determination of the spacecraft position and velocity relative to Pluto and Charon is beyond the charge to the NH Navigation Team, and will fall to the REX Team working in collaboration with other NH investigations. Planned data types bearing on the relative positioning of NH spacecraft relative to Pluto and Charon are the REX occultation times, which are expected to provide positions relative to the centers of figure projected onto the plane determined by the Earth and the NH trajectory during the occultations, optical imaging of Pluto and Charon, and optical navigation data relative to the star field. The results will likely remain ambiguous with respect to uncertainties in the

Table 8 Instrumental characteristics for REX radiometry

Wavelength λ	4.2 cm (7.2 GHz, X-band uplink)
HGA diameter	2.1 m
HGA gain	41.9 dBi
3 dB beam diameter	1.2° (20.9 mrad)
System temperature T_{sys}	150 K
RF Bandwidth	250 MHz
IF Bandwidth	4.5 MHz
Polarization	RCP (Chan A)/LCP (Chan B)
ΔT_{rms}	70 mK (at $\tau = 1$ s)

location of the center of mass with respect to the center of figure, which can be evaluated to some extent based on imaging evaluations of the evolutionary history of the two bodies.

Radio occultation measurements and camera observations of Pluto and Charon will constrain the radii. Due to their slow rotation, both bodies are expected to be close to spherical objects. The derived occultation radii alone is therefore expected to be representative of any chord passing through them. In any event, the best volume measurement will be derived from the combination of occultation and imaging constraints, thereby removing uncertainties in shape as a limitation on our knowledge of the bulk density. The relative accuracy of the bulk density for a radius accuracy as large as 1 km can be on the order of 10^{-3}, which is more than an order of magnitude more accurate than the densities known today and sufficient to distinguish between different models of the interior.

4 Radiometry

4.1 Thermal Emission Observations

At those times when the New Horizons spacecraft high gain antenna (HGA) points toward Pluto or Charon, the REX instrument, operating in a 'radiometry mode,' will receive 7.2 GHz thermal radio emission from the two bodies. Opportunities to observe radio thermal emission occur during the several minutes of radio occultation measurements when the disks of Pluto and Charon obscure the Earth. The REX instrument will detect radiation from the obscuring body as an increase in the radio system noise level in the radiometry channel and also an increase in the noise floor of the occultation channel. These observations will be used to derive the nightside emission temperatures of Pluto and Charon.

For the nightside observations the New Horizons HGA will remain pointed in the Earth direction for the duration of the occultation events, beginning and ending well-above any anticipated sensible atmospheric and ionospheric effects. The motion of the spacecraft will carry the HGA beam across the surface of the planetary disk during the short interruption in uplink signal reception while NH is behind the occulted body. This fortuitous geometry provides opportunities to determine the otherwise inaccessible nightside surface decimeter-emission temperatures. Supplementary observations are desired before the encounter in order to compare the conditions of the sunlit side under near-full-disk illumination with those of the nightside hemisphere; in this connection, it is possible that pointing maneuvers undertaken for other instruments will also sweep the HGA across Pluto, Charon, or both.

The applicable REX instrument radiometric system parameters are given in Table 8. Key geometric and sequencing parameters of this specific observation opportunity at Pluto and Charon are listed in Tables 9 and 10.

Table 9 Pluto encounter: Nominal conditions

Encounter date	14 July 2015 at 11:59:00 SCET
Pluto Closest Approach (PCA)	11:59:00 SCET
Solar distance	32.91 AU
Earth distance	31.91 AU [near opposition]
Pluto distance	11,095 km
Relative velocity	13.78 km s^{-1}
Sun occultation entry	12:43:21 SCET (PCA + 44^m21^s)
Earth occultation entry	12:44:05 SCET (PCA + 45^m05^s)
Pluto distance	38,895 km
Pluto latitude	16.7°
Beam diameter	814 km (0.35 D_p)
Earth occultation midpoint	12:49:34 SCET (PCA + 50^m34^s)
Pluto distance	43,258 km
Pluto latitude	50.1°
Beam diameter	906 km (0.39 D_p)
Sun occultation exit	12:54:10 SCET (PCA + 55^m10^s)
Earth occultation exit	12:55:04 SCET (PCA + 56^m04^s)
Pluto distance	47,666 km
Pluto latitude	−12.7°
Beam diameter	998 km (0.43 D_p)
Duration of occultation	10^m59^s (659 s)

During occultation ingress and egress, noise contributions from the radiative sky background will include the sun at 33 AU, which will be within the beam of the HGA. Assuming that the sun's radiative temperature at $\lambda = 4.2$ cm is about 13,500 K (Zirin et al. 1991), the solar contribution to the system noise is $T_{\text{sun}} \simeq 1.0$ K. Including the cosmic microwave background radiation of $T_{\text{cmb}} \simeq 2.7$ K, the predicted sky temperature becomes $T_{\text{sky}} \simeq 3.7$ K. This is only a small fraction of the estimated overall system temperature $T_{\text{sys}} \approx 150$ K.

The sensitivity of the REX instrument for radiometry ΔT_{rms} can be expressed as,

$$\Delta T_{\text{rms}} = \frac{T_{\text{sys}}}{\sqrt{\Delta f \tau}}, \tag{10}$$

where Δf is the receiver bandwidth and τ is the integration time. Using the applicable IF-bandwidth of 4.5 MHz and $\tau = 1$ s, the above formula yields a senitivity for the emission temperature in the neighborhood of $\Delta T_{\text{rms}} \simeq 70$ mK. Additional analysis of surface emissivity effects will be required to estimate the physical temperature. Knowing the temperature to this degree of accuracy is of interest because of its importance for the vapor pressure of gases in the nightside atmosphere (e.g., Cruikshank et al. 1997).

4.2 Surface Temperatures of Pluto and Charon

As shown in Fig. 7, the HGA boresight follows the approximately diametric subspacecraft track across the disk of Pluto. As a result of the dominant component of spacecraft motion

Table 10 Charon encounter: Nominal conditions

Encounter date	14 July 2015
Charon closest approach	12:12:56 SCET (PCA + $13^{m}56^{s}$)
Solar distance	32.91 AU
Earth distance	31.91 AU [near opposition]
Charon distance	26,925 km
Relative velocity	13.88 km s^{-1}
Sun occultation entry	14:13:35 SCET (PCA + $2^{h}14^{m}35^{s}$)
Earth occultation entry	14:15:11 SCET (PCA + $2^{h}16^{m}11^{s}$)
Charon distance	105,340 km
Charon latitude	−11.0°
Beam diameter	2,206 km (1.86 D_c)
Sun occultation exit	14:16:40 SCET (PCA + $2^{h}17^{m}40^{s}$)
Earth occultation midpoint	14:17:12 SCET (PCA + $2^{h}18^{m}12^{s}$)
Charon distance	106,965 km
Charon latitude	7.8°
Beam diameter	2,240 km (1.89 D_c)
Earth occultation exit	14:19:13 SCET (PCA + $2^{h}20^{m}13^{s}$)
Charon distance	108,591 km
Charon latitude	−33.4°
Beam diameter	2,274 km (1.92 D_c)
Duration of occultation	$4^{m}02^{s}$ (242 s)

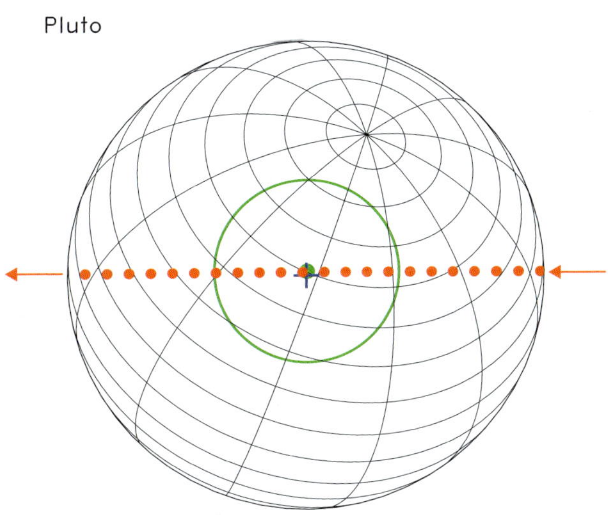

Fig. 7 View of Pluto nightside from the NH spacecraft during Earth occultation on 14 July 2015. The virtual position of Earth, the nominal pointing target for the spacecraft HGA, is shown at 30-s intervals along the trajectory. The subspacecraft point on Pluto at mid-occultation is marked with a *blue cross*. The *green dots* and *concenteric green circle* mark the HGA boresight pointing position and 3-dB beam size on Pluto for the same instant of mid-occultation

away from the Sun, the HGA beam on Pluto increases in size as it sweeps across the disk, but remains much smaller than the diameter of Pluto, taken as 2,330 ± 30 km (Tholen and Buie 1997).

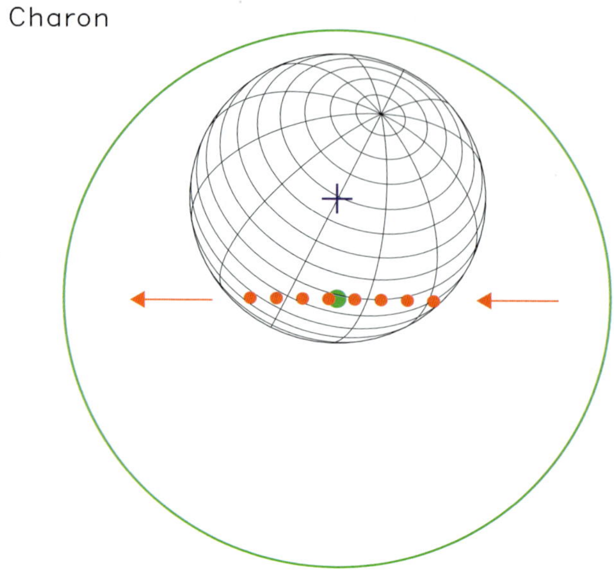

Fig. 8 View of Charon nightside from the NH spacecraft during Earth occultation. See similar sketch of the Pluto occultation geometry for explanations of the various features. A major difference at Charon with respect to the Pluto occultation is that the 3-dB antenna beam is much larger than the Charon disk

Under these conditions, it is possible to use the approximation that the receiver sky temperature approaches that of the Pluto nightside kinetic surface temperature to within a few percent. Minor corrections are required to account for the obscuration of the sun and antenna side lobe effects. Although Pluto is a relatively fast rotator, the high tilt of its rotation axis may lead to significant differences between the microwave surface temperature on the backside as compared with the sunward side. That is, the nightside could be distinctly cooler than the roughly 40 ± 5 K derived in the late 1980s and early 1990s from ground-based mm and far infrared observations of the sunlit hemisphere (Altenhoff et al. 1988; Jewitt 1994; also see Tryka et al. 1994), and the more recent observations in the range of 54–63 K (Lellouch et al. 2000).

Figure 8 shows the geometrical conditions at Charon, which are not as favorable as those for Pluto. As a result of the greater distance and smaller absolute disk size, the HGA beam always covers the entire disk of Charon. Nevertheless, the considerable differences in the surface properties derived for Pluto and Charon would likely be manifested in different inherent emissivities and thus different nighttime temperatures (Cruikshank et al. 1997).

5 REX Implementation

5.1 Implementation-Overview

The Flight Element of REX is based on the shared use of the onboard transceiver in order to avoid functional duplication of spacecraft hardware. Transfer to REX of the uplink signal from a tap near the end of the transceiver amplifier chain provides an efficient method for providing the signals needed for occultation and radiometry with minimal additional hardware.

As the signals presented to REX are the same as those processed in the receiver for purposes of spacecraft tracking and commanding (Fig. 1 and accompanying discussion), the signal bandwidth at the REX input is optimized primarily for transceiver communications and tracking. This bandwidth represents a usable choice for radiometry, but is much greater than is required for the occultation experiment.

Given the input constraints from the transceiver, REX hardware implements two parallel signal processing functions: *i*) for radiometry, determination of the total noise power in the input stream to REX, which is proportional to the linear combination of 4.2 cm-λ noise entering the spacecraft antenna with that internally generated by the receiver, and *ii*) for occultation data analysis and gravity, narrow-band filtering of the input signal to isolate the uplink signal and to reduce the retained bandwidth to the minimum required. In addition, REX handles the necessary housekeeping tasks needed to maintain its interface with the spacecraft system. Required inputs to the REX system are a USO-derived clock signal and the signal feed from the NH transceiver. REX hardware is an integral part of the NH transceiver; as a result REX control functions are implemented as transceiver commands.

The inputs to the REX instrument are samples of a 4.5 MHz bandwidth transceiver IF channel sampled at the rate of 10 Ms s^{-1}. This input stream is processed to accomplish the two primary functions described above. In the first instance the processing consists of simple squaring and accumulation of the IF channel output. In the second instance the processing is based on a series of complex steps which begin with digitally translating the wideband input to baseband using a frequency reference derived from the NH USO. As indicated in Fig. 1, this processing takes place in a single field programmable gate array (FPGA), with spacecraft resource requirements as summarized in Table 2. As a result of the complexity of coherent signal processing, the bulk of the FPGA computational capacity is dedicated to the capture of the uplink occultation signal.

5.2 Extraction of Total Received Signal Power

Referring to the upper path within the dashed box, "FPGA," in Fig. 1 the REX power integrator follows the conversion of the uplink NH transceiver signal to 10 bit digital samples. These data are passed to the REX processor at a rate of 10 Ms s^{-1}, where they are processed to extract the total power in the input stream. This is accomplished by the simple expedient of squaring and adding, as

$$P_{\mathrm{up}}(k) = \frac{1}{N} \sum_{i=N(k-1)+1}^{kN} s_i^2, \tag{11}$$

where, i is the index of the input samples from the quantizer, k is the number of the output sample, and N is the number of squared samples included in the average during each 0.1 s of observation time. In the REX realization P_{up} is represented by 40 bits. These power estimates are interleaved in the REX high-speed output stream and sent to the spacecraft solid state recorder (SSR).

5.3 Multiplierless, Anti-aliasing Filter Design

Referring to the lower path within the dashed box, "FPGA," in Fig. 1, the REX signal processor conditions the baseband signal by reducing the bandwidth to the near-minimum required to support the radio occultation experiment. As indicated above, this is accomplished by heterodyning to zero frequency the uplink carrier signal centered initially at the 2.5 MHz IF center frequency, followed by use of time-invarient baseband filters to reduce the bandwidth. In designing the method of conversion to baseband frequencies from the IF channel, the precise frequencies and bandwidths of 4.5 MHz filter input and the REX processor channel were chosen to permit an efficient digital translation scheme represented schematically

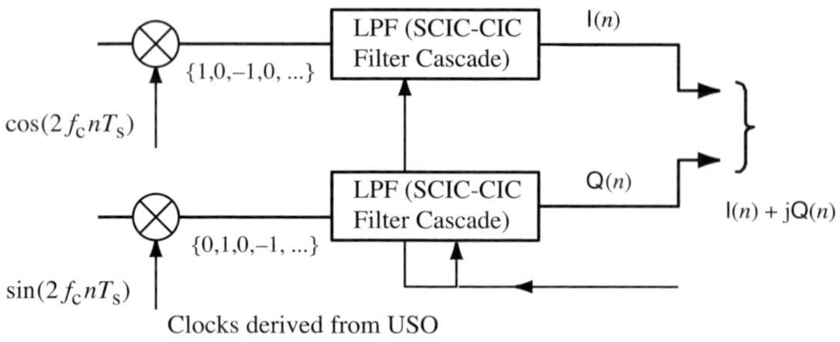

Fig. 9 Block diagram of REX narrow band signal processing. An efficient means of heterodyning the wideband uplink signal to baseband is afforded by the choice of centering the wideband signal at 2.5 MHz, and using a 10 MHz reference frequency. The 1 : 4 ratio of these frequencies results in a multiplier sequence, shown in the figure just to the right of the multipliers, that in effect sends alternate ADC samples successively to the I and Q channels, where the sign is changed of every other sample in each channel. This simple prescription involves only a multiplexer and a switchable inverter, without need for the degree of precision and processor speed, or use of complex multipliers required in a more conventional design

in Fig. 9. The process of down conversion is accomplished by first multiplying the wideband signal by the in-phase reference, i.e., a signal that is proportional to the cosine of the reference frequency, and a parallel multiplication by the quadrature reference, i.e., a signal that is proportional to the sine of the reference frequency.

The downconversion step produces two components, one which is ‘in-phase’ with the USO-derived reference signal, the ‘I’ component, and the other ‘in quadrature’ with the USO-derived reference phase, the ‘Q’ component. At the same time, the center frequency of both components is shifted to 0 Hz, from 2.5 MHz. The I,Q signal components are then filtered simultaneously by identical $\approx$1 kHz lowpass ‘anti-aliasing’ filters in order to suppress all but the central portions of the original IF spectrum, followed by decimation to the lower sampling frequency 2.441 kHz. Each anti-aliasing filter is implemented as a low power decimation filter making use of the signal processing architecture described below. As noted above, all other mixing frequencies in the NH transceiver are also derived from the same USO reference source as the REX processor, so that the fundamental frequency and phase stability of the system is set by the USO, and realized by the quality of the transceiver circuitry.

In a digital implementation certain choices of the reference frequency lead to marked simplifications of digital processes. This characteristic of digital systems is exploited here by choosing the ratio of the center frequency of the wideband channel, $f_c = 2.5$ MHz, and the sampling frequency, f_s, to be 1 : 4, i.e.,

$$\frac{f_c}{f_s} = \frac{1}{4}. \tag{12}$$

With four samples for each cycle of f_c the sequences of values for cosine and sine variations over a single cycle at the sampling frequency become ‘1, 0, -1, 0,’ and ‘0, 1, 0, -1,’ respectively, repeating ad infinitum. Consequently, the result of multiplication by cosine and sine at a frequency f_s can be implemented by simple sample selections and sign changes to the values of the signal at f_c.

The process used here is illustrated in the left hand side of Fig. 9. For our choice of frequency ratios the algorithm is implemented by selecting only the even or only the odd samples and reversing the sign of every other sample in the selected sequences, thus obviating any requirement for multiplication.

The anti-aliasing filter itself is also based on a class of digital, linear phase finite impulse response (FIR) filters comprising a cascade of integrators and comb filters (Hogenauer 1981; Kwentus et al. 1997). In the REX implementation neither the integrators nor the comb filters use hardware multipliers, and for this reason they are referred to as 'multiplierless.' In particular, the transfer function of the integrators is,

$$H_{\mathrm{I}}(z) = \frac{1}{1 - z^{-1}} \tag{13}$$

and comb filters with transfer function

$$H_{\mathrm{C}}(z) = 1 - z^{-1}. \tag{14}$$

Such filters typically are realized by highly efficient and well-understood digital processes. The standard approach, however, results in filter pass band characteristics with levels of out-of-band signal rejection that are deleterious to the REX performance. To understand this, consider that the input noise to the digital filter spans 4.5 MHz, while the desired output bandwidth is only $\approx$1 kHz. As a result of sampling and signal decimation, noise that was initially well controlled at the higher, 10 Ms s^{-1} rate, is increased about 20,000 times by aliasing when the sampling rate is reduced to 2.4 kHz. Thus, narrow band digital filter must have extremely low out-of-band response.

For REX this problem has been solved, uniquely we believe, by modification of the standard 'integrator and comb' filter to further suppress the filter side band responses. This approach reduces the overall filter sideband response by and additional 20 dB over that of a CIC filter along, while also improving the shape of filter pass band. Details of the algorithm employed are explained in the Appendix.

6 REX Commissioning

6.1 Purpose

In the months immediately following launch, the New Horizons project initiated a series of commissioning activities designed to verify instrumental functionality and to measure the performance of the instrument complement. Commissioning of the REX instrument began with an initial test on April 19, 2006, designed to verify basic functionality and performance. This test was conducted while the spacecraft was still spinning following launch. The test comprised an *i*) exercise of calibration test patterns, *ii*) reception of an uplink signal transmitted by a DSN ground station for specific evaluation of the inflight REX system gain, linearity, stability of the composite uplink, USO, and REX receiver chain, *iii*) characterization of the REX bandpass frequency response, and *iv*) a search for spurious signals, if any, in the REX passband. These tests were repeated on June 1, 2006 with the spacecraft attitude stabilized on three axes. Two additional commissioning tests with three-axis attitude stabilization were conducted on June 20 and 29, 2006, for the purpose of measuring the beam pattern of the spacecraft high gain antenna (HGA) and for calibrating the REX system noise figure and radiometric sensitivity.

6.2 Functional Verification

In addition to its operational interface with to the NH transceiver, the input to the REX signal processor can be commanded to process a set of internal test patterns stored in the

REX signal processor. The test pattern input sequences comprise an impulse response, three square waves of different frequencies, two pseudo-random number sequences that differ only in their amplitudes, and a sequence containing only zeros. The response of the REX signal processor to each of these test patterns is fully deterministic, thus the REX output for each of the test pattern inputs can be compared with the expected response sample-by-sample and bit-by-bit. Each time the REX processor is enabled in the NH transceiver, the first part of the test executes a subset of the test patterns for functional verification. For the commissioning tests the subset selected contained six sequences: All zeros, impulse excitation, low-frequency square wave, full-scale pseudo-random sequence, and a repeat of the all-zeros sequence. Each of these test patterns is enabled for six frames, or approximately six seconds, during which the REX output is recorded in the spacecraft solid state recorder (SSR). After the REX contents of the SSR are received on Earth, the REX data are compared with the expected test pattern response on a sample-by-sample and bit-by-bit basis. In all cases, the test pattern outputs matched precisely the expected response for each of the commissioning tests and for each of the test patterns, thereby certifying the functionality of the REX signal processor.

6.3 Evaluation of Spurious Responses, no Uplink

A major concern in the design and integration of the REX system was to avoid the introduction of spurious receiver responses into the REX output. In practical terms the design criterion adopted was that the strength of any spurious signals within the REX passband would be at least −55 dB-Hz below the planned uplink signal transmitted from the ground. The REX commissioning test to certify this level of performance placed the receiver in REX mode, selected the transceiver ADC source for REX, and recorded REX data for 100 s. Subsequently, in post-processing on the ground, frequency power spectra were computed for each of 100 frames, i.e., to 1 Hz resolution, and averaged over the 100 frames. The resulting test spectra were inspected for the presence of spurious responses. No such responses were observed with strength greater than five sigma above that of the spectrum noise floor, where for REX spectra the 5-σ level is −65 dB relative to the uplink signal strength. Thus, the spurious signal design criterion of −55 dB-Hz was met in flight.

6.4 Uplink Signal Acquisition

The radio occultation experiment to probe the structure of Pluto's atmosphere requires that REX instrumentation achieve an uplink signal-to-noise ratio (SNR) of at least 55 dB-Hz, everywhere within an approximately 1-kHz band centered on the planned frequency of the uplink signal. At a nominal distance to Pluto of 33 AU, the SNR requirement is met by transmitting approximately 10 kW at a frequency of 7.2 GHz to the spacecraft from one or more 34-m DSN antennas, resulting in an absolute uplink power in the NH transceiver radiometric channel of −110 dBm. The 55 dB-Hz criterion is achieved when the noise floor in the REX spectrum is −170 dBm-Hz or less, and the spurious-free dynamic range (SFDR), is 55 dB-Hz or better.

To verify that REX has adequate sensitivity the REX commissioning tests emulated the expected uplink signal strength at 40 AU by reducing the transmitted power from the DSN to a level which resulted in a signal strength of −110 dBm in REX's baseband channel. We used data obtained under the conditions just described to quantify the uplink signal strength, the SFDR, and the level of the noise floor. For this purpose we calculated the frequency spectra of REX data with a frequency resolution of 1 Hz. Figure 10 shows an average of 100

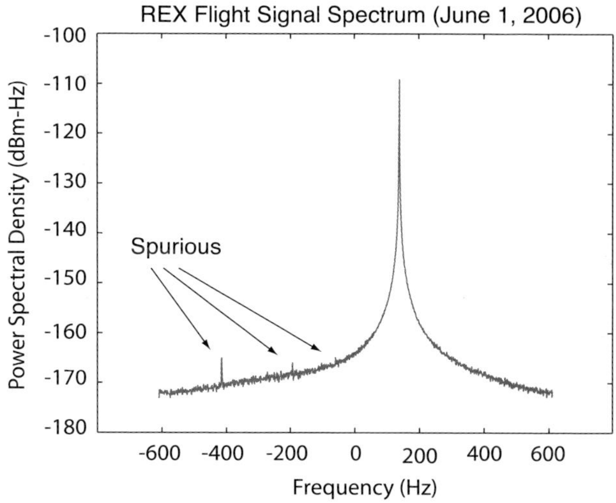

Fig. 10 Example DSN signal acquired during REX radiometric commissioning tests. Spectrum was computed from REX output data to a resolution of 1 Hz, and is the average of 100 unwindowed FFTs. Signal is located at +135 Hz offset from the expected received frequency; broad spectral wings are a feature of the analysis window used. Labeled spurious responses, and others not visible here, are modulation artifacts of only minor consequence. See text

such spectra; the uplink signal strength is close to −105 dBm, somewhat stronger than the −110 dBm level expected at Pluto encounter. The noise floor at −171 dBm-Hz, is somewhat lower than pre-flight tests. At the same time the spectrum is free of spurious signals except for three features located near −50, −200, and −400 Hz and its harmonics. Finer analysis shows that in addition to these features there are also two symmetrically located responses near 30 Hz on either side of the carrier that are not seen in the figure. None of these spurious features are evident in data obtained with the uplink off. The absence of these features in the absence of an uplink suggests that they originate as mixing products within the NH REX-Tranceiver system. In any event the levels of these spurious signals is below that of the required performance.

6.5 USO Stability

Over the course of the REX Commissioning the uplink was transmitted without modulation, with the radiated frequency from the DSN adjusted to compensate for Doppler shifts associated with the relative motions of the Earth and spacecraft. Signal acquisitions by REX spanned periods of typically 100 s, but occasionally lasted as along as 1000 s. For these acquisitions the frequency of the acquired signal in the REX channel can be estimated with a precision limited by the signal SNR, typically 60 dB-Hz, corresponding to an upper bound of about 0.001 Hz. A sample profile of frequency estimates of the uplink signal is presented in Fig. 11. The top plot of the pair shows a series of frequency estimates obtained each 0.2 s during a 100 s span. In this interval the frequency drifted upward at a nearly linear rate over the entire 100 s of the observation. Removing the trend to second order leaves residuals with a standard deviation of 0.008 Hz, for the 0.2 s integration interval. Expressing this deviation relative to the nominal received frequency of $7.2 \cdot 10^9$ corresponds to a fractional deviation over a 0.2 s interval of about 10^{-12}. Repeating the estimates using integration intervals of 1 and 10 s, respectively, results in variations of $5 \cdot 10^{-13}$ in 1 s and $1.4 \cdot 10^{-13}$ in 10 s, suggesting that the USO on transceiver Side A is approaching the stability required for the radio occultation experiment at Pluto. As of this writing no data are available for the second USO on Side B.

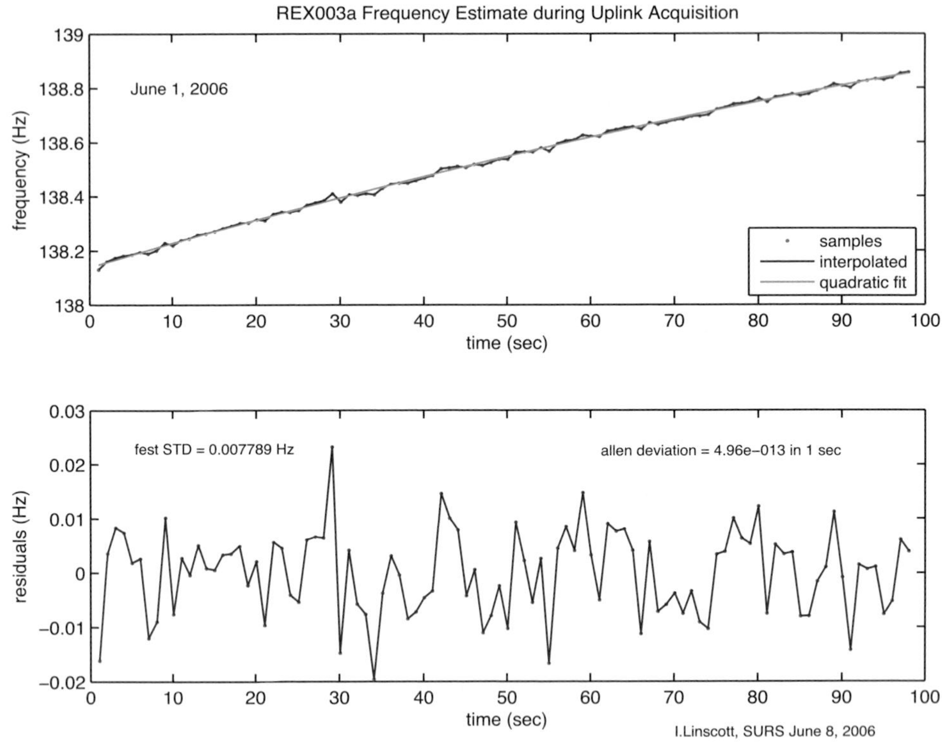

Fig. 11 REX estimates of uplink frequency. *Upper plot* gives the estimated frequency profile obtained using a 0.2 Hz frequency analysis window sequentally over a 100 s duration REX data set, showing a nearly linear upward drift. Removing the trend to second order leaves residuals in *lower plot*, with a standard deviation of 0.008 Hz for the 0.2 s window. Allan deviation at 0.2 s integration interval is $\approx 10^{-12}$, relative to the 7.2 GHz uplink frequency. Estimates obtained over integration time intervals of 1 and 10 s, respectively, result in corresponding Allan deviations of $5 \cdot 10^{-13}$ and $1.4 \cdot 10^{-13}$, respectively. Results are from transceiver Side A

6.6 REX Passband

In order to measure the in-flight response of the REX narrow band filter, the uplink signal frequency was swept slowly across the REX passband during REX Commissioning on April 19 and again on June 20, 2006. In the April 19th test, during which the spacecraft was in its spin-stabilized attitude control mode, a small sinusoidal variation of ± 0.5 dB at the spin frequency was observed in the received power. This effect is attributed to a slight misalignment of the spacecraft spin axis with the HGA boresight. We compensated for this effect by modeling the power variations and removing the effect of the spin from the data. Figure 12 shows the REX bandpass swept-frequency response including both the uncompensated response and the adjusted response using the sinusoidal model. The side lobes of the REX lowpass filter appear at frequencies of about ± 1.8 kHz, where the relative signal level of -44 dB is consistent with the filter design. The swept-frequency response is correctly symmetric, and the noise level drops to a minimum at ± 3 kHz, which is the location of the multiplierless FIR filter nulls.

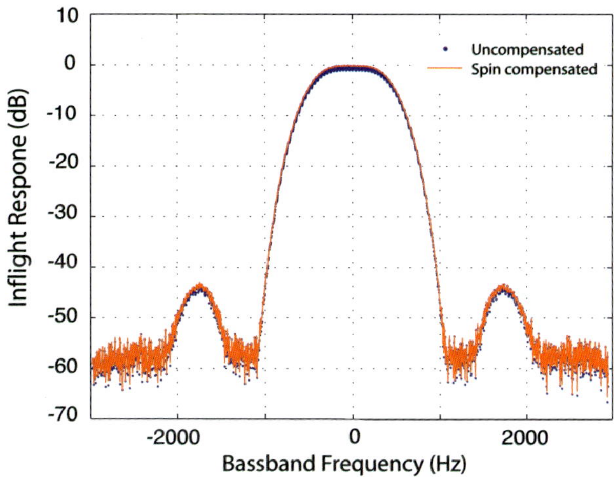

Fig. 12 REX uplink filter passband. Curve gives response of REX to a slow frequency sweep across passband. *Blue curve* is response uncompensated for spacecraft spin; *red* shows the corrected response obtained by use of a sinusoidal modulation model to represent spacecraft spin modulation. Difference in levels between the passband and the two side lobes on outer wings of band shown illustrate the ultimate filter out-of-band noise rejection

6.7 HGA Beam Pattern

The REX commissioning test on July 20, 2006 was dedicated to mapping the beam pattern of the NH spacecraft high-gain antenna. We obtained the beam pattern by continually adjusting the frequency of an unmodulated uplink DSN signal of constant power to arrive at the NH spacecraft with a constant frequency; this signal served as a calibration source. At the same time, we varied the spacecraft attitude with respect to the direction to Earth, thus implementing a scan of the HGA beam over a small range of angles about the Earth direction, centered approximately on the beam maximum. The initial offset of the scan was set at the upper left corner of a $2^\circ \times 2^\circ$ angular box. The beam direction then was made to 'nod and step' parallel to the box edges so as to perform a raster scan bracketing the Earth direction. At the same time the transceiver captured the uplink signal in REX mode and the spacecraft body vectors were time-tagged and logged. The combination of these time sequences allowed us to reference estimates of the uplink signal power to the spacecraft pointing direction.

The resultant HGA response is shown in Fig. 13. An approximately 0.2° offset of the HGA response apparent in the figure is consistent with the ± 0.5 dB variation observed in spin mode. It is believed to be the result of a shift in the center of mass of the spacecraft from the pre-launch condition, a likely consequence of the redistribution of fluid in the fuel tanks.

6.8 Radiometer Calibration

On June 29, 2006, while in REX mode, we obtained a series of five crossed scans of radio astronomy sources together with dwells on cold sky. The spacecraft HGA was initially commanded to point at an offset from the source direction of -1° along the NH body coordinate x, and then scanned across the source at 100 μrad s^{-1} to $x = +1^\circ$, a maneuver that required approximately 350 s. Similar scans were performed for the z-coordinate but with a dwell of 300 s at the origin $x = z = 0$.

The radiometer function in REX integrates the power in data samples from the wideband, 4.5 MHz radiometric channel (*v.* Sect. 5.2). The accumulated growth of the integrated value is placed in the REX high-speed data stream 10 times per second, and the accumulator is

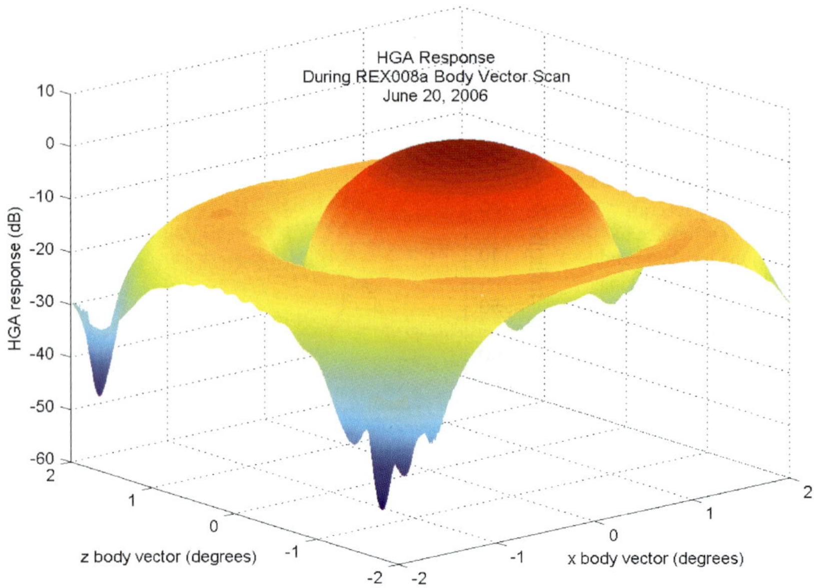

Fig. 13 New Horizons high gain antenna beam pattern at 7.2 GHz. Response of NH HGA to an uplink signal as the spacecraft raster-scanned a $2^\circ \times 2^\circ$ region about the Earth direction. Actual center of beam is offset $\approx 0.2^\circ$ from Earth direction, in $+x$ direction

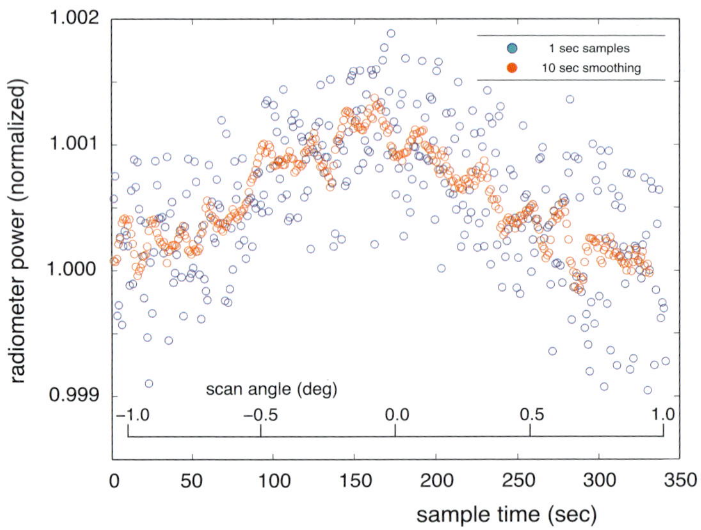

Fig. 14 REX scans of Cygnus-A. Points give values of REX radiometer accumulator vs scan angle. Points with 10 s averaging are correlated by smoothing

reset every 1.024 s, i.e., on the REX frame boundaries. The x-axis scans obtained for the radio source Cygnus-A are shown in Fig. 14, normalized to the cold-sky power level. Scans of Cassiopia-A are similar.

The standard deviation of the 10-s averages indicates that the NH transceiver is radiometrically stable at a level of approximately 5 parts in 10,000, and thus adequate for measuring radiometric temperature to a precision of 0.1 K, or about 1 part in 1000.

7 Summary

Planned New Horizons REX observations of Pluto and Charon, discussed above, address directly our goals for increased understanding of this interesting system: *i*) measurement of atmospheric conditions near and at the surface of Pluto for determination of surface conditions and possible observation of ionospheres at Pluto and Charon; *ii*) an independent determination of the Pluto–Charon system mass, as well as an independent determination of the Pluto and Charon masses; *iii*) nightside observations of the thermal emission from Pluto and Charon with sufficient resolution to determine the brightness distribution across the disk of Pluto; *iv*) expected improvements in the densities of Pluto and Charon from the radio occultation chord in combination with other NH observations, and from expected improvements in the mass of Charon.

In addition, the New Horizons mission has the potential to observe the anomalous accelerations known as the Pioneer effect.

The Radio Experiment is poised to investigate a number of Pluto Science Definition Team Category I, II, and III questions in the Pluto system, and to address additional questions susceptible to radio tracking and propagation methods. Practical implementation of these investigations is made possible by close cooperation between the REX Science Team, the designers and implementers of the spacecraft radio transceiver system, and the NASA Deep Space Network. With the New Horizons Mission successfully in flight operations the science goals set for REX are within reach.

Acknowledgements This work is carried out as part of the New Horizons Radio Science Investigation (REX). In the U.S. REX is supported by the National Aeronautics and Space Administration through the New Horizons Project at the Southwest Research Institute. This paper presents results of a research project partially funded by the Deutsches Zentrum für Luft-und Raumfahrt (DLR). We thank the Precision Time and Frequency Section and the Radio Frequency Engineering Group of the Space Department at APL for their collaboration and critical contributions to the realization and operation of the REX instrument.

Appendix: System Function of the SCIC Filter

A Theory of the CIC-SCIC Filter

A.1 Cascade Integrator Comb (CIC) Filter

Specific details of the design and construction of the CIC and SCIC filters introduced in Sect. 5.3 are discussed here. Hogenauer (1981) proposed use of a cascade of N stages of integrators operating at a sampling frequency, f_s, and comb filters operating at a lower sampling frequency, $f_{s/R}$, as a 'cascaded integrator comb' decimation filter, or CIC. The CIC has the following transfer function,

$$H_{\mathrm{CIC}}(z) = H_{\mathrm{I}}^{N}(z) H_{\mathrm{C}}^{N}(z) = \left(\frac{1 - z^{-RM}}{1 - z^{-1}}\right)^{N} = \left[\sum_{k=0}^{RM-1} z^{-k}\right]^{N}, \tag{15}$$

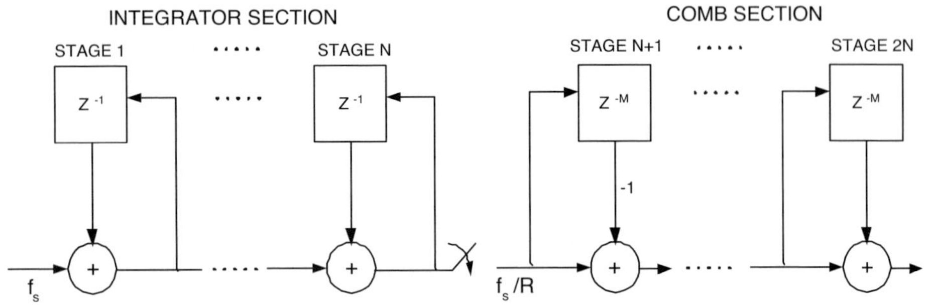

Fig. 15 Schematic diagram of the multiplierless CIC architecture

where the factor R is the decimation ratio of the input to the output sampling frequencies. As revealed in the form of the transfer function, the CIC is a series of N identical FIR filters. A CIC is represented schematically in Fig. 15.

The frequency response of the above filter derived from the transfer function is,

$$H_{\text{CIC}}\left(e^{j\omega}\right) = \left\{ \frac{\sin\frac{\omega RM}{2}}{\sin\frac{\omega}{2}} e^{-j\omega\left[\frac{RM-1}{2}\right]} \right\}^{N}. \tag{16}$$

The CIC frequency response, (16), is characterized by a sequence of nulls at multiples of $f = 1/M$, where M is the integrator depth, chosen to control the placement of the nulls. The design is such that the location of the nulls when aliased fold back onto $f = 0$, so that the aliasing near $f = 0$, or 'DC,' is suppressed substantially. For REX observations the uplink signal will be placed in frequency near $f_s = 2.5 + \Delta$ MHz, where typically $|\Delta| \approx 10^2$ Hz. With $f_c = 10$ MHz and $\frac{f_c}{f_s} = 1/4$, use of the receiver architecture in Fig. 9 results in the uplink signal appearing at frequency Δ relative to DC. For these choices the nulls of a CIC filter strongly suppress aliased out-of-band signal components. In the REX CIC a decimation ratio is $R = 4096$, i.e., 10 MHz/4096 = 2.441 kHz.

The transfer function in (15) indicates that the CIC filter produces a gain of $(RM)^N$ at DC, which has serious implications for precision at high values of R. Additionally, the CIC filter has the disadvantage of affording little choice in the filter response characteristics. These limitations are addressed by splitting the decimation factor in two parts $R = 4096 = 64 \cdot 64$. The first decimation factor is realized in a traditional CIC filter. The second factor, or 'stage,' is implemented in a sophisticated modification of the CIC filter, called a 'Sharpened CIC' filter. As discussed in Hogenauer (1981), the efficiency of the overall design arises from the attributes that: *i*) it requires no multipliers and, *ii*) the second cascade of comb filters operates at the lower data rate, resulting in substantial power savings. Moreover, the simplicity of the corresponding physical structure makes it appropriate for hardware implementation in an FPGA.

A.2 Sharpened CIC Filter

The sharpened CIC, or 'SCIC,' filter was first proposed by Kwentus et al. (1997), based on an earlier suggestion by Kaiser and Hamming (1977), as a method to overcome a rather significant drop off in the passband that is characteristic of the standard CIC filter. As it happens, use of the SCIC architecture also improves the out-of-band response,

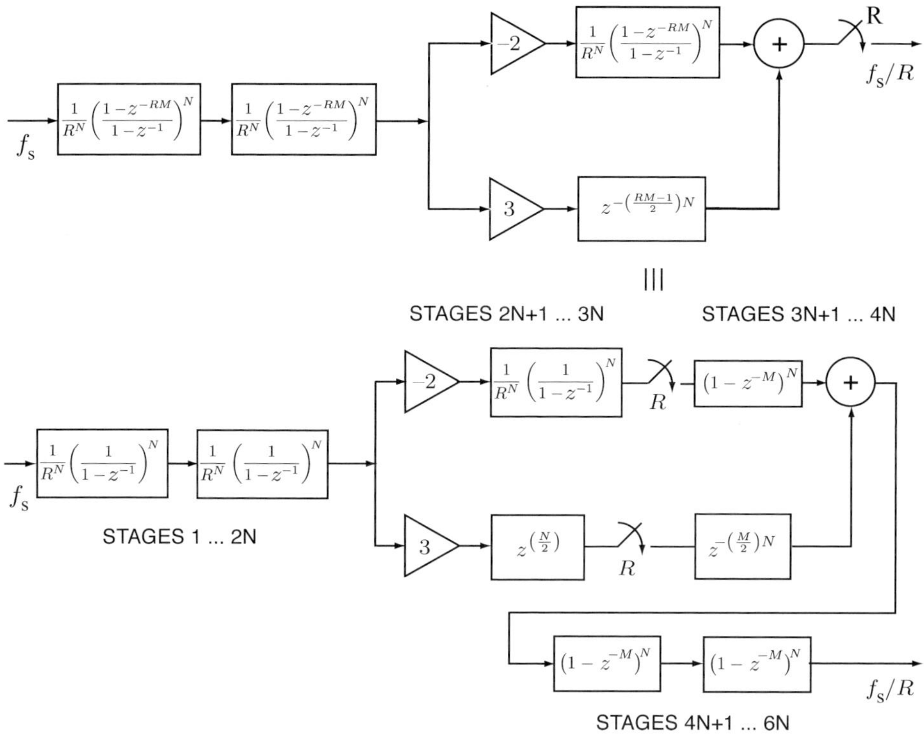

Fig. 16 Schematic diagram of the multiplierless SCIC architecture. *Lower diagram* illustrates the form of implementation in the REX system. *Lower-right-hand branch* contains two delay equalizers

relative to the CIC design. A schematic representation of the SCIC filter is shown in Fig. 16.

The response of the SCIC filter is,

$$
\begin{aligned}
H_{\mathrm{SCIC}}(z) &= H_{\mathrm{CIC}}^{2}(z)\left[3z^{\left(\frac{RM-1}{2}\right)N} - 2H_{\mathrm{CIC}}(z)\right] \\
&= \left(\frac{1-z^{-RM}}{1-z^{-1}}\right)^{2N}\left[3z^{-\left(\frac{RM-1}{2}\right)N} - 2\left(\frac{1-z^{-RM}}{1-z^{-1}}\right)^{N}\right].
\end{aligned} \tag{17}
$$

For the REX implementation, $R = 64$, $M = 1$, and $N = 2$, giving the frequency response of the SCIC filter is shown in Fig. 17.

The CIC-SCIC combination retains the advantage of having nulls at multiples of $f = 1/M$, while achieving a steeper drop at the band edge response than in the CIC filter alone.

A.3 Lowpass, Anti-aliasing Filter Design

A precise implementation of a staged CIC-SCIC filter involves careful design of register length along the data path in the two filters. To do so requires modeling and simulation of the growth of errors due to truncation and rounding along the data path, as there is no

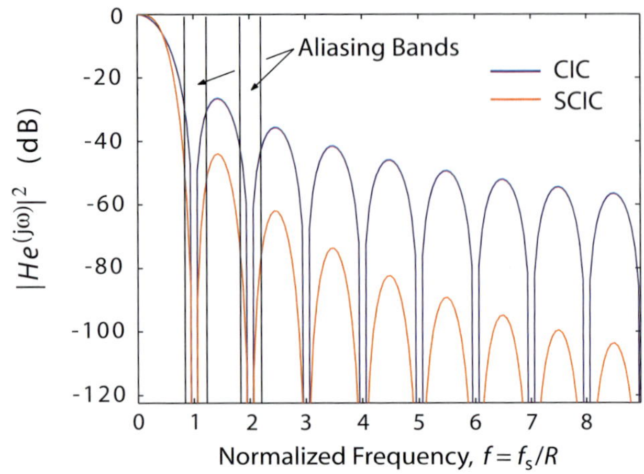

Fig. 17 Frequency response of SCIC and CIC filters. 'Aliasing bands' indicate first two folding frequencies. Note increased sharpness of the SCIC nulls in the aliasing bands as compared with the CIC realization. SCIC filter also has the flatter passband and sharper transition; compare differences between SCIC and CIC curves within the main lobe. For an input sampling frequency, f_S, the output frequency is f, as indicated

analytic solution to the problem. The visibility afforded by this approach makes is possible to optimize the allocation of precision along the processing path, which again improves the efficiency of the physical realization.

The precision in CIC and SCIC filters is determined by the overall register growth, combined with a truncation principle that accounts for the propagation of quantization errors along the data path of the filter. By combining register growth with truncation, a worst-case design principle is established wherein the truncation error at the filter's output uniformly bounds the error incurred at the intermediate stages. Specifically, drawing from the discussion by Hogenauer (1981), if B_j denotes the number of bits truncated at the jth-stage, $j = 1, \ldots, 6N$, for the SCIC filter as shown in Fig. 16, the truncation error has a uniform distribution with a width of

$$E_j = \begin{cases} 0, & \text{in the absence of truncation or rounding} \\ 2^{B_j}, & \text{otherwise.} \end{cases} \tag{18}$$

The mean and variance of the error sources at the jth-stage are given by

$$\mu_j = \begin{cases} \frac{1}{2}E_j, & \text{if truncation is applied} \\ 0, & \text{otherwise} \end{cases} \tag{19}$$

and the variance of the error is

$$\sigma_j^2 = \frac{1}{12}E_j^2. \tag{20}$$

Further, let h_j denote the impulse response of the system function of the jth-stage up to and including the last stage. The functional representation of h_j is derived and discussed in detail in the appendix. The error introduced at the jth-stage propagates to the output shaped by the impulse response, h_j. In particular, the mean of the error at the output of the SCIC filter due to truncation at the jth-stage is

$$\mu_{T_j} = \mu_j D_j, \tag{21}$$

where

$$D_j = \begin{cases} \sum_k h_j(k), & j = 1, \dots, 6N \\ 1, & j = 6N + 1 \end{cases}$$

is the "mean error gain" for the jth error source due to truncation.

Similarly, the variance at the output of the SCIC due to truncation at the jth-stage is

$$\sigma_{T_j}^2 = \sigma_j^2 F_j^2, \tag{22}$$

where the sub-T stands for "total" and

$$F_j^2 = \begin{cases} \sum_k h_j^2(k), & j = 1, \dots, 6N \\ 1, & j = 6N + 1 \end{cases}$$

is the 'variance error gain' corresponding to the jth error source.

Combining the results above, the total mean and variance of the error at the output are given by

$$\mu_T = \sum_{j=1}^{6N+1} \mu_{T_j}, \tag{23}$$

$$\sigma_T^2 = \sum_{j=1}^{6N+1} \sigma_{T_j}^2. \tag{24}$$

Based on (23) and (24), the design criterion for the register length of each of the $6N$ filter stages is:

i) Make the variance due to truncation of the first $6N$ error sources at least as small as the error introduced by truncating the output.
ii) Distribute the accumulating truncation error equally among the first $6N$ sources.

Use of these criteria implies,

$$\sigma_{T_j}^2 \le \frac{1}{6N} \sigma_{T_{6N+1}}^2, \quad j = 1, \dots, 6N \tag{25}$$

and that,

$$\sigma_{T_j}^2 = \frac{1}{12} 2^{2B_j} F_j^2, \tag{26}$$

where B_j is the number of bits truncated at stage j from the maximum register length, $B_{\max}$, which is itself a function of the inherent gain of the filter.

The filter gain for the SCIC filter is

$$G_{\max} = \sum_{k=0}^{3(RM-1)N} |h_1(k)|. \tag{27}$$

It can be seen from (27) that $G_{\max} = (RM)^{3N}$ corresponding to the gain in the path that includes the additional CIC block. Hence,

$$B_{\max} = \lceil 3N \log_2 RM + B_{\text{in}} - 1 \rceil, \tag{28}$$

where B_{in} is the number of bits to the input of the SCIC filter. Combining (26)–(28), the precision of the jth filter stage in the REX implementation is

$$B_j \leq -\log_2 F_j + \log_2 \sigma_{T_{6N+1}} + \frac{1}{2}\log_2 \frac{2}{N}. \tag{29}$$

It can be seen from Figs. 16 and (17) that the transfer function from the jth-stage up to and including the last stage is given by,

$$H_j(z) = \begin{cases} H_{\text{I}}^{2N-j+1}\left[3z^{-\left(\frac{RM-1}{2}\right)N} - 2H_{\text{C}}^{N}H_{\text{I}}^{N}\right]H_{\text{C}}^{2N}, & j = 1,\ldots,2N \\ -2H_{\text{C}}^{3N}H_{\text{I}}^{3N-j+1}, & j = 2N+1,\ldots,3N \\ H_{\text{C}}^{6N-j+1}, & j = 3N+1,\ldots,6N. \end{cases} \tag{30}$$

Expanding the terms,

$$H_j(z) = \begin{cases} \left(\frac{1}{1-z^{-1}}\right)^{2N-j+1}\left(1-z^{-RM}\right)^{2N}\left[3z^{-\left(\frac{RM-1}{2}\right)N} - 2\left(\frac{1-z^{-RM}}{1-z^{-1}}\right)^{N}\right], \\ \quad j = 1,\ldots,2N \\ -2\left(1-z^{-RM}\right)^{3N}\left(\frac{1}{1-z^{-1}}\right)^{3N-j+1}, & j = 2N+1,\ldots,3N \\ \left(1-z^{-RM}\right)^{6N-j+1}, & j = 3N+1,\ldots,6N \end{cases} \tag{31}$$

For an even number of cascade stages, N, let $N = 2P$,

$$H_j(z) = \begin{cases} H_j^0(z) = 3z^{-(RM-1)P}\left(\frac{1}{1-z^{-1}}\right)^{2N-j+1}\left(1-z^{-RM}\right)^{2N} \\ \qquad - 2\left(\frac{1-z^{-RM}}{1-z^{-1}}\right)^{N}\frac{\left(1-z^{-RM}\right)^{2N}}{\left(1-z^{-1}\right)^{2N-j+1}}, \\ \quad j = 1,\ldots,2N \\ H_j^1(z) = -2\left(1-z^{-RM}\right)^{3N}\left(\frac{1}{1-z^{-1}}\right)^{3N-j+1}, & j = 2N+1,\ldots,3N \\ H_j^2(z) = \left(1-z^{-RM}\right)^{6N-j+1}, & j = 3N+1,\ldots,6N \end{cases} \tag{32}$$

We address each of the individual cases, $H^0(z)$, $H^1(z)$ and $H^2(z)$ individually to derive the corresponding impulse response values.

First consider $H^0(z)$, and let $H_j^0(z) = X_j^{00}(z) - X_j^{01}(z)$, where

$$X_j^{00}(z) = 3z^{-(RM-1)P}X_j^{00'}(z), \tag{33}$$

$$X_j^{00'}(z) = \left(\frac{1}{1-z^{-1}}\right)^{2N-j+1}\left(1-z^{-RM}\right)^{2N}, \tag{34}$$

$$X_j^{01}(z) = \left(\frac{1 - z^{-RM}}{1 - z^{-1}}\right)^N \frac{(1 - z^{-RM})^{2N}}{(1 - z^{-1})^{2N-j+1}}. \tag{35}$$

Manipulating the middle expression above for $X_j^{00'}(z)$,

$$\begin{aligned}
X_j^{00'}(z) &= \left(\frac{1}{1 - z^{-1}}\right)^{2N-j+1} \left(1 - z^{-RM}\right)^{2N} \\
&= \left(1 - z^{-RM}\right)^{j-1} \left(\frac{1 - z^{-RM}}{1 - z^{-1}}\right)^{2N-j+1} \\
&= \left(1 - z^{-RM}\right)^{j-1} \left[\sum_{k=0}^{RM-1} z^{-k}\right]^{2N-j+1} \\
&= \sum_{k=0}^{2N(RM-1)+j-1} \left[\sum_{l=0}^{\lfloor \frac{k}{RM} \rfloor} (-1)^l \binom{2N}{l} \binom{2N - j + k - RM\,l}{k - RM\,l}\right] z^{-k}.
\end{aligned} \tag{36}$$

This can also be written as a single sum,

$$x_j^{00'}(k) = \sum_{l=0}^{\lfloor \frac{k}{RM} \rfloor} (-1)^l \binom{2N}{l} \binom{2N - j + k - RM\,l}{k - RM\,l},$$
$$k = 0, \ldots, 2N(RM - 1) + j - 1. \tag{37}$$

Then, since, $X_j^{00}(z) = z^{-(RM-1)P} X_j^{00'}(z)$, we have,

$$x_j^{00'}(k) = x_j^{00}[k - (RM - 1)P]. \tag{38}$$

Repeating this approach for $X_j^{01}(z)$,

$$\begin{aligned}
X_j^{01}(z) &= \left(\frac{1 - z^{-RM}}{1 - z^{-1}}\right)^N \frac{(1 - z^{-RM})^{2N}}{(1 - z^{-1})^{2N-j+1}} \\
&= \left(\frac{1 - z^{-RM}}{1 - z^{-1}}\right)^{3N-j+1} \left(1 - z^{-RM}\right)^{j-1} \\
&= \left[\sum_{k=0}^{RM-1} z^{-k}\right]^{3N-j+1} \left(1 - z^{-RM}\right)^{j-1} \\
&= \left[\sum_{l=0}^{3N} \binom{3N}{l} (-1)^l z^{-RM\,l}\right] \left[\sum_{\nu=0}^{\infty} \binom{3N - j + \nu}{\nu} z^{-\nu}\right] \\
&= \sum_{l=0}^{3N} \sum_{\nu=0}^{\infty} (-1)^l \binom{3N}{l} \binom{3N - j + \nu}{\nu} z^{-(RM\,l+\nu)} \\
&= \sum_{k \geq 0} \left[\sum_{l=0}^{\lfloor \frac{k}{RM} \rfloor} (-1)^l \binom{3N}{l} \binom{3N - j + k - RM\,l}{k - RM\,l}\right] z^{-k}.
\end{aligned} \tag{39}$$

But,

$$X_j^{01}(z) = \frac{(1-z^{-RM})^{3N}}{(1-z^{-1})^{2N-j+1}} = \sum_{k=0}^{3(RM-1)N+j-1} x_j^{01}(k) z^{-k}.$$

Hence, the range of k is established to be $k = 0, 1, \ldots, 3(RM-1)N + j - 1$ giving,

$$x_j^{01}(k) = \sum_{l=0}^{\lfloor \frac{k}{RM} \rfloor} (-1)^l \binom{3N}{l} \binom{3N-j+k-RM\,l}{k-RM\,l}. \tag{40}$$

Putting this all together,

$$h_j^0(k) = \begin{cases} \sum_{l=0}^{\lfloor \frac{k}{RM} \rfloor} (-1)^l \binom{3N}{l} \binom{3N-j+k-RM\,l}{k-RM\,l}, \\ \quad k = 0, 1, \ldots, \frac{(RM-1)N}{2} \\ \sum_{l=0}^{\lfloor \frac{k}{RM} \rfloor} (-1)^l \left[\binom{3N}{l} \binom{3N-j+k-RM\,l}{k-RM\,l} + \binom{2N}{l} \binom{2N-j+k-RM\,l}{k-RM\,l} \right], \\ \quad k = \frac{(RM-1)N}{2} + 1, \ldots, 2(RM-1)N + j - 1 \\ \sum_{l=0}^{\lfloor \frac{k}{RM} \rfloor} (-1)^l \binom{3N}{l} \binom{3N-j+k-RM\,l}{k-RM\,l}, \\ \quad k = 2(RM-1)N + j, \ldots, 3(RM-1)N + j - 1. \end{cases} \tag{41}$$

Similarly,

$$\begin{aligned} H_j^1(z) &= \left(\frac{1}{1-z^{-1}} \right)^{3N-j+1} \left(1 - z^{-RM}\right)^{3N} \\ &= \left(1 - z^{-RM}\right)^{j-1} \left(\frac{1-z^{-RM}}{1-z^{-1}} \right)^{3N-j+1} \\ &= \left(1 - z^{-RM}\right)^{j-1} \left[\sum_{k=0}^{RM-1} z^{-k} \right]^{3N-j+1} \\ &= \sum_{k=0}^{3(RM-1)N+j-1} \left[\sum_{l=0}^{\lfloor \frac{k}{RM} \rfloor} (-1)^l \binom{3N}{l} \binom{3N-j+k-RM\,l}{k-RM\,l} \right] z^{-k}. \end{aligned}$$

Hence,

$$h_j^1(k) = \sum_{l=0}^{\lfloor \frac{k}{RM} \rfloor} (-1)^l \binom{3N}{l} \binom{3N-j+k-RM\,l}{k-RM\,l}, \quad k = 0, 1, \ldots, 3(RM-1)N + j - 1. \tag{42}$$

Finally,

$$H_j^2(z) = \left(1 - z^{-RM}\right)^{6N-j+1} \tag{43}$$

and

$$h_j^2(k) = \begin{cases} (-1)^l \binom{6n-j+1}{l}, & k = RM\,l, \text{ where } l = 0, \ldots, (6N - j + 1) \\ 0, & \text{otherwise.} \end{cases}$$

References

R. Albrecht, C. Barbieri, H.-M. Adorf, G. Corrain, A. Gemmo, P. Greenfield, O. Hainaut, N. Hook, D.J. Tholen, J.C. Blades, W.B. Sparks, High-resolution imaging of the Pluto–Charon system with the Faint Object Camera of the Hubble Space Telescope. Astrophys. J. Lett. **435**, 75 (1994)

W.J. Altenhoff, R. Chini, H. Hein, E. Kreysa, P.G. Mezger, C. Salter, J.B. Schraml, First radio astronomical estimate of the temperature of Pluto. Astron. Astrophys. **190**, L15 (1988)

J.D. Anderson, J.W. Armstrong, J.K. Campbell, F.B. Estabrook, T.P. Krisher, E.L. Lau, Gravitation and celestial mechanics investigations with Galileo. Space Sci. Rev. **60**, 591 (1992)

J.D. Anderson, P.A. Laing, E.L. Lau, A.S. Liu, M.M. Nieto, S.G. Turyshev, Indication, from Pioneer 10/11, Galileo, and Ulysses data, of an apparent anomalous, weak, long-range acceleration. Phys. Rev. Lett. **81**, 2858 (1998)

J.D. Anderson, P.A. Laing, E.L. Lau, A.S. Liu, M.M. Nieto, S.G. Turyshev, Study of the anomalous acceleration of Pioneer 10 and 11. Phys. Rev. D. **65**, 082004 (2002)

S.W. Asmar, J.W. Armstrong, L. Iess, P. Tortora, Spacecraft Doppler tracking: Noise budget and accuracy achievable in precision radio science observations. Radio Sci. **40**, RS2001 (2005). doi:10.1029/2004RS003101

R.S. Bokulic, M.J. Reinhart, C.E. Willey, R.K. Stilwell, J.E. Penn, J.R. Norton, S. Cheng, D.J. DeCicco, R.C. Schulze, Advances in deep-space telecommunications technology at the applied physics laboratory. Acta Astronautica **52**, 467 (2003)

B.J. Buratti, R.S. Dunbar, E.F. Tedesco, J. Gibson, R.L. Marcialis, F. Wong, S. Benett, Modeling Pluto–Charon mutual events. II. CCD observations with the 60 in. telescope at Palomar Mountain. Astron. J. **110**, 1405 (1995)

G.N. Brown Jr., W.T. Ziegler, Vapor pressure and heats of vaporization and sublimation of liquids and solids of interest in cryogenics below 1-atm pressure. Adv. Cryogen. Eng. **25**, 662 (1979)

M.W. Buie, W.M. Grundy, E.F. Young, L.A. Young, S.A. Stern, Orbits and photometry of Pluto's satellites: Charon, S/2005 P1, and S/2005 P2. Astron. J. **132**, 290 (2006)

D.P. Cruikshank, T.L. Roush, J.M. Moore, M.V. Sykes, T.C. Owen, M.J. Bartholomew, R.H. Brown, K.A. Tryka, The surfaces of Pluto and Charon, in *Pluto and Charon*, ed. by S.A. Stern, D.J. Tholen (Univ. of Ariz. Press, Tucson, 1997), p. 221

C. Deboy, C. Haskens, D. Duven, R. Schulze, J.R. Jensen, M. Bernacik, W. Millard, The New Horizons mission to Pluto: Advances in telecommunications system design. Acta Astronautica **57**, 540 (2005)

J.L. Elliot, E.W. Dunham, A.S. Bosh, S.M. Slivan, L.A. Young, L.H. Wasserman, R.L. Millis, Pluto's atmosphere. Icarus **77**, 148 (1989)

J.L. Elliot, L.A. Young, Analysis of stellar occultation data for planetary atmospheres: I. Model fitting, with application to Pluto. Astron. J. **103**, 991 (1992)

J.L. Elliot, A. Ates, B.A. Babcock, A.S. Bosh, M.W. Buie, K.B. Clancy, E.W. Dunham, S.S. Eikenberry, D.T. Hall, S.D. Kern, S.K. Leggett, S.E. Levine, D.-S. Moon, C.B. Olkin, D.J. Osip, J.M. Pasachoff, B.E. Penprase, M.J. Person, S. Qu, J.T. Rayner, L.C. Roberts Jr., C.V. Salyk, S.P. Souza, R.C. Stone, B.W. Taylor, D.J. Tholen, J.E. Thomas-Osip, D.R. Ticehurst, L.H. Wasserman, The recent expansion of Pluto's atmosphere. Nature **424**, 165 (2003)

V.R. Eshleman, Pluto's atmosphere: Models based on refraction, inversion, and vapor-pressure equilibrium. Icarus **80**, 439 (1989)

G. Fjeldbo, A.J. Kliore, V.R. Eshleman, The neutral atmosphere of Venus as studied with the Mariner V radio occultation experiments. Astron. J. **76**, 123 (1971)

G.H. Fountain, D. Kusnierkeiwicz, D.B. Hersman, T. Herder, S.R. Vernon, J.M. Stratton, S. Williams, C. DeBoy, G. Rogers, J. Kinnison, The New Horizons Spacecraft. Space Sci. Rev. (2006, this issue)

D.L. Gresh, E.A. Marouf, G.L. Tyler, P.A. Rosen, R.A. Simpson, Voyager radio occultation by Uranus rings. Icarus **78**, 131 (1989)

W.M. Grundy, M.W. Buie, Distribution and evolution of CH_4, N_2, and CO ices on Pluto's surface: 1995 to 1998. Icarus **153**, 248 (2001)

A.A.S. Gulbis, J.L. Elliot, M.J. Person, E.R. Adams, B.A. Babcock, M. Emilio, J.W. Gangestad, S.D. Kern, E.A. Kramer, D.J. Osip, J.M. Pasachoff, S.P. Souza, T. Tuvikene, Charon's radius and atmospheric constraints from observations of a stellar occultation. Nature **439**, 48 (2006). doi:10.1038/nature04276

E.M. Gurrola, Interpretation of radar data from the icy Galilean satellites and Triton, Ph.D. Dissertation, Stanford University, 1996

C.J. Hansen, D.A. Paige, Seasonal nitrogen cycles on Pluto. Icarus **12**, 247 (1996)

D.P. Hinson, F.M. Flasar, A.J. Kliore, P.J. Schinder, J.D. Twicken, R.G. Herrera, Jupiter's ionosphere: Results from the first Galileo radio occultation experiment. Geophys. Res. Lett. **24**, 2107 (1997)

D.P. Hinson, J.D. Twicken, E.T. Karayel, Jupiter's ionosphere: New results from Voyager 2 radio occultation measurements. J. Geophys. Res. **103**, 9505 (1998)

D.P. Hinson, R.A. Simpson, J.D. Twicken, G.L. Tyler, F.M. Flasar, Initial results from radio occultation measurements with Mars Global Surveyor. J. Geophys. Res. **104**, 1999 (1999)

D.P. Hinson, G.L. Tyler, J.L. Hollingsworth, R.J. Wilson, Radio occultation measurements of forced atmospheric waves on Mars. J. Geophys. Res. **106**, 1463 (2001)

E.B. Hogenauer, An economical class of digital filters for decimation and interpolation. IEEE Trans. Acoust., Speech, Signal Process. **ASSP-29**, 155 (1981)

W.B. Hubbard, D.M. Hunten, S.W. Dieters, K.M. Hill, R.D. Watson, Occultation evidence for an atmosphere on Pluto. Nature **336**, 452 (1988)

W.B. Hubbard, R.V. Yelle, J.I. Lunine, Nonisothermal Pluto atmosphere models. Icarus **84**, 1 (1990)

J.F. Jensen, R.S. Bokulic, Experimental verification of noncoherent doppler tracking at the deep space network. IEEE Trans. Aero. Elect. Syst. **36**, 1401 (2000)

D.C. Jewitt, Heat from Pluto. Astron. J. **107**, 372 (1994)

J. Kaiser, R. Hamming, Sharpening the response of a symmetric nonrecursive filter by multiple use of the same filter. IEEE Trans. Acoust., Speech, Signal Process. **25**, 415 (1977)

E.T. Karayel, D.P. Hinson, Sub-Fresnel-scale vertical resolution in atmospheric profiles from radio occultation. Radio Sci. **32**, 411 (1997)

V.A. Krasnopolsky, D.P. Cruikshank, Photochemistry of Pluto's atmosphere and ionosphere near perihelion. J. Geophys. Res. **104**, 21979 (1999)

A.Y. Kwentus, Z. Jiang, A.N. Willson Jr., Application of filter sharpening to cascaded integrator-comb decimation filters. IEEE Trans. Signal Process. **45**, 457 (1997)

E. Lellouch, R. Laureijs, B. Schmitt, E. Quirico, C. de Bergh, J. Crovisier, A. Coustenis, Pluto's non-isothermal surface. Icarus **147**, 220 (2000)

G.F. Lindal, G.E. Wood, H.B. Hotz, D.N. Sweetnam, V.R. Eshleman, G.L. Tyler, The atmosphere of Titan: An analysis of the Voyager 1 radio occultation measurements. Icarus **53**, 348 (1983)

E.A. Marouf, G.L. Tyler, P.A. Rosen, Profiling Saturn's rings by radio occultation. Icarus **68**, 120 (1986)

W.B. McKinnon, J.I. Lunine, D. Banfield, Origin and evolution of Triton, in *Neptune and Triton*, ed. by D.P. Cruikshank (Univ. of Ariz. Press, Tucson, 1995), p. 807

R.L. Millis, L.H. Wassermann, O.G. Franz, R.A. Nye, J.L. Elliot, E.W. Dunham, A.S. Bosh, L.A. Young, S.M. Slivan, A.C. Gilmore, P.M. Kilmartin, W.H. Allen, R.D. Watson, S.W. Dieters, K.M. Hill, A.B. Gills, G. Blow, J. Priestley, W.M. Kissling, W.S.G. Walker, B.G. Marino, D.G. Dix, A.A. Page, J.E. Ross, H.P. Avey, D. Hickey, H.D. Kennedy, K.A. Mottram, G. Moyland, T. Murphy, C.C. Dahn, A.R. Klemola, Pluto's radius and atmosphere: Results from the entire 9 June 1988 occultation data set. Icarus **105**, 282 (1993)

T.C. Owen, T.L. Roush, D.P. Cruikshank, J.L. Elliot, L.A. Young, C. de Bergh, B. Schmitt, T.R. Geballe, R.H. Brown, M.J. Bartholomew, Surface ices and the atmospheric composition of Pluto. Science **261**, 745 (1993)

M. Pätzold, B. Husler, A. Wennmacher, K. Aksnes, J.D. Anderson, S.W. Asmar, J.-P. Barriot, H. Boehnhardt, W. Eidel, F.M. Neubauer, O. Olsen, J. Schmitt, J. Schwinger, N. Thomas, Gravity field determination of a Comet Nucleus: Rosetta at P/Wirtanen. Astron. Astrophys. **375**, 651 (2001)

M.J. Person, J.L. Elliot, A.A.S. Gulbis, J.M. Pasachoff, B.A. Babcock, S.P. Souza, J. Gangestad, Charon's radius and density from the combined data sets of the 2005 July 11 occultatation. Astron. J. **132**, 1540 (2006)

K. Reinsch, V. Burwitz, M.C. Festou, Albedo maps of Pluto and improved physical parameters of the Pluto-Charon system. Icarus **108**, 209 (1994)

B. Sicardy, T. Widemann, E. Lellouch, C. Veillet, J.-C. Cuillandre, F. Colas, F. Roques, W. Beisker, M. Kretlow, A.-M. Lagrange, E. Gendron, F. Lacombe, J. Lecacheux, C. Birnbaum, A. Fienga, C. Leyrat, A. Maury, E. Raynaud, S. Renner, M. Schultheis, K. Brooks, A. Delsanti, O.R. Hainaut, R. Gilmozzi, C. Lidman, J. Spyromilio, M. Rapaport, P. Rosenzweig, O. Naranjo, L. Porras, F. Diaz, H. Calderon, S. Carrillo, A. Carvajal, E. Recalde, L. Gaviria Cavero, C. Montalvo, D. Barria, R. Campos, R. Duffard,

H. Levato, Large changes in Pluto's atmosphere as revealed by recent stellar occultations. Nature **424**, 168 (2003)

B. Sicardy, A. Bellucci, E. Gendron, F. Lacombe, S. Lacour, J. Lecacheux, E. Lellouch, S. Renner, S. Pau, F. Roques, T. Widemann, F. Colas, F. Vachier, R. Vieira Martins, N. Ageorges, O. Hainaut, O. Marco, W. Beisker, E. Hummel, C. Feinstein, H. Levato, A. Maury, E. Frappa, B. Gaillard, M. Lavayssiere, M. Di Sora, F. Mallia, G. Masi, R. Behrend, F. Carrier, O. Mousis, P. Rousselot, A. Alvarez-Candal, D. Lazzaro, C. Veiga, A.H. Andrei, M. Assafin, D.N. da Silva Neto, C. Jacques, E. Pimentel, D. Weaver, J.-F. Lecampion, F. Doncel, T. Momiyama, G. Tancredi, Charon's size and an upper limit on its atmosphere from a stellar occultation. Nature **439**, 52 (2006)

D.E. Smith, M.T. Zuber, H.V. Frey, J.B. Garvin, J.W. Head, D.O. Muhleman, G.H. Pettengill, R.J. Phillips, S.C. Solomon, H.J. Zwally, W.B. Banerdt, T.C. Duxbury, M.P. Golombek, F.G. Lemoine, G.A. Neumann, D.D. Rowlands, O. Aharonson, P.G. Ford, A.B. Ivanov, P.J. McGovern, J.B. Abshire, R.S. Afzal, X. Sun, Mars orbiter laser altimeter: Experiment summary after the first year of global mapping of Mars. J. Geophys. Res. **106**, 23689 (2001)

J.R. Spencer, J.A. Stansberry, L.M. Trafton, E.F. Young, R.P. Binzel, S.K. Croft, Volatile transport, seasonal cycles, and atmospheric dynamics on Pluto, in *Pluto and Charon*, ed. by S.A. Stern, D.J. Tholen (Univ. of Ariz. Press, Tucson, 1997), p. 435

E.M. Standish Jr., No dynamical evidence in the optical observations. Astron. J. **105**, 2000 (1993)

J.A. Stansberry, J.I. Lunine, W.B. Hubbard, R.V. Yelle, D. M Hunten, Mirages and the nature of Pluto's atmosphere. Icarus **111**, 503 (1994)

D.F. Strobel, X. Zhu, M.E. Summers, M. H Stevens, On the vertical thermal structure of Pluto's atmosphere. Icarus **120**, 266 (1996)

M.E. Summers, D.F. Strobel, G.R. Gladstone, 1997, Chemical models of Pluto's atmosphere, in *Pluto and Charon*, ed. by S.A. Stern, D.J. Tholen (Univ. of Ariz. Press, Tucson, 1997), p. 391

D.J. Tholen, M.W. Buie, Further analysis of Pluto–Charon mutual event observations. Bull. Am. Astron. Soc. **22**, 1129 (abstract) (1990)

D.J. Tholen, M.W. Buie, Bulk properties of Pluto and Charon, in *Pluto and Charon*, ed. by S.A. Stern, D.J. Tholen (Univ. of Ariz. Press, Tucson, 1997), p. 193

K.A. Tryka, R.H. Brown, D.P. Cruikshank, T.C. Owen, T.R. Geballe, C. DeBergh, Temperature of nitrogen ice on Pluto and its implications for flux measurements. Icarus **112**, 513 (1994)

G.L. Tyler, Radio propagation experiments in the outer solar system with Voyager. Proc. IEEE **75**, 1404 (1987)

G.L. Tyler, D.N. Sweetnam, J.D. Anderson, S.E. Borutzki, J.K. Campbell, V.R. Eshleman, D.L. Gresh, E.M. Gurrola, D.P. Hinson, N. Kawashima, E.R. Kursinski, G.S. Levy, G.F. Lindal, J.R. Lyons, E.A. Marouf, P.A. Rosen, R.A. Simpson, G.E. Wood, Voyager radio science observations of Neptune and Triton. Science **246**, 1466 (1989)

G.L. Tyler, G. Balmino, D.P. Hinson, W.L. Sjogren, D.E. Smith, R.A. Simpson, S.W. Asmar, P. Priest, J.D. Twicken, Radio science observations with Mars Global Surveyor: Orbit insertion through one Mars year in mapping orbit. J. Geophys. Res. **106**, 23,327 (2001)

R.V. Yelle, J.L. Elliot, Atmospheric structure and composition, in *Pluto and Charon*, ed. by S.A. Stern, D.J. Tholen (Univ. of Ariz. Press, Tucson, 1997), p. 347

E.F. Young, R.P. Binzel, A new determination of radii and limb parameters for Pluto and Charon from mutual event lightcurves. Icarus **108**, 219 (1994)

H. Zirin, B.M. Baumert, G.J. Hurford, The microwave brightness temperature spectrum of the quiet Sun. Astrophys. J. **370**, 779 (1991)

The Solar Wind Around Pluto (SWAP) Instrument Aboard *New Horizons*

D. McComas · F. Allegrini · F. Bagenal · P. Casey · P. Delamere · D. Demkee · G. Dunn · H. Elliott · J. Hanley · K. Johnson · J. Langle · G. Miller · S. Pope · M. Reno · B. Rodriguez · N. Schwadron · P. Valek · S. Weidner

Originally published in the journal Space Science Reviews, Volume 140, Nos 1–4, 261–313.
DOI: 10.1007/s11214-007-9205-3

Abstract The Solar Wind Around Pluto (SWAP) instrument on *New Horizons* will measure the interaction between the solar wind and ions created by atmospheric loss from Pluto. These measurements provide a characterization of the total loss rate and allow us to examine the complex plasma interactions at Pluto for the first time. Constrained to fit within minimal resources, SWAP is optimized to make plasma-ion measurements at all rotation angles as the *New Horizons* spacecraft scans to image Pluto and Charon during the flyby. To meet these unique requirements, we combined a cylindrically symmetric retarding potential analyzer with small deflectors, a top-hat analyzer, and a redundant/coincidence detection scheme. This configuration allows for highly sensitive measurements and a controllable energy passband at all scan angles of the spacecraft.

Keywords Solar wind · Pluto plasma interaction · Pickup ions

1 Introduction

The *New Horizons* mission (Stern et al. 2007) will make the first up-close and detailed observations of Pluto and its moons. These observations include measurements of the solar

D. McComas (✉) · F. Allegrini · P. Casey · D. Demkee · G. Dunn · H. Elliott · J. Hanley · K. Johnson · J. Langle · G. Miller · S. Pope · M. Reno · B. Rodriguez · N. Schwadron · P. Valek · S. Weidner
Southwest Research Institute, San Antonio, TX 78228-0510, USA
e-mail: DMcComas@swri.edu

F. Bagenal · P. Delamere
University of Colorado, Boulder, CO 80309, USA

J. Langle
Midwest Research Institute, Kansas City, MO 64110, USA

N. Schwadron
Boston University, Boston, MA 02215, USA

wind interaction with Pluto. The Solar Wind Around Pluto (SWAP) instrument is designed to measure the tenuous solar wind out at ~32 AU and its interaction with Pluto. SWAP directly addresses the Group 1 science objective for the *New Horizons* mission to "characterize the neutral atmosphere of Pluto and its escape rate". In addition, SWAP makes measurements critical for the Group 2 objective of characterizing Pluto's ionosphere and solar wind interaction. Finally, SWAP makes observations relevant to two Group 3 science objectives for *New Horizons*, both in support of characterizing the energetic particle environment of Pluto and Charon and in searching for magnetic fields of Pluto and Charon.

Pluto's thick atmosphere escapes the planet's weak gravity and streams away as neutral particles. The escaping neutral atoms are ionized by solar photons. As the solar wind approaches Pluto it picks up these ions and electrons and carries them downstream. This pickup process slows the solar wind, and the type of interaction varies greatly depending on the atmospheric escape rate, all the way from comet-like for larger escape rates to Venus-like for low escape rates. SWAP measurements should provide the best estimate of the overall atmospheric escape rate at Pluto and allow the first detailed examination of its plasma interactions with the solar wind.

The SWAP observations are extremely challenging because the solar wind flux, which falls off roughly with the square of heliocentric distance, is approximately three orders of magnitude lower at Pluto compared to typical solar wind fluxes observed near Earth's orbit. In addition, because the solar wind continues to cool as it propagates through the heliosphere, the solar wind beam becomes narrow in both angle and energy.

The SWAP design was strongly driven by three constraints: (1) very low use of spacecraft resources (mass, power, telemetry, etc.), as this is a noncore instrument on a relatively small planetary mission; (2) very high sensitivity to measure the solar wind and its interaction with Pluto out at ~32 AU, where the density is lower by a factor of ~1000 compared to that at 1 AU; and (3) the need to make observations over a very large range of angles, as the spacecraft constantly repoints its body-mounted cameras during the flyby. Given these not-entirely-consistent design drivers, we developed an entirely new design that combines elements of several different previous plasma instruments. Together these components comprise the SWAP instrument, which will measure the speed, density, and temperature of the distant solar wind and its interaction with Pluto.

2 Scientific Background and Objectives

Only partly in jest, Dessler and Russell (1980) suggested that Pluto might act like a colossal comet. The 1988 stellar occultation showed that Pluto's tenuous atmosphere could indeed be escaping (Hubbard et al. 1988; Elliot et al. 1989). Applying basic cometary theories to Pluto, Bagenal and McNutt (1989) showed that photoionization of escaping neutral molecules from Pluto's atmosphere could significantly alter the solar wind flow around Pluto for sufficiently large escape rates. For large atmospheric escape rates, the interaction may be best described as "comet-like", with significant mass-loading over an extensive region; for small escape rates the interaction is probably confined to a much smaller region, creating a more "Venus-like" interaction (Luhmann et al. 1991), where electrical currents in the gravitationally bound ionosphere deflect the solar wind flow. Figure 1 compares these two types of interactions schematically. At aphelion (50 AU), should Pluto's atmosphere completely collapse and freeze onto the surface, then the interaction becomes "Moon-like" with the solar wind suffering minimal deflection and directly bombarding the bare, icy dayside surface.

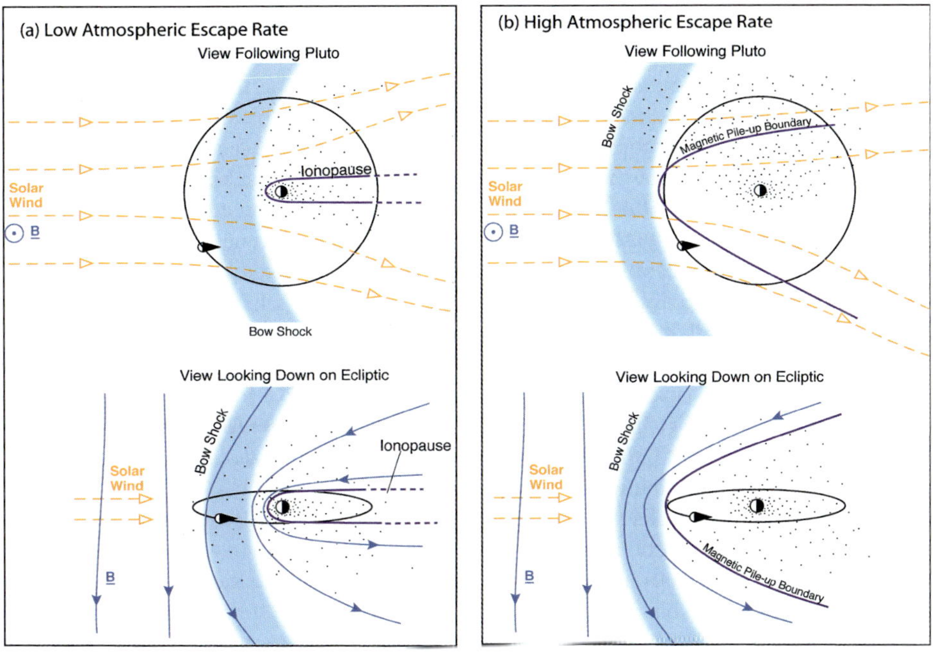

Fig. 1 Solar wind interaction with Pluto for (**a**) low atmospheric escape rate and (**b**) high atmospheric escape rate. The *dots* indicate ions produced by the photoionization of Pluto's escaping atmosphere. These pickup ions will move perpendicular to both the magnetic field (B) and solar wind (*upward in the top two diagrams*). Note that although the interplanetary magnetic field (IMF) tends to be close to tangential to the Sun direction, the sign of the direction varies on time scales of days. The asymmetries of the interaction will flip as the magnetic field changes direction. The *lower panels* show a plane parallel to the ecliptic plane though at the time of the *New Horizons* flyby Pluto will be well above the ecliptic plane

Table 1 Typical solar wind interaction properties at Pluto (at 30 AU), taken from Bagenal et al. (1997) using $R_P = 1150$ km

Magnetic field strength	0.2 nT
Proton density	0.01 cm^{-3}
Solar wind speed	450 km/s
Proton temperature	1.3 eV
Alfvén Mach number	~45
Sonic Mach number	~40
Proton gyroradius	23,000 km (~20 R_P)
N_2^+ gyroradius	658,000 km (~550 R_P)
Ion inertial length	2280 km (~2 R_P)
Electron inertial length	53 km

Not having a detectable atmosphere, Charon almost certainly has such a "Moon-like" interaction, remaining primarily in the solar wind if Pluto's interaction is weak but becoming totally engulfed if Pluto's interaction is strongly "comet-like" and extends beyond Charon's orbit at 17 R_P (Pluto radii, ~1150 km). For a review of early studies of the solar wind interaction with Pluto see Bagenal et al. (1997), which also reviews the implications of the unlikely possibility of Pluto having an intrinsic magnetic field.

The solar wind is supersonic (actually superfast-mode magnetosonic) so that when it impinges on a magnetic obstacle (such as the magnetosphere of the Earth or another planet) an upstream bow shock must form to slow and deflect the supersonic plasma. The weak interplanetary magnetic field (IMF) at 30 AU (see Table 1 for typical solar wind properties near Pluto) and heavy ions formed by photoionizing the heavy molecules of Pluto's escaping atmosphere have very large gyroradii ($\sim$500 R_P). The net results of these nonfluid or kinetic effects are to make the bow shock a thick transition region and to make the shape of the interaction region asymmetric, where the direction of asymmetry is governed by the direction of the IMF. Recent simulations of the solar wind interaction with a strongly escaping atmosphere have necessarily been 3D and have either taken a multifluid approach (solar wind proton fluid and pickup ion fluid) or a hybrid approach (electron fluid with ion particles) (Harnett et al. 2005; Delamere and Bagenal 2004).

In the following we discuss how the current understanding of Pluto's atmosphere leads us to expect a more comet-like interaction at the time of the *New Horizons* flyby in 2015. Through measurements of bulk properties of the solar wind (flow, density, and temperature) as well as the energy distribution of solar wind and pickup ions, the SWAP instrument will not only characterize the solar wind interaction with Pluto but will also allow us to determine the global rate of atmospheric escape.

In the comet-like scenario, variations on the scale of the interaction region can be substantial over periods of days, and a factor of $\sim$10 variations in the solar wind flux can change the size of the interaction region from a few to more than 20 R_P. It is therefore critical to measure the solar wind for several solar rotations ($\sim$26 days per rotation) before and after the flyby to characterize the most likely external solar wind properties during the actual encounter period. Furthermore, because the strong asymmetry of the interaction depends on the direction of the IMF, our analysis of SWAP data will need assistance from increasingly capable models of solar wind structure based on plasma and magnetometer data from spacecraft elsewhere in the solar system.

2.1 Atmospheric Escape

The exact nature of Pluto's plasma interaction is critically dependent on the hydrodynamic escape rate of the atmosphere from its weak gravity. Escaping neutrals are photoionized by solar UV (or, less likely, suffer an ionizing collision). Freshly ionized particles experience an electric field owing to their motion relative to the IMF (which is carried away from the Sun by the solar wind) and are accelerated by this motional electric field, extracting momentum from the solar wind flow. This electrodynamic interaction modifies the solar wind flow. Estimates of Pluto's atmospheric escape rate, Q, vary substantially: McNutt (1989) estimated 2.3–5.5 $\times 10^{27}$ s^{-1} for CH_4-dominated outflow, Krasnopolsky (1999) found a hydrodynamic outflow of N_2 of 2.0–2.6 $\times 10^{27}$ s^{-1}, whereas Tian and Toon (2005) derived values for N_2 escape as high as 2×10^{28} s^{-1}. The nature of the plasma interaction varies considerably over this range of escape rates since the scale of the interaction is proportional to Q (Bagenal and McNutt 1989).

Hydrodynamic escape of an atmosphere occurs when the atmospheric gases in the upper atmosphere (in the vicinity of the exobase) are heated significantly (so that the thermal speed becomes comparable to the local sound speed). Krasnopolsky and Cruikshank (1999) reviewed the photochemistry of Pluto's atmosphere and Krasnopolsky (1999) summarized approaches taken in modeling the complexities of hydrodynamic escape at Pluto. Earlier models approximated all the heating that occurs in a thin layer of the atmosphere. Recently, however, Tian and Toon (2005) developed a model that includes a distributed heating function appropriate for EUV absorption by the dominant molecule N_2 relatively high in Pluto's atmosphere, which leads to larger escape rates. These authors derive an exobase height of 10–13 R_P and a transonic point of $\sim$30 R_P. The fact that even with significant heating the escape speed ($<$100 m/s) is subsonic at the exobase is consistent with what Krasnopolsky (1999) called "slow hydrodynamic escape". Nevertheless, an exobase at 10–13 R_P implies a very extended atmosphere, and the present *New Horizons* trajectory with a currently planned closest approach of $\sim$9 R_P, may well briefly dip below the exobase.

Tian and Toon (2005) modeled the effects of variations in (i) EUV flux over the solar cycle and (ii) Pluto's distance from the Sun to estimate the variability of atmospheric escape over the full Pluto orbit. They found escape rates varying from 2×10^{28} s^{-1} (for Pluto at 30 AU and solar maximum activity) to 1×10^{28} s^{-1} (for Pluto at 40 AU and solar minimum). These authors, however, were unable to find a stable solution for atmospheric escape at the low heating levels appropriate for aphelion (50 AU). Whether Pluto retains a stable atmosphere through aphelion or not remains an open issue. Recent occultation measurements suggest that Pluto's atmosphere has been expanding rather than collapsing as Pluto recedes from the Sun (Elliot et al. 2003). The possibility that the atmospheric pressure could be increasing on Pluto even after perihelion was predicted (Stern et al. 1988; Hansen and Paige 1996). These authors suggested that there would be a phase lag between perihelion and the maximum atmospheric pressure. There is also the possibility that the polar obliquity is now putting some new, fresh frost into sunlight, further driving atmospheric expansion.

If we assume that the current estimates for the escaping atmosphere have persisted over the age of the solar system, Pluto could have lost hundreds of meters to kilometers of material to space via "escape erosion". In this case, the ancient topography made of volatiles (N_2 and CO particularly, but also CH_4) would have escaped to space. This fantastic possibility appears to be something wholly unique to Pluto, since even Triton is in the Jeans escape regime. Thus, escape erosion might have caused Pluto to have a young surface, and if so, imaging from *New Horizons* will give us information on the recent Kuiper Belt impactor size–frequency distribution, in contrast to the time-integral distribution provided by imaging of Charon, where such erosion is not occurring.

2.2 Solar Wind Interaction at High Atmospheric Escape

Galeev et al. (1985) derived a size for the interaction region (over which significant momentum would be extracted from the solar wind) for comets that is proportional to Q and inversely proportional to the solar wind flux, $n_{sw} V_{sw}$. Bagenal and McNutt (1989) applied this simple scaling law to Pluto and found that for typical solar wind conditions and photoionization of CH_4 or N_2, the scale size of the dayside interaction region, R_{SO} (the "standoff" distance) is $R_{SO}/R_P = Q/Q_0$), where Q_0 is 1.5×10^{27} s^{-1}. Applying the upper range of atmospheric escape rates discussed above, one finds $R_{SO} = 6$–13 R_P for 1–2×10^{27} s^{-1} (assuming an escape speed of 100 m/s). Note that in these simple calculations one assumes that the exobase is close to the planet. Tian and Toon (2005) suggest that the exobase may be as high as 10–13 R_P, in which case the standoff distance would be expanded accordingly. Nevertheless, the linear dependence on solar wind flux means that larger (factor of $\sim$10) variations in solar wind density would produce a similar variation in R_{SO}, often extending the interaction region beyond the orbit of Charon at 17 R_P.

This simple scaling is based on a fluid approach that is appropriate for cometary interactions in the inner Solar System. The weak IMF at 30 AU (see Table 1) means that ion

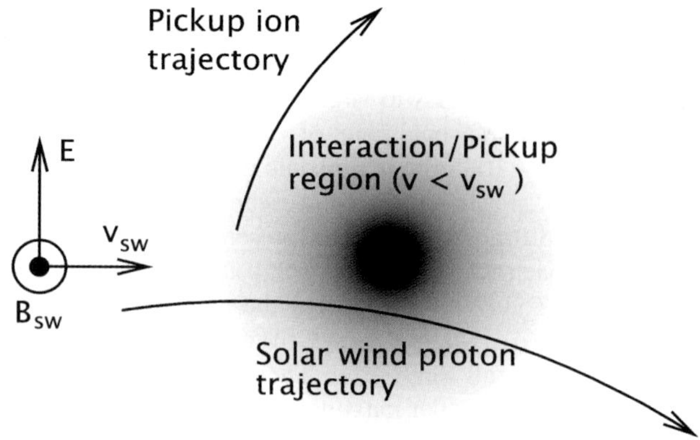

Fig. 2 Schematic of ion motion near Pluto's interaction region. Pickup ions move initially in the direction of the solar wind convection electric field ($+y$) and the solar wind flow is deflected in the $-y$ direction, consistent with momentum conservation

kinetic effects are crucial at Pluto because of the large pickup-ion gyroradius (Kecskemety and Cravens 1993) and the large turning distance for solar wind protons (Bagenal et al. 1997).

Previous models of weakly outgassing bodies used two-dimensional, two-ion fluid and/or hybrid (i.e., fluid electrons with kinetic ions) simulations (Bogdanov et al. 1996; Sauer et al. 1997; Hopcroft and Chapman 2001). All of these models showed the formation of asymmetric plasma structures. Sauer et al. (1997) specifically investigated the Pluto plasma interaction with two-dimensional models for $Q > 10^{27}\ \mathrm{s}^{-1}$. Although the two-dimensional models provide a good qualitative description of the interaction, the quantitative details of the plasma coupling (i.e., momentum transfer) require three-dimensional models. More recently, Lipatov et al. (2002) performed a three-dimensional hybrid code simulation of the solar wind interaction with weak comets and illustrated the dependence of gas production rates ($6.2 \times 10^{26} < Q < 10^{28}\ \mathrm{s}^{-1}$) on plasma structure. However, these results are applicable to 1 AU, where the ions are more strongly magnetized, compared to the situation at 30 AU, where the IMF is very weak and the solar wind is tenuous.

Delamere and Bagenal (2004) modeled the kinetic interaction of the solar wind with Pluto's escaping atmosphere using a hybrid simulation that treats the pickup ions and solar wind protons as particles and the electrons as a massless fluid. A hybrid code is a reasonable approach for a system with scale sizes ranging from the ion inertial length of the solar wind protons ($\sim 2\ R_P$) to the pickup ion (N_2^+) gyroradius ($\sim 500\ R_P$). Figure 2 shows a schematic of the ion motion near Pluto in this kinetic regime. Note the asymmetry of the interaction imposed by the direction of the magnetic field in the solar wind ($\mathbf{B}_{sw}$).

Figures 3, 4, and 5 show the density, temperature, and bulk flow resulting from hybrid simulations for $|\mathbf{B}| = 0.2$ nT, an atmospheric escape rate of $Q = 1 \times 10^{27}\ \mathrm{s}^{-1}$ and an atmospheric escape speed of 100 m/s (Delamere and Bagenal 2004) in the same geometry depicted in Fig. 2 (upstream solar wind flow from the left and IMF pointing out of the page). The main features of the interaction are a broad region of proton heating extending $\sim 20\ R_P$ upstream of Pluto, which is consistent with the expected location of a bow shock. The kinetic energy of the solar wind protons is not fully converted to thermal energy, and the shock structure forms only in the upper half of the interaction region. In the lower half of the interaction region the protons are compressed (illustrated by higher density in Fig. 3, higher temperature in Fig. 4, and slightly reduced and deflected flow in Fig. 5) behind what is effectively a magnetic pileup boundary. Since the gyroradii of the pickup ions is huge compared with the scale of the interaction region, they effectively move upward in the geometry illustrated in these figures. The density of pickup ions is relatively small but,

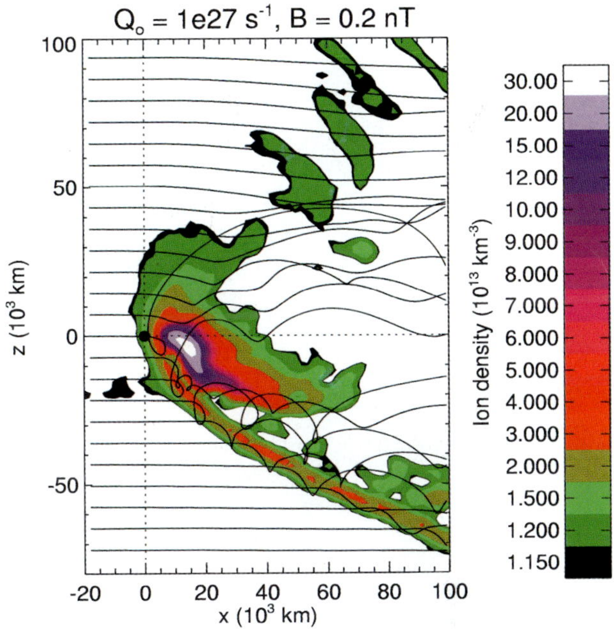

Fig. 3 Sample trajectories of solar wind protons near Pluto (origin of plot) through the interaction region with total ion density (N_2^+ plus solar wind protons). The trajectories are altered at the magnetic pileup boundary, but only partially thermalized as illustrated in Fig. 4. The magnetic pileup boundary is very abrupt in the $-z$ half-plane

because of their high mass, the momentum they extract from the solar wind is significant. The solar wind is slowed down (leading to compression and enhanced magnetic field) and deflected in the opposite direction to the pickup ions. Superimposed on the density contours in Fig. 3 are trajectories of solar wind protons. The large gyrations reflect the weak magnetic field's inability to turn the protons. The smaller gyrations indicate stronger magnetic fields in the pileup region. Harnett et al. (2005) showed that multifluid simulations produced qualitatively similar, though less pronounced, asymmetries in the interaction region.

Clearly, the upstream direction and magnitude of the IMF are critical drivers for the solar wind interaction with Pluto. Unfortunately, *New Horizons* did not have the resources to carry a magnetometer, so we will need to use our best knowledge of solar wind physics to infer the magnetic properties. Generally speaking, the IMF becomes increasingly tightly wound spirals on conical surfaces (for different latitudes) with increasing distance from the Sun. Thus, we expect the average field direction to be nearly perpendicular to the flow by 32 AU. Because of the alternating sector structure of the IMF, it is equally likely that the IMF will be pointed in either direction. Measurements of pickup ions, either by SWAP or by PEPSSI (McNutt et al. 2007) will provide the critical indication of the sign of the magnetic field at the time of the encounter.

2.3 Solar Wind Interaction at Low Atmospheric Escape

If Pluto's escape rate is less than $\sim 10^{27}$ s^{-1}, then the solar wind is expected to penetrate close to Pluto and impinge directly onto its ionosphere. Strictly speaking, the "obstacle" that deflects the solar wind in this interaction is produced by the electrical currents induced in the ionosphere. Such an interaction would be similar to the solar wind interaction with Venus or the saturnian magnetospheric interaction with Titan (Luhmann et al. 1991). The photochemistry of Pluto's upper atmosphere and ionosphere modeled by both Krasnopolsky and Cruikshank (1999) and by Ip et al. (2000) show peak ionospheric densities of a few

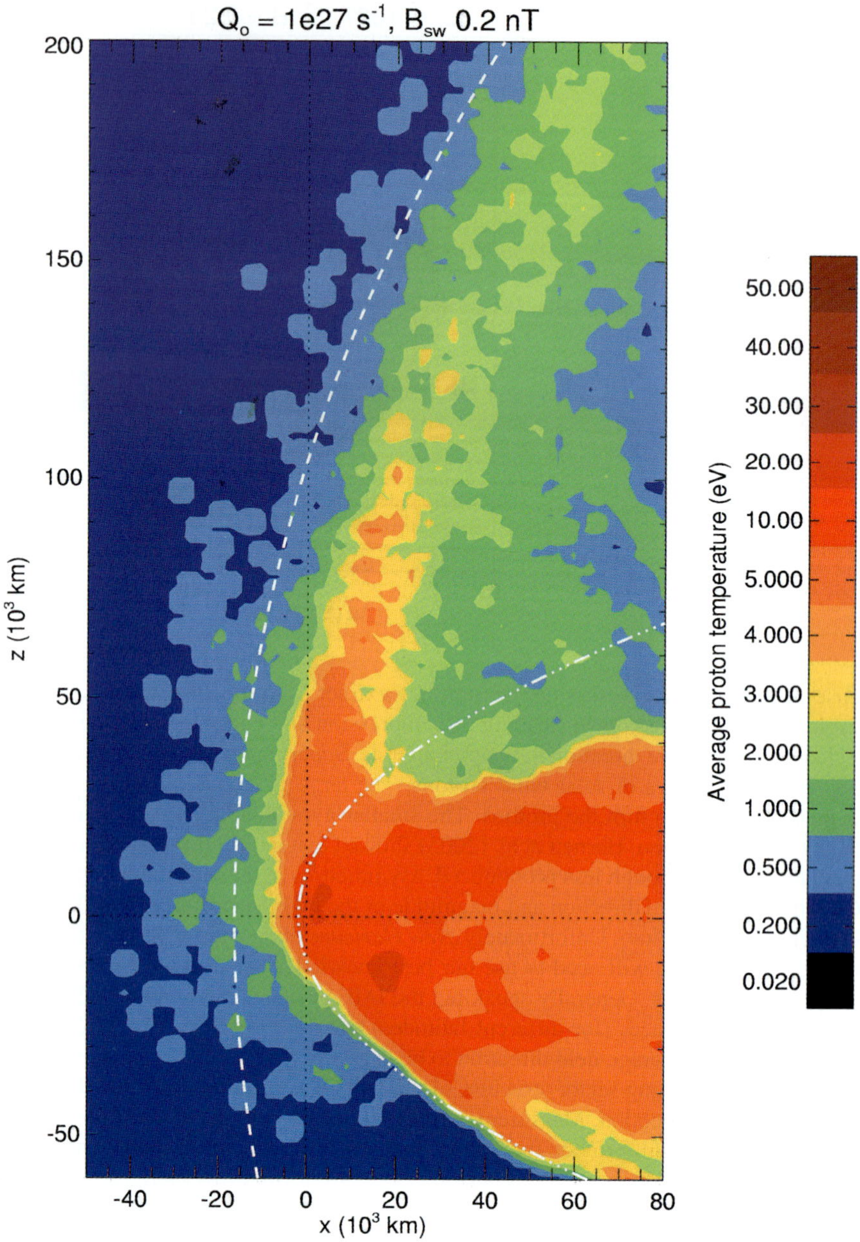

Fig. 4 Column-averaged proton temperature in the plane perpendicular to $B_{\rm SW}$. The maximum temperature for the solar wind protons (i.e., fully thermalized) moving initially at 340 km/s is 360 eV. The interaction only partially thermalizes the solar wind with maximum temperatures of a few tens of electron volts. The dashed line is the expected location (i.e., based on upstream standoff distance) of the bow shock and the dash-dot-dot-dot line is the expected location of the magnetic pileup boundary based on calculations for solar wind stagnation near comets for $Q = 10^{27}$ s^{-1} and $B_{\rm SW} = 0.2$ nT (Biermann et al. 1967; Galeev et al. 1985). Given the extreme asymmetry of the interaction, it is not possible to fit the bow shock and the magnetic pileup boundary with symmetric parabolic functions. We based our fit for the bow shock on the $+z$ half-plane and the fit for the magnetic pileup boundary on the $-z$ half-plane

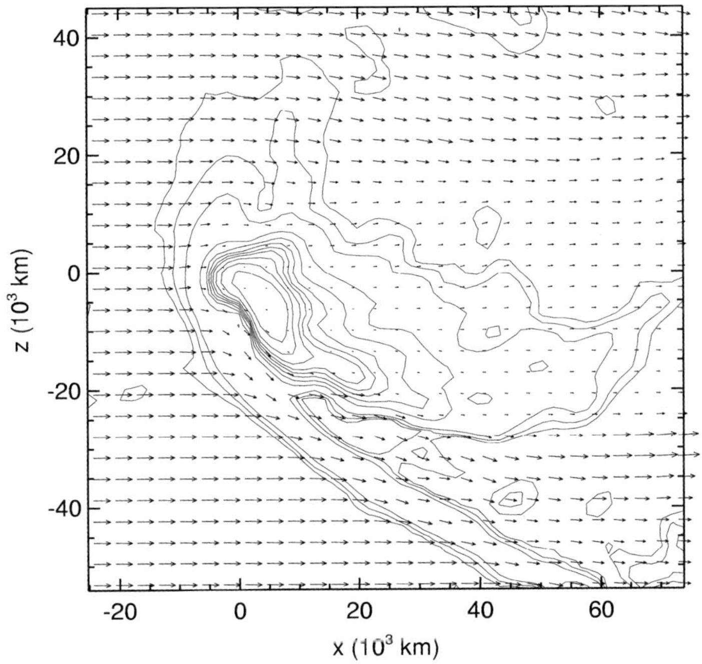

Fig. 5 Ion bulk flow velocity (pickup ions plus solar wind protons) with contours of total ion density illustrating the high degree of asymmetry in the plasma flow

thousand cm^{-3} at altitudes of $\sim$1000 km above the surface of Pluto. They find the main ionospheric ion to be H_2CN^+.

The solar wind deflection and any pickup ions will be significantly harder to measure on the *New Horizons* flyby in the case of a Venus-like interaction. However, the broad nature of the bow shock and the large gyroradii of any ions beyond the ionosphere should still produce detectable signatures several radii away from Pluto.

2.4 Solar Wind Interaction at Aphelion

It is not clear whether the atmosphere of Pluto completely collapses onto the surface when Pluto reaches aphelion. In the absence of a significant atmosphere Pluto will have a "Moon-like" interaction, as is expected for its moon Charon. In this type of interaction, the solar wind is absorbed on the sunlit side and the IMF diffuses through the nonconducting bodies, generating an extremely hard vacuum in the cavity behind. It seems unlikely at temperatures below 40 K that Pluto's icy outer layers could be electrically conducting. But should they have significant conductivity, the plasma interaction may be similar to the solar wind/asteroid interaction described by Wang and Kivelson (1996), Omidi et al. (2002), and Blanco-Cano et al. (2003). If the dimensions of the obstacle are small compared to the ion inertial length, then the interaction is mediated by the whistler mode. Comparison of *Galileo* observations of asteroid-associated perturbations with numerical models confirm the whistler-mode interaction. Magnetohydrodynamic shock waves are absent with small obstacles, as compressional waves can only exist on size scales larger than the ion inertial length. Pluto represents a possible intermediate case if the interaction region is limited to an area close to the planet since the solar wind proton inertial length is roughly 2 R_P.

2.5 Heliospheric Pickup Protons

In addition to the primary science of measuring the solar wind interaction with Pluto, SWAP may afford an excellent opportunity to measure heliospheric pickup protons on its way out through the heliosphere, *en route* to Pluto (and beyond).

The solar system moves continually through the local interstellar medium (LISM)—the part of the galaxy nearest to our solar system—causing a complex set of interactions between the outflowing solar wind and the matter in the LISM (see Fig. 6). These interstellar interactions set up the plasma boundaries of our heliosphere: the termination shock where the solar wind abruptly slows, becoming subsonic, prior to bending back into an extended tail in the interstellar downstream direction; the heliopause separating the subsonic solar wind from the ionized plasma of the interstellar medium; and possibly a bow shock where the LISM plasma becomes subsonic prior to being deflected around the heliosphere.

Interstellar atoms, predominantly H, continually drift into and through the heliosphere and, because of their neutrality, are unimpeded by the solar wind. The H atoms on trajectories toward the Sun move into regions of increasingly dense solar wind and higher levels of solar radiation, enhancing the probability of ionization through charge exchange or photoionization. The vast majority of interstellar H atoms that penetrate inside of 4 AU become ionized and incorporated into the solar wind flow. Upon ionization, newborn ions begin essentially at rest in the reference frame of the spacecraft and Sun. Just like for planetary pickup ions, as described above, because the moving solar wind carries frozen-in magnetic

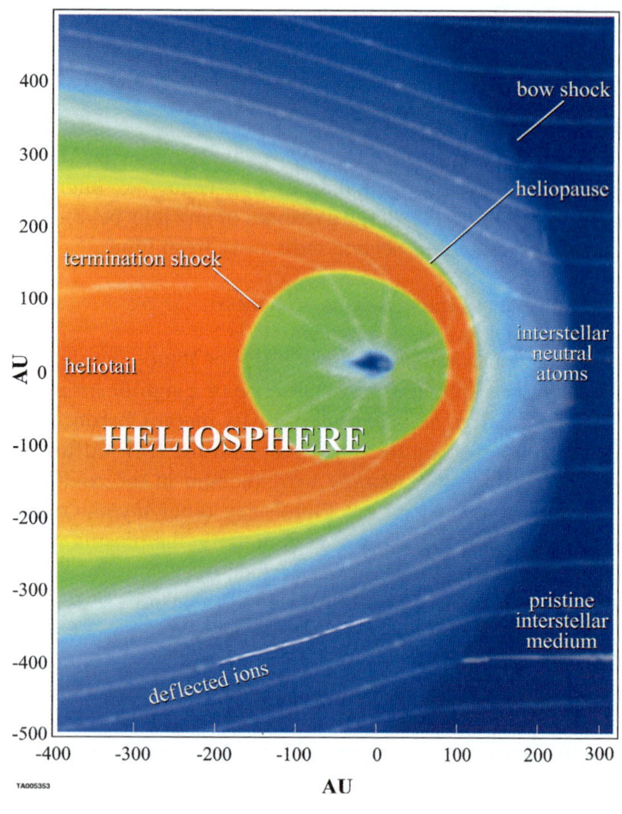

Fig. 6 The SWAP instrument will measure protons produced from neutral matter that drifts into the heliosphere from the LISM

field lines, the newborn ions encounter a motional electric field ($-\mathbf{v}_{sw} \times \mathbf{B}$) exerted by the solar wind and get picked up and carried outward with the solar wind. As they move outward, magnetic inhomogeneities in the IMF scatter the pickup ions, forming an almost spherical ring distribution in velocity space. As measured by spacecraft, these interstellar pickup-ion distributions are essentially flat for speeds less than $2v_{sw}$, then drop off sharply at speeds above this limit (Gloeckler et al. 1995).

The first interstellar pickup ions discovered (Möbius et al. 1985) were He^+ created by ionization of interstellar neutral He that penetrated within 1 AU of the Sun. The composition and velocity-space-resolved measurements by the SWICS experiment on *Ulysses* (Gloeckler et al. 1992) made it possible to explore pickup ions from 1.35 to 5.4 AU in great detail. The review by Gloeckler and Geiss (1998) provides an excellent summary of these pickup-ion observations. The observations include the most abundant pickup ion, H^+, second most abundant, He^+, and several other interstellar pickup-ion species, N^+, O^+, and Ne^+ (Geiss et al. 1994). The SWICS team demonstrated the existence of pickup distributions of He^{++}, which is produced largely by double charge exchange of atomic He with solar wind alpha particles (Gloeckler et al. 1997) and rare $^3He^+$ pickup ions. In addition to the interstellar pickup ions, SWICS distributions showed that the majority of the C^+ and a fraction of the O^+ and N^+ are produced by an additional "inner source" of neutral atoms located near the Sun (Geiss et al. 1995). The inner-source, velocity-space distributions are significantly modified as they cool over the solar wind's transition from the near-Sun source to several AU where they were observed (Schwadron et al. 2000).

Ulysses/SWICS also observed the ubiquity of pickup-ion "tails" in slow solar wind (Gloeckler et al. 1994; Schwadron et al. 1996; Gloeckler 1999). These tails do not correlate strongly with the presence of shocks but do correlate with compressive magnetosonic waves, suggesting that pickup ions are subject to strong statistical acceleration through processes such as transit-time damping of magnetosonic waves in slow solar wind (Schwadron et al. 1996; Fisk et al. 2000).

Prior to *Ulysses*, it was expected that pickup-ion distributions should be fairly isotropic because of pitch-angle scattering from background turbulence and self-generated waves (Lee and Ip 1987). Instead, pickup-ion distributions were observed to be highly anisotropic (Gloeckler et al. 1995) with scattering mean free paths $\sim$1 AU; the most likely cause is the inhibition of scattering through 90° pitch angle (Fisk et al. 1997). Although this lack of scattering is not fully understood, it has been shown that the turbulence of the wind has a strong 2D component (Matthaeus et al. 1990; Bieber et al. 1996) that is ineffective for pickup-ion scattering (Bieber et al. 1994; Zank et al. 1998). Pickup-ion models have been devised that take into account the long scattering mean free path (e.g., Isenberg 1997; Schwadron 1998).

Interestingly, the same self-generated turbulence associated with pickup-ion scattering has been considered an energy source for heating the solar wind (Zank et al. 1996; Matthaeus et al. 1999). These theories for the turbulent heating of solar wind take into account both the pickup-ion-driven turbulence, which is predominant outside of $\sim$8 AU, and wind shear. Smith et al. (2001) and Smith et al. (2006) have rigorously tested the heating models using data from *Voyager 2* and *Pioneer 11*, and improved models of the pickup-ion scattering and associated wave excitation (Isenberg 2005) have led to remarkable agreement between data and theory.

Charge exchange between interstellar hydrogen and solar wind protons leads to a complex interaction near the nose of the heliosheath, where a so-called hydrogen wall is formed from slowed interstellar hydrogen atoms and charge-exchanged solar wind protons. These interactions cause the removal or filtration of a fraction of the penetrating interstellar hydrogen atoms (e.g., Baranov and Malama 1995). The solar radiation pressure and the rates of photoionization and charge exchange vary with solar latitude and over the solar cycle. Sophisticated models of interstellar neutral atoms have been developed to take these effects into account (e.g., Izmodenov et al. 1999).

Cassini was the first spacecraft to measure pickup ions in the interstellar medium downstream region of the heliosphere beyond 1 AU (McComas et al. 2004a). Both interstellar pickup H^+ and He^+ were identified with the familiar cutoff in velocity space at twice the solar wind speed. Observed enhancements in the pickup He^+ were consistent with gravitational focusing by the Sun. Further, McComas et al. also reported observations of the interstellar hydrogen shadow from depletion of H atoms in the downstream region caused by the outward force of radiation pressure (which exceeded the gravitational force at the time of observation) and the high probability of ionization for atoms that must pass close to the Sun to move behind it.

The trajectory of *New Horizons* takes the spacecraft out to Pluto, which is currently toward the nose of the heliosphere. Figure 7 shows simulated distribution functions of pickup protons and solar wind protons in the spacecraft frame. We have used a kappa distribution for the solar wind protons (e.g., Vasyliunas 1968; Collier 1993) (with kappa $\sim$3, which is typical), and a density that falls off as R^{-2} from a 1-AU density of 5 cm^{-3} and temperatures consistent with observations (e.g., Smith et al. 2001). For the pickup protons, we have taken a steady-state distribution that includes convection and adiabatic cooling in the radially expanding flow (Vasyliunas and Siscoe 1976). The proton-pickup rate at each location is the local ionization rate times the neutral density solved by using the "hot" model (Fahr 1971; Thomas 1978; Wu and Judge 1979), which accounts for gravitational focusing by the Sun, ionization loss, and the finite temperature of incoming neutral atoms. We have taken an interstellar density near the termination shock of 0.1 cm^{-3}, a neutral temperature of 11,000 K, a neutral inflow speed of 22 km/s, and an ionization rate of 7×10^{-7} s^{-1} (referenced at 1 AU). We have also taken the force of radiation pressure to be comparable to that of gravity. The top panel in Fig. 7 shows that solar wind protons are typically comparable to pickup protons near 5 AU. The pickup protons are observable near the knee of the distribution, provided the solar wind is not significantly hotter or has significantly broader tails than those shown. Since the plasma near the ecliptic plane is highly variable near 5 AU, the detection of pickup protons is possible, but is often obscured by the solar wind proton distribution. As *New Horizons* moves out toward the nose of the heliosphere, the situation changes. The solar wind density falls off with the square of heliocentric distance (whereas the pickup proton density falls off only as R^{-1}) and cools. Both of these effects make it much easier to detect "clean" pickup-proton distributions.

The unique trajectory of *New Horizons* combined with SWAP's capability of pickup-proton measurement should enable it to address some fascinating issues. The distribution functions measured by SWAP will allow investigation of possible sources (other than interstellar) of pickup ions. For example, it is thought that an outer source of pickup ions may be caused by the interaction of solar wind with dust grains from the Kuiper Belt (Schwadron et al. 2002). The outer source is likely to be far more variable in space and time than the interstellar source.

3 Instrument Description

The SWAP electro-optics consists of (1) a retarding potential analyzer (RPA), (2) a deflector (DFL), and (3) an electrostatic analyzer (ESA). Collectively, these elements select the angles

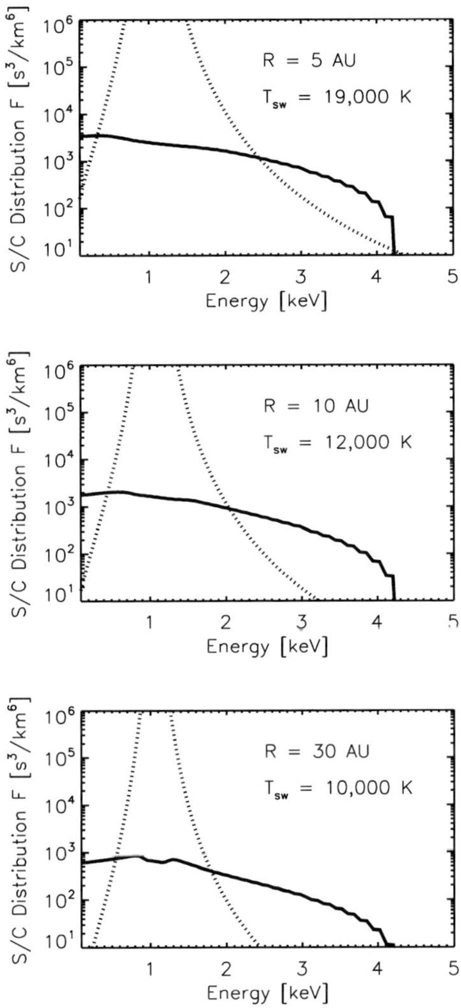

Fig. 7 Pickup-proton (*solid curves*) and solar wind (*dashed curves*) distribution functions shown with distance from the Sun toward the nose of the heliosphere. The pickup-proton distribution functions become more and more prominent compared to the cooled solar wind distributions further out in the heliosphere

and energies of the solar wind and pickup ions to be measured. Ions selected by the electro-optics are then registered with a coincidence detector system. Figure 8 schematically depicts the SWAP principle electro-optics. Ions enter through the RPA with all ions having energy per charge (E/q) less than the RPA voltage being rejected by this high-pass filter. Ions entering at angles from above horizontal in this figure can be deflected into the subsequent electro-optics by applying a voltage to the deflector ring. Ions with E/q greater than the RPA voltage are then selected by the ESA, which rejects ions outside the selected E/q range as well as UV light and neutral particles.

Figure 9 schematically shows how SWAP's ESA and RPA are used together to select the E/q passband. When the RPA is off, the passband is determined solely by the ESA, which has an 8.5% FWHM resolution (top panel). At increasing RPA voltages for a given ESA setting, the passband is cut off in a variable "shark-fin" shape, allowing roughly two decades decreased sensitivity (middle panel). Finally, differentiating adjacent RPA/ESA combinations, or better yet deconvolving multiple combinations, provides high-resolution differential measurements of the incident beam.

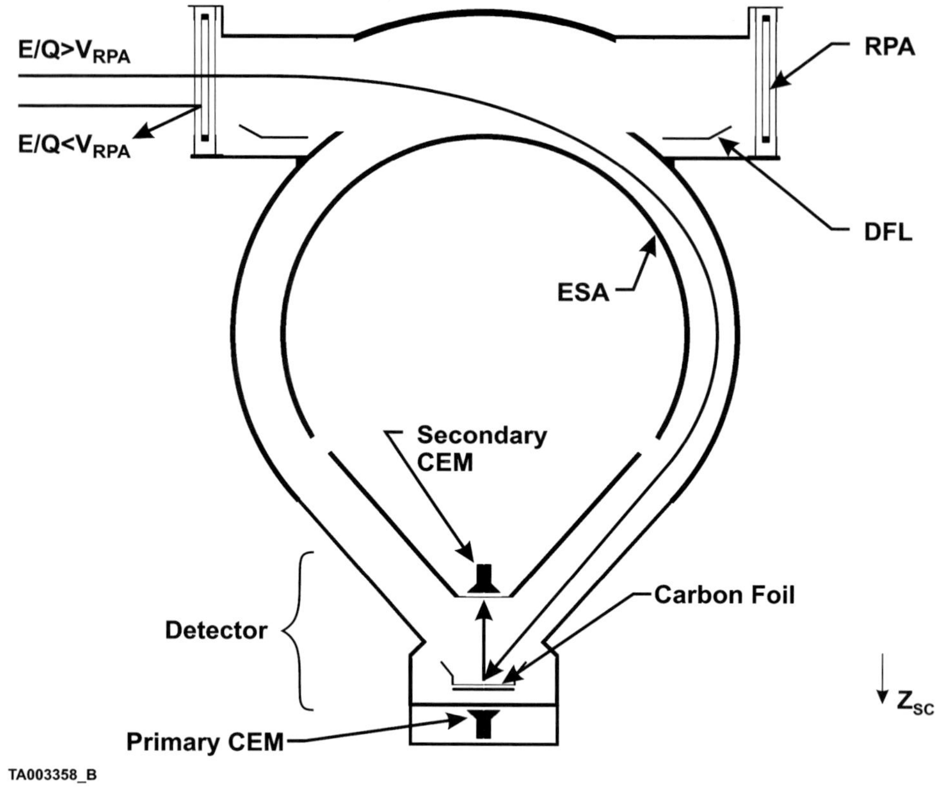

Fig. 8 Schematic diagram of the SWAP electro-optics including RPA, deflector, ESA, and detector section

As an example of how the RPA can be used to measure very small changes in beam energy, and hence solar wind speed, Figure 10 shows the count rate as a function of RPA voltages for a range of different ion beam energies taken with the flight electro-optics during instrument calibration (described in more detail in Sect. 4.1). A small subset of the full energy range of SWAP (990 to 1010 eV) is shown to highlight the energy resolution possible by this instrument. For each scan, the count rate is normalized to the rate when the RPA voltage is set at zero to take into account the differences in the ion beam flux. Differences in the ion energy as small as 1–2 eV are distinguishable at typical solar wind energies of ~1000 eV.

The transmitted ions are postaccelerated into the detector section, which employs an ultra-thin carbon foil and two channel electron multipliers (CEMs) to make a coincidence measurement of both the primary particle and the secondary electrons generated when the primary particle passes through an ultra-thin carbon foil. Charge amplifiers (CHAMPs) service the two CEMs and transmit digital pulses when events are detected. High-voltage power supplies (HVPSs) provide power for the CEMs and sweep the voltages on the electro-optics. The control board processes the pulses from the CHAMPs, controls the sweeping of the high voltages, digitizes the housekeeping data, creates telemetry packets for transmission, accepts commands, and converts spacecraft power into the secondary voltages required by the instrument. Key properties of the SWAP instrument are given in Table 2.

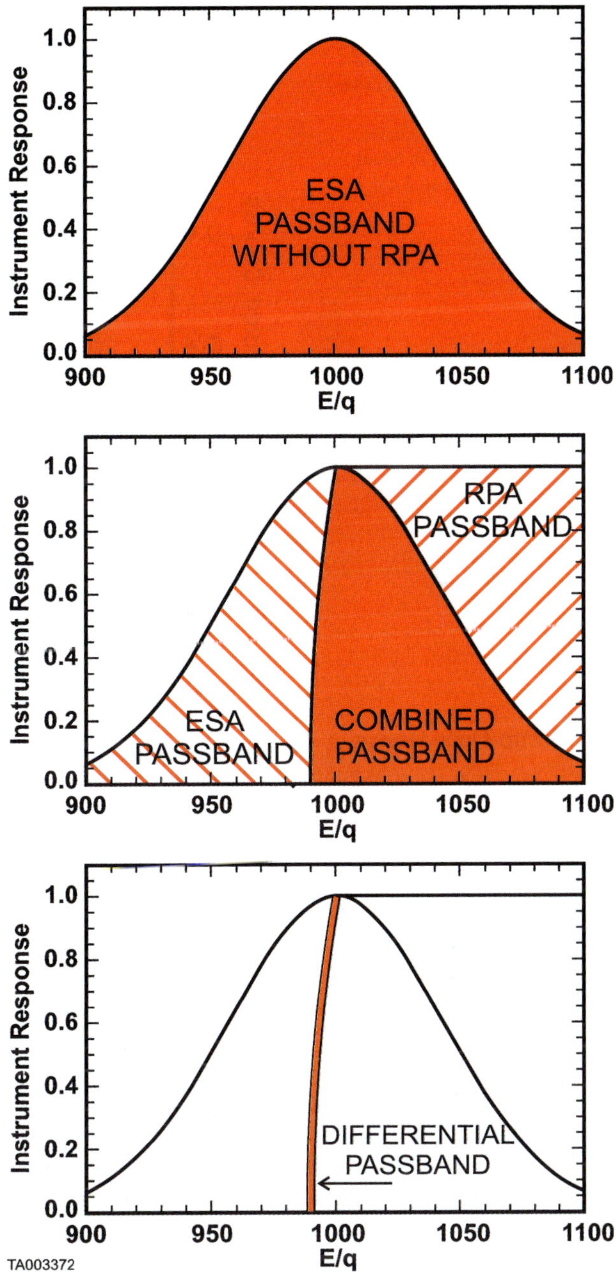

Fig. 9 SWAP principle of operation. Different combinations of RPA and ESA settings provide for a variable E/q passband. Multiple combinations of settings can be differentiated to produce very high resolution measurements

The *New Horizons* spacecraft points its optical instruments by rotating the spacecraft about its Z-axis. To make solar wind measurements largely independent of this scan angle, SWAP was designed to be essentially symmetric about Z. SWAP is mounted into a cylindrical bracket as depicted in Fig. 11. By placing it on the $-Z$ corner of the spacecraft, the instrument's field of view is clear of any structures or obstructions for all scan angles.

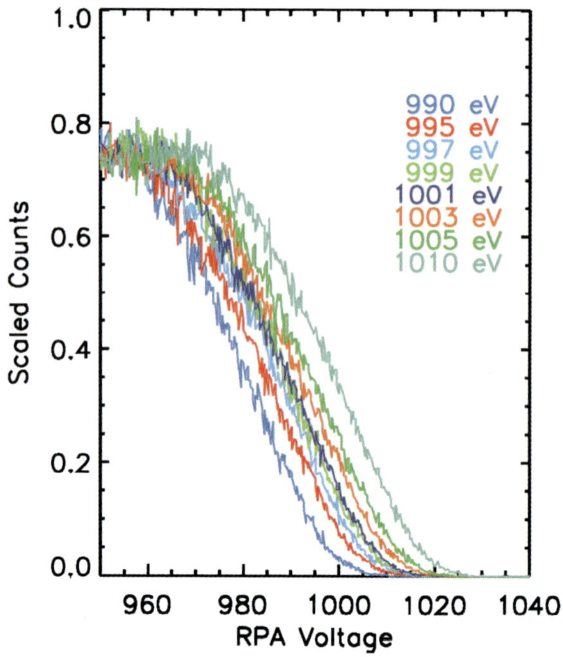

Fig. 10 RPA resolution of ion beams with energies from 990 to 1010 eV. Each RPA scan shown was taken with the beam incident normal to the RPA. Deconvolution of such curves should make changes in beam energy as small as 1 to 2 eV possible

3.1 Electro-Optic Design

The electro-optic design is driven by the unique needs of the *New Horizons* mission to Pluto. A large-aperture instrument is required since the density of the solar wind falls off as the square of the distance from the Sun. At ~32 AU, where we will encounter Pluto, the mean solar wind density is roughly 3 decades lower than the solar wind at 1 AU. Also, the *New Horizons* spacecraft does not have a scan platform for the optical instruments. Instead, the spacecraft slews about the Z-axis to provide the required scanning motion. Because SWAP's focus is to measure the solar wind, we required an electro-optic design that was largely unaffected by this rotation about the Z-axis. Therefore, we selected a top-hat design with an ESA that has a large angular and energy acceptance to maximize the effective aperture. Since the driving measurement is to look for small changes in the solar wind speed as *New Horizons* passes Pluto, we included an RPA to allow finer resolution energy measurements and to provide a variable electrostatic passband. This is accomplished by "crossing" the ESA and RPA such that the RPA admits particles only in the highest energy portion of the ESA passband as described in the previous section. Also, because the spacecraft will sometimes be required to tip out of the ecliptic plane, we incorporated a deflector to adjust the look direction of the instrument. Finally, because the mission is designed so the spacecraft rarely points its $-Y$-axis toward the Sun, we can provide an unobstructed field of view of $>270°$ in the X–Y plane, centered on the $+Y$-axis, which is aligned with the high gain antenna (HGA). Nominally, the HGA points toward the Earth within a few degrees of the Sun for a large fraction of the Pluto encounter. All supports for the top-hat top-plate, RPA, ESA, and the aperture doors for protection during launch are restricted to the remaining 90° aligned with the $-Y$-axis. Figure 12 shows the key dimensions of the SWAP electro-optics.

Table 2 Measured properties of the SWAP instrument

Field of view	$276° \times 10°$ (deflectable $>15°$ toward $-Z$)
Energy range	
ESA (bin centers)	35 eV to 7.5 keV
RPA	0 to 2000 V
Energy resolution	
ESA ($\Delta E/E$)	0.085 FWHM
RPA	0.5 V steps (high resolution requires deconvolution)
ESA factor (beam energy/ESA voltage)	1.88
Dynamic range	$\sim 10^6$
Geometric factor (hot plasma)	Coincidence: 2.1×10^{-3} cm^2 sr eV/eV Total: 1.3×10^{-2} cm^2 sr eV/eV
Cold beam effective area* (normal incidence)	Coincidence: 3.3×10^{-2} cm^2 Total: 1.9×10^{-1} cm^2
Time resolution	Full energy range and (1) detailed peak measurements or (2) additional full energy sweep every 64 s (128 steps with 0.39 s accumulation times)
Mass	3.29 kg
Volume	0.011 m^3
Power	2.84 W
Telemetry	$<$1–280 bps

*Calculation of expected count rates is presented in Sect. 5.1

3.1.1 RPA Design

The RPA consists of four concentric aluminum cylinders or screens, each of which is machined with ~90,000 close-packed holes to create a self-supporting grid structure. Each cylinder, approximately 65% transmissive, is 0.762 mm thick and has been machined into a gridlike structure by drilling 0.343-mm-diameter holes through nominal 0.394-mm-thick aluminum in a close-packed hexagonal configuration (Fig. 13). From outermost to innermost, the outer diameters of the four RPA screens are 174.4, 169.6, 166.4, and 161.6 mm. The outside and inside cylinders are at ground potential (0 V). The two central cylinders are biased from 0 to +2000 V in 0.49-V steps and isolated from the rest of the structure by ceramic insulators. There is a 2.032-mm gap between the biased and grounded RPA grids to provide sufficient high-voltage clearance for the −2000 V potential. The two biased RPA grids are separated by 1.016 mm to stop penetration of lower energy equipotentials through the grid holes, thus creating a much more uniform electric field in the RPA region. The unobstructed field of view (FOV) is 276° in the roll direction.

The RPA provides a low-pass filter with a relatively sharp energy cutoff so that we can make a fine sweep across the solar wind beam after locating it with a coarse ESA energy scan. Ions that have sufficient energy to climb the electrostatic "hill" set by the voltage on

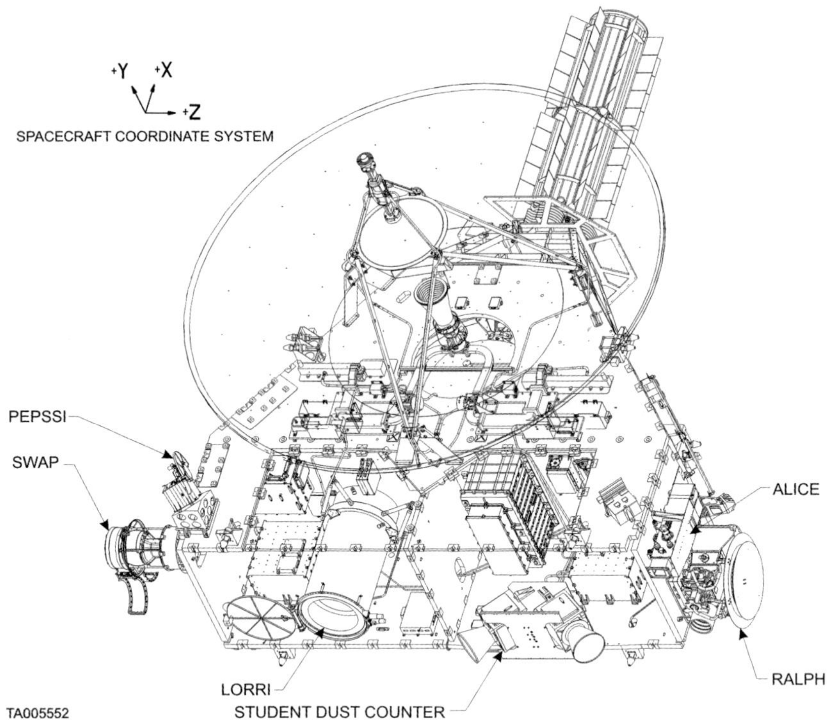

Fig. 11 Location of SWAP on the *New Horizons* spacecraft

the inner RPA grids are reaccelerated to their original energy as they pass from the inner RPA grids to the final grounded RPA grid.

3.1.2 Deflector

SWAP incorporates a very simple deflector that is used to deflect particles from above the central plane of the instrument (from further out in the $-Z$ axis of the spacecraft) into the ESA. The deflector is located just inboard of the RPA. The voltage on the deflector ring is varied from 0 to +4000 V. It deflects up to 7000 eV/q particles up to 15° into the ESA (with larger deflections for lower energies).

3.1.3 ESA

The ESA provides coarse energy selection and protects the detectors from UV light. The dimensions of the top-hat ESA are shown in Fig. 12. The outer ESA is serrated and blackened with Ebanol-C, a copper-black process that greatly reduces scattering of light and particles. The inner ESA, which is blackened but not serrated, is supported on insulators that attach it to a cantilevered support structure (Fig. 14). On the other side of this structure a grounded cone completes the ESA design by providing a field-free region for particles to enter into the detector region. The voltage on the ESA is varied from 0 to −4000 V. It has a ratio of central E/q of particle transmitted to ESA voltage, or "K-factor", of 1.88 and selects up to 7.5 keV/q particles (passband central energy) with a $\Delta E/E$ of 8.5% FWHM.

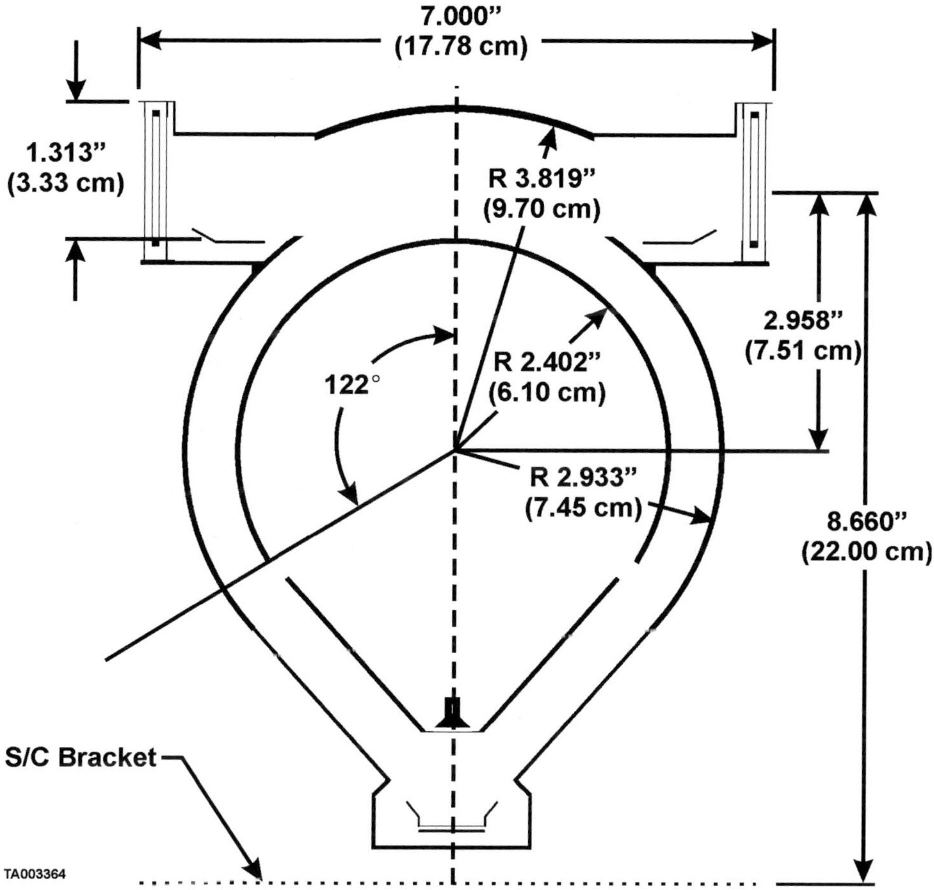

Fig. 12 Key dimensions of the electro-optics

3.2 Detector Design

SWAP employs a coincidence system to detect incoming ions. After ions of the desired energy and angle have been selected by the electro-optics system, they pass through a field-free region between the ESA and detector region. Once a particle enters the detector region, it is accelerated by the CEM high voltage toward the focus ring, on which is suspended an ultra-thin carbon foil (McComas et al. 2004b). The carbon foil is nominally 1 μg/cm^2 thick and is suspended on a 64% transmissive grid. The particle passes through the focus ring, which is at the primary CEM (PCEM) HVPS output voltage, and travels on to the PCEM (Fig. 15). Forward-scattered electrons from the carbon foil are also accelerated to the PCEM by the ~100 V potential created by the PCEM strip current and a resistive divider. Backward-scattered electrons are directed by the focus ring toward the secondary CEM (SCEM), which collects them.

Counts from these two CEMs are registered by CHAMPs and their associated electronics. A count from either the PCEM or the SCEM starts a 100-ns coincidence window timer. Given the particle trajectories and the electron trajectories, it is possible for either the PCEM or the SCEM to trigger first, so no specific arrival order is required by the electronics. The

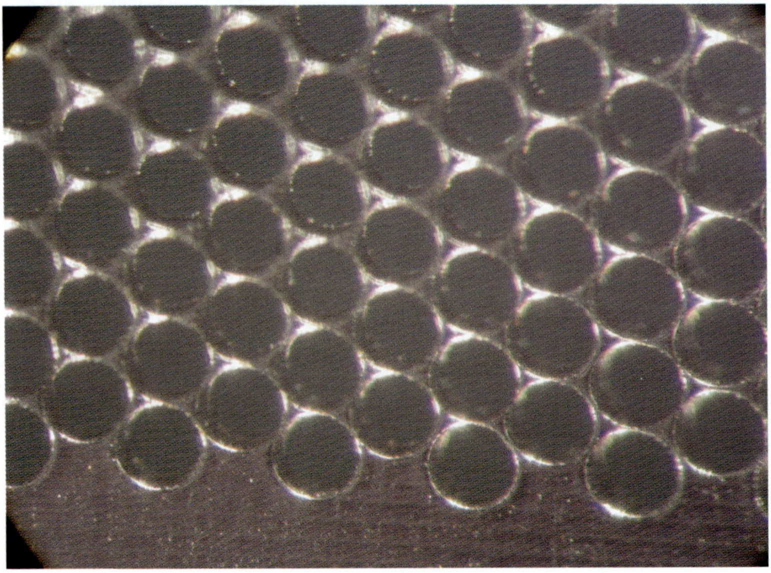

Fig. 13 Photograph of 0.343-mm RPA holes on 0.381 centers

CEMs (Fig. 16) have quad spiral channels with a resistance of 300 Mohm and dark counts less than 0.04 s^{-1}.

The coincidence detector system reduces the background from CEM dark counts, penetrating energetic particles, and UV noise, allowing SWAP to have a very low noise floor. Two detectors further provide the redundancy needed for a long-duration mission. SWAP can still make its primary science measurement using either one of its CEMs. This requirement for redundancy drove us to add a −1-kV focus ring supply that is slaved to the SCEM and diode OR-ed with the PCEM supply to the focus ring. If the PCEM is turned off, the −1-kV focus ring supply will accelerate back-scattered electrons from the carbon foil into the SCEM.

The long CEM lifetime needed for this mission required us to select ultra-low outgassing materials everywhere near the detector. We used only glass, metal, and ceramic materials for all parts of the detector that had venting access to the CEM detectors. High-voltage and low-voltage cabling was brought to bulkheads, and the signals were conducted through ceramic feedthroughs to the exact location they were required (Fig. 17).

3.3 Mechanical Design

The SWAP mechanical design consists of three main subassemblies: optics and detector, aperture door, and electronics packaging and cabling (Fig. 18). In addition to these instrument subassemblies, we also discuss SWAP's structural and thermal design in the following sections.

3.3.1 Optics and Detector Mechanical Design

The critical requirements driving the optics and detector mechanical design were to mount the electro-optical components as specified by the ray-traced model, produce optically black

Fig. 14 Cantilevered inner-ESA dome with (*bottom panel*) and without (*top panel*) blackening

surfaces inside the instrument, ensure contamination control for the detectors, provide for easy refurbishment, and accurately align all critical components.

We used a SIMION ray-tracing model to define the electro-optics component (Fig. 19) sizes and locations (RPA, ESA, and CEMs) and required that the mounting of the various elements should minimally impact particle trajectories. We accommodated this by mounting the inner ESA and SCEM assemblies from a cantilevered support hidden in the 90° region where SWAP does not require particle viewing. The location of this cantilever matches up with the hinge assembly on the door.

To accommodate the cleanliness requirement, the optics and detector areas included a split between ultra-clean and electronics volumes. We used ceramic for the electronics near the primary CEM, and we fed the cabling into the primary CEM area using passthroughs in the cantilever support so that it would not pass through the ESA gap.

We refurbished SWAP after spacecraft environmental testing, replacing the CEMs and the ultra-thin carbon foil with ones that had been stored in a clean environment after burn-in. To minimize the risk of dismantling the instrument, the CEM assemblies were designed so that they could be easily replaced. To meet the alignment requirements for the electro-optics, we designed features that would control the concentricity and placement of the optics and detectors in relation to each other both before and after refurbishment.

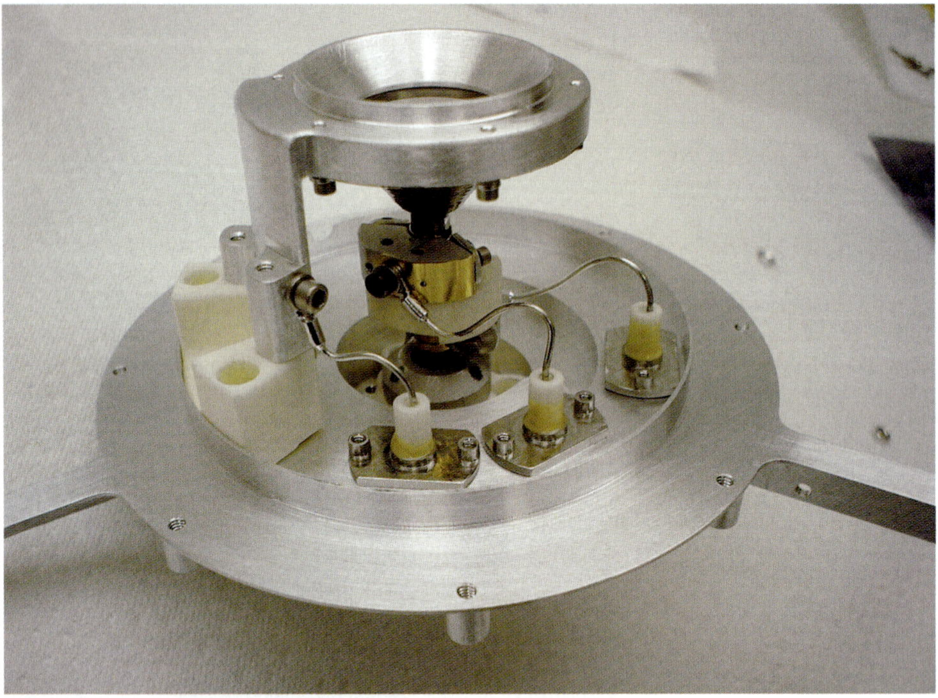

Fig. 15 PCEM and focus ring with carbon foil (partially assembled)

Fig. 16 Quad spiral channel electron multiplier

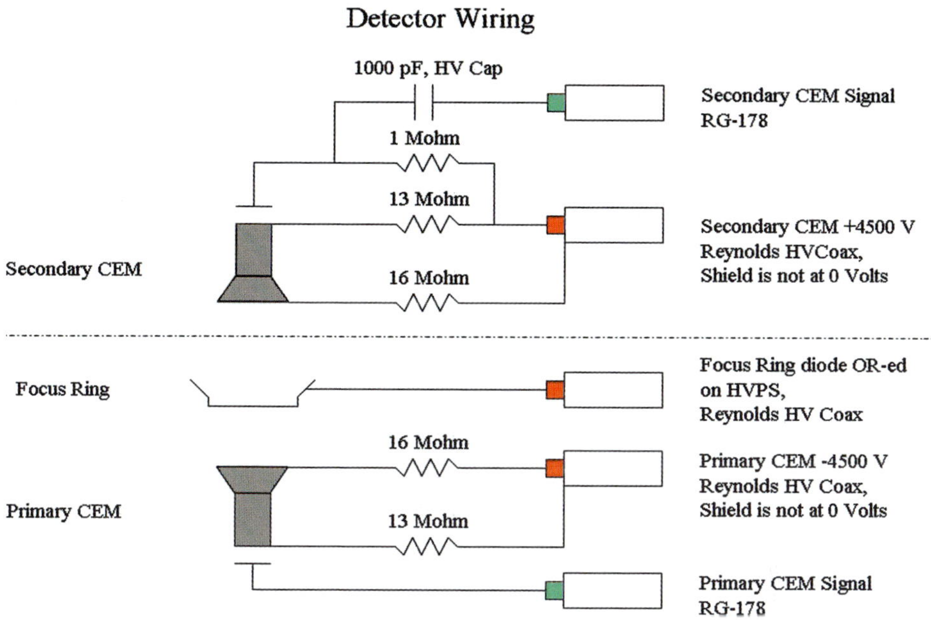

Fig. 17 Detector wiring diagram

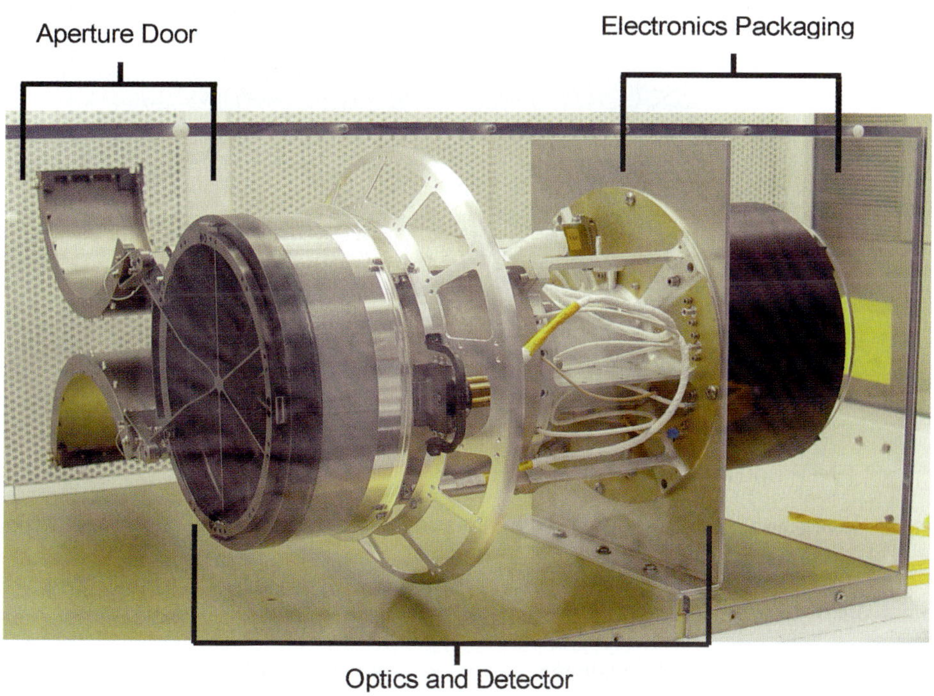

Fig. 18 Photograph of SWAP instrument highlighting subassemblies

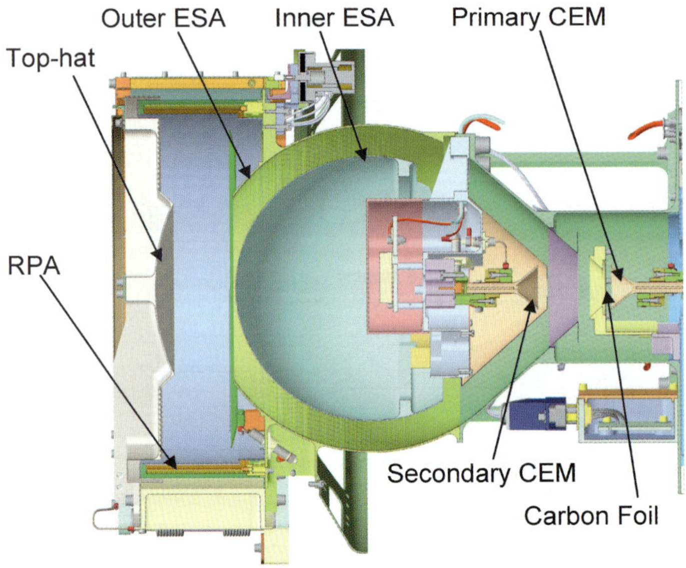

Fig. 19 Mechanical configuration of the optics and detector assembly

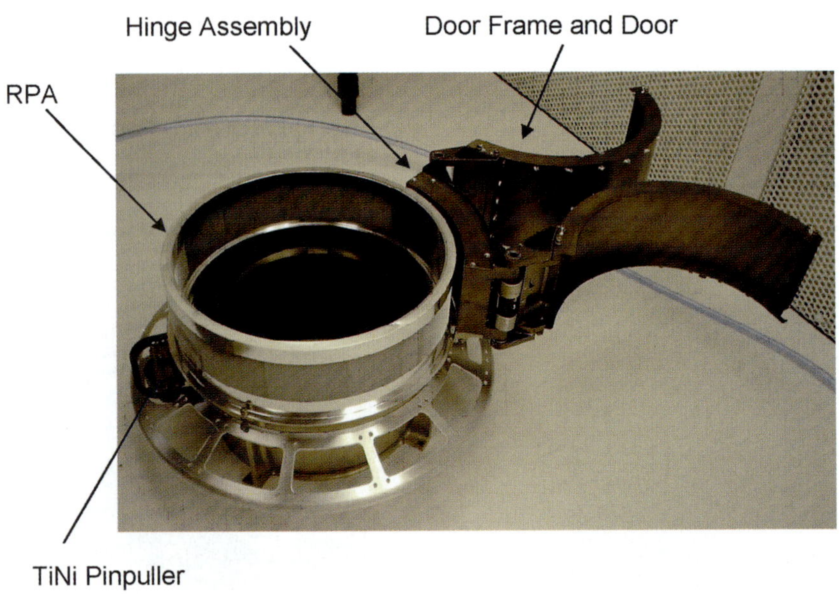

Fig. 20 Door assembled on the SWAP outer ESA. The TiNi pinpuller releases the door so it can spring open

3.3.2 Aperture Door Design

The aperture door's (Fig. 20) main requirement is to protect the SWAP RPA from contamination and damage during ground and launch operations. Although we designed the door to open one time after launch, it was important that it be easily reset for testing on the spacecraft. We opened the door multiple times as part of the instrument verification and during the environmental test flow at the spacecraft level. The door could not be a glinting source because it stays with SWAP during the entire mission to Pluto. Spacecraft requirements dictated that external surfaces be conductively coupled to the instrument. Therefore, we coated the door surfaces with black nickel and provided for ground straps built into the design. The *New Horizons* Component Environmental Specification required verifying through testing that the door's torque margin was greater than 2.25. Testing demonstrated a torque margin of 3.10, and the door successfully opened in flight.

3.3.3 Electronics Packaging and Cabling Design

Figure 21 shows the as-built electronics packaging and cabling that are part of the SWAP instrument. Panel (a) shows the HVPS and control board with the electronics before the

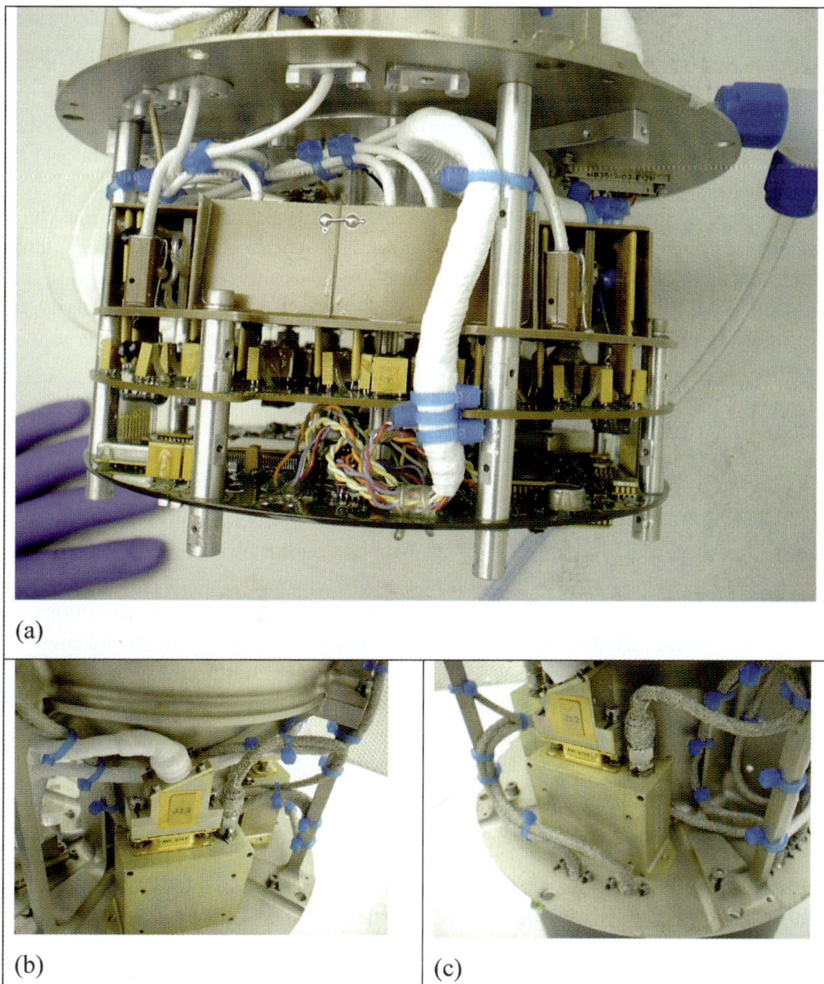

Fig. 21 As-built electronics packaging and cabling

electronics housing was installed. Panels (b) and (c) show different views of the CHAMP electronics and cables that bring the HV to the RPA and ESA.

To save weight and volume, the electronics are compactly packaged below the ESA in an "electronics volume", as shown in panel (a). The electronics are laid out on three boards, which are connected via pin/socket connections. The top board contains the HV, the middle board includes the drivers for the HVPS, and the bottom board receives power from and provides telemetry to the spacecraft. The CHAMP boards were originally intended for the lower electronics, but engineering testing showed that they needed to be closer to the detectors to decrease noise. We routed the HV and signal cables through the electronics housing to their appropriate location and included cable routing tie points in the outer ESA design to prevent the cables from moving during launch loads or transportation. The standoffs between the three bottom boards structurally attached the boards to the housing and allowed heat to conductively pass to the housing.

3.3.4 Structural Design

The Component Environmental Specification defined the following critical structural requirements for SWAP: (1) quasi-static load of 30g force applied separately along three orthogonal axes, (2) first mode structural frequency constraints of >70 Hz (thrust direction) and >50 Hz (lateral direction), (3) sine vibration up to 20g (22–24 Hz), and (4) random vibration with an overall amplitude of 10.4 Grms.

We performed a structural analysis during the final design phase to verify that the SWAP structure would meet these requirements. The analysis focused on the critical items such as the cantilevered ESA and outer ESA. The analysis also showed that the first natural frequency of the SWAP instrument, 180 Hz, was well above the requirement. Finally, we performed vibration testing during environmental testing to ensure that SWAP would survive the prescribed levels.

3.3.5 Thermal Design

SWAP has flight temperature limits of 0 to +40°C (operating) and −20 to +50°C (nonoperating). We initially performed a thermal analysis to show that the electronics and other temperature-sensitive parts of the SWAP instrument could survive these temperature extremes. The analysis also showed that the heat exchange with the spacecraft was less than 5 W (the limiting case was when the instrument was at its hot limit with the door open), as required by the spacecraft. During environmental testing, we performed a thermal vacuum test to validate the functionality of SWAP at all temperatures within these extremes; hot and cold turn-ons of the instrument were also performed.

3.4 Electronics

3.4.1 CHAMPs

The CHAMPs for SWAP convert a charge pulse from the CEMs into a logic pulse that can be registered and processed on the control board. The CHAMPs (Fig. 22) are located as close to the detectors as practical in separate enclosures mounted to the top of the strong back. The SWAP design incorporates high-speed commercial hybrid CHAMPs, which work reliably to >1 MHz rates and have voltage-adjustable thresholds. The output from the CEMs is brought to the CHAMP through a short coaxial cable through a bulkhead connector that

Fig. 22 Charge amplifier

ties the shield to chassis ground. A 100-ohm resistor and back-to-back diodes protect the input from discharges, and the input is AC coupled to the charge amp.

Injecting test pulses into the front end demonstrates the end-to-end integrity of the microcircuits and cabling for CEM pulse processing. Test pulses are driven from the control board to the CHAMP, which is twisted with a return line.

The threshold voltage for the CHAMP is set by a digital-to-analog converter (DAC) on the control board, and this value can be set through standard software commands, allowing the threshold to be updated at any time. A resistor sets the output pulse width to 70 ns and the amplifier dead time to 100 ns. The output pulses from the amplifier are buffered by two Schmitt trigger buffers and transmitted through a back-terminated series resistor to the cable that connects to the control board.

3.4.2 HVPS

The HVPSs (Fig. 23) set the voltages on the optical surfaces (RPA, DFL, ESA, and focus ring) as well as supply power to the PCEM and SCEM. Table 3 shows the primary HVPS properties. The two detector supplies are single string, but the two independent detectors provide redundancy. The focus ring is diode-ORed with the output of the PCEM and a −1000-V focus ring supply that is created and controlled in parallel with the SCEM. The optical power supplies are redundant and diode-OR-ed together. The secondary ground on the HVPS is "zap-trapped" to chassis through back-to-back diodes for protection. Mechanically, the HVPSs are built onto two interconnected boards: the driver board and the multiplier board.

Table 3 Key HVPS specifications

	PCEM	SCEM	Focus ring	RPA	DFL	ESA
Voltage range [V]	0 to −4500	0 to +4500	0 to −1000	0 to +2000	0 to +4000	0 to −4000
Ripple	0.5 V_{rms}	0.5 V_{rms}	10 V_{rms}	0.1 V_{rms}	0.5 V_{rms}	0.5 V_{rms}
Settling time	N/A	N/A	N/A	100 ms to 0.1%	100 ms to 0.1%	100 ms to 0.1%
Accuracy	5 V	5 V	20%	0.5 V	4 V	4 V

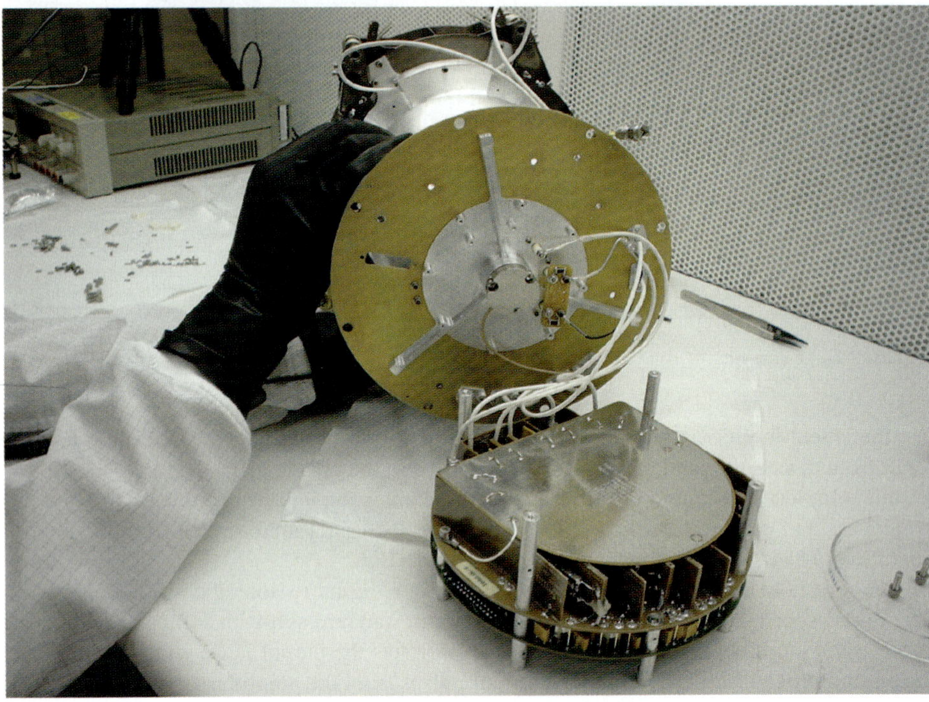

Fig. 23 HVPS being assembled into the SWAP sensor

3.4.3 Control Board

The SWAP control board provides the electrical interface between the *New Horizons* spacecraft and the SWAP instrument. All command, telemetry, power, safe/arm, and actuation interfaces reside on the control board. Software on an 8051 microcontroller responds to commands, controls the operation of the instrument, sequences the high-voltage power supplies, collects the data, and formats telemetry for downlink.

SWAP communicates to the spacecraft through two redundant (A and B side) asynchronous RS-422 interfaces. SWAP accepts serial commands, produces serial telemetry, and synchronizes communication with the spacecraft through a pulse per second (PPS) line. The command and telemetry data are transmitted at 38,400 baud.

The control board contains the EMI filters and DC–DC converters, which create the +5 and −5 V secondary power rails that are isolated from the spacecraft primary bus. Two separate 1.5-W DC–DC converters provide power for the instrument. Since the DC–DC converters do not have provisions for synchronization, each converter has an independent EMI filter to eliminate low-frequency noise from the beat between the two converter oscillators. The control board has a set of power MOSFETs, which allow us to switch low-voltage power to the PCEM HVPS, SCEM HVPS, optical HVPS bank A, optical HVPS bank B, and the housekeeping circuits.

The microcontroller provides all of the onboard processing required to operate the instrument, responds to commands, and produces telemetry. The CPU executes at 4.9152 MHz, and a minimum of 12 clocks are required to execute a single instruction, so the top speed of the processor is 0.4 million instructions per second (MIPS). From this low rate, the clock can be divided down further to reduce the instruction rate to 0.05 MIPS. Because of spacecraft-level power constraints every effort was made to reduce the operating power required for SWAP. This maximized SWAP's ability to remain on during the Pluto encounter sequence when the spacecraft is power-limited.

Boot code for the microcontroller resides in a radiation-hardened 32k 8-bit programmable read-only memory (PROM). This ensures that the instrument can always boot and establish basic communication with the spacecraft even if other memory devices have endured temporary upsets or even permanent degradation. Two separate 128k 8-bit electrically erasable programmable read-only memory (EEPROM) devices provide redundant storage for a 64k 8-bit storage area for program code and a 64k 8-bit storage area for look-up tables (LUTs). Finally, a 128k 8-bit static random access memory (SRAM) is used to provide 64k $\times$ 8 bits of code memory and 64k $\times$ 8 bits of data memory. During normal operations, if the program code in one of the EEPROM banks has a valid checksum, then it is loaded into RAM by the boot PROM and executed. A field-programmable gate array (FPGA) controls the memory mapping and memory windows that the microcontroller needs to access these devices.

Signals received from the primary and secondary CHAMPs are processed on the control board. When the electro-optics have been set to the appropriate levels and sufficient settling time has elapsed, the software opens up an acquisition window and totals all of the primary and secondary CEM pulses that occurred during the acquisition window. Whenever a secondary or primary event occurs, a 100-ns coincidence window is started. If a pulse is received from the other charge amplifier during the 100-ns coincidence window, then a coincidence counter is incremented.

The control board sets the RPA, DFL, and ESA high-voltage optical settings using a set of independent DACs. The analog settings, along with digital enables, and the low-voltage power for the primary and redundant supplies are carried to the HVPS over a dedicated board-to-board connector. The HVPS returns the analog current and voltage monitors to the control board for analog housekeeping and real-time monitoring.

The health and safety of the instrument in general and the CEMs in particular are monitored extensively by the control board. The count rate of each detector is totaled and checked against a software limit every 0.5 s. The supply voltage and strip current for each CEM is monitored and verified by the software every 1 s. The strip current is also compared against a threshold level with 32 values. If the strip current exceeds this threshold, the software receives an interrupt and may immediately take action. This information, along with temperatures in the instrument and the current and voltage monitors of the low-voltage power supplies are transmitted to the ground in housekeeping telemetry packets.

Table 4 Operational phases and SWAP modes

SWAP operational phase	Description	BOOT	LVENG	LVSCI	HVENG	HVSCI
Commissioning: low-voltage	LV functional was performed and memory load and dump capabilities were tested.	X	X	X		
Commissioning: door opening	SWAP door was opened using the S/C actuator bus. SWAP was off.					
Commissioning: initial high-voltage	HVPSs were turned on for the first time. High-voltage functional test was performed.	X		X	X	
Commissioning: nominal science	Fully functioning using HVSCI mode.	X	X		X	X
Jupiter encounter	Fully functioning using HVSCI mode.	X	X		X	X
Cruise	Functional test to be performed; CEM gain test and EEPROM refresh.	X	X		X	X
Pluto encounter	Fully functioning using HVSCI mode.	X	X		X	X

3.5 Modes of Operation

SWAP flight activities are conducted during the following operational phases: (1) commissioning, (2) Jupiter encounter, (3) cruise/annual checkout, and (4) Pluto encounter. To accommodate these different activities, SWAP has various modes that are controlled by the SWAP onboard flight software (FSW). A mapping of the operational phases to SWAP modes is described in Table 4, and a description of the modes and corresponding power and telemetry resource consumption is shown in Table 5.

Mode changes are triggered by command with the exception of the BOOT to Low Voltage Engineering Mode transition, which is based on time since power up and is automatic. Flight operations, and therefore the modes, drive the requirements for the FSW.

3.6 Flight Software

3.6.1 Overall Capabilities

The FSW runs on the 80C51 microprocessor at 4.9152 MHz, equating to approximately 400,000 instructions per second. It is written almost entirely in the C programming language. The FSW consists of two different code images stored in nonvolatile memory. At power on, the FSW runs directly from a PROM code image. This code gathers diagnostic information on the SWAP hardware, provides the ability to reload EEPROM code or tables, and determines which copies of EEPROM code or table to use for science acquisition. Once

Table 5 SWAP modes and resources

Name	Description	+30 V S/C power (W)	SWAP internal (+5 V/−5 V) power (W)	Telemetry output[a] (bits/s)
OFF	No power being applied to SWAP.	0	0	0
BOOT	Bootup run from PROM image. Checksum of nonvolatile area and read/write tests of RAM performed. This mode has the ability to upload and commit new code and table images to EEPROM. Transition to LVENG is automatic based on time since power up.	0.70	0.44	1200
LVENG	Low-voltage engineering run from EEPROM image. Safe mode (no HV), no sweeping takes place, for engineering and testing.	0.70	0.44	600
LVSCI	Low-voltage science run from EEPROM image. The instrument will begin sending test pulses through the CHAMPs in a pattern that will simulate a normal science sweep. This mode emulates HVSCI without HV.	0.74	0.48	291
HVENG	High-voltage engineering run from EEPROM image. HV can be set according to commands rather than being controlled by the sweep algorithm; used during initial turn-on, calibration, and ramping of HVPS prior to nominal science use in HVSCI mode.	1.33	0.98	290
HVSCI	High-voltage science run from EEPROM image. HV is on; ESA/RPA/DFL supplies are swept according to tables.	1.33	0.93 (avg)	291

[a]These are typical values for these modes. Other rates can be commanded as needed

the EEPROM code and tables have been chosen, the EEPROM code is copied into RAM for execution.

The FSW addresses the following requirements areas: data interfacing and synchronization with the spacecraft, instrument engineering and safety, and science support. The software is synchronized with the 1 PPS signal from the spacecraft. The spacecraft sends a mission-elapsed time (MET) to each of the instruments for a common time tag. The 1-s period is subdivided into twenty 50-ms time periods, which allow for scheduling of software processes and managing the S/C timing requirements for command and telemetry.

The software processes for science provide support in (1) managing the setting of HVPS optical power supply levels during science-data acquisition (during HVSCI mode, the ESA, RPA and DFL are set by means of lookup tables), (2) starting and collection of counter data from the PCEM, SCEM, and coincidence electronics, and (3) performing calculations onboard using the counter data and telemeter science-data products.

In nominal operations, commands are required for the high-voltage ramp after a power on, but once the ramp has been completed, a small number of commands are used for initiating science acquisition that set a configuration for acquisition, the deflection angle, and the telemetry rates.

The FSW outputs three types of telemetry: instrument state information, which can be read and acted upon by the spacecraft software for onboard anomaly recovery; engineering telemetry (housekeeping, messages, and memory dump); and science telemetry (real-time, summary, and histograms). The telemetry is summarized in Table 6.

3.6.2 Science-Data Collection

Science data are collected in the HVENG and HVSCI modes. HVENG was used extensively during commissioning for initial HV ramp-up. HVSCI is the primary SWAP science mode. In HVSCI, the optical power supplies are stepped every 0.5 s and the PCEM, SCEM, and coincidence counter data are collected with each step. During each 0.5-s period, approximately 100 ms are allowed for the optical power supply settling time and 390 ms are allocated to the counter sample period. The duration of the latter is set with digital hardware. An overall cadence of 64 s consisting of 128 0.5-s steps defines the 64-s science-acquisition frames and hence all science FSW activities. Two different user-selectable methods of sweeping during the 64 s can be performed: coarse-fine and coarse-coarse sweeps. A typical coarse-fine sweep consists of a 32-s coarse sweep, which covers the entire energy range with 64 logarithmically spaced optical power supply values, followed by a 32-s (also 64 0.5-s steps) fine sweep. A coarse-coarse sweep consists of two 32-s coarse sweeps performed in one 64-s period. For both sweep types, the optical power supply values are read from one of several user-selectable tables.

For the coarse-fine sweep, the peak value of the coincidence counter during the coarse sweep is calculated to determine the center of the fine sweep so that finer resolution sweeping around the peak response can be performed.

A 64-s acquisition frame period running a coarse-fine sweep from SWAP calibration is shown in Fig. 24, which plots the PCEM, SCEM, and coincidence count rates and the ESA and RPA voltages. A relationship between energy and coincidence rate can be derived from such a plot given that with each 0.5-s step, a different incident energy is defined for each ESA–RPA voltage combination. The figure shows how the energy is swept from high to low, with the coarse sweep taking place from time = 0 to 32 s and a fine sweep from time = 32 to 64 s.

To control acquisition while minimizing commanding to SWAP, science-data collection instructions are stored in user-selectable tables in two identical EEPROMS. The hierarchy of tables is as follows:

- Plan (64 tables) schedules which sweep table is active for each 64-s acquisition frame.
- Sweep (16 tables) defines which energy step and ESA/RPA table is used for each 0.5-s sample. This determines, for example, whether a sweep is coarse-fine or coarse-coarse.
- ESA (4 tables) provides a mapping between energy step and the DAC value to program the ESA optical power supply.
- DFL values (1 table) map energy step and angle to DFL DAC setting.
- RPA values (4 tables) map energy and angle to RPA DAC setting.

With these tables, the user chooses a plan number and angle setting to define science acquisition.

The summary is composed of a series of ongoing calculations based on the fine portion of a coarse-fine sweep. An array of coincidence counts is collected and processed during each 64-s acquisition frame. Figure 25 shows an idealized coincidence count rate response

Table 6 SWAP telemetry

Name	Description	Typical packet rate	Typical bit rate (bits/s)
Instrument state	SWAP information used onboard by S/C autonomy software. Instrument state includes heartbeat and safety flags.	1 Hz	Varies—not considered part of SWAP
Housekeeping (engineering)	Instrument status is contained in this packet (e.g., opcode echo, FPGA status, non-safety-critical monitors, software variable status, etc.).	0.5 Hz in LVENG; once every 64 s in other modes	292 in LVENG; 9.1 in other modes
Memory dump (engineering)	Memory dump for diagnostics.	Low	N/A
Message (engineering)	Warning message that is issued to provide additional detail if anomalous activities occur.	Low	N/A
Real-time (Science)	Contains the most detailed temporal data from SWAP consisting of RPA, ESA and DFL commanded DAC levels corresponding to primary CEM, secondary CEM, and coincidence counts.	1 Hz during commissioning; 1 set of 64 consecutive packets per hour otherwise	280 during commissioning and Pluto Encounter; 5.0 otherwise
Summary (Science)	Summary accumulation over a set period of time, typically an hour. A calculation is performed using the coincidence data from the previous 64-s period to generate values that are related to density, velocity, and temperature and output in telemetry. The minimum, maximum, and variance of these parameters are also output in telemetry.	1 per hour	0.19
Histogram (Science)	All of the count data are accumulated over a set period of time, typically a day. The data are accumulated in a normalized energy array of 2048 elements. The normalization occurs such that for each 64-s acquisition period, the center of the array corresponds to the solar wind peak during those 64 s. The entire array is brought down in telemetry. Because of its size, the array must be trickled out in 64 packets. This packet contains the histogram header (APID 0x586) and the beginning of the histogram data. It is followed by 63 Science Histogram Data (APID 0x587) packets to create a complete histogram set.	1 set of 64 (1 header + 63 data) packets per day	0.15

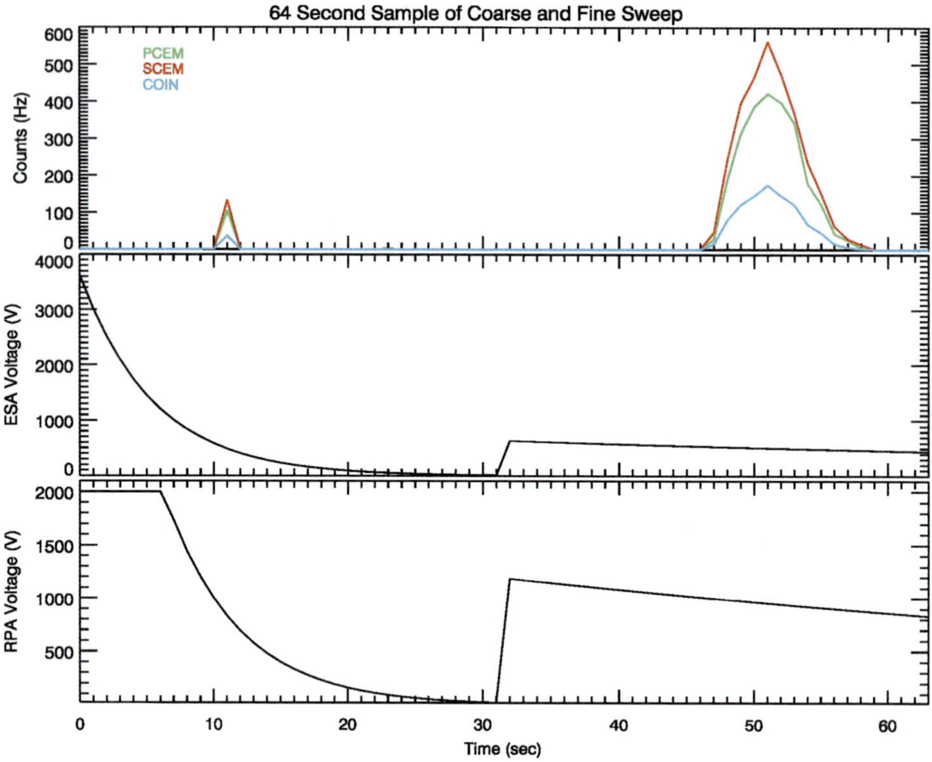

Fig. 24 A coarse and fine sweep acquired during SWAP ground calibration activities

collected during a 32-s fine sweep in which the values related to pseudo-velocity, pseudo-density, and pseudo-temperature of a fine sweep are defined. For each of the summary values (angle, pseudo-density, pseudo-velocity, and pseudo-temperature), a running log of the sum, minimum, maximum, and variance are calculated. While the sum, minimum, and maximum are represented in telemetry as 32-bit numbers, the variance is calculated by summing the squares of each of the values and accumulated into a 64-bit number. Because the summary packet defaults to transmitting once every hour, SWAP can use the summary to collect data even during low-telemetry rate periods. The rate can be altered via SWAP command for testing or other purposes.

Figure 26 summarizes the histogram calculation. The top half of the figure shows a time line of two consecutive 64-s frames. The red and blue plots show the coincidence counters during the coarse sweep and the black plots show the response from the fine sweeps. The red plot shows that the peak was found at an energy index of 512, which results in the FSW running a subsequent fine sweep from energy indices 480 to 543 to center the coarse peak in the middle of the fine sweep.

The coarse sweep data are shifted and accumulated into a normalized histogram array that is composed of 2048 32-bit energy bins. It is "normalized" because each bin corresponds to a relative energy index where 1024 always corresponds to the peak of the coarse array. In Fig. 26, for Frame i, the fine peak energy index is 512. This number is used to calculate the shift required to move the wider spaced coarse sweep so its energy index 512 aligns with histogram array index 1024. Note that the wide spacing of the coarse array is transferred

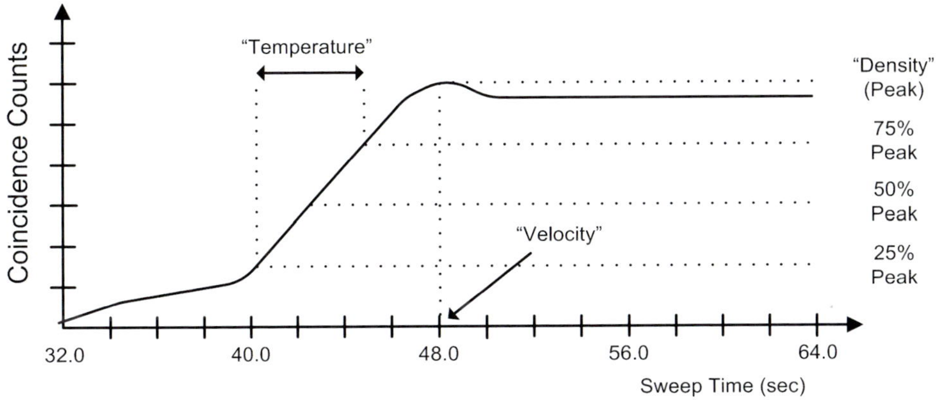

Fig. 25 Illustration of SWAP summary calculation

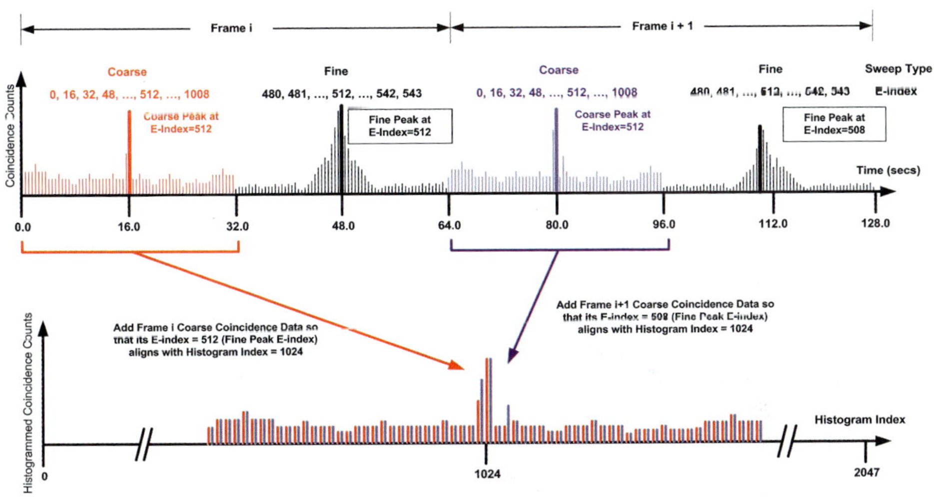

Fig. 26 Illustration of histogram calculation. The *bold vertical lines on the upper plot* show the maximum coincidence value for that 64-sample (32-s) span

also, so there are gaps between consecutive coarse samples accumulated into the array. In Frame $i + 1$, the fine peak energy index is 508, so a slight shift between the frames is noted and results in the second coarse data set not aligning exactly over the first coarse data set.

The number of times a particular bin (also with indices 0 through 2047) has been accumulated is telemetered along with the histogram array so that the activity in each energy bin can be reported. The histogram packet defaults to being sent out approximately once per day. Because of the size of one histogram array, the packet is divided and sent out over the course of 64 s. The histogram is used so that SWAP can collect data even during times where only low-telemetry rates are allowed. Like the summary packet, the rate can be altered via SWAP command.

4 Instrument Performance

We characterized SWAP's performance through both laboratory testing and electro-optics modeling. The purpose of the laboratory testing was to calibrate the instrument and determine a ground truth. The computer models allowed us to examine more combinations of incident-beam and instrument parameters than possible in laboratory testing and to allow for analysis of other parameter combinations as needed after launch.

4.1 Laboratory Testing

SWAP was extensively calibrated at the SwRI Ion Calibration Facility. This facility produces ion beams from <500 eV to 51 keV. The ion species can be selected for mass per charge from 1 to >40. Figure 27 shows the SWAP instrument mounted on the facility's four-axis positioning system, with the doors that protect the sensor for launch in their open configuration. Full functionality of the instrument was demonstrated during calibration, including its ability to open the doors while under high vacuum. The four-axis positioning system allows vertical and horizontal displacement in the plane normal to the ion bean and rotation about two orthogonal axes that cross in the beam path. The outer rotation is about a vertical axis, and the inner, nested, rotation is about a horizontal axis.

The SWAP instrument was mounted with its center axis aligned to the inner-stage rotation axis, which is equivalent to roll maneuvers the *New Horizons* spacecraft will perform. Because of SWAP's 17-cm-diameter aperture, outer-stage rotations resulted in translation in the input aperture. We used the horizontal stage to return the aperture to its original position. The positioning system's range of motion is large enough to fully illuminate the instrument's FOV.

In addition to testing performed on individual components and on a high-fidelity prototype, the full flight instrument underwent a number of calibrations. Calibrations were per-

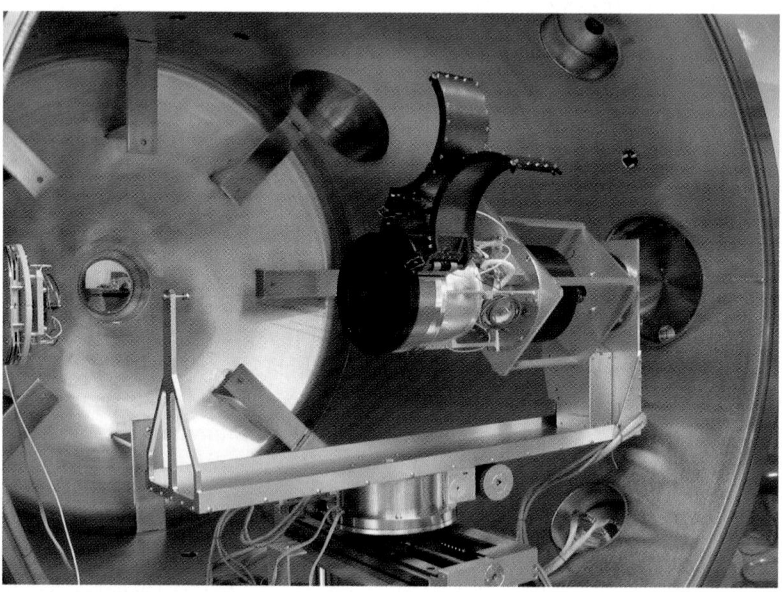

Fig. 27 SWAP instrument mounted on four-axis positioning system

formed before and after instrument-level environmental testing, after refurbishment activities following spacecraft environmental testing, and a limited engineering beam test immediately prior to delivery to the spacecraft for final integration. The instrument was calibrated over the full angular response and energies from 500 eV to 8 keV. The nature of electrostatic optics allowed the response of the instrument to other energies to be easily scaled. Because the solar wind protons have a nominal energy of ~1 keV, a large fraction of the calibration data was taken at this energy. Results presented here are from the instrument in the final delivered configuration.

4.1.1 RPA

Figure 28 shows the measured energy response of the RPA to a beam of 1-keV protons. The primary, secondary, and coincidence rates are shown in blue, green, and orange, respectively. The RPA grids are curved, rounding the cutoff of the transmission as compared to a planer RPA. This is seen as the reduction of counts from 950 to 1000 V. However, as shown in Fig. 10, by making small steps in the RPA voltage, we are able to differentiate beams that are only ~1–2 eV apart at 1 keV.

Figure 29 shows the general shape of the RPA response over the full voltage range. Normalized coincidence is plotted as a function of the RPA voltage/beam energy. Because the individual RPA grids have a finite thickness, they act as a series of electrostatic lenses. The features in this figure, including maximum near 0.945 of the 1-keV beam energy, are due to the focusing properties, as will be discussed in Sect. 4.2.

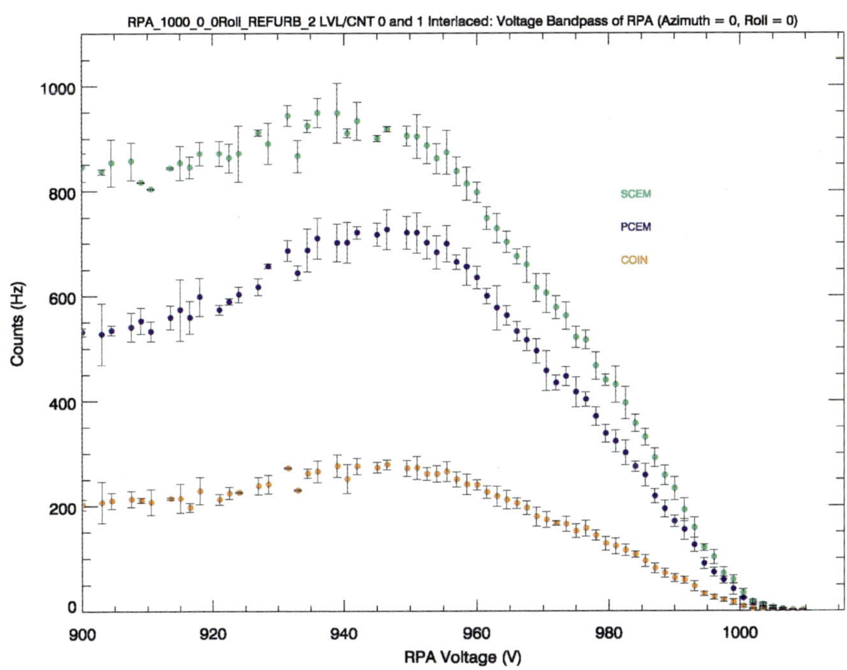

Fig. 28 Measured energy response of RPA to a beam of 1-keV protons

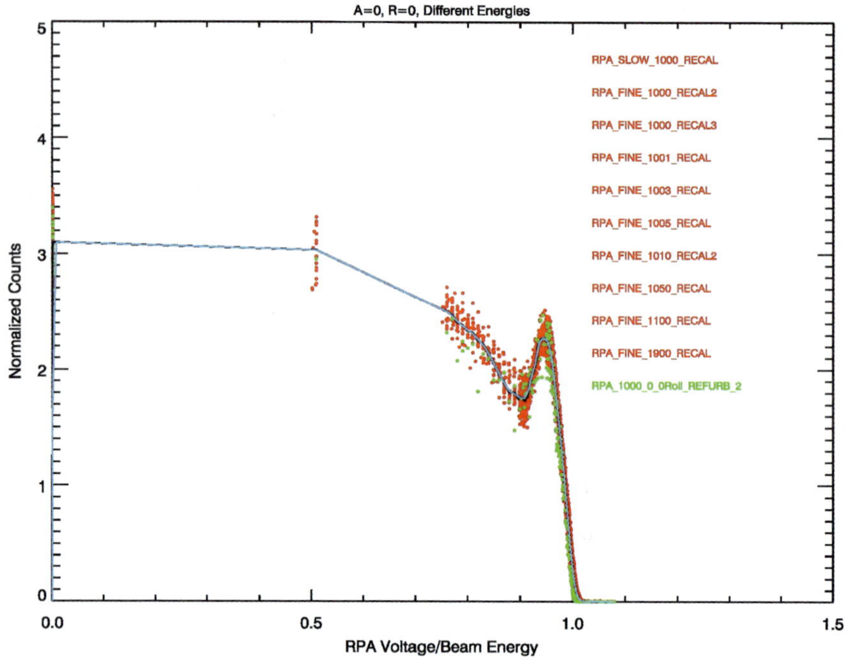

Fig. 29 General shape of RPA response over full voltage range

4.1.2 Deflector System

The FOV of the SWAP instrument is 276° about the spacecraft roll axis by 10° in the plane normal to the roll axis. SWAP uses an electrostatic deflection (DFL) plate to increase the FOV out of the plane by up to 15° in azimuth (α). The required DFL voltage, normalized to the beam energy, can be expressed as

$$\frac{V_{\text{deflector}}}{E_{\text{beam}}} = 0.0007\alpha^2 + 0.0209\alpha + 0.1009, \tag{1}$$

where α is the angle from the plane.

For non-normal incident particles, the RPA response needs a calibration correction. For an ideal RPA, non-normal incident particles with $E/q > (V_{\text{RPA}}/\cos^2\alpha)$ would be passed. The unique focusing properties of the SWAP RPA lead to a more complicated response function. Figure 30 shows the normalized response of the RPA as a function of the azimuth angle. The RPA voltage is scaled by the incident beam energy. Protons with energy of 1000, 1010, and 1900 eV are shown in red, green, and blue, respectively. We set the DFL voltage according to (1), leading to an empirically derived calibration function for SWAP:

$$\begin{aligned} f &= \cos^2(2\alpha), \quad \alpha \le 4.0, \\ f &= \cos^2(3.5\alpha - 20) - 0.03, \quad 4.0 < \alpha \le 10, \\ f &= \cos^2(5.5\alpha - 65) - 0.085, \quad \alpha > 10. \end{aligned} \tag{2}$$

For the SWAP RPA, non-normal incident particles with $E/q > (V_{\text{RPA}}/f(\alpha))$ would be passed.

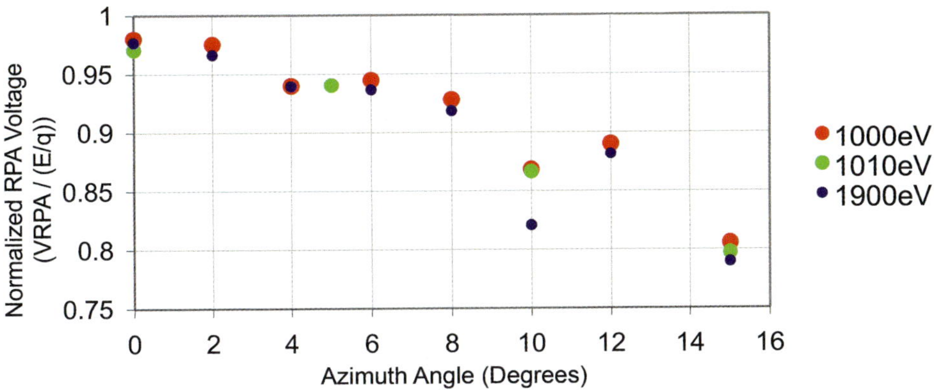

Fig. 30 The normalized peak response of the RPA shown as a function of the incident azimuth angle. The RPA voltage is scaled by the incident beam energy. Protons with energies of 1000, 1010, and 1900 eV are shown in red, green, and blue, respectively. The DFL voltage was set according to (1). For an ideal RPA, non-normal incident particles with $E/q > (V_{\text{RPA}} \cos^2 \alpha)$ would be passed. The unique focusing properties of the SWAP RPA lead to a more complicated response function, requiring an empirically derived calibration function as shown in (2)

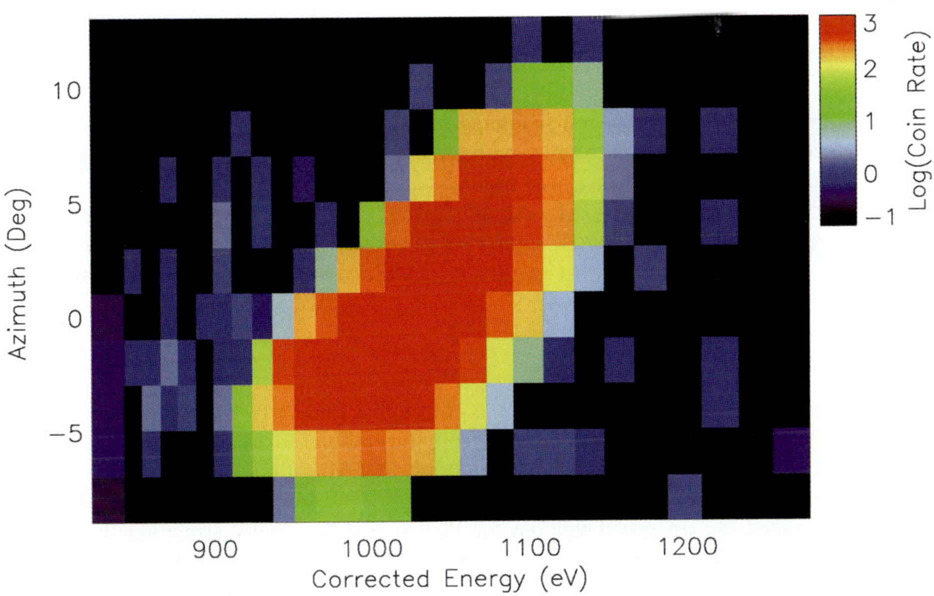

Fig. 31 The energy–angle response of the SWAP instrument shown for a 1-keV proton beam

4.1.3 ESA

Figure 31 shows an example of the energy–angle response of the ESA to a 1-keV proton beam. The RPA and deflector system's voltages were at zero for the data shown here. The ESA's analyzer constant is 1.88, the energy resolution $\Delta E/E$ is 8.5%, and the angular FOV, undeflected, is 10°.

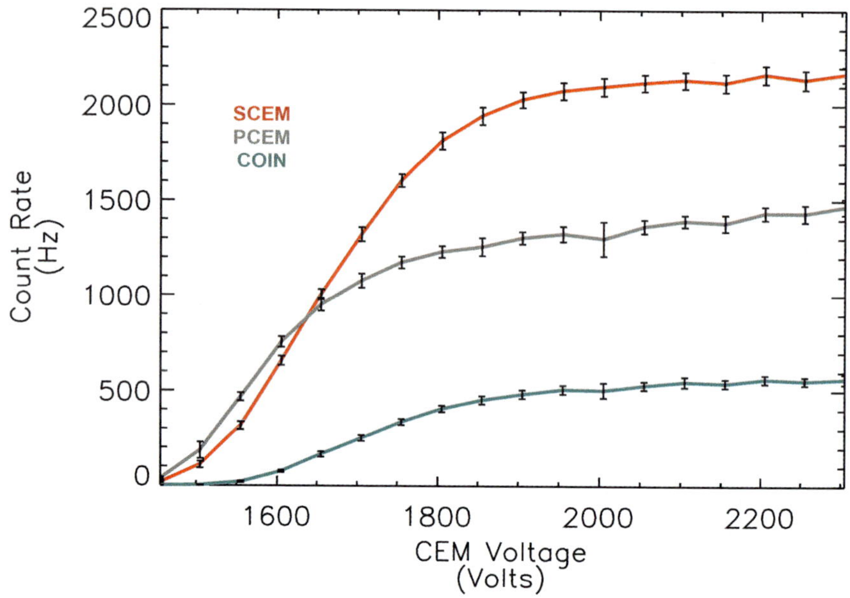

Fig. 32 PCEM (*gray*), SCEM (*red*), and coincidence (*blue/green*) rates shown as a function of CEM voltage. The magnitude of the PCEM voltage is set equal to the SCEM voltage for the data shown here

In addition to performing the coarse energy measurement, the ESA also blocks UV light to the detectors. A krypton line source, with emission at 123.6 nm and approximately the same intensity of the Sun at 1 AU, was used to test SWAP for rejection of scattered light. This is a worst-case test since SWAP operates further out in the heliosphere where the UV flux is lower by a factor of R^2. For all angles tested with the UV source, the count rate never exceeded 1 Hz for either primary or secondary detectors, and no coincidence events were observed.

4.1.4 Detectors

Ions that make it through the sensor then pass through a thin carbon foil and are measured by the PCEM. Secondary electrons liberated from the carbon foil are attracted to the SCEM. Events that are measured by both the PCEM and SCEM within 100 ns are recorded as coincidence (COIN) events.

We determined the operation voltage for PCEM and SCEM by sweeping the voltage while illuminating the instrument with a constant intensity 1-keV proton source. Figure 32 shows the response of the PCEM, SCEM, and the COIN rate as a function of voltage. For the data shown here, the voltage applied to the PCEM and SCEM were equal. From these tests, we selected a nominal operation voltage of 2100 V at the start of mission, which is well into the saturation part of the CEM gain curves. The high-voltage power supplies for the CEMs were designed to output up to 4.5 kV, allowing us to increase the voltage to the CEMs if their gain degrades over the mission lifetime.

4.2 Instrument Modeling

We obtained instrument-modeling results with ion-optics simulations using SIMION 7.0 software (Dahl 2000). SIMION is an ion-optics simulation program that models problems with 2D or 3D electrostatic potential arrays. It creates potential arrays based on a user-given geometry input (usually in the form of a text file containing commands). The potentials of points outside the electrodes are determined by solving the Laplace equation by finite difference methods with a maximum of 50-million array elements. The time-step integration for the calculation of ion trajectories uses a fourth-order Runge–Kutta method, with ions flown individually or in groups. Parameters such as position, velocity components, and energy can be recorded at different locations in the instrument model.

The SWAP model consists of a 3D geometry with cylindrical symmetry. The limitation of 50-million array elements prevents our reproducing the finer structure of the RPA design with the instrument design in the same potential array. Therefore, the RPA grids are modeled separately with a much finer resolution. We discuss the SWAP model (instrument with "ideal" RPA) in Sect. 4.2.1 and describe the RPA model in Sect. 4.2.2.

4.2.1 Instrument Model with Ideal RPA

Our model geometry (Fig. 33) fully conforms to the as-built mechanical design of the flight instrument, with as much detail as the resolution permits. This mechanical detail helps us account for effects such as fringing fields. Figure 34 shows the focusing properties of the ESA for normal incidence ions in two orthogonal planes. In both planes, the ions are focused inside the ESA part and then accelerated and refocused as they get closer to the detectors. However, the impact positions on the primary CEM are nicely spread in the funnel to avoid a hot spot.

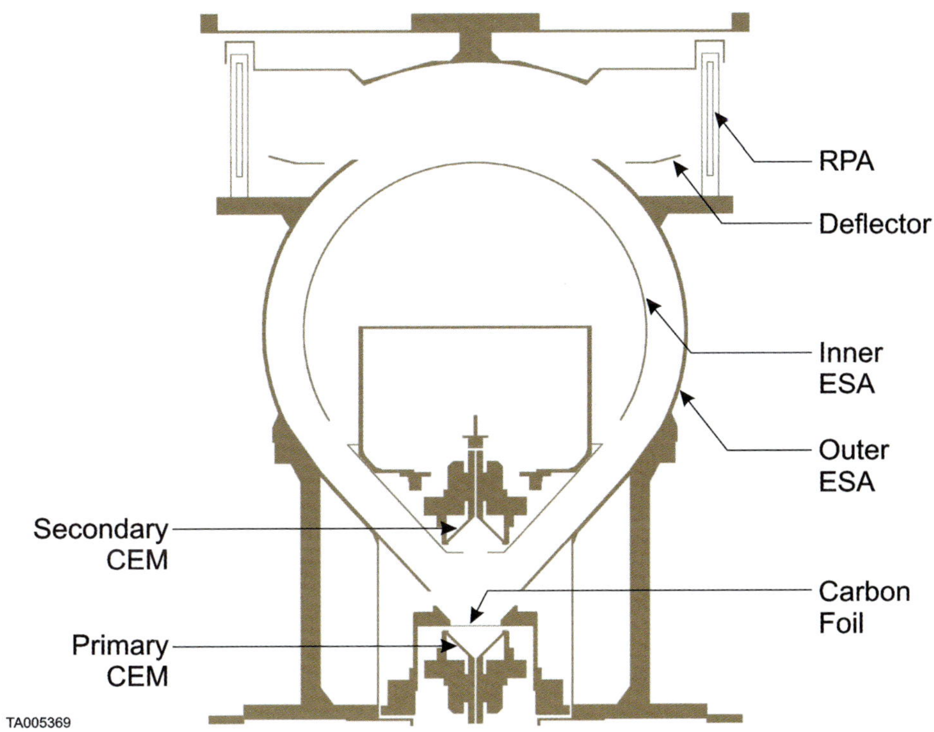

Fig. 33 Cross section of the electro-optics model for the ESA. The model has a main axis of symmetry

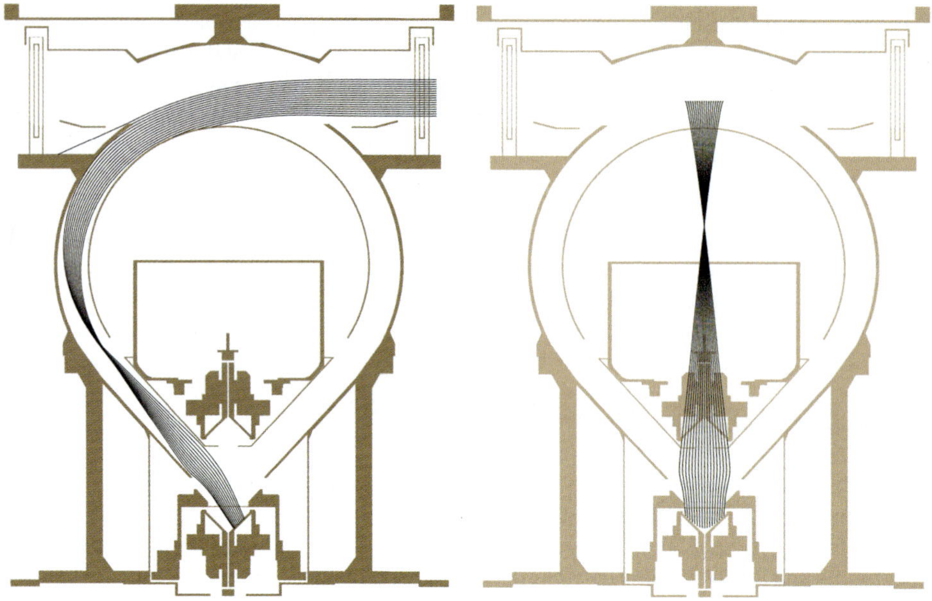

Fig. 34 Focusing properties of the ESA shown in two orthogonal projections

Table 7 Comparison of the instrument characterization between measurements with ion beam and model

	Calibration with ion beam	Instrument model
E/q resolution ($\Delta E_{\mathrm{FWHM}}/E$)	8.5%	9.5%*
k [V/eV]	1.88	1.86
Angular acceptance at HM [°] (undeflected)	−4.9 to +5.2	−4.6° to 8.1°*

*Model is broader because collimating effects of RPA are not included

The instrument was thoroughly characterized with ion beam calibrations (above) and the results agree very well with those from the modeling. Table 7 compares the energy-per-charge resolution, the k factor, and the angular acceptance.

Figure 35 shows the angle–energy response. The angle and energy binnings have been set to the same values for better comparison. The right portion of the figure shows the normalized transmission, which is color-coded using a logarithmic scale. Again, there is excellent agreement between the modeling and the calibration measurements with an ion beam.

4.2.2 RPA Model

We modeled the RPA grids separately with much higher resolution, matching the thickness and hole pattern of the mechanical drawings. The model includes the curvature of the grids and drilled holes, which are aligned for the four grids. This configuration does not account for a possible misalignment of the hole patterns, which may cause some differences between the model and actual hardware. Figure 36 shows an isometric view of part of the RPA.

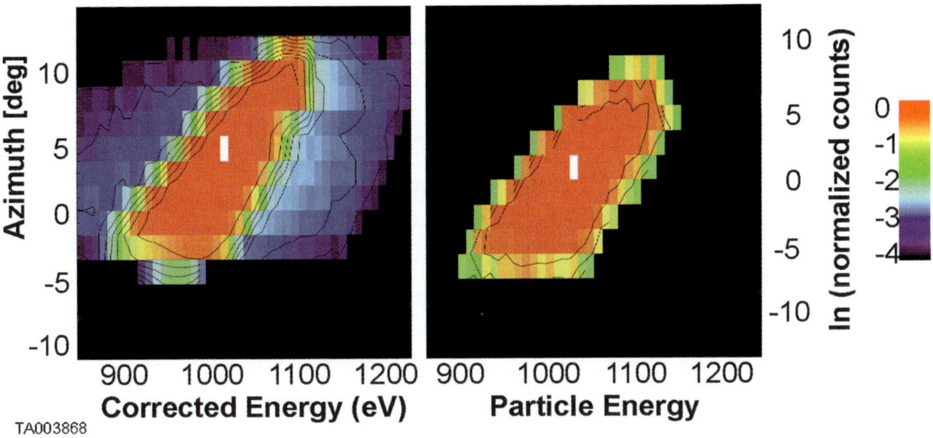

Fig. 35 Comparison between the angle–energy transmission function of the instrument (*left*) and the model (*right*)

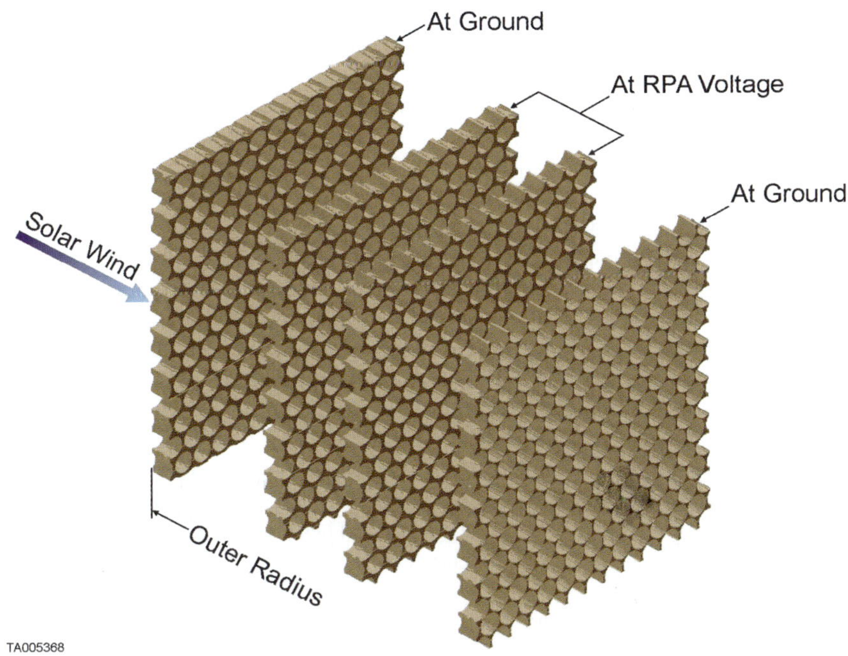

Fig. 36 Isometric view of a part of the RPA. The voltage is applied on the two middle grids. The holes are aligned so that their axes are radial. The thickness of the grids and the diameter of the holes are in accordance with the mechanical drawings

The ions are flown through the RPA grids, and the Cartesian coordinates and velocity-vector components are recorded at the exit of the RPA for those ions that make it through. Those ions are then flown inside the instrument model described in Sect. 4.2.1, and an ion that makes it to the carbon foil is considered to be detected.

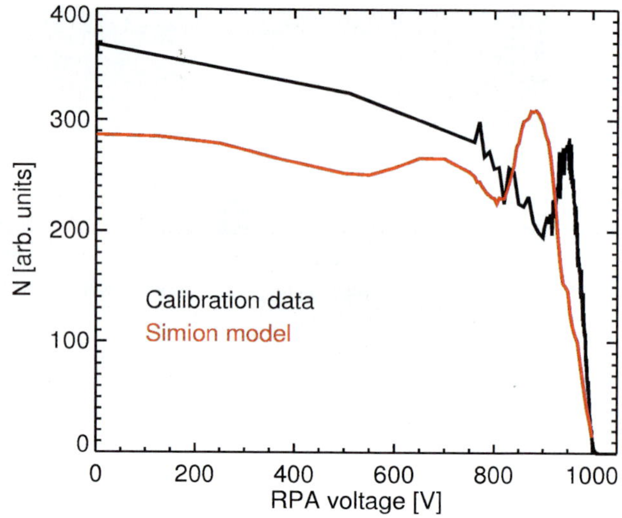

Fig. 37 Relative transmission function as a function of the RPA voltage for a 1 keV/e ion beam and an ESA voltage of −524 V

In the following discussion, we present results using a 1 keV/q parallel beam, which is uniformly distributed over the aperture of the instrument at normal incidence (0° in azimuth and elevation). Figure 37 shows a transmission curve as a function of the RPA voltage for an ESA voltage of −524 V, with the ordinate in arbitrary units. The red curve depicts the results of the RPA model and is compared to the calibration data under the same conditions (black curve).

The RPA does not give a sharp cutoff at 1 kV because each hole of the RPA grids acts like a tiny lens, focusing or defocusing the ion trajectories. At a given E/q, the deflection increases with the RPA voltage. Because of the grid thickness, some ions hit the inside of the holes, which can also cause a nonuniform response below the cutoff. Because the model holes are radially aligned, the E/q and the RPA voltage have to be within a certain ratio for the trajectory to deviate from a row of holes to another neighboring row. Therefore, transmission varies as a function of RPA voltage for a particular E/q. Although the general responses are similar, a discrepancy exists between the calibration data and the simulations. The likely cause of this discrepancy is that the RPA grid holes are not aligned along a radial direction. In addition, the ion beam in calibration entered the instrument with a small angle with respect to the normal of the RPA grids (offset in azimuth and/or elevation angle) and the ion beam had a small divergence ($\sim$1.0° at 1 keV/e). These results highlight the importance of using the flight calibration data in concert with the instrument model for understanding the detailed response function.

We have presented the simulation results for one particular case; however, the ESA, deflector, and RPA responses were simulated with numerous different initial conditions. The model aids in understanding the instrument response and provides additional interesting information that can be used for the analysis and interpretation of the data.

5 Science Operations and Data

This section describes the expected fluxes and predicted count rates, instrument operations, and data pipeline. We used the fluxes measured from previous missions and the instrument model just described to estimate the anticipated count rates in the solar wind and at Jupiter.

The subsequent operations discussion describes the voltage settings, which we vary according to the anticipated count rates. Finally, the data pipeline section describes the data products and level 2 data analysis.

5.1 Fluxes and Predicted Rates

We predict SWAP count rates in the solar wind scaling from fluxes observed by *Voyager 2* (Richardson et al. 1996) over a large range of heliocentric distances. The proton flux decreases with distance from the Sun (Fig. 38) primarily because the density decreases with distance. For our calculations we use a constant speed of 1 keV ($\sim$440 km s^{-1}). Fit functions for the density and density standard deviations are

$$n_p = 5.22r^{-1.93}\ [\mathrm{cm}^{-3}] \tag{3a}$$

and

$$\sigma(n_p) = 7.4r^{-2.15}\ [\mathrm{cm}^{-3}], \tag{3b}$$

where the standard deviation for the speed is based on Fig. 6 of Richardson et al. (1996).

We used the solar wind fluxes from Fig. 38 and the instrument model described here to estimate primary and secondary CEM count rates in the solar wind as a function of heliocentric distance. The results are shown in Fig. 39. The instrument model includes the instrument effective area and the ESA and RPA response curves. The model also includes the deflector and angular response function and assumes a 1-keV beam centered in the ESA passband. The red line at 1 MHz indicates a reasonable maximum count rate for the detectors. Although the count rates are too high in the inner heliosphere with these voltage settings, the RPA and ESA voltages can be adjusted to reduce passband.

As already described, the ESA and RPA can be crossed to provide a reduced energy passband. Figure 40 shows the estimated count rates as a function of the center energy of

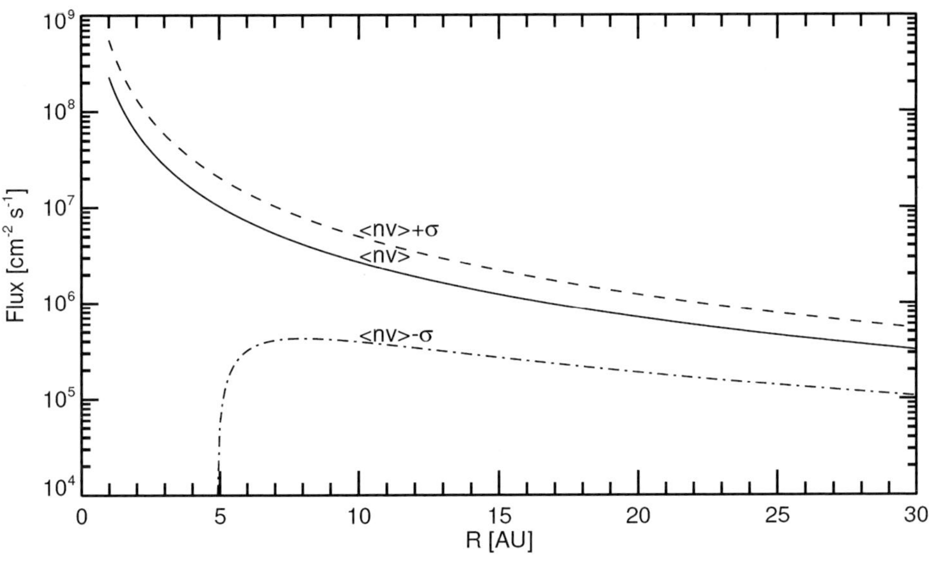

Fig. 38 Flux of solar wind protons versus heliocentric distance. The *solid curve* shows average values; the *other two* show $\pm1\sigma$ probabilities

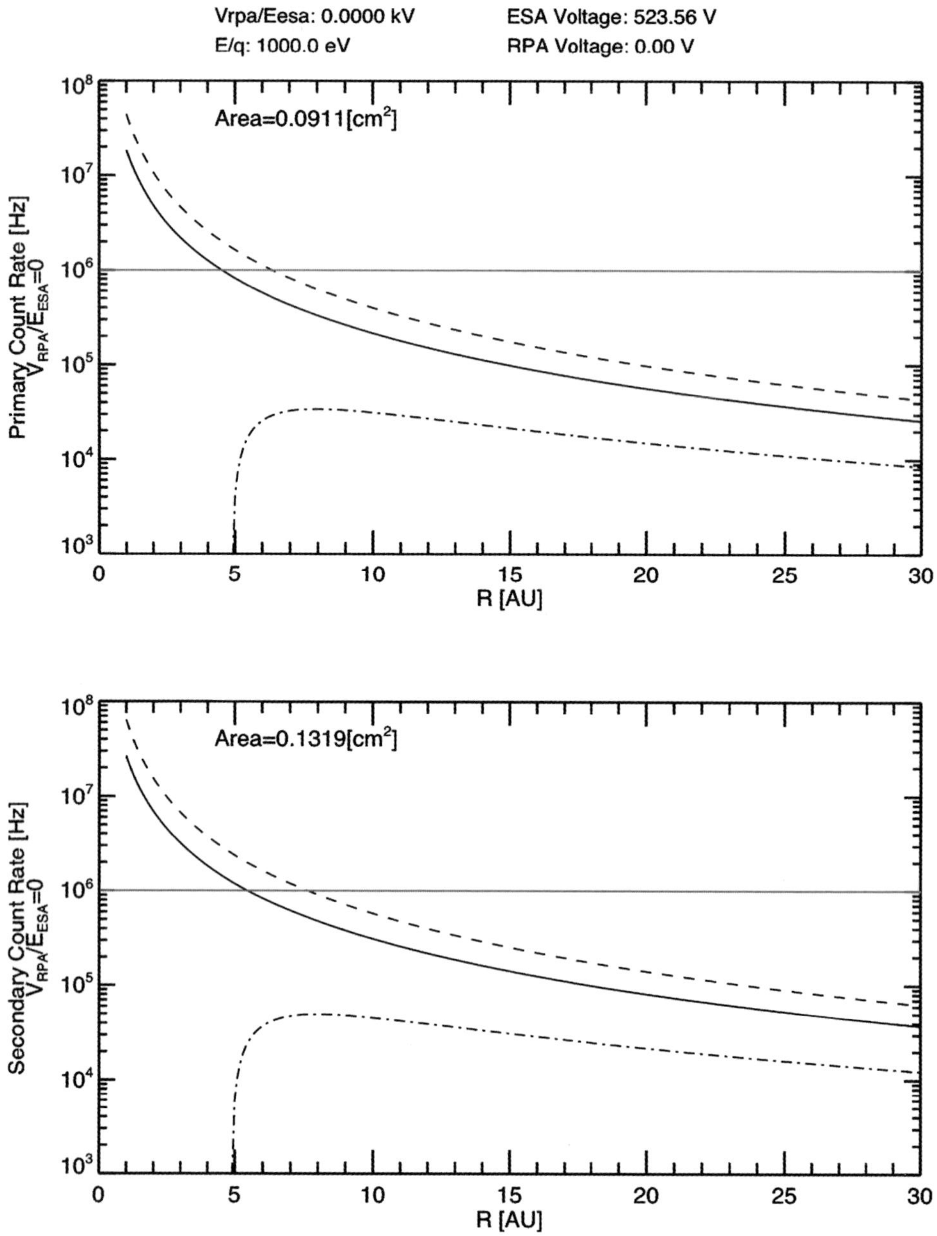

Fig. 39 Estimated count rates on the primary (*top*) and secondary (*bottom*) CEM as a function of heliocentric distance for the fluxes given in Fig. 38

the ESA (kV_{ESA}) passband for a 1-keV beam with a flux that is one standard deviation higher than the average flux at 5 AU. Each curve corresponds to sweeping the RPA and ESA voltages while keeping the RPA voltage to ESA energy ratio fixed. At a ratio of 1.03 the expected count rates at 5 AU are reduced to near maximum allowable levels, and for a ratio

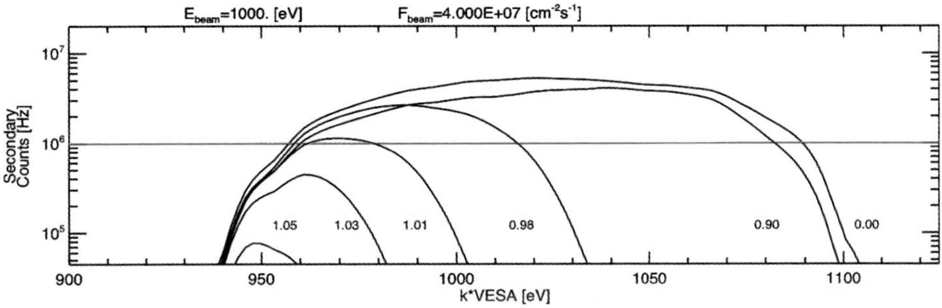

Fig. 40 Estimated count rates on the secondary CEM as a function of the center of the ESA bandpass. *Each curve* corresponds to sweeping ESA and RPA voltages such that the ratio of the RPA voltage to the center energy of the ESA bandpass ($V_{\text{RPA}}/E_{\text{ESA}}$) is fixed. The flux used is higher than normal for 5 AU

of 1.05 the expected rates are $<10^5$ Hz. For the last phase of commissioning we used a ratio of 1.05 for safety.

5.2 Instrument Operations

For the Jupiter phase of the mission, we determined that the radiation levels during the encounter should be low enough to allow SWAP to operate for the entire encounter. SWAP's shielding is similar to *Ulysses*/SWOOPS, and the *New Horizons* spacecraft does not get as close to Jupiter's radiation belts as *Ulysses* did. Phillips et al. (1993) showed that *Ulysses*/SWOOPS operated as close as 16 R_{J}, compared to the *New Horizon*'s closest approach of about 32 R_{J}. Inbound to Jupiter, we anticipate that real-time science mode data will be taken twice per hour, with 20 measurements recorded per hour in Jupiter's tail. The Jupiter tail observations continue until about 100 days after closest approach, which corresponds to about 2200 R_{J} downstream. Although SWAP is not designed for measuring Jovian ion distributions, we are hopeful that some new observations will be possible.

Past Jupiter, we plan cruise phase science to study how the solar wind and pickup ions evolve with radial distance from the Sun. For cruise phase, real-time data will be recorded at a low rate and summary and histogram data at a moderate rate. In addition, the gain on the CEMs will be measured during each annual check out.

For the *New Horizons* primary mission, we will use SWAP's maximum sensitivity and operate at its highest rate throughout the Pluto and Kuiper Belt Object flybys. These observations give a spatial resolution of <3000 km, adequate to determine whether Pluto has a bow shock and, if so, locate it with sufficient accuracy to make a good measure of the atmospheric escape rate for the likely orientation of the IMF at flyby.

5.3 Data Pipeline

SWAP data are provided at three levels in the data pipeline. Level 1 is the raw decommuted data. Level 2 involves converting the level 1 data to more physical units, correlating data with thruster firings to look for any enhanced backgrounds, calculating some orbital and pointing information, and producing E/q spectrograms. For level 3, we further process the data to obtain speed, temperature, and relative density.

5.3.1 Level 1

Level 1 data are comprised of real-time science, summary, histogram, and housekeeping packets. We decommute and place the raw data into flexible image transport system (FITS) files. Most of the SWAP data consist of binary tables with columns for given instrument parameters and rows for each measurement time. Since the histogram data consist of arrays, it is stored as an image in the FITS file.

We place all the real-time data for a given day into one file. A complete real-time observation consists of 64 real-time packets. To make sure that an instrument cycle is not split across two daily files, a parameter in the real-time packet is used, which indicates the beginning and end of the 64-s instrument cycle. We also create daily files for summary data, but the histogram data are handled differently since the histogram arrays are divided into 64 packets. A set of 64 histogram packets are combined and placed into one file.

We divide each FITS file into extensions. The first two real-time data extensions contain header information and the real-time data as a binary table. Additional table extensions contain housekeeping and thruster-firing data. We use this same organization for the summary data. The first two histogram extensions contain the array (image) with the number of samples placed into each bin, and another image holds the array of counts detected in each bin. The housekeeping, thruster, source sequence count (used to check for data gaps), and checksum (used to check for corrupted data) information is contained in additional extensions. For real-time and summary files, the source sequence and checksum information is not stored as separate extensions but stored directly inside the tables.

5.3.2 Level 2

We also store SWAP level 2 data as binary tables and images in FITS files. The header information from level 1 is duplicated into level 2, since the header contains level 1 processing information and orbit-attitude information for the start time of the each file. The second extension for the real-time and summary level 2 files contains the corresponding data from those packets converted from raw to engineering units (e.g., converting the RPA DAC step number to voltage). The real-time files contain several additional extensions that hold spectrogram information. These extensions consist of the coarse scan (first 32 steps) 2D spectrogram count arrays, the corresponding 2D error spectrograms, the y-axis labels (energy label) for the spectrograms, and the x-axis time label information. The real-time and summary files have housekeeping, quality, and thruster-firing extensions. The histogram level 2 file format mirrors the histogram level 1 format, but it includes quality flags.

We base the flags that assess the quality of the data on housekeeping data, orbit and attitude, and how well the software is working. Currently, these flags are based on standard operating ranges for housekeeping data, but additional ones can be created later as needed.

We are developing a SPICE code to calculate the time in UTC, position and velocity of the spacecraft in a variety of coordinate systems, field of view of the instrument, and the Sun location in instrument coordinates. The boresight of SWAP is nearly aligned with the $+Y_{\mathrm{sc}}$ axis; therefore the SWAP frame kernel consists of only two small rotations from spacecraft to instrument coordinates. We are integrating the SPICE code into level 2, and an orbit and attitude table extension with all SPICE calculations for each measurement time is planned for the level 2 files. Flags will be created to indicate whether or not the Sun is in SWAP's FOV, since accurate solar wind properties can only be determined if the bulk of the solar wind beam enters the instrument.

We will produce energy–time spectrograms from real-time science mode data. These count rate spectrograms provide a way to examine the data at high time resolution over the

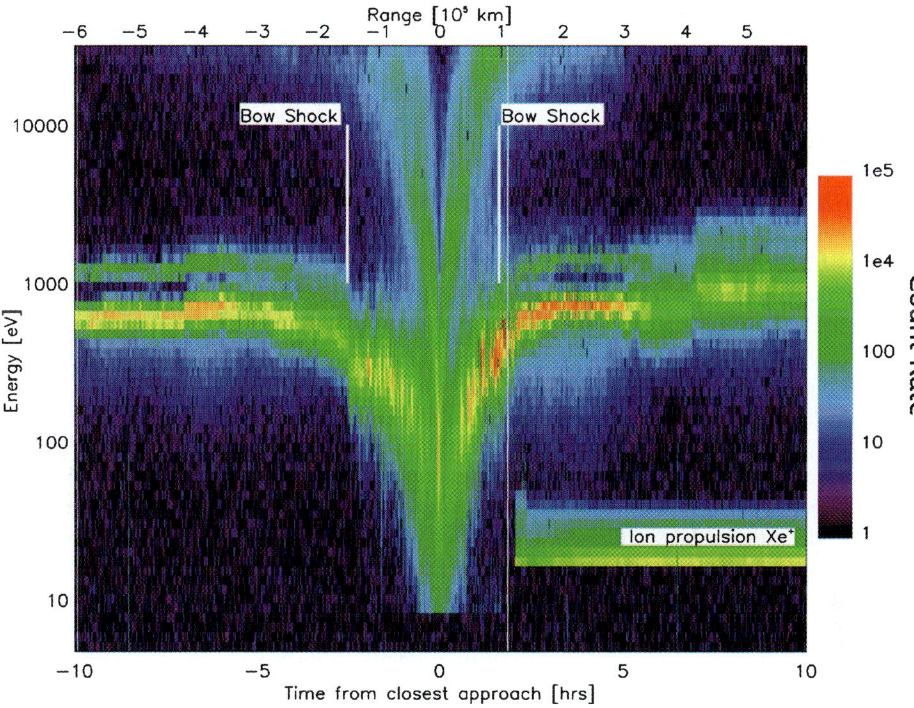

Fig. 41 Comet Borrelly flyby E/q–time spectrogram (Fig. 1 of Young et al. (2004))

full energy range of the instrument. Such spectrograms have proven useful for analyzing a wide variety of distribution types. They have been used to analyze the solar wind, comets, plasma boundaries, and shocks. As an example, Fig. 41 shows a spectrogram used to analyze ion measurements for the Comet Borrelly flyby. In this figure it is easy to locate the bow shock and see the slowing of the solar wind caused by mass-loading.

5.3.3 Level 3

Speed, temperature, and relative density depend heavily on the calibration, orbit, and attitude information. We have planned these data products for level 3 since the energy distribution (spectrograms) and the orbit and attitude information, both of which are accomplished in level 2 analysis, must be determined first. Speed, temperature, and relative density from the energy distribution/spectrum for the solar wind measurements will be derived during solar wind intervals. When only pickup ions are observed, the energy distribution/spectrum is a more appropriate data product. We will use ESA and RPA response functions, the angular response function, the instrument solid angle, detector efficiencies, and the effective area determined during calibration to establish plasma properties and distributions from the detected count rates. The effective area (Table 2) is used for a cold beam such as the solar wind when the entire solar wind beam is in the field of view. Note that the effective area is for a normal incident beam and the transmission varies as a function of angle; therefore, all of the angular-dependent calibration functions discussed in Sect. 4 are also required to determine the flux. Precise speeds and temperatures will be obtained from the fine RPA scans. Lab calibration results already indicate the ability to accurately determine the solar wind beam energy and speed with the RPA fine scans. For times when the deflector is operated, we will use the deflector's calibration data to analyze the data.

6 Early SWAP Observations

New Horizons was launched toward Jupiter and on to Pluto on January 19, 2006. Early operations included commissioning of all of the instruments onboard. The SWAP cover doors were opened, low and high voltages were turned on, and all aspects of the SWAP instrument were demonstrated to be functioning nominally in space. Because of the closeness to the Sun and high fluxes possible (Figs. 38 and 39), we ran SWAP with a very high crossing ratio of 1.05% between the ESA and RPA. This crossing effectively limits the counts to ~1% of what would be otherwise measured for all E/q values less that 2 keV/q; above that value, the RPA is fixed at 2 kV, so the energy passband of the ESA is effectively fully open.

Figure 42 shows an initial color-coded spectrogram of the background-subtracted coincidence measurements of solar wind ions as a function of E/q for January 8–20, 2006, when *New Horizons* was ~4.9 AU from the Sun (~0.4 AU upstream from Jupiter). The lower trace is produced by solar wind protons, and the upper trace comes from the alpha particles (He^{++}), which travel at approximately the same speed as the protons and thus have twice the E/q. Alpha particle densities in the solar wind are typically only ~4% that of the protons, so SWAP's enhanced sensitivity above 2 keV/q is nearly ideal for making high-sensitivity measurements of these ions while still limiting counts from protons to safe levels. Solar wind speed is a function E/q, with 1-keV protons corresponding to typical, ~440 km/s solar wind with larger (smaller) E/q representing faster (slower) wind speeds.

A forward interplanetary shock passed over the *New Horizons* spacecraft at ~1800 on January 11 and a reverse shock passed by at ~1300 on January 14; the speed immediately following the latter shock was in excess of 600 km/s. Such forward–reverse shock pairs, followed by rarefaction regions (slowly decreasing speeds and consequent falling E/q of the proton and alpha beams) are typical of corotation interaction regions (CIRs) in the solar wind at these distances (Hundhausen and Gosling 1976; Smith and Wolfe 1976). These and the continuing upstream measurements of the solar wind are critical for understanding the external plasma environment as *New Horizons* entered the jovian magnetosphere late in February 2007.

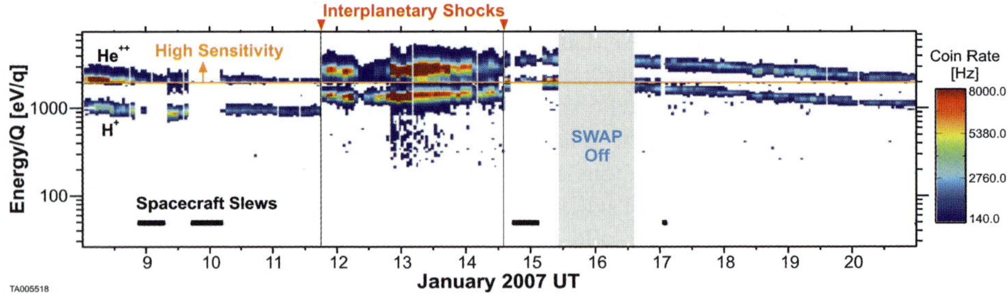

Fig. 42 Early observations from SWAP in the solar wind at a heliocentric distance of ~4.9 AU. The *upper* (*lower*) trace shows solar wind alpha particles (protons) with the sensitivity significantly reduced below 2 keV/q for safety reasons (see text). Times when the proton and alpha particle beams are greatly reduced or disappear correspond to times when the instrument was temporarily off or the spacecraft pointed SWAP's aperture away from the Sun. The observed stream structure and forward–reverse shock pairs are consistent with a corotation interaction region in the solar wind at these heliocentric distances

7 Conclusions

The SWAP instrument on *New Horizons* will provide the best measurement of the global atmospheric escape rate from Pluto. This escape rate determines the type of interaction that Pluto has with the solar wind. For a large escape rate, with many ions leaving the atmosphere, the interaction should be comet-like, where ionization of atmospheric material mass-loads the solar wind, causing slowing of the overall bulk flow of the solar wind and draping of the IMF. For lower escape rates, the interaction is likely more like the interaction that the solar wind has with Venus. There, the interaction size is much smaller and the solar wind interacts directly with ions closely bound to the planetary atmosphere, causing a much smaller region of draped magnetic field. In either case, SWAP should also be able to identify and locate a bow shock, if one exists.

In addition to measuring the overall escape rate from Pluto's atmosphere and allowing us to develop an understanding of the solar wind interaction with Pluto for the first time, SWAP should also be able to measure hydrogen pickup ions around Pluto. In addition, in the unlikely event that Pluto has any significant internal or remnant magnetic field, SWAP should measure the effects that such a field would have on the solar wind plasma interaction. Finally, *en route* to Pluto and Kuiper Belt Objects beyond, SWAP should make unique observation of the variation of heliospheric pickup hydrogen ions and sample the jovian magnetosphere.

The SWAP instrument was designed to meet the unique and challenging requirements of measuring the solar wind and its interaction with Pluto out at 30 AU, from a spacecraft that scans its orientation over large angles to image Pluto and Charon with body-mounted cameras as it flies by. In addition, because observations of the plasma interaction is not primary science for the mission, resources for SWAP were highly constrained and we had to remain focused on developing a simple instrument optimized only for measurements of the solar wind interaction with Pluto, while working within the detailed constraints of the *New Horizons* spacecraft and mission plan.

To satisfy all of these requirements and constraints, we designed the SWAP instrument to be cylindrically symmetric over 270° of rotation angle, and combined an RPA that acts as a high-pass filter, a deflector ring, and an ESA. Different combinations of the RPA and ESA produce a variable passband, and deconvolution of measurements made while microstepping the RPA allows the measurement of very small changes in the solar wind speed. Ions that pass through the SWAP electro-optics are then measured in a coincidence detector section that both provides redundancy for making the measurements and suppresses the background generated by anything other than real ions transiting through the instrument. The SWAP instrument has been well characterized by both ground calibrations and a detailed instrument model. SWAP has been successfully turned on in space and is making routine science observations.

SWAP will make unique and exciting measurements *en route*, at Pluto, and beyond. SWAP will measure the slowing of the solar wind flow, mass-loading of the solar wind plasma, the escape rate of atmospheric material from Pluto and the fascinating solar wind interaction with Pluto for the first time.

Acknowledgements We gratefully acknowledge all of the scientists, engineers, technicians, and support personnel who have made the SWAP instrument and *New Horizons* mission a reality. Although it is not possible to list everyone who has contributed to the SWAP instrument, a partial list includes Maureen Ahr, Barry Eggington, George Geleta, Pat Gonzales, Evan Guy, Helen Hart, Marc Johnson, Matt Maple, Greg Palacios, Robert Rendon, Syrrel Rogillio, Carlos Urdiales, Kenny Whitton, and Mike Young. We also gratefully acknowledge the work of Wendy Mills, who helped assemble and edit this paper, and two reviewers and Alan Stern, who provided helpful comments on this manuscript. This work was carried out under the *New Horizons* mission, which is part of NASA's New Frontiers program.

References

F. Bagenal, R.L. McNutt, Geophys. Res. Lett. **16**, 1229–1232 (1989)

F. Bagenal, T.E. Cravens, J.G. Luhmann, R.L. McNutt, A.F. Cheng, *Pluto and Charon: Pluto's Interaction with the Solar Wind* (University of Arizona Press, Tucson, 1997), p. 523

V.B. Baranov, Y.G. Malama, J. Geophys. Res. **100**(A8), 14755–14761 (1995)

J.W. Bieber, W. Wanner, W.H. Matthaeus, J. Geophys. Res. **101**, 2511 (1996)

J.W. Bieber, W.H. Matthaeus, C.W. Smith, Astrophys. J. **420**, 294 (1994)

L. Biermann, B. Brosowski, H.U. Schmidt, Sol. Phys. **1**, 254–284 (1967)

X. Blanco-Cano, N. Omidi, C.T. Russell, J. Geophys. Res. **108**, 1216 (2003)

A. Bogdanov, K. Sauer, K. Baumgärtel, K. Srivastava, Planet. Space Sci. **44**, 519–528 (1996)

M.R. Collier, Geophys. Res. Lett. **20**, 1531–1534 (1993)

D.A. Dahl, Int. J. Mass Spectrom. **200**, 3–25 (2000)

P.A. Delamere, F. Bagenal, Geophys. Res. Lett. **31**, L04807 (2004)

A.J. Dessler, C.T. Russell, Eos **61**, 691 (1980)

J.L. Elliot, E.W. Dunham, A.S. Bosh, S.M. Sliven, L.A. Young, L.W. Wasserman, R.L. Millis, Icarus **77**, 148 (1989)

J.L. Elliot, A. Ates, B.A. Babcock, A.S. Bosh, M.W. Buie, K.B. Clancy, E.W. Dunham, S.S. Eikenberry, D.T. Hall, S.D. Kern, S.K. Leggett, S.E. Levine, D.S. Moon, C.B. Olkin, D.J. Osip, J.M. Pasachoff, B.E. Penprase, M.J. Person, S. Qu, J.T. Rayner, L.C. Roberts, C.V. Salyk, S.P. Souza, R.C. Stone, B.W. Taylor, D.J. Tholen, J.E. Thomas-Osip, D.R. Ticehurst, L.H. Wasserman, Nature **424**, 165 (2003)

H.J. Fahr, Astron. Astrophys. **14**, 263 (1971)

L.A. Fisk, G. Gloeckler, T.H. Zurbuchen, N.A. Schwadron, in *Acceleration and Transport of Energetic Particles Observed in the Heliosphere*, ed. by R.A. Mewaldt, J.R. Jokipii, M.A. Lee, E. Möbius (American Institute of Physics, Melville, 2000), pp. 229–233.

L.A. Fisk, N.A. Schwadron, G. Gloeckler, Geophys. Res. Lett. **24**, 93 (1997)

A.A. Galeev, T.E. Cravens, T.I. Gombosi, Astrophys. J. **289**, 807 (1985)

J. Geiss, G. Gloeckler, U. Mall, R. von Steiger, A.B. Galvin, K.W. Ogilvie, Astron. Astrophys. **282**, 924 (1994)

J. Geiss, G. Gloeckler, R. von Steiger, H. Balsiger, L.A. Fisk, A.B. Galvin, F.M. Ipavich, S. Livi, J.F. McKenzie, K.W. Ogilvie, B. Wilken, Science **268**, 1033–1036 (1995)

G. Gloeckler, N.A. Schwadron, L.A. Fisk, J. Geiss, Geophys. Res. Lett. **22**, 2665 (1995)

G. Gloeckler, J. Geiss, H. Balsiger, P. Bedini, J.C. Cain, J. Fisher, L.A. Fisk, A.B. Galvin, F. Gliem, D.C. Hamilton, Astron. Astrophys. Suppl. Ser. **92**, 267–289 (1992)

G. Gloeckler, J. Geiss, Space Sci. Rev. **86**, 127 (1998)

G.L. Gloeckler, A. Fisk, J. Geiss, Nature **386**, 374–377 (1997)

G. Gloeckler, J. Geiss, E.C. Roelof, L.A. Fisk, F.M. Ipavich, K.W. Ogilvie, L.J. Lanzerotti, R. von Steiger, B. Wilken, J. Geophys. Res. **99**, 17637 (1994)

G. Gloeckler, Space Sci. Rev. **89**, 91–104 (1999)

C.J. Hansen, D.A. Paige, Icarus **120**(2), 247–265 (1996)

E.M. Harnett, R.M. Winglee, P.A. Delamere, Geophys. Res. Lett. **32**, L19104 (2005)

M.W. Hopcroft, S.C. Chapman, Geophys. Res. Lett. **28**, 1115 (2001)

W.B. Hubbard, D.M. Hunten, S.W. Dieters, K.M. Hill, R.D. Watson, Nature **336**, 452 (1988)

A.J. Hundhausen, J.T. Gosling, J. Geophys. Res. **81**, 1436 (1976)

W. Ip, A. Kopp, L.M. Lara, R. Rodrigo, Adv. Space Res. **26**, 1559 (2000)

P.A. Isenberg, J. Geophys. Res. **102**, 4719 (1997)

P.A. Isenberg, Astrophys. J. **623**, 502–510 (2005)

V.V. Izmodenov, R. Lallement, J. Geiss, Astron. Astrophys. **344**, 317–321 (1999)

K. Kecskemety, T.E. Cravens, Geophys. Res. Lett. **20**, 543–546 (1993)

V.A. Krasnopolsky, J. Geophys. Res. **104**, 5955–5962 (1999)

V.A. Krasnopolsky, D.P. Cruikshank, J. Geophys. Res. **104**, 21979 (1999)

M.A. Lee, W. Ip, J. Geophys. Res. **92**, 11041 (1987)

A.S. Lipatov, U. Motschmann, T. Bagdonat, Planet. Space Sci. **50**, 403–411 (2002)

J.G. Luhmann, C.T. Russell, K. Schwingenschuh, E. Eroshenko, J. Geophys. Res. **96**, 11199 (1991)

W.H. Matthaeus, M.L. Goldstein, D.A. Roberts, J. Geophys. Res. **95**, 20673 (1990)

W.H. Matthaeus, G.P. Zank, C.W. Smith, S. Oughton, Phys. Rev. Lett. **82**, 3444–3447 (1999)

D.J. McComas, N.A. Schwadron, F.J. Crary, H.A. Elliott, D.T. Young, J.T. Gosling, M.F. Thomsen, E. Sittler, J.J. Berthelier, K. Szego, A.J. Coates, J. Geophys. Res. **109**, A02104 (2004a)

D.J. McComas, F. Allegrini, C.J. Pollock, H.O. Funsten, S. Ritzau, G. Gloeckler, Rev. Sci. Instrum. **75**(11), 4863–4870 (2004b)

R.L. McNutt, Geophys. Res. Lett. **16**, 1225–1228 (1989)
McNutt et al., Space Sci. Rev. (2007), this issue
E. Möbius, D. Hovestadt, B. Klecker, M. Scholer, G. Gloeckler, Nature **318**, 426–429 (1985)
N. Omidi, X. Blanco-Cano, C.T. Russell, H. Karimabadi, M. Acuna, J. Geophys. Res. **107**, 1487 (2002)
J.L. Phillips, S.J. Bame, M.F. Thomsen, B.E. Goldstein, E.J. Smith, J. Geophys. Res. **98**(A12), 21189–21202 (1993)
J.D. Richardson, J.W. Belcher, A.J. Lazarus, K.I. Paularena, P.R. Gazis, in *Proceedings of the Eighth International Solar Wind Conference*, ed. by D. Winterhalter, J.T. Gosling, S.R. Habbal, W.S. Kurth (American Institute of Physics, Melville, 1996), pp. 483–486
K. Sauer, A. Lipatov, K. Baumgärtel, E. Dubinin, Adv. Space Res. **20**, 295 (1997)
N.A. Schwadron, J. Geiss, L.A. Fisk, G. Gloeckler, T.H. Zurbuchen, R. von Steiger, J. Geophys. Res. **105**(A4), 7465 (2000)
N.A. Schwadron, L.A. Fisk, G. Gloeckler, Geophys. Res. Lett. **23**, 2871–2874 (1996)
N.A. Schwadron, J. Geophys. Res. **103**, 20643–20650 (1998)
N.A. Schwadron, M. Combi, W. Huebner, D.J. McComas, Geophys. Res. Lett. **29**, 1993 (2002)
C.W. Smith, W.H. Matthaeus, G.P. Zank, N.F. Ness, S. Oughton, J.D. Richardson, J. Geophys. Res. **106**, 8253–8272 (2001). doi: 10.1029/200JA000366
C.W. Smith, P.A. Isenberg, W.H. Matthaeus, J.D. Richardson, Astrophys. J. **638**, 508–517 (2006)
E.J. Smith, J.H. Wolfe, Geophys. Res. Lett. **3**, 137 (1976)
A.S. Stern, L.M. Trafton, G.R. Gladstone, Icarus **75**, 485–498 (1988)
Stern et al., New Horizons mission. Space Sci. Rev. (2007), this issue
G.E. Thomas, Annu. Rev. Earth Planet. Sci. **6**, 173 (1978)
F. Tian, O.B. Toon, Geophys. Res. Lett. **32**, L18201 (2005)
V.M. Vasyliunas, J. Geophys. Res. **73**, 2839–2884 (1968)
V.M. Vasyliunas, G.L. Siscoe, J. Geophys. Res. **81**(7), 1247–1252 (1976)
Z. Wang, M.G. Kivelson, J. Geophys. Res. **101**, 24479–24494 (1996)
F.M. Wu, D.L. Judge, Astrophys. J. **231**(2), 594–605 (1979)
D.T. Young, F.J. Crary, J.E. Nordholt, F. Bagenal, D. Boice, J.L. Burch, A. Eviatar, R. Goldstein, J.J. Hanley, D.J. Lawrence, D.J. McComas, R. Meier, D. Reisenfeld, K. Sauer, R.C. Wiens, Icarus **167**, 80–88 (2004)
G.P. Zank, W.H. Matthaeus, J.W. Bieber, H. Moraal, J. Geophys. Res. **103**, 2085 (1998)
G.P. Zank, H.L. Pauls, I.H. Cairnes, G.M. Webb, J. Geophys. Res. **101**, 457 (1996)

The Pluto Energetic Particle Spectrometer Science Investigation (PEPSSI) on the New Horizons Mission

Ralph L. McNutt Jr. · Stefano A. Livi · Reid S. Gurnee · Matthew E. Hill · Kim A. Cooper · G. Bruce Andrews · Edwin P. Keath · Stamatios M. Krimigis · Donald G. Mitchell · Barry Tossman · Fran Bagenal · John D. Boldt · Walter Bradley · William S. Devereux · George C. Ho · Stephen E. Jaskulek · Thomas W. LeFevere · Horace Malcom · Geoffrey A. Marcus · John R. Hayes · G. Ty Moore · Mark E. Perry · Bruce D. Williams · Paul Wilson IV · Lawrence E. Brown · Martha B. Kusterer · Jon D. Vandegriff

Originally published in the journal Space Science Reviews, Volume 140, Nos 1–4, 315–385.
DOI: 10.1007/s11214-008-9436-y

Abstract The Pluto Energetic Particle Spectrometer Science Investigation (PEPSSI) comprises the hardware and accompanying science investigation on the New Horizons spacecraft to measure pick-up ions from Pluto's outgassing atmosphere. To the extent that Pluto retains its characteristics similar to those of a "heavy comet" as detected in stellar occultations since the early 1980s, these measurements will characterize the neutral atmosphere of Pluto while providing a consistency check on the atmospheric escape rate at the encounter epoch with that deduced from the atmospheric structure at lower altitudes by the ALICE, REX, and SWAP experiments on New Horizons. In addition, PEPSSI will characterize any extended ionosphere and solar wind interaction while also characterizing the energetic particle environment of Pluto, Charon, and their associated system. First proposed for development for the Pluto Express mission in September 1993, what became the PEPSSI instrument went through a number of development stages to meet the requirements of such an instrument for a mission to Pluto while minimizing the required spacecraft resources. The PEPSSI instrument provides for measurements of ions (with compositional information) and electrons from 10 s of keV to ~1 MeV in a $160^\circ \times 12^\circ$ fan-shaped beam in six sectors for 1.5 kg and ~2.5 W.

E.P. Keath, B. Tossman, W. Bradley, and P. Wilson IV are retired.

R.L. McNutt Jr. (✉) · R.S. Gurnee · M.E. Hill · K.A. Cooper · G.B. Andrews · E.P. Keath · S.M. Krimigis · D.G. Mitchell · B. Tossman · J.D. Boldt · W. Bradley · W.S. Devereux · G.C. Ho · S.E. Jaskulek · T.W. LeFevere · H. Malcom · G.A. Marcus · J.R. Hayes · G.T. Moore · M.E. Perry · B.D. Williams · P. Wilson IV · L.E. Brown · M.B. Kusterer · J.D. Vandegriff
Applied Physics Laboratory, The John Hopkins University, 11100 John Hopkins Road, Laurel, MD 20723, USA
e-mail: Ralph.mcnutt@jhuapl.edu

S.A. Livi
Southwest Research Institute, 6220 Culebra Road, San Antonio, TX 78228, USA

S.M. Krimigis
Academy of Athens, 28 Panapistimiou, 10679 Athens, Greece

F. Bagenal
The University of Colorado, Boulder, CO 80309, USA

Keywords New Horizons · PEPSSI · Pluto · Energetic particle instrument

Abbreviations

1PPS	One Pulse Per Second
ADC	Analog-to-digital converter
APL	Applied Physics Laboratory
ASIC	Application specific integrated circuit
C&DH	Command and Data Handling
CCSDS	Consultative Committee for Space Data Systems
CFD	Constant Fraction Discriminator
CSA	Charge Sensitive Amplifier
eV	Electron Volt
FITS	Flexible Image Transport System
FOV	Field of View
FWHM	Full Width Half Maximum
GSE	Ground support equipment
GSFC	Goddard Space Flight Center
HDU	Header Data Unit
HV	High Voltage Section of HVPS
HVPS	High Voltage Power Supply (HV and Bias Supply Sections)
IEM	Integrated Electronics Module
IGSE	Instrument Ground Support Equipment
ICD	Interface Control Document
ITF	Instrument Transfer Frame
LED	Leading Edge Discriminator
MCP	Micro-channel plate
MIDL	Mission Independent Data Layer
MDM	Master Data Manager
MET	Mission Elapsed Time
MOI	Moment of inertia
NA	Not applicable
NASA	National Aeronautics and Space Administration
NH	New Horizons
ns	nanosecond $= 10^{-9}$ s
PDS	Planetary Data System
PEPSSI	Pluto Energetic Particle Spectrometer Science Investigation
PFF	Pluto Fast Flyby
PHA	Pulse height analysis
PIDDP	Planetary Instrument Definition and Development Program
ps	picosecond $= 10^{-12}$ s
psi	Pounds per square inch
RTG	Radioisotope Thermoelectric Generator
SQL	Structured Query Language
SSD	Solid-state detector
SSR	Solid-state recorder
STP	Supplemented Telemetry Packet
SwRI	Southwest Research Institute
TDC	Time-to-digital chip
TOF	Time of flight

TRIO Temperature remote input/output
T-V Thermal-vacuum
UART Universal asynchronous receive and transmit

1 Introduction

The Pluto Energetic Particle Spectrometer Science Investigation (PEPSSI) is one of seven scientific instruments/experiments (Weaver et al. 2008) on board the New Horizons spacecraft (Fountain et al. 2008), now on its way to Pluto (Stern 2008). While it is doubtful that Pluto has an intrinsic magnetic field and magnetosphere that accelerates charged particles to high energies, Pluto does have (or has had in the very recent past) a substantial atmosphere (Elliot et al. 1989, 2003, 2007; Brosch 1995; Sicardy et al. 2003; Fink et al. 1980) that is escaping into the solar wind in a comet-like interaction (Bagenal and McNutt 1989; McNutt 1989; Kecskemety and Cravens 1993; Bagenal et al. 1997; Trafton et al. 1997; Krasnopolsky 1999; Delamere and Bagenal 2004; Tian and Toon 2005). Measured interactions at comets show that the outgassing cometary neutral atoms and molecules charge-exchange with the solar wind and are accelerated in the process (Galeev et al. 1985; Mendis et al. 1986; Galeev 1987; Neugebauer 1990; Coates et al. 1993a, 1993b; Huddleston et al. 1993; Motschmann and Glassmeier 1993). By measuring the in situ energetic particle population, identifying those ions from the emitting body, and noting their variation with distance to the emitting body, the outgassing source strength may be deduced (Gloeckler et al. 1986). It is also important to note that energization beyond what one would naively expect from pick-up alone is also observed at comets (McKenna-Lawlor et al. 1986; Richardson et al. 1986; Sanderson et al. 1986; Somogyi et al. 1986). For example, shock acceleration of particles at Venus can elevate some of the particles to substantial ($\sim$100 keV) energies (Williams et al. 1991). Making these measurements to determine the "outer boundary" of the influence of Pluto's atmosphere is the primary objective of the PEPSSI instrument. The extent of the interaction with the solar wind will be determined by comparing the PEPSSI measurements with those obtained of the solar wind by the SWAP instrument (McComas et al. 2008). A deduced atmospheric profile from the surface to the edge of Pluto's atmosphere will be assembled from combining PEPSSI, the New Horizons ultraviolet imaging spectrometer Alice (Stern et al. 2008) and the New Horizons radio experiment (REX) (Tyler et al. 2008) measurements.

PEPSSI combines energy and time-of-flight measurements in a low-mass (<1.5 kg), low-power (<2.5 W) unit. At the same time, the instrument has a relatively large total geometric factor of $\sim$0.1 cm^2 sr and enables directional information of the particle distribution across a $\sim 12^\circ \times 160^\circ$ swath in six $\sim 25^\circ$-wide angular bins. The instrument also discriminates between electrons and ions without the use of (relatively) heavy permanent magnets (or power-hungry electromagnets). While the use of silicon solid state detectors (SSDs) for measuring energetic particles date back to the near-beginning of the space program, the additional species discrimination made possible by including time-of-flight measurements (thus independently, but simultaneously, measuring both an ion energy and speed) is a relatively newer development (McEntire et al. 1985; Lanzerotti et al. 1992; Williams et al. 1992; Krimigis et al. 2004). Further, packaging of this capability into such a compact instrument provided a variety of engineering and programmatic challenges.

1.1 Previous Similar Instrumentation

PEPSSI combines time-of-flight (TOF) and total-energy measurements in six angular sectors packaged into a compact arrangement (Fig. 1). Unlike plasma instruments incorporating

Fig. 1 PEPSSI mounted on the New Horizons spacecraft via its complex-angle bracket in the clean room at APL prior to being covered with its thermal blanket

a TOF system at lower energies (Gloeckler et al. 1985), no pre-acceleration, high-voltages ($\geq$10 s of kilovolts, kV) are used in this device.

PEPSSI is a direct outgrowth of a variety of internal research and development activities within the Johns Hopkins University Applied Physics Laboratory (APL) going as far back as 1980 (De Amicis 1988) as well as NASA's Planetary Instrument Definition and Development Program (PIDDP). The original goal of "a total instrument mass and power of less than 0.5 kg and 0.5 W" resulted in a design known colloquially as "the hockey puck" due to the cylindrical aperture/detector/time-of-flight (TOF) section that has about the same proportions and size as a regulation ice hockey puck. Little from a second PIDDP effort was available for incorporation into the MESSENGER flight hardware given the timing of MESSENGER's selection for flight (McNutt et al. 2006). Some improvements were possible for PEPSSI due to work under this second grant prior to the selection of New Horizons for flight (McNutt et al. 1996; Andrews et al. 1998). As had been predicted in 1992, the most difficult aspect was the development of a scientifically useful energetic particle instrument with the small mass and power limits that had to be met to allow for inclusion on a Pluto mission (Neugebauer et al. 1993; Lunine et al. 1995).

2 Scientific Background and Objectives

2.1 The Interaction of Pluto with the Solar Wind

The interaction of the various planets of the solar system with the solar wind goes back to the initial forays outside of the Earth's magnetosphere in the early 1960s. As part of the studies that eventually led to the Voyager program (Dryer et al. 1973) made an initial scoping study of what a solar wind interaction with Pluto might look like. They noted that for the anticipated scale lengths a kinetic approach was more proper for Pluto as for Mercury and that for a vanishingly small ionospheric scale height at Pluto, a long, induced magnetotail is, nonetheless, expected. With plans for a Pluto mission using grand-tour-like trajectories eventually canceled, the then-future Voyager encounter with Neptune and its large moon Triton (Morrison et al. 1982) would potentially provide the best insight into Pluto's interaction with the solar wind (McNutt 1982).

Interest in Pluto's interaction with the solar wind was reignited by the stellar occultation of 1988 (Elliot et al. 1989) and its interpretation of showing significant atmospheric escape (Bagenal and McNutt 1989; McNutt 1989). Current thinking is that the interaction of Pluto

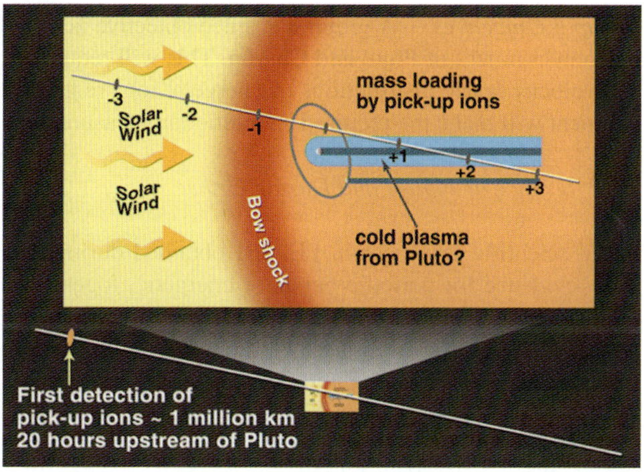

Fig. 2 Schematic of expected interaction of Pluto with the solar wind for the strong-interaction limit. Based upon scaling from cometary interactions, the first detection of pickup ions from Pluto are expected as early as ~20 hours before closest approach

with the solar wind is something between that of an un-magnetized planet, such as Venus and that of a comet, depending upon the strength of the atmospheric outflow. If Pluto were to have even a weak intrinsic magnetization, then the interaction would be more akin to that of the magnetized planets due to the weak solar wind ram pressure at ~30 AU and beyond. Estimates of the overall outgassing rate of the atmosphere Q_0 are between $\sim 10^{27}$ and 10^{28} molecules s^{-1}. Such rates have effects ranging from just shielding the surface from the solar wind to producing a well-formed magnetosphere encompassing the orbit of Charon (Fig. 2). Detailed discussions of past and current thinking about the interaction is given by (Bagenal et al. 1997) and (McComas et al. 2008), respectively, and more detailed simulations have been carried out as well (Delamere and Bagenal 2004; Harnett et al. 2005).

2.2 PEPSSI Science Objectives

2.2.1 PEPSSI Objectives at Pluto

The PEPSSI sensor is designed to perform *in situ* measurements of the mass, energy spectra, and distributions of moderately energetic particles in the near-Pluto environment and in the Pluto-interaction region. The instrument measures particle velocity and energy, and derives particle mass. It discriminates between electrons, protons, alphas, and carbon-nitrogen-oxygen (CNO—taken as a closely-spaced group in atomic weight), and heavier ions. The direction of particles is also determined within one of six sectors. PEPSSI objectives, within the context of New Horizons science-mission-group objectives include:

Group 1 Objective. A group 1 objective is characterization of the neutral atmosphere of Pluto and its escape rate. To support this objective, PEPSSI will detect heavy ions and measure associated energy spectra and spatial variation along the trajectory. By analogy with cometary measurements, these measurements will be used to determine the neutral particle escape rate, which along with UV spectral measurements made in the upper atmosphere (Stern et al. 2008), will be used to put together a fully self-consistent model of Pluto's upper atmosphere to satisfy this group 1 objective.

Group 2 Objectives. A group 2 science objective is characterization of Pluto's ionosphere and interaction with the solar wind. This characterization will be aided by PEPSSI measurements of the spatial extent and composition of pickup ions; these measurements are complementary to those that will be made by SWAP (McComas et al. 2008).

Group 3 Objectives. A group 3 science objective is characterization of the energetic particle environment of Pluto and Charon. This will require measurement of the spatial extent and velocity-space distributions of energetic ions (e.g., H^+, N^+, and N_2^+). The PEPSSI instrument will make the required energetic ion measurements.

2.2.2 Science at Jupiter and During Cruise

Initial scientific results from PEPSSI obtained during the New Horizons flyby of Jupiter show evidence for periodic bursts of energetic, Iogenic particles down the magnetotail of the planet (McNutt et al. 2007). In addition, this magnetospheric passage provided an opportunity for significant in-flight calibration activities that have been used to inform ongoing rehearsal activities for the Pluto flyby.

There is no official "cruise science" on the New Horizons mission. This is by design to help save the instruments for the Pluto flyby as well as minimize operational costs by putting the spacecraft in hibernation for the majority of the cruise from Jupiter to Pluto. Data on the interplanetary medium will be collected as activity during the annual checkouts allows. The general trajectory of New Horizons into the direction toward the incoming interstellar wind will allow for important comparisons of energetic particles and transients in the interplanetary medium as they propagate toward and into the inner heliosheath, now being explored by Voyagers 1 and 2.

2.3 Measurement Requirements

2.3.1 Measurement Ranges

Energy thresholds and energy ranges depend upon the energy measurement mode, i.e., TOF-only, energy (SSD) only, or coincidence measurements through the entire system (Table 1).

2.3.2 Derived Instrument Specifications

2.3.2.1 Mass Resolution (Mass Uncertainty) Particle mass is derived from energy and TOF measurements. The uncertainty in the derived mass, i.e., the mass resolution, is determined by (a) energy measurement resolution, (b) TOF measurement resolution, (c) particle mass, and (d) calibration accuracy. For Energy-plus-TOF measurements, the mass resolution for three species of particles (spanning light, medium, and heavy mass) is specified as <2

Table 1 Energy measurement ranges for PEPSSI

Species	Energy measurement range		
	Energy + TOF measure	Energy-only measure	TOF-only measure
Energetic electrons	Not applicable	25 keV to 500 keV	Not applicable
Protons	25 keV to 1 MeV	Not applicable	$\sim$ 1 keV to 1 MeV
Atomic ions, e.g., CNO group, Mg^+, Si^+, Ne^+	60 keV to 1 MeV	Not applicable	15 keV to $\sim$1 MeV/nucleon
Molecular ions e.g., N_2^+, O_2^+	100 keV to 1 MeV	Not applicable	30 keV to $\sim$1 MeV/nucleon

atomic mass units (AMU) for H^+ (25 keV to 1 MeV), <5 AMU for $C^+/N^+/O^+$ (60 keV to 1 MeV), and <15 AMU for Fe^+ (60 keV to 1 MeV).

For TOF-Only measurements, the means to ascertain particle mass is less precise, and this requirement, with respect to mass resolution, is to distinguish between H^+ and CNO group particles. To support derivation of species mass, for particle energies in the ~1 keV to 1 MeV range, PEPSSI was specified to be capable of measuring particle TOF over a range of 1 to 320 ns. As TOF measurements respond to the energy-per-mass of an incident particle, these measurements can, in principle, respond to heavy ions up to ~30 MeV total energy (~1 MeV/nucleon). However, such extremely high energies are not expected in the Pluto environment and the current capability of the instrument is only to deal with events with up to ~1 MeV total energy.

2.3.2.2 Species Mass Range The PEPSSI instrument is constrained in downlink capability from Pluto as well as in the mass and power available for the instrument. Hence, prudent choices had to be made to meet all of the constraints while still enabling the collection of appropriate data from the vicinity of Pluto and its transmission to Earth following the flyby. Species resolution for the various energy spectra is limited by the counting statistics and the physical size of the detector that limits the TOF drift space. To enable the discrimination of solar wind particles (primarily protons and alpha particles, i.e. doubly-ionized helium nuclei) from pickup particles from Pluto, including atomic "debris" as well as ionized molecules of nitrogen, methane, and carbon monoxide, and allow for discovery science within the confines of the requirements, energy spectra are output for proton events, electron events, CNO events, and heavy particle events (>24 AMU, typified by Fe).

2.3.2.3 Sensitivity and Geometric Factor Requirements With PEPSSI mounted on the spacecraft, including installation of the RTG power source, and with the PEPPSI covers closed, the background ion particle count rate was specified not to exceed one particle per second. Expected fluxes at Pluto are relatively low (~ 100 events per second), so to stay within low power limits of operation, the PEPSSI instrument was specified to be capable of processing at least 10^3 particle events per second, where this event rate is applicable to the total of all classes of measurements, i.e., Energy-plus-TOF, Energy-Only, and TOF-Only. The PEPSSI instrument has the potential to measure particle events at a much higher rate; this rate should be established once the analysis of data from the Jupiter flyby is fully analyzed.

2.3.2.4 Geometric Factor Expected count rates at Pluto are unknown but expected to be low. Hence, the geometric factor was required to be as large as possible, consistent with the targeted low mass of the instrument of ~1.5 kg.

On the basis of these trades, the PEPSSI geometric factors, for electron and ion detection, were specified to meet or exceed the values given in Table 2. The geometric factors for electron and ion detection are different because of the difference in numbers of ion and electron detectors. The values that follow apply to the entire aperture acceptance angle of 160° by 12°, i.e. the geometric factor per "pixel" is less.

2.3.2.5 Integration Interval Nominally, energy-plus-TOF measurements, used to determine particle species and associated energy spectra, are integrated over as low as a 1-second interval (based upon consideration of spacecraft speed, and hence spatial resolution, telemetry rates and data volume playback during the Pluto encounter). TOF measurements, used to determine particle velocity distribution, are integrated over the identical time interval. By command, the integration interval may be adjusted from 1 to 65,535 seconds.

Table 2 PEPSSI sensor specifications

Ion detection geometric factor	≥0.1 cm^2 steradian
Electron detection geometric factor	≥0.033 cm^2 steradian
Acceptance angle	160° by 12°, 6 sectors of 25° by 12° each, 2° gaps between sectors
Aperture area	0.6 cm × 1.2 cm per sector
TOF length	6 cm nominal between entry, exit foils
Number of Detectors Per Sector	2
Detector Area	>0.4 cm^2, ion and electron detectors
Number of Ion Detectors	9
Number of Electron Detectors	3 (located in sectors 0,2,5)

2.3.3 Measurement Resolution Requirements

These specifications flow, in turn, to the next level of implementation requirements that drove the system design and implementation.

2.3.3.1 Energy Resolution As a goal, instrument energy measurement resolution, which includes the effects of all noise sources including analog-to digital converter (ADC) quantization noise, was 5 keV full-width at half maximum (FWHM) or less. As a requirement, instrument energy measurement resolution for ions was ~7 keV or better, for electrons 8 keV or better; ADC quantization is equivalent to 1 keV. The Energy resolution is determined to a large degree by the performance of the energy-peak detector chips for the given SSD detector capacitance and leakage currents. The energy resolution of the energy-peak detector chips is a trade-off between power dissipation, mainly in the charge sensitive amplifier, and integration time of the shaper. Both parameters are fixed on the energy board with two resistors (common for all 12 channels) as described in more detail in the electronics section.

The overall energy resolution is also determined by the digital noise (mainly the ADC noise) on the energy board. The clock frequency of the temperature remote input/output (TRIO) application specific integrated circuit (ASIC) chip for the ADC can be set comparatively low, to 150 kHz, and, given the very low digital power dissipation of the chip (1 mW digital power), the system noise is within the specification levels.

2.3.3.2 TOF Resolution Time measurement resolution, which includes the effects of time jitter, time walk, path dispersion, and time quantization, are specified as 1 ns FWHM or less.

The time resolution is dependent on the time walk and time jitter of the constant fraction discriminator (CFD) ASIC chip and the time jitter of the time-to-digital converter (TDC) ASIC chip. The CFD time walk is ~200–300 ps for a 1:100 input energy dynamic range; the time jitter is in the range 50–100 ps depending on the power dissipation. The time jitter of the TDC chip is ~50 ps. Thus the overall electronics time resolution is <500 ps. The optimum is achieved with proper delay line selection and optimization of the power dissipation within the power budget limits.

2.3.4 Platform Requirements

For proper interpretation of the PEPSSI data, the instrument alignment with respect to the spacecraft and the knowledge of the spacecraft attitude are both required to 1.5°. For deciphering the measurements to be made at Pluto during the Pluto flyby, knowledge of spacecraft distance to Pluto is required to an accuracy of 225 km. This knowledge of position figure is equal to roughly 1/10 the diameter of Pluto, and is achieved with the spacecraft timing system and navigation of the spacecraft at the system level.

2.3.5 Time Resolution Requirement

2.3.5.1 Timetagging Requirements For proper interpretation and processing of science data, knowledge of the data intervals over which PEPSSI science data is collected must be known to within to ±1 second of spacecraft mission elapsed time (MET). To this end, data packets sent from PEPSSI to the command and data handling (C&DH) system are time tagged with MET time to 1-second resolution (Fountain et al. 2008). Based on spacecraft time-keeping requirements, knowledge of MET time relative to UTC time to ~10 ms accuracy is known after the fact.

2.3.5.2 Science Data Synchronization PEPSSI science data (species energy spectra, velocity, and pulse-height analysis (PHA)) are collected over fixed time intervals. These data intervals are synchronized by, and time aligned with, one pulse-per-second (1PPS) timing epochs input from the spacecraft that are coincident with the MET one-second time increments.

2.3.6 Calibration Requirements

2.3.6.1 Required Ground Calibration Calibration of the PEPSSI instrument is required in order to meet instrument performance specifications. Ground calibration tests were planned using both linear accelerator facilities at APL and Van De Graff facilities located at GSFC. These facilities provide a calibrated source of electrons and a variety of ions (e.g., protons, helium, CNO, iron) at a variety of energies. Time and availability issues reduced the calibration plan to using the APL facility only. Calibration was performed with the PEPSSI instrument in vacuum at a temperature of 25°C for each sector. Calibration was performed prior to installation on the spacecraft. A calibrated alpha-particle source is installed in the collimator assembly and described below. This source was used during thermal vacuum tests to verify instrument performance has not changed.

2.3.6.2 In-flight Calibration Characterization Following launch, the PEPSSI instrument calibration had to be characterized during initial instrument test and prior to all major data collecting operations. To characterize the instrument, background particle data was collected and downlinked to the PEPSSI payload operations center. Archived energy and TOF measurement data in the PHA data packet, as well as particle species data, have been processed to evaluate instrument calibration. To correct for some drift in measurements, new 'look-up' tables have been configured and uplinked. Further characterization occurred during the close approach to Jupiter, using that planet's magnetospheric plasma as a calibration source. The characterization will be monitored during the yearly checkouts during the cruise to Pluto. The installed alpha-particle calibration source will also be monitored to look for any changes in instrument performance prior to arrival in Pluto-space in 2015.

2.3.6.3 Energy Board Temperature Monitor Energy measurements vary slightly as a function of energy board component temperature. A temperature sensor is installed on the energy board, and this temperature data is telemetered in instrument housekeeping telemetry. Energy board temperature information allows the instrument ground processing software to account for any slight errors in energy measurement due to energy-board temperature variation.

3 Technical Description

3.1 Instrument Overview

Given the requirements, as well as ongoing lessons from the near-concurrent development of the EPS detector for the MESSENGER mission (Andrews et al. 2007), the required functionality was achieved within the prescribed constraints. These properties are given in Tables 3 and 4. Some of the numbers are still being refined based upon the final analysis of instrument performance during the New Horizons flyby of Jupiter (28 Feb 2007).

PEPSSI consists of a collimator and sensor assembly, referred to as the sensor module, which is mounted atop an electronic board stack (Figs. 3 and 4). The electronic stack consists of six metal-framed electronic boards. The stack is a cube measuring approximately 10 cm on a side. The sensor module is approximately 2 cm high and protrudes out from the stack an additional 6 cm. A simplified block diagram of the PEPSSI instrument is shown in Fig. 5. The corresponding simplified schematic for operation is shown in Fig. 6.

Table 3 PEPSSI performance parameters

Geometric Factor (Start Foil)		0.15 cm^2 sr
Geometric Factor (each Electron Pixel)		0.011 cm^2 sr
Geometric Factor (each Ion Pixel)		0.013 cm^2 sr
Electron Efficiency vs. Energy		Est. roughly at 100% at mid energies
Low Energy Ion Efficiency vs. $En \times M$		>4% details in Sects. 4.3.3.6 and 4.4.2
High Energy Ion Efficiency vs. $En \times M$		>4% details in Sects. 4.3.3.6 and 4.4.2
Energy Coverage		See Table 4
$\Delta E/E$ capability at 50 keV		14%
$\Delta E/E$ capability at >100 keV		10%
Energy/TOF Channel resolution		250 eV/167 ps Granularity
TOF dispersive spread		2.5 to 3.5 ns (expected)
Mass Species Separation		See Table 4
Angular Coverage (total)		$160° \times 12°$
Angular Coverage (Electrons)		$147° \times 12°$
Angular Coverage (Low Energy Ions)		$160° \times 12°$
Angular Coverage (High Energy Ions)		$147° \times 12°$
Angular Coverage (Diagnostic Ions)		$147° \times 12°$
Angular Pixels (Electrons)		$12° \times 27°$ (13° with 13.6° gap)
Angular Pixels (Low Energy Ions)		$12° \times 30°$ (full ~27° coverage)
Angular Pixels (High Energy Ions)		$12° \times 27°$ (13° with 13.6° gap)
Angular Pixels (Diagnostic Ions)		$12° \times 27°$ (13° with 13.6° gap)
Electron Scattering Angle Contrast		Several percent
Energy Coverage:		Technique
Electrons	25 keV–500 keV	Singles/Foil Technique
Low Energy Ions	1 keV–1 MeV	TOF Only
Low Energy CNO	15 keV–1 MeV	PH-based mass discrimination
High Energy Protons	40 keV–1 MeV	TOF vs. Energy
High Energy CNO	100 keV–1.2 MeV	TOF vs. Energy

Table 4 Final PEPSSI flight parameters and assessment against requirements

Parameter	Requirement	Measured or derived from measurements	Comments
Engineering Parameters			
Mass	<1.5 kg	1.475 kg	Measured at delivery
Power Average	2.55 W	2.49 W	Not including heater power
Power Peak	same	2.51 W	Additional power needed one time to deploy cover
Data Volume	<135 bps	91 bps	With FAST Compression
Size	<1200 cm^3	352 cm^3	Includes sunshade and mounting tabs/structures
Performance Parameters			
Geometric Factor	>0.1 cm^2 sr	0.15 cm^2 sr	Calculated: PDR
Ion Energy Range	15 keV/n to 1 MeV/n	~1 keV/n to 1 MeV/n	Lowest energies with TOF-only, as expected
Electron Energies	25 to 500 keV	25 to 500 keV	Not verified in calibration
Energy Resolution	10 keV	<5 keV	
Species	H, He, CNO, e^-	H, He, CNO, Fe, e^-	
Field-of-View	160° × 12°	160° × 12°	
Angular Resolution	25° × 12°	25° × 12°	
TOF Range	1–320 ns	1–320 ns	Verified electronically
TOF Resolution	<5 ns	<4 ns (FWHM)	
Integration period	10 s adjustable	10 s default	Programmable from 1 to 65,535 s

As shown in the block diagram, the sensor module includes a time-of-flight (TOF) section 6 cm long feeding a solid-state Si detector (SSD) array. The SSD array is connected to the energy board and measures particle energy. Secondary electrons, generated by ions passing through the entry and exit foils, are detected to measure ion TOF. Event energy and TOF measurements are combined to derive mass and to identify particle species.

While PEPSSI uses the same type of energy and time-of-flight measurement scheme as the previous such instruments mentioned above, it also employs multiple (six) angular apertures in a swath in order to provide angular information without relying upon either a mechanical scanning mechanism or the rotation of the spacecraft. The PEPSSI acceptance angle is fan-like and measures 160° by 12° with six 25° segments. Each segment is separated by a 2° gap. Its total ion geometric factor is greater than 0.1 cm^2 sr. Particle direction is determined by the particular 25° sector in which it is detected. The angular resolution of the instrument was required to determine the incoming direction to better than 40° for the ions in each of the six sectors, i.e., within a 25° by 12° window. Electron particle direction is limited to sector 0, sector 2, and sector 5.

N.B. The expected (atmospheric) species to dominate at Pluto are N_2^+, CH_4^+, and associated breakup products, e.g. N^+. Accelerated protons from the solar wind (H^+) are also likely to be seen. Fe^+ (at 56 AMU) s not expected from Pluto, but indicates the potential mass coverage of the instrument. Ar^+ (40 AMU) may be present, albeit at low concentrations, but is within the mass coverage; see discussion of Sect. 2.3.2.2.

The entrance apertures for the axially symmetric TOF section are 6 mm wide. A 50 Å aluminum/350 Å polyimide/50 Å aluminum composite foil covers each aperture. These

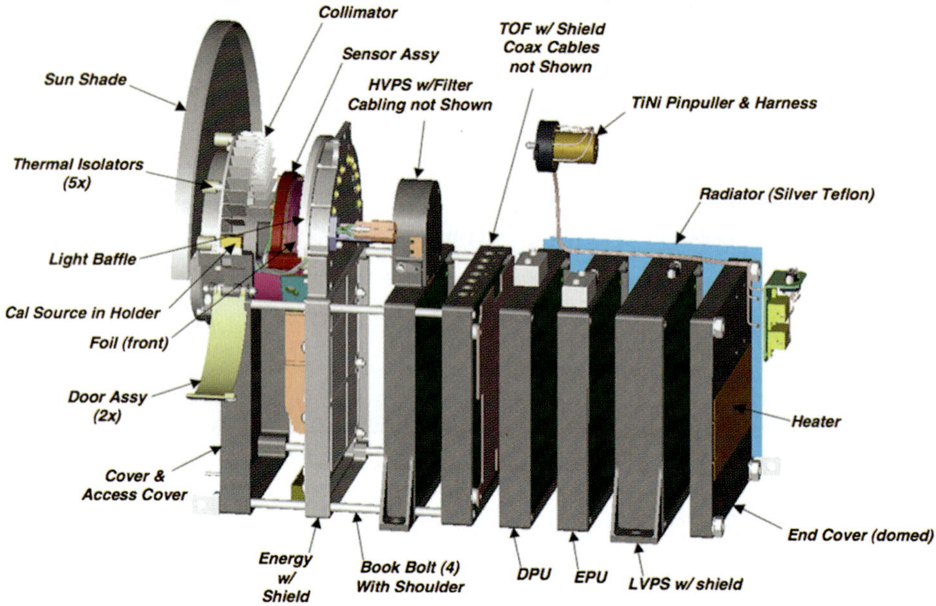

Fig. 3 Expanded view of the PEPSSI instrument showing, *left* to *right*: sunshade, collimator assembly (with acoustic doors open—flight configuration), sensor assembly, and electronics boards stack

Fig. 4 Assembled view of the sensor, again in flight configuration

foils reduce the TOF UV Lyman-alpha photon background. A 50 Å palladium/500 Å polyimide/50 Å palladium foil covers the exit apertures. Both entrance and exit aperture foils are mounted on a high-transmittance stainless steel grid supported on a stainless steel frame.

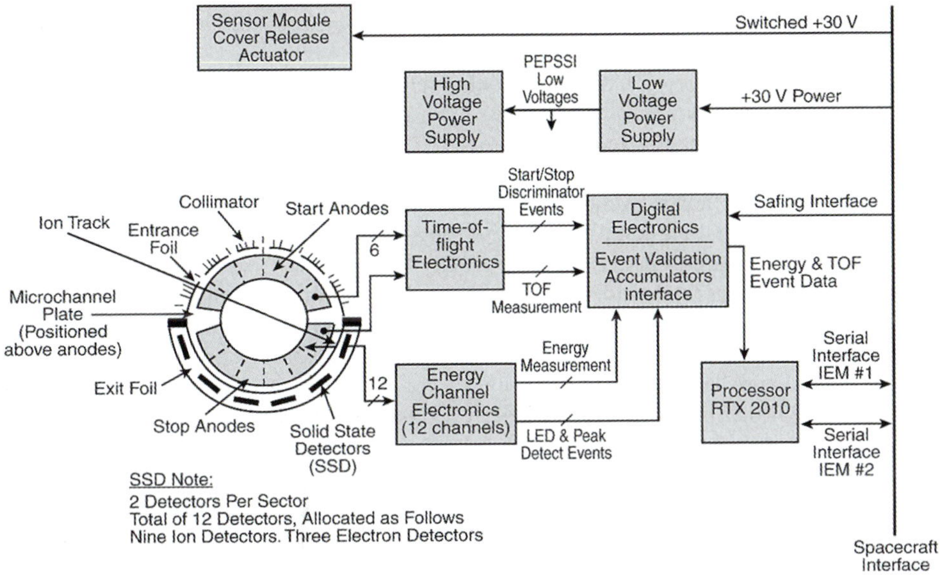

Fig. 5 Simplified block diagram of the PEPSSI instrument. Flow of information from the sensor to the spacecraft interface is shown

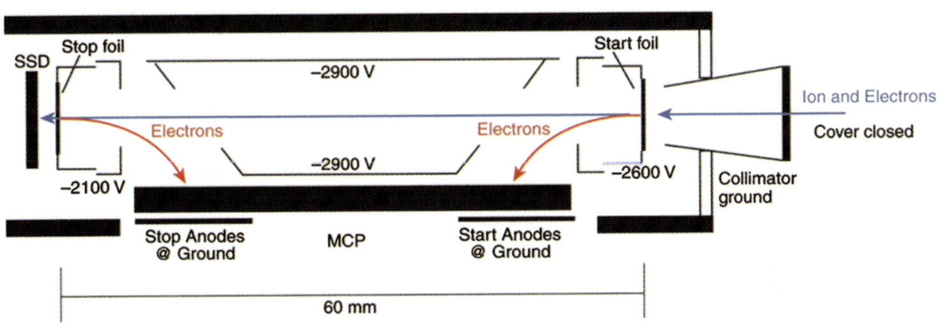

Fig. 6 Simplified operation of the PEPSSI sensor (1 of 6 shown). Incoming ions and electrons pass through the collimator assembly hitting the "start" foil and ejecting secondary electrons. The start electrons hit the start anode flagging the time-of-flight start. The incoming particles cross the 6-cm drift space and then hit the stop foil, again ejecting secondary electrons that hit the stop anode. The incoming particle then hits the SSD where its total energy is recorded. The electronics detect and classify both nominal events and valid "triples" (that have valid start, stop, and energy signals within appropriate electronic windows such that the incoming particle can be fully categorized

3.1.1 Differences from EPS on MESSENGER

The MESSENGER EPS unit was the first flight unit of this miniaturized "hockey-puck" design. PEPSSI presented different requirements with respect to available power and maximum count rates. While larger rates were anticipated for the encounter with Jupiter and could be used to good advantage for some calibration activities, PEPSSI requirements have always been focused at the conditions near Pluto. In addition, the launch of New Horizons over two years after that of MESSENGER allowed for some upgrades to be included in the PEPSSI unit.

The combination of lower expected counting rates and the need to save power led to the implementation of (1) lower-current MCPs, (2) a new high-efficiency, low-voltage power supply (LVPS) with a complete custom converter design to save power, (3) a new energy board using new application-specific-integrated-circuit (ASIC) chips to save power, and (4) a redesigned TOF board with different amplifiers and delay lines to achieve higher gain.

With the need for long-term reliability during the cruise to Pluto, PEPSSI enabled a high-voltage (HV) safing system, which is usually active (such a system is also on EPS but causes significant cross-talk with the SSDs if active; the safing system on PEPSSI has not tripped to the date). This system samples the HV current 1000 times per second, and shuts down the supply if the HV current exceeds a threshold; it is inactive only during HV ramp up and ramp down, and then only momentarily for each step. A different HVPS driver is used on PEPPSI that provides improved performance (not incorporated on EPS due to the MESSENGER launch schedule). The PEPSSI unit also incorporates a buffer amplifier and a filter, neither of which are present in the EPS unit.

The EPS instrument has 24 detectors: six large detectors for ions, six large detectors for electrons, and corresponding small detectors as well. There are six processing chains for the ions and six for the electrons, with small or large detector usage dictated by ground command. The particle flux in the outer heliosphere in general, and near Pluto in particular is expected to be much lower than near Mercury and less variable. To ensure detection of the lowest fluxes PEPSSI uses only 12 detectors (small and large pixels as laid out on EPS and electrically tied together) and uses only three electron detectors to increase the geometric factor for the ions.

The front-end collimator is more open on PEPSSI than on EPS to increase the geometric factor, and PEPSSI also has an added sunshade to be able to look back near the solar direction (EPS is mounted well away from the Sun-looking direction on the MESSENGER spacecraft due to the high solar luminosity at Mercury's perihelion).

EPS is thermally isolated from the MESSENGER spacecraft with a separate thermal radiator to reject internal heat to space, while PEPSSI is quasi-thermally coupled to the spacecraft-mounting bracket (conductively isolated but radiatively coupled). Furthermore, the EPS thermal design uses heaters and mechanical thermostats to keep the instrument from getting too cold, while PEPSSI relies on the spacecraft to keep it sufficiently warm (a "hibernation" operational heater was installed as a precaution).

3.2 Mechanical Design

Top-level mechanical design requirements and design factors of safety are in accord with those used throughout the New Horizons project. In particular, the PEPSSI instrument was designed to withstand a quasi-static load limit of 30 g along three orthogonal axes (applicable to the primary structure design and as multiplied by the appropriate factor of safety). In addition, first mode structural frequencies were specified to be above 70 Hz in the spacecraft thrust direction and above 50 Hz in the lateral directions and the instrument was designed to withstand a maximum pressure rate change of 1.0 psi/s. All of these specifications were verified during environmental tests.

3.2.1 Dimensions and Mounting

The PEPSSI envelope is 19.7 cm × 14.7 cm × 21.6 cm as installed and 25.1 cm × 14.7 cm × 21.6 cm following the one-time opening of the acoustic doors. The instrument itself is

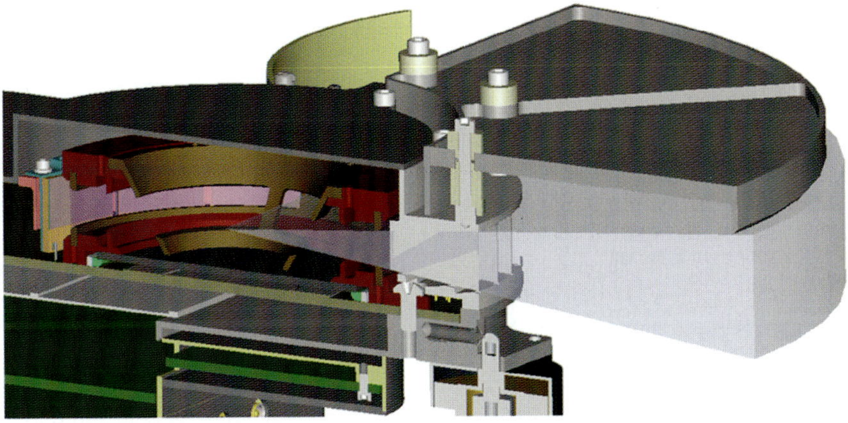

Fig. 7 Cutaway view of the PEPSSI sensor showing the FOV defined by the internal sensor structure, collimator assembly, and Sun-shade (*top*)

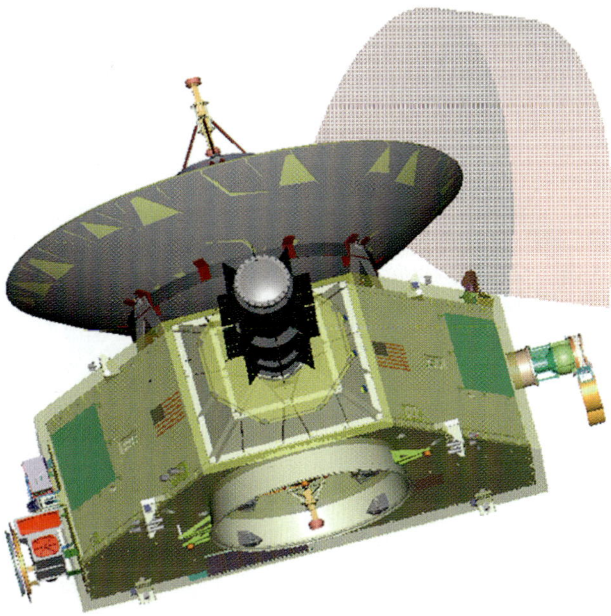

Fig. 8 Location of PEPSSI on the New Horizons spacecraft. The *lightly shaded area* denotes the PEPSSI Field-of-View (FOV)

mounted on a bracket that provides for a complex angular offset of the field of view (FOV) of the viewing fan from any of the spacecraft decks (Figs. 1, 7, and 10).

The complex angular offset was employed to optimize the viewing of freshly ionized pick-up ions in the vicinity of Pluto due to charge exchange with neutrals from Pluto's atmosphere. This allows for the instrument to be mounted to the spacecraft deck while looking past the high gain antenna (HGA) but not being obscured by it (Fig. 8).

Alignment for PEPSSI's FOV on the observatory's top deck was determined by the design and location of its mounting bracket. Alignment control was to be kept within 1.5° of the Euler angle rotations defining the bracket's mounting surface. Knowledge of PEPSSI's FOV was specified to be within 1.5° of the observatory's coordinate system, as referenced

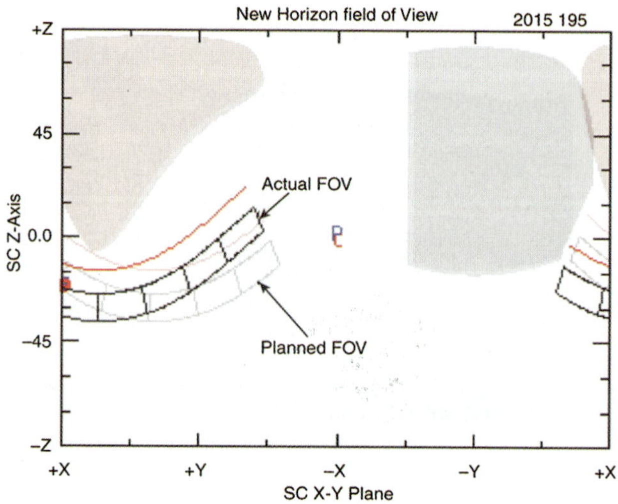

Fig. 9 The PEPSSI FOV in angular coordinates referenced to the main deck of the spacecraft. The *shaded area* above the FOV traces indicates the HGA and the *shaded area* to the *right* indicates the body of the spacecraft. The *red line* indicates the Sun shield, the *red dot* the location of the Sun and the P and C (near the *center*) the locations of Pluto and Charon in August 2006

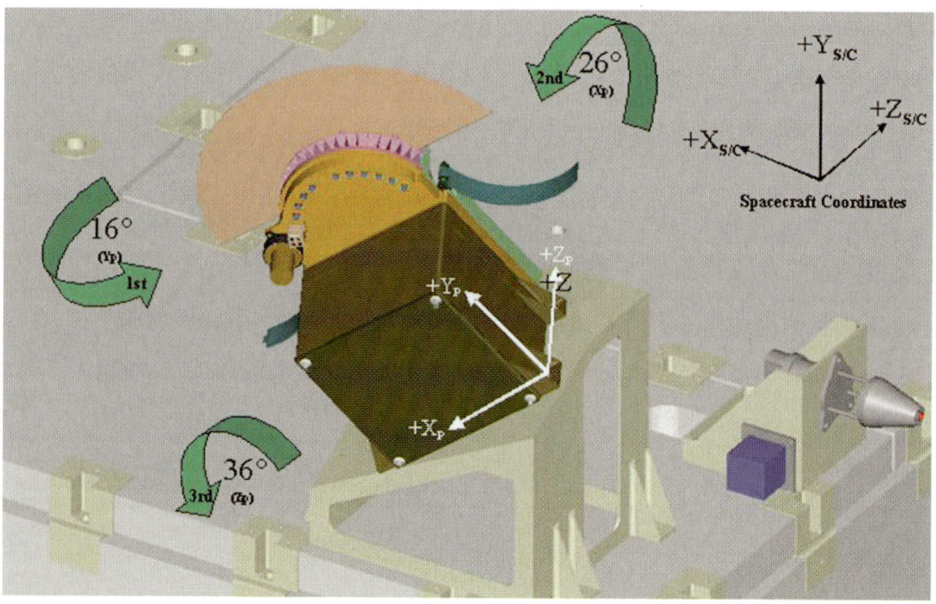

Fig. 10 PEPSSI mounting orientation with respect to the New Horizons deck

to the orthogonal plane surfaces of the PEPSSI instrument. An error in interpretation of the one of the interface control documents (ICDs) led to a misalignment in the bracket surface as manufactured. Figure 9 shows the planned and actual fields of view in the coordinate system of the observatory, i.e., the New Horizons spacecraft.

With this FOV PEPSSI points at an off-angle to the Sun during spacecraft downlink operations and most of Pluto flyby operations. (The HGA points towards the Earth, along the $+Y$ axis, during downlink operations). This allows PEPSSI to detect pickup ions, coming from the general direction of the Sun, while keeping the Sun out of the FOV.

There are no rigid requirements to keep the Sun out of the instrument FOV; however, for normal operation the Sun should be $\sim 10^{\circ}$ from the instrument boresight. In the event the Sun is in the instrument FOV, no damage to the instrument will occur. However, the flood of energetic photons may cause a degradation of both energy and TOF measurements. An initial approach for mitigating these effects, in the event the Sun was in the FOV, was to disable individual start and stop that would be affected. This approach was abandoned during development due to reliability concerns with the switches required for its implementation. Enhanced count rates due to the presence of the Sun in the FOV was what led to the mapping of the actual FOV. For heliocentric distances beyond ~ 10 AU, the low solar intensity is no longer an instrument health and safety concern.

PEPSSI is mounted to the spacecraft's top panel with four (4) #8-32 stainless-steel screws into threaded inserts on a dedicated spacecraft-designed mounting bracket. The instrument is thermally isolated from the bracket by thermal washers between the bracket and PEPSSI's mounting tabs. PEPSSI's alignment and repeatability of mounting with respect to the observatory's coordinate system was satisfied by dimensional tolerancing and position of the mounting holes in the bracket and on the spacecraft.

3.2.2 Mass Properties

The mass budget for the PEPSSI instrument was 1500 grams, including 120 grams of margin. This mass included all sensor pieces as well as the mounting bolts and was worked to on a board-by-board basis (including the metal framing for the boards). The measured sensor mass at delivery to the spacecraft was 1475 g.

3.2.3 Deployable Cover

The PEPSSI design includes two cover doors installed over the instrument aperture. One purpose of the instrument covers is to prevent acoustic (air pressure) damage to the thin entry/exit foils within the sensor module during launch. Another is to keep out air-born dust and contaminants during instrument ground test. During all ground operations and storage, except functional test of cover release, the aperture covers were maintained in place to prevent damage to the entry/exit foils and to prevent contamination of the MCP.

The covers are crescent shaped and each covers one-half of the 160° aperture angle. Each cover is mounted at one end on a hinge bracket assembly that includes a Mandrel with torsion spring. In the closed position an actuator pin holds the covers in place. To open the covers, the actuator is fired to retract the retaining pin, allowing the covers to spring open. A spacecraft command must be sent to fire the actuator. Once open, the covers are maintained in the open position by torsion-spring action. To allow for ground tests (e.g., EMI tests) with the covers open, the actuator used to pin the doors closed was capable of at least 100 cycles of operation.

Cover opening was tested for correct actuator function in a vacuum chamber prior to flight. The covers were opened in flight on 3 May 2006.

3.2.4 Instrument Purge

The PEPSSI instrument interface included a purge manifold. Once PEPSSI was assembled with the MCP, it required constant purge with at least MIL-P-27401D Grade C (99.995% pure) nitrogen, at a flow rate of 0.1 liters/minute.

The purge system was capable of being disabled for a short period of time (up to 30 minutes) while in a class 10,000 or better clean room. Records were maintained of all periods when purge had either been removed or failed.

3.2.5 Handling Requirements

The PEPSSI instrument was kept under purge at all times, except during test in vacuum chambers. The PEPSSI door remained closed to protect the entry and exit foils. When open, all airflow toward the foils, including breath, was prevented. As long as the cover is closed and the unit is being purged, it could be in any normal room environment.

Proper procedures and equipment to eliminate the risk of electrostatic discharge were followed and used. The collimator was not cleaned or handled in any manner, and the cleaning of the outside cover was minimized.

A high-voltage power supply (HVPS) safing plug was included and removed prior to launch. In addition, a cover-release arming-plug was installed before launch to enable cover release commands in space.

3.2.6 Transportation and Storage

The PEPSSI instrument required a shipping container to provide protection during transportation and storage. The instrument was constrained and padded within the container, which is airtight and was purged with dry nitrogen and sealed, to minimize vibration and shock during transport.

3.2.7 Vacuum and Outgassing Requirement

The PEPSSI HVPS was required to be turned on to full voltage only in a vacuum environment $<3 \times 10^{-6}$ Torr. As a consequence, the HVPS could not be activated until sufficient time in vacuum had elapsed for spacecraft and instrument outgassing. A high concentration of outgassing products and particles could contaminate the MCP under high-voltage conditions, so outgas products must be at a sufficiently low level for safe turn-on of the HVPS.

Before launch, during both instrument-level and spacecraft-level thermal-vacuum tests, appropriate detectors were placed inside the test chamber to monitor outgas products. During spacecraft-level testing, 24 hours of time in vacuum (at $<3 \times 10^{-6}$ Torr) was required before the HVPS could be safely turned on. During instrument-level test, the test conductor determined when it was safe to turn on the HVPS. Prior to HVPS activation, the PEPSSI instrument could be (and was) powered on for low-level engineering level checkout. At full atmosphere, during instrument and spacecraft level testing, the HVPS voltage (applied to the MCP) was kept less than 700 volts. After the New Horizons launch, the requirement was for at least two weeks to pass prior to first activation of the HVPS. The PEPSSI instrument was powered earlier (with HVPS off) to allow for initial checkout (20–22 Feb 2006) and engineering-level checks (1–2 Mar, 27 Apr, and 2 May 2006).

3.3 Detectors and Electronics

Power as initially allocated to the instrument took into account the limited resources on the spacecraft. Referenced to the +30 V input the peak power specifications were $<$2000 mW in science mode and $<$1750 mW for the high voltage power supply (HVPS) disabled. During development, these numbers were found not to be realizable without extensive additional development. To remain within financial resources a request was granted to raise the allowable power required by 550 mW.

3.3.1 Energy Measure

Each SSD sector in the sensor module has 2 detectors. In sectors 0, 2, 5, one detects ions, the other electrons. The electron detectors are covered with a 1-μm Al layer, to block low-energy protons and heavier ions. The detector pairs in sectors 1, 3, 4 are just ion detectors. A key to the electronics and functional layout is given in Fig. 11. The electron detectors measure energetic electrons from 25 keV to 500 keV. The Al layer blocks protons and ion particles with energies less than 100 keV, and particle energy levels above 100 keV are expected to be rare in the near-Pluto environment. For those rare events, where ion energy levels exceed 100 keV, coincident TOF measurements will be used to discriminate between ions and energetic electrons.

Ion energy measurements using the ion detectors are combined with coincident TOF measurements to derive particle mass and identify particle species. Particle energy for protons greater than 25 keV and heavy ions (such as the CNO group) greater than 60 keV are measured, up to a maximum of 1 MeV. Lower-energy ion fluxes are measured using TOF-Only measurements; detection of micro-channel plate (MCP) pulse height provides a coarse indication of low-energy particle mass.

3.3.2 Time-of-Flight Measure

Before an ion passes through the TOF head, it is first accelerated by a 2.6-kV potential. Secondary electrons from the foils are electrostatically separated on the MCP, providing start and stop signals for TOF measurements. The segmented MCP anode, with one start

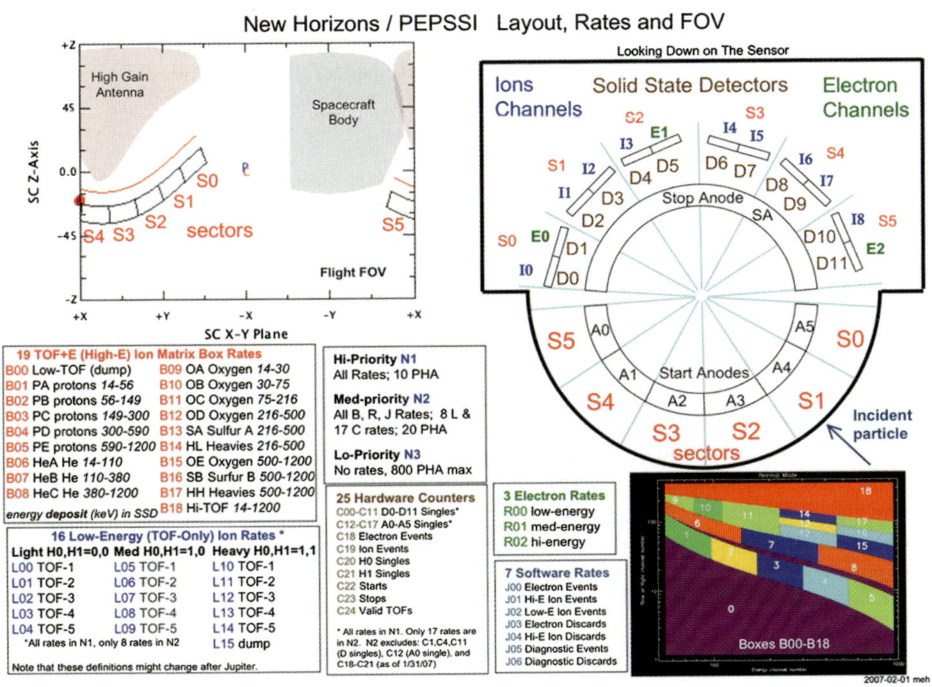

Fig. 11 Schematic layout of PEPSSI showing the nominal FOV and detectors and rates for the various modes and channels

Table 5 Electronics boards in the PEPSSI instrument

PEPSSI board	Board function(s)
Energy Board	Amplify SSD event signals (12 channels—signal proportional to particle energy) and digitize (10 bits). The measurement is achieved with low-power, low-noise Energy (CSA + Shaper)—Peak Detector ASICs
TOF Board	Amplify start (six) and stop anode signals, compute TOF (time duration between start and stop anode signals). The measurement is achieved with the low-power, low-jitter, and low-walk TOF chip (CFD + TDC)
HVPS Board	Provide high voltage to entry/exit foils (−2600 V), MCP (−2100 V and −100 V), deflector plates (−2900 V), and SSD bias voltage (−100 V)
Digital (Interface) Board	Event validation logic; energy and TOF event counters; interface functions
Events Processor Board	Command and Telemetry communications to spacecraft; events processing
LVPS Board	Supply ±15 V, ±5 V, 2.5 V to electronic boards

anode for each of the six angular segments, determines the direction of travel, particularly for lower-energy ions that do not yield an SSD signal above threshold. A nominal 500-volt accelerating potential between the foil and the MCP surface controls the electrostatic steering of secondary electrons. The dispersion in transit time is less than 400 picoseconds (ps).

3.3.3 Electronics

The six electronics boards in the electronics stack (cf. Fig. 3) each provide a specific function for the instrument (Table 5). The common board size has also been used on other APL spacecraft subsystems, notable the electronics for the (redundant) digital processing units and several of the instruments on the MESSENGER spacecraft (Gold et al. 2001).

3.3.4 Operation

3.3.4.1 Science Mode In the science mode, three basic classes of measurements are concurrently made. The first class, called TOF-plus-Energy, uses SSD measurements in coincidence with TOF measurements to determine energetic ion particle mass (species) and associated energy spectra. Measurements are collected from six different sectors and particle direction to a particular 25° by 12° sector is determined by the start sector.

The second class of measurement is referred to as TOF-Only. It is a measure of ion particle velocity based on the TOF measurement and is made if no coincident energy event is detected at the SSDs. Particle composition is categorized as light (protons), medium (mass $<$ CNO), or heavy (mass $\geq$ CNO). These measurements occur if the ion particle energy is below the energy threshold of the SSD detector.

The third class of measurements consists of Energy-Only measurements of particles where the SSD measurement event is not coincident with any TOF measurement event. These measurements include electron energy measurements, where transit time of electrons through the TOF chamber is, for practical purposes, zero, as well as ion energy measurements made when the ion fails to generate any secondary electrons (for TOF measurement).

3.3.4.2 HVPS Activation The high voltage (HV) supply of the HVPS can only be turned on in vacuum ($<3 \times 10^{-6}$ Torr). If the HV supply had been turned on accidentally during ground test in air, the MCP could have been degraded. Therefore, when the PEPSSI instrument was powered up on the ground, the HVPS (consisting of HV and bias power supply) was disabled and no science data was collected. A sequence of instrument commands is required to enable the HVPS, fine-tune voltages, and tweak HVPS clock frequency for optimum efficiency. Valid science data is output after the HVPS command sequence has been executed and science data enabled.

3.3.4.3 Cover Door Open At first power-up on orbit, the PEPSSI covers are in the closed position. A spacecraft command must be sent to fire the actuator and open the front covers. Once open, the covers remain open indefinitely, and no cover release commands are required at any subsequent turn-on of the instrument.

3.3.4.4 Test Mode The instrument can implement a test mode that allows for data input and output through a test port instead of the C&DH communication ports. This is strictly for use during instrument software development pre-flight and is not for spacecraft level test or for use in space operations.

3.3.4.5 Calibration Ground-based calibration was performed with the instrument in a dedicated calibration mode ("diagnostic mode"). Based on an evaluation of telemetered energy and TOF data, instrument calibration tables were configured and subsequently loaded into the instrument.

3.3.5 Electrical Interface

There are five electrical interfaces from PEPSSI back to the spacecraft. Only the power and command and telemetry interfaces are still of use. The others were used for ground testing, safing, and the deployment of the cover that occurred as part of the overall commissioning.

The PEPSSI instrument does not require a survival heater for operation. However, PEPSSI does require a spacecraft-provided and controlled heater that can be can be powered during cruise phase when observatory power is available. The heater provides thermal margin, if required, and dissipates 1 watt when switched on.

3.3.5.1 PEPSSI Power Interface Switched power is provided to PEPSSI at 30 volts nominal and is regulated to within ±1 volt of nominal. Primary power is supplied to the PEPSSI support electronics through a separate dedicated connector that has no signal or secondary lines. Total power consumed by PEPSSI was initially specified to be 2 watts or less, but was later increased during the design process.

There are specifications regarding input voltage characteristics (operating and survival) regarding regulation, input voltage ripple, source voltage transients, source impedance, and voltage turn-on/turn-off rates. Inrush current transients at power turn-on and power turn-off imposed on the instrument are also specified. The instrument was designed and tested to operate over and survive respective voltage operating conditions as required.

3.3.5.2 PEPSSI Command and Telemetry Interface PEPSSI implements a low-speed command and telemetry interface circuit that conforms to the EIA RS-422 standard for serial data transmission. The low-speed buses for PEPSSI each consist of three circuits: a 1-pulse/second (1PPS) sync signal, a command circuit, and a telemetry circuit.

PEPSSI implements two RS-422 bi-directional universal asynchronous receive and transmit (UART) ports to the spacecraft. The first port connects to the spacecraft C&DH system located in the integrated electronics module (IEM) #1, and the second port connects to the spacecraft C&DH system located in IEM #2. Normally, only one of the two IEMs is active, and PEPSSI responds to and interacts with the active IEM (Fountain et al. 2008).

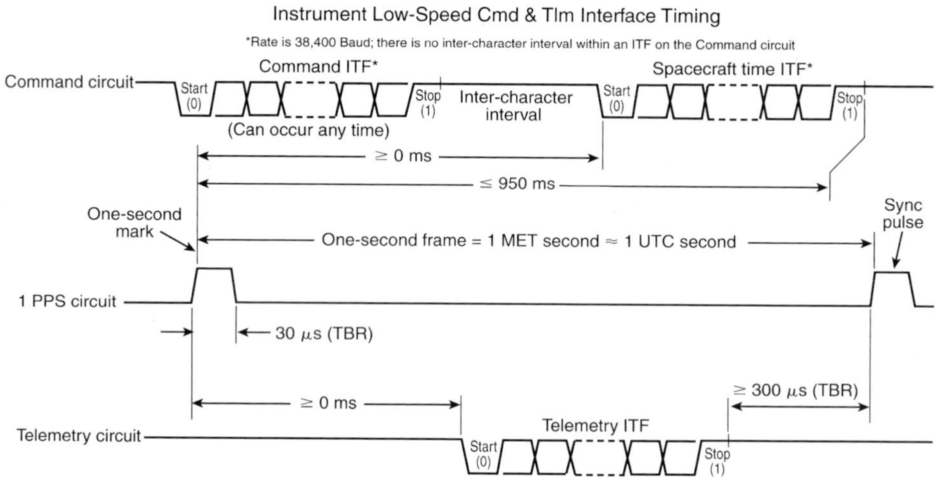

Fig. 12 PEPSSI command and telemetry timing

The serial ports are bi-directional and PEPSSI is capable of simultaneously sending and receiving at 38.4 k baud rate with 8 data bits, no parity, 1 start bit, and 1 stop bit. The least significant bit of each byte is sent first. For multi-byte values, "big endian" format is used, where the most significant byte is sent first; the least significant bit is referred to as "b0". PEPSSI receives instrument commands and MET time from the spacecraft over this port, and sends instrument science and housekeeping data to the spacecraft. Data sent to the spacecraft is normally recorded in the SSR of the respective IEM for latter downlink to the ground station.

The active IEM and PEPSSI exchange messages over the UART using the standard UART protocol to govern the lower-level aspects of the transfer. A higher-level construct, an Instrument Transfer Frame (ITF) protocol, is used for higher-level synchronization and error control. The IEM provides PEPSSI with command messages and spacecraft time messages. PEPSSI provides the IEM with TLM messages that contain instrument state data as well as science data. PEPSSI does not exchange messages with any other instrument and every message to or from PEPSSI is contained within an ITF.

In addition to Transmit (to spacecraft) and Receive (from spacecraft) signal lines, each port includes the 1PPS signal input to PEPSSI. This signal provides nominal one-second timing information; command and telemetry transfer is synchronized to 1PPS epochs as illustrated in Fig. 12.

Commands, instrument telemetry, and MET time are packed in the ITF format that is described below. This frame is defined by the time between the rising edges of any two adjacent 1PPS signals. Command and telemetry data frames are transferred as a serial 8-bit byte stream.

MET time is transferred from the spacecraft to PEPSSI over each one-second interval. MET time is transferred to 1-second precision and is valid when the 1PPS signal is asserted. MET time is transferred within 0 to 950 ms following each 1PPS.

PEPSSI telemetry data may be transferred anytime between 1PPS epochs except when the 1PPS rising edge occurs. All telemetry transmissions terminate 300 μs prior to the rising edge of the 1PPS. All three signals (Transmit, Receive, 1PPS) are implemented as complementary RS-422 electrical standard interfaces.

PEPSSI sends a number of different science and housekeeping data products to the spacecraft. These products are packed in CCSDS telemetry packet format. A number of CCSDS telemetry packets may be included in each telemetry ITF sent to the spacecraft.

3.3.5.3 Test Port Interface For early hardware and software development, PEPSSI implemented an RS-232 protocol test interface to the instrument processor. However this was intended only for early development and test and was not accessed at the spacecraft level of integration and test.

3.3.5.4 HVPS Safing The PEPSSI HVPS could be safely turned on to full level only when the instrument is in vacuum. Damage to the MCP may result if high voltage power is applied at atmospheric or partial atmospheric pressures. To prevent accidental turn-on of the HVPS, the PEPSSI instrument was designed to accept a safing plug connection—installation of the safing plug was required to disable HVPS operation. The safing plug was removed before launch to enable on-orbit HVPS operation.

3.3.5.5 PEPSSI Cover Release The PEPSSI design includes a cover for the instrument aperture. One purpose of this cover is to protect the instrument foils during ground vibration and acoustic shock tests. Prior to launch, the cover is opened only during T-V, calibration, and special cover tests; the cover remained closed at all other times. The instrument cover is opened by spacecraft command. The spacecraft command system and associated spacecraft ground support equipment (GSE) included implementation of an arming plug. The arming plug was removed during ground system test to prevent accidental opening of the cover and then installed before launch to enable on-orbit cover release operation.

The spacecraft provided a one-time activation signal to open the cover. An electrical non-explosive shape memory alloy (SMA) pin-puller is incorporated in the instrument and was used for the release. The SMA pin-puller employs redundant circuitry. Each circuit requires a current load of between 0.5 to 2.0 amps at 30 VDC. Depending upon the operational load selected, the spacecraft power distribution unit could apply this power for a minimum of 200 ms. The pin-puller employs an auto shut-off switch removing power after it has activated, approximately 100 ms after power is applied. The cover has been successfully opened, and this feature will no longer be used.

3.4 Telemetered Data Products

3.4.1 Proton and Electron Energy Spectra

In the energy spectra data for detected electron, proton, and heavy ion particles each energy bin corresponds to a particular particle energy range (defined in the software data base look-up Table 6), and the telemetered value represents the number of particles detected in the respective energy bin. The initial look-up table that defines the energy bin coordinates for each particle was determined and loaded during pre-launch instrument calibration.

Ion particle data is output for each of six sectors; electron data is output for each of three sectors. The integration interval, i.e., the time interval over which the data is collected, is nominally 10 seconds, but data integration interval may be changed by command from 1 second to 65,535 seconds.

For those sectors that include an electron detector (sectors 0, 2, 5), the number of ion events counted in each energy bin represents the number of ion particles impinging on the respective single ion detector. For those sectors that include ion detector pairs (sectors 1, 3, 4), the number of ion events counted in each energy bin represents the sum of ion particles impinging on the respective ion detector pair.

Table 6 Proton/electron/heavy ion energy spectra data products

Species	Number of energy bins	Bits per bin	Scaling	Number of sectors	Number of bits per integration interval
Electrons	3	10	Log compressed, 5-bit mantissa, 5-bit exponent, base 2	3	90
Protons	6	10		6	360
Heavy Ions (>1 AMU)	4	10		6	240
Total bits per integration interval→					690

Table 7 Heavy ion energy spectra

Species	Number of energy bins	Bits per bin	Scaling	Number of sectors	Number of bits per integration interval
CNO	6	10	Log compressed, 5-bit mantissa, 5-bit exponent, Base 2	6	360
N_2	3	10		6	180
Total bits per integration interval→					540

3.4.2 Heavy Ion Energy Spectra

The energy spectra data for heavy ions, output from each of six sectors, is conditional; it is output only if the particle event rate exceeds a predefined threshold (Table 7). The event rate threshold includes hysteresis, so that once heavy ion spectra data is output, the event rate must drop to a somewhat lower value for output to cease. The particle event rate threshold is programmable by instrument command and is typically set at 100 particle events per second. The data integration interval, nominally 10 seconds, is identical to the integration time interval for all other spectra type data. It may be adjusted by instrument command over a range of 1 to 65,535 seconds.

For those sectors that include an electron detector (sectors 0, 2, 5), the number of heavy ion events counted in each energy bin represents the number of ion particles impinging on the respective single ion detector. For those sectors that include ion detector pairs (sectors 1, 3, 4), the number of heavy ion events counted in each energy bin represents the number of ion particles impinging on the respective pair of SSD ion detectors.

3.4.3 TOF-Only Velocity Spectra

Particle velocity spectra data, derived from TOF-only measurements are comprised of three sets of particle velocity data corresponding to particle mass categorized as light, medium, or heavy (Table 8) The two stop-anode discriminators are employed to differentiate between light, medium, and heavy particles. The discriminator threshold settings normally are set so that light particles correspond to protons, medium particles correspond to the CNO group, and heavy particles correspond to particles with mass greater than the CNO group. Similar to heavy ion spectra, particle velocity spectra data is output only if the particle event rate exceeds a predefined threshold. As with the other products just mentioned, the data integration interval, nominally 10 seconds, is identical to the integration time interval for all

Table 8 TOF-only velocity spectra

Mass category	Number of spectral bins	Bits per bin	Scaling	Range	Number of sectors	Number of bits per integration interval
Light	5	10	Units = km/s	200 km/s	6	300
Medium	5	10	Log compressed, 5-bit mantissa,	to 2000 km/s spanning	6	300
Heavy	5	10	5-bit exponent, base 2	5 bins	6	300
Total bits per integration interval→						900

Table 9 Singles event data (event validity check)

Event description	Number of event counters	Bits per counter	Number of count bits	Scaling
SSD Leading Edge Discriminator —Ion Event	9	10	90	Log compressed, 5-bit mantissa, 5-bit exponent, base 2
SSD Leading Edge Discriminator —Electron Event	3	10	30	
SSD Peak Detector—Ion Event	9	10	90	
SSD Peak Detector—Electron Event	3	10	30	
Start Anode Discriminator Event	6	10	60	
CFD Start Event	1	10	10	
CFD Stop Event	1	10	10	Log compressed, 5-bit mantissa, 5-bit exponent, base 2
Valid TOF Events	1	10	10	
Ion Events	1	10	10	
Electron Events	1	10	10	
Software Counters	3	10	30	
Total data bit count per integration interval →			380	

other spectra type data. It may be adjusted by instrument command over a range of 1 to 65,535 seconds.

3.4.4 Singles-Event Data (for Event Validity Check)

To allow validation of instrument science data, PEPSSI counts and telemeters a number of instrument single events (Table 9). Single-event data is collected over the same identical data integration intervals as ion-energy spectra and velocity spectra data. It is nominally 10 seconds, but may be adjusted by instrument command over a range of 1 to 65,535 seconds.

3.4.5 PHA (Pulse Height Analysis) Event Data

Detailed particle event data are retained for a sub-set of the total number of events due to data volume limitations on the SSR (Table 10). A rotating priority scheme is used to save the most important data and done for each of the data types using integration intervals,

Table 10 PHA event data

Data item	Description	Number of bits
TOF [units = ns]	Particle TOF Measurement. All zero fill for energy-only events. Log Compressed Scaling	8
Energy [units = keV]	Particle energy measurement. All zero fill for TOF-only events. Log Compressed Scaling	8
Ion sector	SSD channel/sector identifier, 9 channels	4
Start sector	Identifies the sector where the TOF start anode signal is detected	3
Heavy Discriminator	TOF Stop Anode Discriminator States (Flags). Identifies event particle as light, medium, or heavy	2
Quality flag	Set if multiple-ion hits detected	1
Reserved	Reserved	6
Bits/PHA event→		32
Bits/40 PHA events →		1280

N1, N2, and N3. A separate buffer is maintained for the PHA data that is output in the high-, medium-, and low-telemetry packets. The maximum number of PHA events saved per integration period is 10 for N1 data, 20 for N2 data, and 800 for the N3 data. PHA events are distributed among the buffers in a round-robin fashion: the first detected event is stored into the high-priority (N1) packet buffer, the next event is stored into the medium-priority (N2) packet buffer, and the next event is stored into the low-priority (N3) packet buffer, etc. Each event allocated to a particular buffer is simply stored into the next slot within the buffer until the buffer fills up. Thereafter, the rotating priority PHA replacement scheme is used in deciding which events may be displaced from the filled buffer.

All PHA events are assigned to a priority group based on the event type 1, 2, or 3 (Energy-only or "electron", TOF-plus energy or "high-energy ion", and TOF-only or "low-energy ion," respectively) and the bin number, determined from the appropriate lookup table, for the event. Eight such priority groups have been defined for the PEPSSI system with an 8-element priority array maintained for each event type. Group numbers are held in the array elements in priority order, i.e., the first element contains the group number of the highest priority group, the second element contains the group number of the next lower priority group, and so on. The priority of a particular group can be determined by finding which array element contains that group number.

As PHA events are stored into their respective buffers, the assigned priority and group numbers are stored along with each event. As long as each buffer contains a vacant slot, the next detected event is simply stored into the vacant slot. Once the buffer becomes full, the assigned PHA group priority for a new event will determine its relative ability to displace events that are already in the buffer. For the data in the N1 and N2 telemetry packets, the lowest priority event already in the buffer is identified and displaced by the new event, if the priority of the new event is higher.

A record is maintained of the frequency of occurrence of each group during each integration interval. The most frequently occurring group number is used at the conclusion of the interval to "rotate" the PHA-priorities by pushing the corresponding most frequently occurring group to the bottom of the priority list. This rotated priority list is then used in determining the priority of PHA events in the next integration interval. For the N3 data with up to 800 PHA events, a modified scheme is used. Operationally, the N3 integration rate is set to a long interval (2 hours) to minimize the N3 data, which we are not relying on for PHA data.

Table 11 PEPSSI non-packetized housekeeping data

Parameter	Name	No. of bits	Units
HV Input Current Monitor	CDH_PEPSSI_HV_CURR	8	MICROAMPS
HV Output Voltage Monitor	CDH_PEPSSI_HV_VOLT	8	VOLTS
Bias Input Current Monitor	CDH_PEPSSI_BIAS_CURR	8	MICROAMPS
Bias Output Voltage Monitor	CDH_PEPSSI_BIAS_VOLT	8	VOLTS
Energy Board Temperature	CDH_PEPSSI_ENE_BD_TEMP	8	DEGREES C
HVPS Board Temperature	CDH_PEPSSI_HVPS_BD_TEMP	8	DEGREES C
LVPS Board Temperature	CDH_PEPSSI_LVPS_BD_TEMP	8	DEGREES C
Total bits		56	

3.4.6 Non-Packetized Housekeeping Data

PEPSSI telemeters non-packetized housekeeping data. These include voltage and current measurements that are monitored during activation of the instrument energy measurement system and TOF measurement system. PEPSSI non-packetized housekeeping data are placed in the instrument-defined status field of Instrument Transfer Frames sent from PEPSSI to the spacecraft C&DH system. Housekeeping data is updated at a regular 1-second rate. The total number of bits in each housekeeping frame is 56 bits, resulting in an average output data rate of 56 bps (Table 11).

3.4.7 Quick look Diagnostic Data

For a quick look analysis of instrument status and measurement results, PEPSSI can output a diagnostic set of data at a fixed 300-second rate. It is intended that PEPSSI diagnostic data be downlinked as soon as possible following the near Pluto encounter, i.e., on a first-look basis. This will allow for examination of instrument performance in the quickest and most timely manner. It is estimated, over a 24-hour timeline, roughly the time duration of the near-Pluto encounter, about 480 kbits of PEPSSI quick-look data will be available for downlink.

The diagnostic data is identical to the electron/proton/heavy-ion energy spectra and rate-counter data previously defined, except that it is collected over 300-second integration time intervals. In addition, 600 bits of extended housekeeping data are included. The diagnostic integration time interval is constant at 300 seconds, and the subsequent average output data rate is less than 6 bits per second.

3.4.8 Instrument Data Rate Summary

The telemetry system of the PEPSSI instrument may be configured in a number of different ways by instrument command. The data integration interval for a number of products may be adjusted, heavy ion and TOF velocity data products may be enabled or disabled, and the maximum PHA event rate may be changed. The volume of data telemetered from PEPSSI, and the resulting average data rates, will vary accordingly.

Table 12 lists PEPSSI data products and computes telemetry data rates for a number of representative telemetry system configurations. The first four products listed (proton/electron energy spectra, heavy-ion spectra, TOF velocity spectra, and rate-counter data)

Table 12 Example PEPSSI telemetry data rates as function of telemetry configuration

Telemetry data product	Number of data bits	Data integration interval	Example bit rate (bps) In-space, low-rate configurations 300-second data interval	10-second data interval	In-space, typical configurations 300-second data interval	10-second data interval	Ground test only 1-second data interval
Proton/Electron Energy Spectra	690	10 seconds typical Programmable 1 to 7200 seconds	2.3	69	2.3	69	690
Heavy Ion (CNO, N2) Energy Spectra[a]	540		1.8	Disabled	1.8	54	540
TOF-Only Velocity Spectra[a]	900		3.0	Disabled	3.0	90	900
Singles Events (Rate Counters)	380		1.3	38	1.3	38	380
PHA	32 per PHA event	10 second (fixed)					
Event Rate ≤40 events/s					≤1280	≤1280	≤1280
Event Rate ≤10 events/s			≤320	≤320			
"Non-Packetized" Housekeep Data[b]	64	1 second (fixed)	64	64	64	64	64
Quicklook Diagnostic	1670	300 seconds (fixed)	5.6	5.6	5.6	5.6	5.6
Proton/Electron Spectra							
Rate Counters							
Extended Housekeep							
CCSDS Frame Overhead			8.5	16.5	8.5	16.5	88.3
Average total data bit rate, bps →			≤407	≤513	≤1367	≤1617	≤3948

[a]Heavy ion spectra and TOF velocity spectra outputs are conditional. They are output only if the particle event rate exceeds a predefined threshold. The particle event rate threshold is programmable by instrument command. The output may also be disabled by instrument command regardless of particle event rate

[b]Non-Packetized housekeeping data is packed in the Instrument Transfer Frame instrument status field and is intended for inclusion in spacecraft telemetry packets

represent data collected over a data integration time interval that is changeable and may be set by instrument command. The remaining data products (PHA data, non-packetized housekeeping, and quick look diagnostic data) are collected and output at fixed rates.

The data integration interval, for the first four products listed, is nominally set to 10 seconds and is never set any lower during in-space operation. For anticipated in-space operations, where the heavy-ion spectra and TOF velocity spectra are enabled, and the PHA limit is set to 40 PHA events per second, the maximum output data rate is about 1620 bits per second. For certain ground test operations, prior to spacecraft integration, where data is integrated over short, one-second time intervals, the output data rate averages about 3950

Table 13 Average data rates and volumes

Estimates of nominal PEPSSI telemetry data rates and volumes (prior to compression by Fast algorithm). (The 152 bit ITF overhead is not included)

Packet	Bit length (including CCSDS header)	Nominal output rate (seconds)	Average data rate (bits/second)	24-hour data volume (Megabits)
High priority packet	3640	300	$\approx$12	$\approx$1.0
Medium priority packet	2464	10	$\approx$250	$\approx$21
Low priority packet (first)	30,720[a]	300	$\approx$100	$\approx$8.8
Low priority packet (second)	30,720[a]	300	$\approx$100	$\approx$8.8
Common status packet	456	300	$\approx$1.5	$\approx$0.13
Average bit rate			$\approx$720	

[a]Assumes a maximal length CCSDS packet of 4096 data bytes

bits per second. If required by mission operation or spacecraft constraints, as illustrated, the instrument telemetry system may be configured to telemeter less data.

The PEPSSI instrument outputs four types of telemetry packets on regular time intervals:

- High Priority (N1)
- Medium Priority (N2)
- Low Priority (N3), and
- Common Status ("housekeeping")

The time intervals may be modified by instrument command. Table 13 lists the packet types and provides pre-launch estimates of average data output rates and 24-hour data storage volume for the Pluto encounter. The data rate and volume figures are raw values and do not include reductions due to use of the Fast data compression algorithm. The amount of reduction, in data volume and rate, depends upon the variableness of the raw data stream and therefore, is not fixed. It is anticipated the reduction of data volumes and rate will be 50% or more; therefore, the actual output data rates at the PEPSSI command and telemetry interface depend upon mission phase and ongoing planning. The data rates and volumes given in the table should be considered upper bounds.

The table also indicates that the 24-hour volume for High Priority data is on the order of 1 Mbit. However, the Fast compression algorithm can reduce this volume by up to 50% leaving the remaining volume less than 2% of the spacecraft 24-hour 'first-look', day one downlink capability following the Pluto encounter.

The instrument is quite flexible; for example, the actual bit rate at Jupiter was ~60 bps.

3.4.9 Calibration

For instrument calibration prior to launch, PEPSSI test software included a capability to read out raw instrument measurements. Raw measurement data was evaluated in order to configure the look-up calibration table that was subsequently loaded into the instrument software database.

Table 14 PEPSSI energy measurement system commands

Command description	LED threshold	
	Granularity	Range
Energy discriminator threshold commands		
Set leading edge discriminator (LED) threshold [Chan 1: Chan 12]	≈1 keV	≈15 to 255 keV

3.4.10 Memory Image Dump

PEPSSI implements a capability to load selected portions of processor memory, e.g., look-up tables for instrument calibration. The PEPSSI instrument includes a capability to telemeter an image of the newly loaded segment of memory, to allow load verification in the spacecraft command center.

3.5 Commands

PEPSSI uses a variety of commands to provide functionality to the instrument and set various parameters that affect how the data is collected and processed before being sent to the ground.

3.5.1 Energy Commands

Commands related to energy measurement are given in Table 14. These include commands to adjust energy detection thresholds of leading edge discriminators. The leading edge discriminator thresholds operate in current mode so that a threshold current essentially corresponds to an energy threshold.

3.5.2 HVPS Commands

High-voltage PEPSSI operation requires a vacuum environment. At power turn-on, the HVPS is disabled by default. PEPSSI includes a set of HVPS and bias-voltage-supply commands to enable high-voltage operation, adjust high-voltage power supply clocks for optimum efficiency, and fine-tune the high-voltage outputs. A summary of these commands is given in Table 15.

3.5.3 TOF Commands

In concert with energy threshold commands, PEPSSI implements a set of commands to adjust the thresholds of the start and stop constant fraction discriminators (CFD). With the threshold set to zero, about 5×10^5 electrons are required to generate a start or stop pulse respectively (and thus initiate a TOF measurement). The threshold can be adjusted upward, requiring more electrons and proportionally higher energy levels to initiate a TOF measurement event. A summary TOF command list is given in Table 16.

3.5.4 Process Control Commands

PEPSSI implements a number of commands to manage instrument operations, including energy measurement integration times, enable/disable event pile-up checks, etc. A summary list follows in Table 17.

Table 15 PEPSSI HVPS commands

Description	Scaling Resolution	Range
High voltage power commands		
Enable/disable HVPS commands (safety feature)	NA	NA
Enable/disable HVPS & bias supply clocks	NA	NA
Adjust high voltage clock rate	≈750 Hz	≈75 kHz ± 20%
Set high voltage command limit	≈ − 15 V	≈ − 3500 V
Set high voltage level	≈ − 15 V	≈ − 3500 V
Set high voltage alarm level	≈ − 15 V	≈ − 3500 V
Enable/disable high voltage alarm	NA	NA
Bias voltage power commands		
Set bias voltage level	≈ − 0.5 V	≈ − 125 V
Adjust bias voltage clock rate	≈350 Hz	≈47.25 kHz ± 20%

Table 16 PEPSSI TOF measurement system commands

TOF measurement system commands		
	Units scaling	
Description	Resolution	Range
Start anode leading edge discriminator thresholds		
Set TOF sector threshold [Sector 1:Sector 6]		
Stop anode leading edge discriminator thresholds		
Set heavy 0 threshold		
Set heavy 1 threshold	≈10 mV	≈0 to 2500 mV
TOF constant fractional discriminators		
Set CFD start threshold		
Set CFD stop threshold		
Anode enable/disable		
Enable/disable start anode [anode 1: anode 6]	NA	NA
Enable/disable stop anode [anode 1: anode 6]	NA	NA
Miscellaneous		
Set TOF chip mode	NA	NA
Reset TOF chip	NA	NA

3.6 Telemetry and Command Format

The C&DH system receives all telemetry from PEPSSI as non-packetized critical housekeeping data and as CCSDS packets. The CCSDS formatted packets can be common packets with standard formats, common packets with PEPSSI-specific formats, and PEPSSI-specific packets.

Table 17 PEPSSI process control commands

Description	Scaling Resolution	Range
Measurement process control		
Load look-up tables	NA	NA
Load PHA rotating priority structure	NA	NA
Set integration time for energy, TOF spectra (seconds)	1	65,535
Set maximum telemetered PHA event rate (events per second)	1	1 to 100
Set event rate threshold to enable telemetry of heavy ion & velocity spectra (events per second)	1	0 to 1000 (0 ⇒ enable unconditionally)
Enable/disable telemetry of heavy ion & velocity spectra	NA	NA
Set coincidence window delay	Determined during design phase	
Set coincidence window width		
Mask SSD discriminator firings for selected channel	NA	NA
Enable/disable electron multiple-hit check	NA	NA
Enable/disable ion multiple-hit check	NA	NA
Miscellaneous		
Reset event FIFO	NA	NA
NO-Op command	NA	NA
Memory load command	NA	NA
Memory dump command	NA	NA

3.6.1 Data Rate and Volume

As just noted, the raw data rate, with the instrument telemetry system configured in the preferred manner for space operations, is a maximum of about 2500 bps, representing a total raw data volume collected over 24 hours of about 200 Mega-bits. These are raw figures that apply prior to application of the Fast data compression, which may reduce data volume by up to one-half. If necessary, the instrument telemetry system may be configured for lower data rates.

Telemetry data sent from the PEPSSI instrument is compressed using the Fast algorithm. Fast is an APL-developed data-compression algorithm that is imbedded in instrument common code used by a number of MESSENGER and New Horizon instruments. Fast is a lossless compression algorithm; the reduction of data volume is variable, data volume may be reduced by up to one-half.

3.6.2 Telemetry Formatting

3.6.2.1 Instrument Transfer Frame (ITF) All messages exchanged between the spacecraft C&DH and PEPSSI are transported using the ITF format. The format applies to all telemetry data sent from the instrument and to commands and MET time received from the spacecraft. The 48-bit frame header associated with this format is outlined in Table 18.

Byte 0 is first byte transmitted/received by PEPSSI. The instrument resynchronizes to the incoming sync pattern on each ITF to limit error propagation in case of bit errors. An ITF contains a single message, (S/C time, Command, or Telemetry) and will have variable

Table 18 Instrument transfer frame format

Major field	Field	Byte #	Description
ITF Header	sync pattern 1	0	$0 \times$ fe
	sync pattern 2	1	$0 \times$ fa
	sync pattern 3	2	0×30
	Message type	3	S/C TIME, COMMAND, or TELEMETRY
	Checksum	4	Byte-wise exclusive-OR of all bytes following the checksum
	Message length	5–6	Number of bytes following the length field
Message	Message data	7 to $N-1$	*New Horizons* formatted S/C time, command, or instrument state/telemetry message. $N =$ length of ITF in bytes. max $N = 12$ for S/C time message max $N = 279$ for Command, max $N = 1250$ for Telemetry

length depending on the length of the contained message. The maximum length for command messages is selected to accommodate memory load commands. The maximum length for telemetry messages is selected to provide adequate margin for the one-second timing requirements for telemetry transfer. S/C time messages are fixed length. The time messages contain information in addition to time that vary from instrument to instrument. The ITF header includes a checksum (byte 4) on the data that follows. In the case of Command and S/C Time messages transferred to PEPSSI, the C&DH computes and adds the checksum to the ITF when the telecommand packet from the spacecraft is parsed into PEPSSI commands. PEPSSI computes and adds the checksum to all telemetry messages sent to the C&DH. PEPSSI does not receive or send messages to any other instrument or subsystem other than the C&DH.

3.6.2.2 Telemetry Interface Once per second, the PEPSSI instrument formats a telemetry message containing data in an ITF for transfer to the C&DH. This frame is defined by the time between the rising edges of any two adjacent 1PPS signals. The message includes one or more CCSDS-formatted telemetry packets. The format is shown in Table 19.

3.6.2.3 CCSDS Packetization The header of the CCSDS-formatted packet is defined in Table 20. The length of the packet is variable and is specified in the CCSDS header. The data field of each CCSDS telemetry packet is compressed using the lossless Fast algorithm.

The application process id (ApID) in the primary header identifies the type of packet. The eleven-bit id is divided into a four-bit source and a seven-bit data. This gives an instrument up to 128 possible packet types that can be produced. The four-bit source for the PEPSSI instrument is shown in the table below. For memory dump packets, an instrument should set the lower seven bits of the ApID to a value of 1 (0000001). All other ApID assignments are at the discretion of the instrument. The mapping of PEPSSI instrument data products to specific application process identification (ApID) packets is specified in the PEPSSI software specifications document.

3.6.2.4 Memory Dump Packet New Horizons uses a single memory dump telemetry format for all onboard processors. The format of the telemetry packet is specified in Table 21.

Table 19 Standard telemetry instrument transfer frame format

Name	Length (bits)	Value	Description
Sync pattern 1	8	$0 \times$ fe	
Sync pattern 2	8	$0 \times$ fa	
Sync pattern 3	8	0×30	
Message type	8	0×04	Telemetry
Checksum	8	$\rightarrow$	Byte-wise exclusive-OR of all bytes following the checksum
Message length remaining	16	$\rightarrow$	Number of bytes following the message length field
Heartbeat	1	$\rightarrow$	Alternating bit
Boot/application (optional)	1	$0 =$ Running boot code $1 =$ Running appl. code	Flag describes if boot or application code is running
Turn-off request (optional)	1	$0 =$ No turn-off requested $1 =$ Turn-off requested	Flags describes if instrument requests spacecraft to turn off power or reset
Reserved standardized instrument status	5	$\rightarrow$	Other bits not yet assigned.
Valid command count	8	$\rightarrow$	Module 2^8 count of commands accepted
Invalid command count	8	$\rightarrow$	Module 2^8 count of commands rejected
Instrument defined status	$8 \cdot N$	N/A	Instrument non-packetized status or housekeeping. The C&DH processor will place this data in spacecraft-produced telemetry packets, and will use it in the evaluation of autonomy rules.
Offset to first packet header	16	$0 \times 0000 =$ A packet begins in first byte of telemetry packets field $0 \times$ FFFE $=$ There is no telemetry data in this frame $0 \times$ FFFF $=$ There is telemetry data in this ITF, but no packet begins in this frame	Offset within telemetry data field to start of first data packet header in frame. If an instrument never has more telemetry than can fit in a single ITF, this field will always be 0.
Telemetry data	$8 \cdot M$	$\rightarrow$	New Horizons formatted CCSDS telemetry packets

3.6.3 Command Formatting

3.6.3.1 Command Message The format of a command ITF is defined in Table 22. The command message may be variable in length and length is specified in the header. One command is included in each ITF; however multiple command ITFs may be received from

Table 20 CCSDS telemetry packet header format

Item	Bit Field	Length (Bits)	Value (Binary)	Description
Primary header	Version number	3	000	Designates a source packet
	Type indicator	1	0	Designates a telemetry packet
	Secondary header flag	1	1	Secondary header flag present
	Application process identifier	11	1101 xxxxxxx	PEPPSI packet Identification
	Grouping flags	2	01	First packet in group
			00	Intermediate packet
			10	Last packet in group
			11	Not part of group
	Source count	14		Continuous sequence count (mod 16384) of packets for specific ApID
	Packet length	16		Number of bytes in [packet secondary header + packet data field] − 1
Secondary header	MET	32		Spacecraft MET at time the source Packet is constructed
Data		$8 \cdot M$		Additional data bytes, M, can range from 1 to TBD

the C&DH in any one-second frame. For command verification, command acceptance and command rejection counts are included in the PEPSSI telemetry ITF.

3.6.3.2 Command Format The format of the message data within a command ITF is specified in Table 23.

This format is identical to the NH Spacecraft formatted commands and allows for multiple commands to be packed into a single CCSDS packet. However, only a single instrument command will be sent via ITF at a time.

3.6.3.3 Memory Load Command New Horizons has adopted a single memory load command format for all onboard processors. The format of the command is specified in Table 24.

3.6.3.4 Memory Dump Command The format for the New Horizons memory dump command is specified in Table 25.

3.6.4 MET Time Message

An MET time message is sent from the C&DH to the instrument once per second. The MET time is the predicted time of the next 1PPS. The MET ITF is defined in Table 26. The MET ITF is transferred within 950 ms following each 1PPS epoch.

Table 21 Memory dump telemetry packet format

Item	Length (bits)	Value	Description
Packet primary header	48		For memory dump packets, the lower seven bits of the 11-bit ApID should be binary '000 0001'.
Packet secondary header	32		MET
Address	32		Start Address
Byte count	16	1–256	Number of bytes in this dump packet (not including pad) New Horizons has chosen a maximum dump size of 256 bytes.
Memory type	8	TBD	This field is available for memory architectures that cannot determine the memory type from the address. This field is spare (0×00) for architectures that can.
Spare	8	0	Unused (0×00)
Data	$N \cdot 8$		$1 \leq N \leq 256$ (From 1 to 256 8-bit bytes)
Pad	$M \cdot 8$	0	$0 \leq M \leq 3$ Pad up to 32-Bit boundary.

Table 22 Command message in instrument transfer frame

Name	Byte #	Value	Description
sync pattern 1	0	$0 \times$ fe	$0 \times$ fe
sync pattern 2	1	$0 \times$ fa	$0 \times$ fa
sync pattern 3	2	0×30	0×30
Message type	3	0×02	*COMMAND*
Checksum	4		*byte-wise exclusive-OR of all bytes following the checksum*
Message length	5–6		*number of bytes following the length field (field sized for length <= TBD bytes; messages do NOT span this data structure)*
Message data	7 to N		*New Horizons formatted instrument command max N is TBD*

Table 23 Message data

Name	Length (bits)	Value	Description
Opcode	16		Instrument splits this into 2 8-bit fields (dest, opcode).
Macro field	2	00 = Execute real-time	Instrument use only.
		01 = Invalid	
		10 = Add to Inst. Macro 11 = Add to DPU Macro	Unused or spare (0 × 00) for non-instrument commands.
Command word count	14	$N+2$	Number of 32-bit command words $(N)+2$
Command words	$32 \cdot N$		
Checksum	32		32-bit XOR of the fields defined above

Table 24 Memory load command message

Item	Length (bits)	Value	Description
Opcode	16		
Macro Field	2	→	00 = Execute real-time
			01 = Invalid
			10 = Add to Inst. Macro
			11 = Add to DPU Macro
			Unused (00) for non-APL instruments
Length	14	5–68	Command length in 32-bit words $(4+(N+M)/4)$.
Address	32	→	Start address
Byte Count	16	1–256	Number of bytes to be loaded (not including pad). *New Horizons* has chosen a maximum load size of 256 bytes.
Memory Type	8	TBD	This field is available for memory architectures that cannot determine the memory type from the address. This field is spare (0 × 00) for architectures that can.
Spare	8	0	Unused (0 × 00)
Data	$N \cdot 8$		$1 \leq N \leq 256$ (From 1 to 256 8-bit bytes)
Pad	$M \cdot 8$	0	$0 \leq M \leq 3$
			Pad up to 32-bit boundary. This data will not be loaded but will be included in XOR calculation.
Checksum	32	→	32-bit XOR

Table 25 Memory dump command message

Item	Length (bits)	Value	Description
Opcode	16	→	
Macro Field	2	→	00 = Execute real-time 01 = Invalid 10 = Add to Inst. Macro 11 = Add to DPU Macro Unused (00) for non-APL instruments
Length	14	5	Command length in 32-bit words
Address	32	→	Start Address
Byte Count	32	→	Number of bytes to be dumped.
Memory Type	8	TBD	This field is available for memory architectures that cannot determine the memory type from the address. This field is spare (0 × 00) for architectures that can.
Spare	24	0	Unused (0 × 00)
Checksum	32	→	32-bit XOR

Table 26 C & DH-to-instrument spacecraft time message in instrument transfer frame

Name	Length (Bits)	Value	Description
sync pattern 1	8	0 × fe	
sync pattern 2	8	0 × fa	
sync pattern 3	8	0 × 30	
Message type	8	0 × 01	S/C-Time
Checksum	8		Byte-wise exclusive-OR of all bytes following the checksum
Message length	16	$0 \times 0004 + N$	Expressed in number of bytes of time information plus any additional information sent to that instrument.
S/C-Time	32	MET	Mission elapsed time corresponding to the predicted S/C time at the next SYNC pulse
Additional information sent to the instrument	$8 \cdot N$	N/A	Additional information sent to an instrument can include a closest approach countdown, whether that instrument is allowed to produce a memory dump packet in the next second, scan rate, etc. This field will be customized for PEPSSI, and will be fixed from one second to the next.

3.7 Instrument Environmental Design Requirements

3.7.1 Thermal Interface

The PEPSSI-to-spacecraft thermal interface includes a conduction component via the spacecraft-mounting surface and a radiation component derived by a ratio between blanketed and painted surfaces. Two spacecraft-monitored temperature sensors are continuously monitored via the spacecraft Command and Data Handling (C&DH) system. These sensors are used to ascertain the temperature of the instrument. The spacecraft thermal control system maintains the temperature on the spacecraft side of the PEPSSI mounting plate and the support electronics box mounting points. The spacecraft has no specific requirement for control of the temperature gradient across the mounting interface from any one point to another. The spacecraft also has no specific requirement for control of the interface temperature rate of change. Internal temperature measurements for PEPSSI (two in the instrument and one on the mounting bracket) are included in PEPSSI's housekeeping telemetry data.

3.7.2 PEPSSI Thermal Design Requirements

Design and test limits give the temperature ranges over which PEPSSI was cycled during thermal vacuum testing. The spacecraft interface temperatures include: Operating/Test temperatures 0°C to +40°C, Non-Operating Survival −20°C to +40°C, and Internal Operating Limits with baseplate control for the instrument itself with −45°C to +45°C for the electronics boards and −35°C to +35°C for the solid state detectors. The non-operating survival limits assume the PEPSSI cruise heater is not on. The heater may be turned on during the cruise phase when PEPSSI is off and when observatory power is available to provide margin with respect to the non-operating −20°C limit.

3.7.3 Radiation Shielding Requirements

The New Horizons radiation environment includes exposure to solar protons, Jovian electrons, galactic cosmic rays, and neutrons and gamma rays emitted from the radioisotope thermoelectric generator (RTG) power source. PEPSSI is designed to survive a total ionizing dose of 15 kilorads without failure and to be immune to latch-up. Additional spot shielding to parts was added as required to bring the instrument to this overall shielding level.

3.7.4 Electrostatic Requirements

The electrostatic requirement for PEPSSI is that the observatory's total potential be less than 1000 volts. Analysis by the New Horizons project for the Jupiter flyby showed a worst case ($35 R_J$ closest approach) potential of $\sim +5$ to $+10$ volts on isolated solar-illuminated surfaces, and ~ -200 volts on isolated surfaces in darkness. There is no hard information of the plasma environment of Pluto, but it is not expected to possess a strong intrinsic magnetic field and, therefore, is not expected to have a mechanism to produce a population of particles sufficiently energetic to charge the observatory to a level of concern. The worst Jupiter case is less than the maximum allowable for PEPSSI by a margin of five; and the electrostatic charge at Pluto is expected to be substantially less.

Table 27 PEPSSI thermal test limits

Condition	Limits
Typical	+30°C to +40°C
Operating	0°C to +40°C
Operating test	−10°C to +50°C
Non-operating survival	−30°C to +50°C (Includes 10°C margin)

3.8 Test Requirements

3.8.1 Vibration

Instrument test vibration levels are specified in project documentation for levels appropriate to the launch vehicle stack used for New Horizons. Testing philosophy was based on a protoflight approach, where flight hardware is tested to qualification levels for flight acceptance durations. Tests included sine vibrations test at levels applicable at the spacecraft-mounting interface, with vibration test of three orthogonal axes, including the thrust axis, performed, as well as random vibration tests with the same geometry.

3.8.2 Thermal

The PEPSSI instrument is mounted on a bracket on a side deck of the spacecraft and is thermally conductive through the mounting interface. The instrument temperature is controlled by a tailored thermal coupling at the mounting interface and use of instrument radiating surfaces. No heaters are required. The ranges of mounting interface temperatures, at the bracket-to-spacecraft deck interface, are given in Table 27.

3.8.3 Acoustics and Shock

All instruments, including PEPSSI, were mounted on the spacecraft during separation shock and acoustic test. In addition, acoustic tests were performed on the PEPSSI engineering model instrument, using protoflight acoustic levels, to verify the integrity of the design with respect to acoustic noise. No instrument level acoustic test was carried out for the flight PEPSSI hardware.

3.8.4 EMI/EMC

PEPSSI electromagnetic interference (EMI) and electromagnetic contamination (EMC) level requirements are set forth in the New Horizons project documents. Full conducted emissions and conducted susceptibility as well as radiated emissions, radiated susceptibility, and transients measure tests were run prior to integration with the spacecraft.

4 Performance

4.1 Data Conversion to Physical Units

PEPSSI measurements are intended to generate the information needed to derive the charged ion and electron differential intensities (**I** [$cm^{-2}\ sr^{-1}\ s^{-1}\ keV^{-1}$]) and phase space densities (**PSD** [$s^3\ cm^{-6}$]). The significance of **I** and **PSD** are that, for many particle transport processes within planetary magnetospheres and interplanetary space, they satisfy the collisionless Boltzmann equation, derived from Liouville's Theorem for collision-less transport, for energy preserving, and for non-energy preserving processes, respectively. Our purpose here is to develop the quantitative procedures for converting the count rates (**R** [counts s^{-1}]) reported by PEPSSI in the many predefined channels described in preceding sections, or in channels defined on the ground using the pulse-height-analysis (PHA) data, into estimates of **I** and **PSD** for the various defined ranges of energies, particle species, and arrival angles. The approach follows classical procedures.

4.1.1 Flux, Differential Intensity and Phase Space Density

The rate **R** [particles/s] that traverse an area A can be given by the Flux **F** [$\mathrm{cm}^{-2}\,\mathrm{s}^{-1}$] with:

$$\mathbf{R} = A\mathbf{F} \tag{1}$$

or by the intensity **i** [$\mathrm{cm}^2\,\mathrm{s}^{-1}\,\mathrm{sr}^{-1}$]

$$\mathbf{R} = A\int \mathbf{i}\cos(\hat{u})d\Omega \tag{2}$$

where Ω is solid angle and $\hat{u}$ is the angle to the area normal.

Often used is the quantity differential intensity **I** [$\mathrm{cm}^2\,\mathrm{s}^{-1}\,\mathrm{sr}^{-1}\,\mathrm{keV}^{-1}$], defined as the number of particles of a given species S with energy between E and $E+\Delta E$ that traverse the area A during the time t, where

$$\mathbf{R}(S,E) = \mathbf{I}(S,E)A\Delta\Omega\Delta E \tag{3}$$

In three dimensions, with θ being the polar angle and ϕ the azimuthal angle of a polar reference system:

$$d^3\mathbf{R} = \mathbf{I}(S,E,\theta,\phi)A(\theta,\phi)\cos(\hat{u})tdE\cos\theta d\theta d\phi \tag{4}$$

where we note that the area A and the unit vector $\hat{u}$ pointing normal to A are both functions of θ and ϕ.

Note that for non-relativistic energies, $\mathbf{I}(S,E,\theta,\phi)$ is related to the phase space density **PSD** (number of particles in the configuration space element d^3R and with velocity between v and $v+d^3v$) by the simple relationship

$$\mathbf{PSD}\ [\mathrm{s}^3\,\mathrm{cm}^{-6}] = \mathbf{I}\ [\mathrm{cm}^2\,\mathrm{s}^{-1}\,\mathrm{sr}^{-1}\,\mathrm{keV}^{-1}]m/v^2 \tag{5}$$

For relativistic energies, the phase space density (identified with lower case letters here: **psd**) is defined in terms of particle momentum **p** (**psd** [$\mathrm{s}^3\,\mathrm{cm}^{-6}\,\mathrm{kg}\,\mathrm{m}^{-3}$]), and the relationship to differential intensity **I** is:

$$\mathbf{psd}\ [\mathrm{s}^3\,\mathrm{cm}^{-6}\,\mathrm{kg}\,\mathrm{m}^{-3}] = \mathbf{I}\ [\mathrm{cm}^2\,\mathrm{s}^{-1}\,\mathrm{sr}^{-1}\,\mathrm{keV}^{-1}]/\mathbf{p}^2 \tag{6}$$

4.1.2 Definition of Sensor Transfer Function and Geometric Factor

Using (4), the rate **R** [particles/s] of particles of species S, in the energy band ΔE around mean energy E, angular band $\Delta\theta$ around mean polar direction θ, and angular band $\Delta\phi$ around the mean azimuthal direction ϕ, measured by the instrument can be expressed as:

$$\mathbf{R}(S,E) = \int_{\Delta E}\int_{d\theta}\int_{d\phi}\mathbf{I}(S,E,\theta,\phi)A(\theta,\phi)\cos(\hat{u})dE\cos(\theta)d\theta d\phi \tag{7}$$

One may take the kernel approach to relating **R** to **I** by allowing $\mathbf{I}(S, E, \theta, \phi)$ to be a Dirac δ-function (mono-energetic, infinitely narrow beam). Then:

$$d\mathbf{R}(S, E, \theta, \phi) = d\mathbf{I}(S, E, \theta, \phi)\mathbf{T}(S, E, \theta, \phi) \tag{8}$$

where $\mathbf{T}(S, E, \theta, \phi)$ [cm^2 sr keV] is the kernel, also known as the "transfer function" of the instrument. Then:

$$\mathbf{R}(S, E) = \int_{\Delta E} \int_{d\theta} \int_{d\phi} \mathbf{I}(S, E, \theta, \phi)\mathbf{T}(S, E, \theta, \phi) \cos(\theta) d\theta d\phi \tag{9}$$

It is typical for the first estimate to assume that the particle intensities are constant over the energy-species-angle bandwidths of any one channel. The estimated intensity $\mathbf{I}_{kj}$ associated with a specific detector **k**, which at a specific time views a range of angles centered in a specific direction (θ_k, ϕ_k), and which is associated with a specific energy channel **j**, which measures particle species $\mathbf{S}_j$ over an energy range that stretches from E_j to $E_j + \Delta E_j$, is given by (9) as:

$$\mathbf{R}_{kj} = \int_{\Delta Ej} R_k(S_j, E_j) dE = \int_{\Delta Ej} \mathbf{I}_k(S_j, E)\mathbf{T}(S_j, E)) dE \sim \langle \mathbf{T}_{kj} \rangle \mathbf{I}_k(S_j, E_j) \Delta E_{kj} \tag{10}$$

where $\mathbf{R}_{kj}$ is the channel count rate and where $\langle \mathbf{T}_{jk} \rangle$ is called here the Transfer Factor $\mathbf{H}_{kj}$. It is convenient to separate $\mathbf{H}_{kj}$ into a counting efficiency "$\boldsymbol{\varepsilon}_{kj}$" and a factor that relates strictly to geometry, the so-called geometric factor $\mathbf{G}_{kj}$. That is:

$$\mathbf{I}_k(S_j, E_j) \sim \mathbf{R}_{kj} / (\boldsymbol{\varepsilon}_{kj} \mathbf{G}_{kj} \Delta E_{kj}) \tag{11}$$

Therefore, the first goal of calibration is to obtain detector-channel-averaged transfer factors $\mathbf{H}_{kj}$ (or the equivalent efficiency factor $\boldsymbol{\varepsilon}_{kj}$ since $\mathbf{G}_{kj}$ is easy to calculate), the detector-channel energies E_{kj} and the detector-channel energy band pass ΔE_{kj}.

More quantitative work may require a full inversion of the kernel equation given in (9). Here we are usually aided by the fact that to some level of approximation the Transfer Function **T** is separable in the form:

$$\mathbf{T}(S, E, \theta, \phi) = K(S, E)\mathbf{P}(\theta, \phi) \tag{12a}$$

and hopefully even

$$\mathbf{T}(S, E, \theta, \phi) = \mathbf{K}(S, E)\mathbf{C}(\theta)\mathbf{D}(\phi) \tag{12b}$$

Note that even with these separations, an exact inversion of the integral is rarely possible, and we can compute only the coefficients of some tailored expansion of **K**, **C**, and **D**, as for example in spherical harmonics (using Legendre polynomials). The accuracy of these coefficients depends on both the raster coverage of the measurements and on the calibration.

4.1.3 Goals of the PEPSSI Characterization and Calibration Efforts

The goal of PEPSSI characterization and calibration efforts, then, is to relate the rates, $\mathbf{R}_{kj}$, of the on-board or ground-defined channels to the *in situ* particle intensities $\mathbf{I}(S, E, \theta, \phi)$ by developing a hierarchy of information about the transfer function $\mathbf{T}(S, E, \theta, \phi)$. Specifically, we wish to derive for each channel:

$$E_{kj}, \Delta E_{kj}, \boldsymbol{\varepsilon}_{kj}\mathbf{G}_{kj}, \mathbf{C}_{kj}(\theta), \mathbf{D}_{kj}(\phi), \mathbf{K}_{kj}(S, E), \mathbf{P}_{kj}(\theta, \phi), \quad \text{and finally,} \quad \mathbf{T}_{kj}(S, E, \theta, \phi)$$

Clearly, the further to the right we move into this list, the higher the fidelity the characterization and calibration efforts must be.

Typical calibration procedure consists of stimulating the instrument with energetic particles, first from radiation sources, and then from accelerator beams (fair approximation of a delta function) and recording the response of the different channels. The ideal calibration for PEPSSI meant a scan of the following variables:

Species and Mass: e, H, He, O, Ar (proxy for heavier species, e.g., Fe)
Energy: 1 keV–1 MeV ($\sim$ 30 energies for 10 points/decade)
Polar angle: $\pm$10 deg (12 degrees nominal FOV)
Azimuthal angle: $\pm$90 deg (160 degrees nominal FOV)

Note that even given this level of discreteness, complete characterization to the level of establishing $\mathbf{T}_{kj}(S, E, \theta, \phi)$ for all channels would have required: 5 masses $\times$ 30 energies $\times$ 21 polar angles $\times$ 180 azimuthal angles $=$ 567,000 calibration points! Clearly it was (and is) not possible to run this many beam values without significant infrastructure and automation. Therefore, we are dependent to a substantial degree on the separability of the transfer function as described in (12). An important goal of the characterization and calibration efforts was, in fact, to establish that degree of separability. Simulations play an important role in establishing expectations for the instrument, as well.

4.2 Simulations

Numerical simulation provides an efficient and effective means of verifying and quantifying the flight instrument response. Hence, a range of simulation tools are developed in order to characterize PEPSSI. The results from these simulation tools allow us to calculate the end-to-end expected performance of the instrument which one can use to compare with in-flight and calibration data. Different sections of PEPSSI are simulated using different tools to gain better understanding of each part of the instrument.

4.2.1 Geometric Factor

The number of data points needed to measure the geometric factor is very large, and the time needed to perform these measurements was well beyond the available time. Instead, estimates of the geometric factor rely on computer modeling and analysis. The very same analysis formed the base for the selection of the best collimator configuration: 23 blades, each 25 mm long. This geometry for the collimator defines the maximum acceptance angles through the collimator and the geometric factor $GF = 0.15$ cm^2 sr, which is the total geometric factor for the start system.

4.2.2 Ion Measurements

4.2.2.1 Electron Optics As discussed previously, secondary electrons generated by primary ions in a start and a stop foil are used to generate timing signals to measure the ion's time-of-flight (TOF) within PEPSSI. They are also used to determine the directionality of the low energy ions that do not stimulate the SSDs.

Here, a pie-shaped cutout of the cylindrical PEPSSI sensor was analyzed (cylindrical symmetry is assumed for the electrostatic fields). The electron optics for both the start foil/MCP and stop foil/MCP are identical, hence the analysis yields a good representation for both start and stop electron trajectories. Electrons are generated at the foil when ions

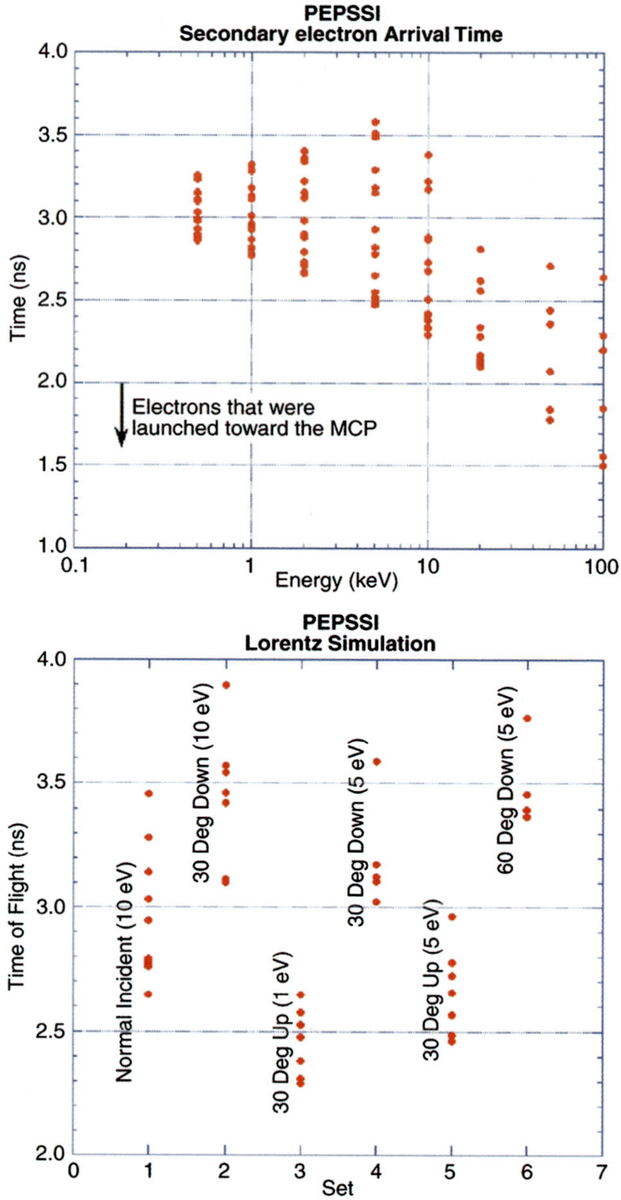

Fig. 13 Time dispersion associated with varying angles for electrons emitted from the center of the foil

transverse through the foil. These low energy electrons (typically <10 eV) are immediately affected by the accelerating voltages within PEPSSI (the high energy ions are not significantly affected by the accelerating voltages) and are steered towards the MCP. An important issue is the dispersion in the arrival times of the secondary electrons as a function of foil position and the angle of emittance of the secondary electrons from the foil. Such dispersion adds to the error in the measurement of time-of-flight. Figure 13 (top) shows the time dispersion associated with varying angles for electrons emitted from the center of the foil. In Fig. 13 (bottom) the dispersion shown for any one column is associated with varying the emission position over the ∼6 mm vertical extent of the foil. The combination of position and angle yields a time dispersion of roughly 1.5–2 ns. Combining the dispersions associated with the start and stop detections in a root-mean-square sense yields a total time dispersion error of roughly 2.1–2.8 ns. Combining that value in a root-mean-square sense with the ∼1.5 ns electronic dispersion yields a total time dispersion that resides between 2.5 and 3.2 ns.

4.2.2.2 Ion Energy Losses As ions transverse the TOF section of PEPSSI through the front foil (∼10 μg/cm^2), back foil (∼ 19 μg/cm^2), dead layer of the solid state detector (∼550 Å), and eventually stop and deposit their total energy in the SSD, they lose energy and scatter depending upon the ion's initial energy, mass, and the medium through which they pass. We have incorporated data from the SRIM particle-interaction-with-matter code (Biersack and Haggmark 1980) in our simulation to simulate realistically the energy loss and scattering as ions transverse through different materials in the PEPSSI sensor.

4.2.2.3 TOF Measurements When ions penetrate through the front foil, a distribution of ion velocities is created, as calculated using SRIM. This distribution of ions is then used to calculate the distributions of ion TOFs. The uncertainty due to secondary electron dispersion and electron noise together with SRIM data are all incorporated in our simulation. Figure 14 shows the simulated response of PEPSSI TOF spectra as a function of particle initial energy for four species (H, ^{4}He, ^{16}O, and ^{56}Fe). At low energy (∼10 s of keV), ions lose significant amounts of energy and scatter significantly when going through the front foil. These effects explain the spread in TOF measurement at low energy. However, at higher energy (∼100 s of keV), the TOF spreads are mostly consequences of the uncertainties in the TOF measurement from both the electronics jitter (∼1.5 ns), and secondary-electron dispersion in the TOF optics as discussed in previous section (except for very heavy ions, e.g., Fe).

4.2.2.4 Total Energy Measurement If an ion has sufficient energy left once it transits the front and back foil, it ends up in the SSD. Depending on the ion's final energy and mass when it reaches the SSD, it can penetrate through the dead layer of the SSD and produce an electronic signal to be measured. Figure 15 shows the simulated measured energy by PEPSSI as a function of the ion's initial energy before entry into PEPSSI. The measured energy shows the cumulative effect of the energy spread due to energy lost throughout the PEPSSI detector (start foil, stop foil, dead layer, and electron hole pair production). The energy loss is greatest for ions with the greatest nuclear charge, here Fe.

4.2.2.5 TOF Versus Energy Finally, the energy measurement is combined with the TOF measurement on a particle-by-particle basis. Even with a ∼3 ns time dispersion, the major elemental species of interest (H, He, CNO, Fe) are discriminated except at the lowest energies for CNO and Fe, where the error in the energy measurements dominates.

4.2.2.6 Efficiencies The efficiency for detection of an ion within the SSDs is roughly ∼100% (except at the very lowest energies where energy straggling can position the energy below the low energy threshold). The efficiency for obtaining a TOF measurement is estimated using the efficiency of generating secondary electrons in both the front and the rear foil. To emit a secondary electron, such electrons must be generated close enough (distance "ρ") to the surface of the foil so that the electron can escape before it is re-assimilated. Thus, very roughly, it is expected that the efficiency for the generation of a secondary electron is proportional to the amount of energy per unit distance (dE/dx [keV/micron]) that

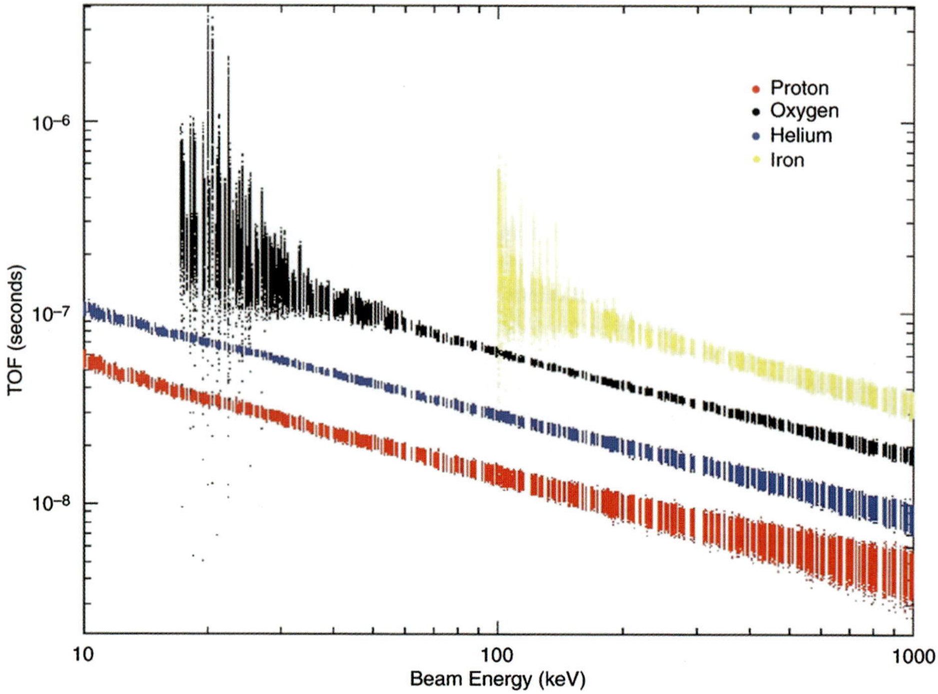

Fig. 14 Simulated TOF spectra as a function of accelerator beam energy for protons and oxygen, helium and iron nuclei

an ion deposits as it goes through this outer thin layer of the foil. The canonical number of secondary electrons generated out of each surface as a proton with 10's of keV energy encounters the foil is between 0.5 and 1 on average (Frischkorn et al. 1983); R.W. McEntire, private communication, 2004; estimated here as 0.75). Since we require a simultaneous start and stop electron, the efficiency of proton detection at 10's of keV is $(1 - e^{-0.75})^2$, where $(1 - e^{-0.75})$ is the Poisson probability of having at least 1 or more electrons emitted when the mean emission number is 0.75. Thus, the efficiency is roughly 28%. Other energies and species may be roughly scaled with this number using tabulated dE/dx values. For example, at 50 keV total energy the dE/dx values for protons and oxygen ions are 120 and 250 keV/micron, respectively (we ignore for now the energy losses suffered by the ion in getting to the position in either foil where the secondary electron is generated). Thus, since the average number of secondary electrons for oxygen will by $\sim 0.75 \times (250/120)$, or 1.56. Poisson statistics tells us that the probability of detecting an oxygen ion with 50 keV energy is $(1 - e^{-1.56})^2$ or 62%.

4.2.3 Electron Measurements

The PEPSSI electron measurement strategy depends on the use of the aluminum flashing on the electron SSD. We therefore need to understand the effect of that flashing on both the ion and electron measurements within the electron SSDs. Based on simulations with GEANT-4 (Agostinelli et al. 2003), Fig. 16 shows the effect of Al flashing on proton measurements, and Fig. 17 shows the effect on electron measurements, for a varying thickness of Al flashing. The baseline spectrum assumed for both protons and electrons is a power-law spectrum with

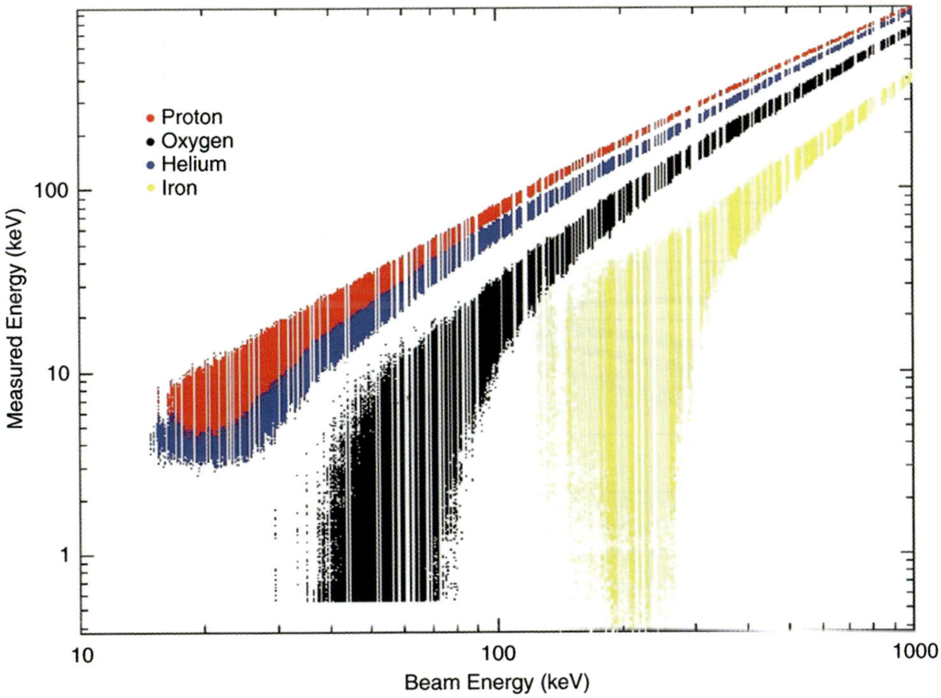

Fig. 15 Simulated measured energy versus incident energy for the same ions as in the previous figure

a spectral index of 3; that is, Intensity $= A \times E^{-3}$, where "A" is a constant. With the flashing thickness utilized with PEPSSI, 1 micron Al, the proton intensities are severely depleted for energies <250 keV. With that flashing thickness, the electron intensities are maintained to energies down to about 20 keV. The drop off of electron intensities at the highest energies occurs because of the electron penetration of the 500-micron detectors. That drop-off is rounded because of electron scattering within the SSD, as demonstrated with an examination of the detailed trajectories of individual electrons with the GEANT-4 simulations. There may be periods of time when the ion counts overwhelm the electron measurements. The most important issue is not that the electrons may be occasionally contaminated, but whether or not the electrons are contaminated in a way that the ion counts are confused with what are really electron counts. This problem is intrinsic to the technique used here to measure the electrons. Extensive experience with similar measurement made by the EPIC instrument on Geotail (Williams et al. 1994) demonstrates that, because of the clean proton measurements (below and above 250 keV) made in roughly the same direction, it is unlikely that ions and electrons will be confused.

4.3 PEPSSI Flight Unit Calibration

4.3.1 The JHU/APL Calibration Facility

Calibration of the PEPSSI instrument was conducted at the JHU/APL instrument testing facility through a combination of long-duration, radiation-source exposures (Am-241 degraded alpha source) and discrete exposures from the 170-kV particle accelerator beam. The

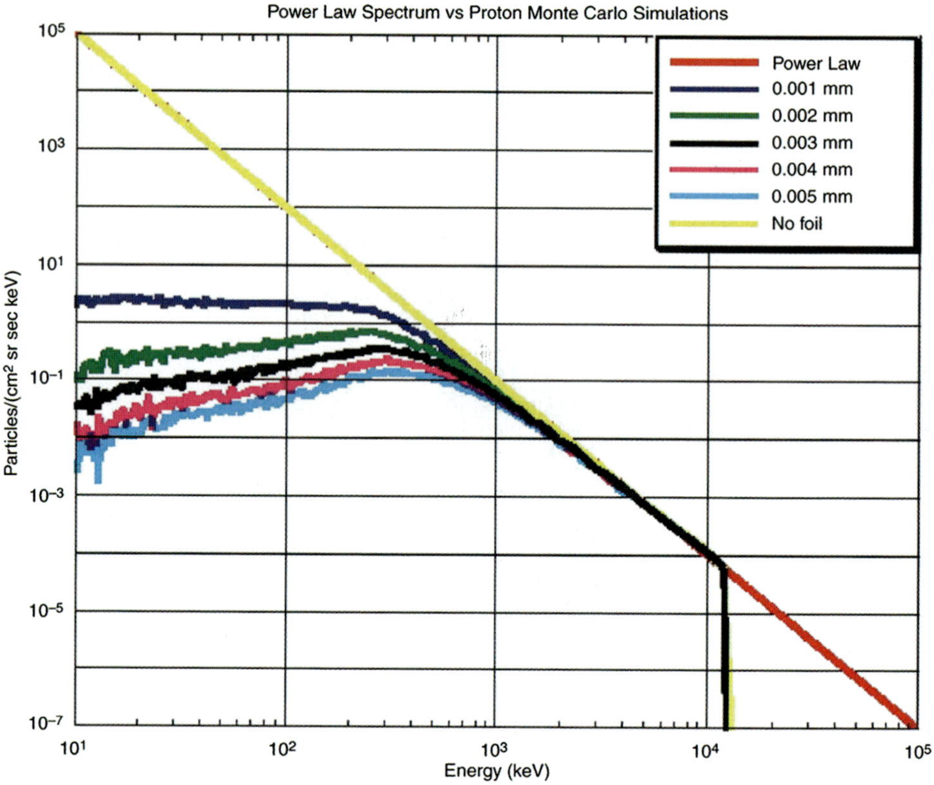

Fig. 16 Simulated measured proton spectra for different Al flashing thicknesses from a known incident spectrum (*red*) on a 500 μm thick silicon detector

Air Insulated Accelerator, obtained from Peabody Scientific, is a versatile system capable of producing a broad range of ion species to energies of 170 keV. The system produces beams of H, O, and noble gas ions with intensities over the range of $\sim 10^2$ to 10^6 particles$/\text{cm}^2/\text{s}$ at the target position in an energy range of 3 to 170 keV. Figure 18 shows the picture of the accelerator HV terminal.

A one-meter diameter and one meter tall cylindrical vacuum chamber is attached to the end of the accelerator beam line with a small gate valve to serve as the shutter (shown in Fig. 19). The chamber has two rotation stages ($\pm 0.1°$) and two translation stages (± 2 μm) that are controlled by stepper motors with digital read-out capability. Within the beam line and in the chamber itself there are various mechanical shutters that are used to control the ion beam size from 1 mm^2 to 1 cm^2. Three different beam monitors are used in the accelerator and chamber for diagnostics. A Faraday cup in the beam line before the beam enters the chamber wall is used for initial diagnostics of the beam intensity. In addition, within the chamber itself there is a stepper-motor-controlled moveable fixture that is mounted with both a channeltron and an SSD to serve as the primary beam monitor. The moveable fixture allows measurement of the species, energy and intensity of the beam inside the chamber itself.

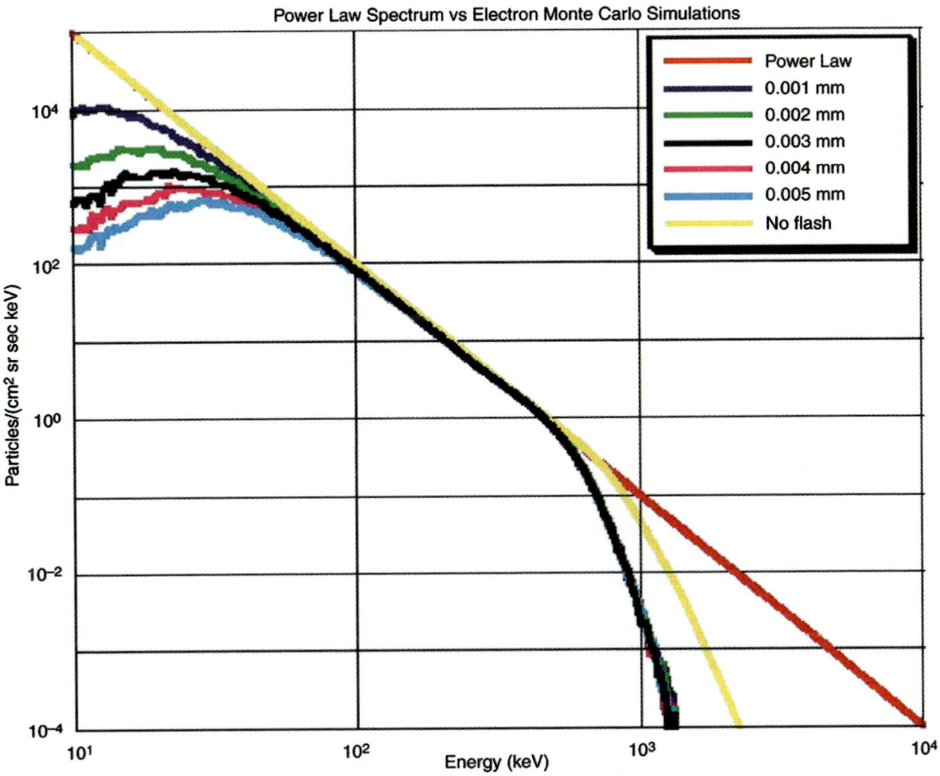

Fig. 17 Simulated measured electron spectra for different Al flashing thicknesses from a known incident spectrum (*red*) on a 500 μm thick silicon detector

4.3.2 Test Set Up

For both the alpha particle and beam calibration test, the PEPSSI unit was mounted inside the vacuum chamber. An emulator box was connected to PEPSSI to act as the New Horizons spacecraft in providing commands and power to the instrument. The emulator box was controlled by a Window-PC running GSEOS to operate and collect data from PEPSSI. All the calibration data were recorded under GSEOS as a recorded file in binary format.

PEPSSI was in full flight configuration during calibration, including the collimator, flight SSDs and foils. The PEPPSI instrument was mounted to a rotating turntable that allowed the beam to be scanned in full 160° in the azimuth direction. The response of the instrument to the vertical angle was not investigated with the accelerator beam.

4.3.3 Ground Calibration

Numerous proton, helium, and oxygen beams, along with radioactive sources, were used to calibrate and characterize PEPSSI performance over the course of two weeks during the ground calibration period. Protons at energies from 30 to 170 keV and oxygen ions at energies from 50 to 170 keV were used to stimulate PEPSSI.

4.3.3.1 Directionality Calibration runs like those depicted in Fig. 20 document the ability of the PEPSSI sensor to determine the directionality of the incoming charged particles. Fig-

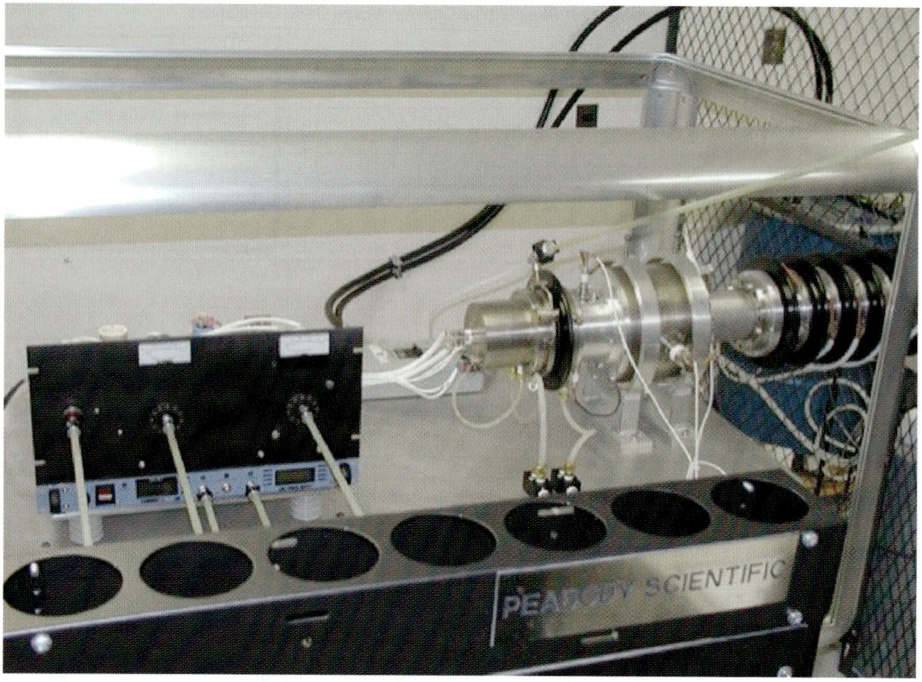

Fig. 18 The JHU/APL particle accelerator showing the high-voltage terminal

Fig. 19 The vacuum chamber used in calibrating the PEPSSI unit. The beam line enters the chamber horizontally from *left* to *right* in the picture

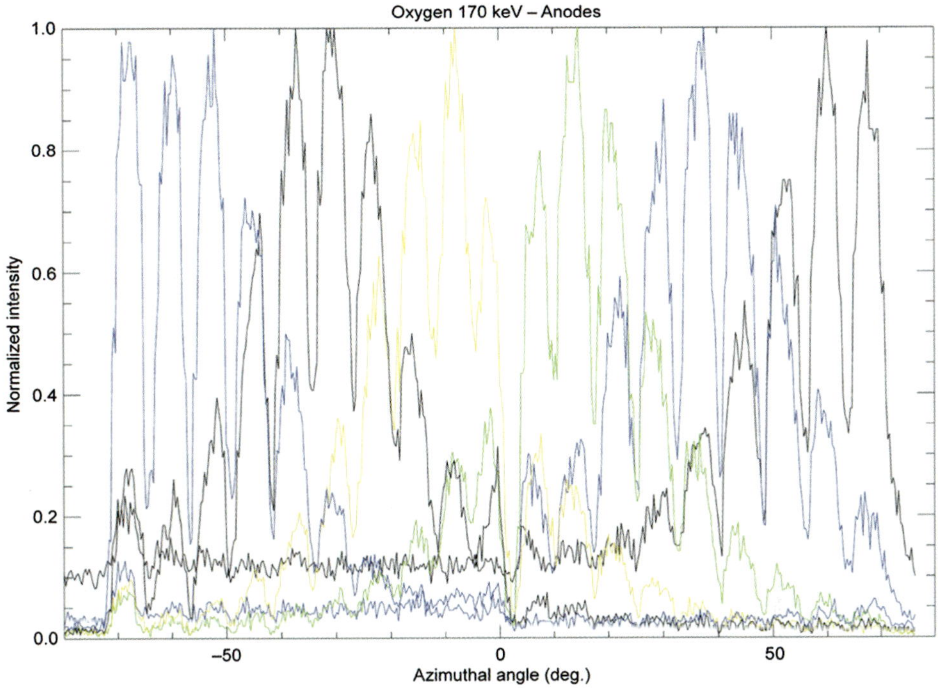

Fig. 20 Angular scan in the azimuth angle direction showing response of discrete start anode (colors) and the internal fins of the collimator to a narrowly collimated oxygen-ion beam (170 keV). All 20 internal fins are seen in the scan. The oxygen beams used have angular spreads on the order of ∼1 degree. For these anode scans, from the left (see identification and locations in Fig. 11): A5—*blue*, A4—*black*, A3—*yellow*, A2—*green*, A1—*blue*, A0—*black*; note the collimator features (see next figure as well), secondary electrons spread and/or electronic cross-talk, i.e., response into two channels at the same incidence angle

ure 20 shows the ability of the discrete "start" anodes to determine the directionality of the incoming particles for the low-energy ion events (no SSD signals). During this test, a constant, collimated oxygen beam is used, and PEPSSI rotated in the full 160° in the azimuth direction. All twelve internal collimator fins can be seen clearly in the figure overlaying the discrete anodes count rate. The discrete anodes perform the imaging function for low-energy ions. The proton beams used have angular spreads on the order of ∼1°. The angular resolution achieved is adequate for the science to be performed. Figure 21 shows the normalized response of the SSDs to the same angular scan. All six pairs of SSD are accounted for except sector 0, 2, and 5 where the electron SSDs are located.

4.3.3.2 Time-of-Flight Resolution The TOF spectrum of an 80 keV proton beam is shown in Fig. 22 with a 2.0 ns FWHM of the TOF peak. Intrinsically the TOF spread ranges between ∼2 and 4 ns for the PEPSSI instrument. There are two causes that contribute to the TOF spread: (1) the dispersive spread in the secondary electrons (∼2 ns), and (2) the performance of the TOF electronics (most notably the CFDs which is <1 ns). The measured TOF spread agrees with the expected PEPSSI performance (2.5–3.5 ns). Figure 23 shows the conversion coefficients taken from simulation to relate the measured TOF bin number into actually TOF in nanoseconds. Figure 24 summarizes the TOF performance of PEPSSI during beam calibration; the lowest energy that has a resolvable TOF spectrum for protons (oxygen ions) is 30 keV (50 keV).

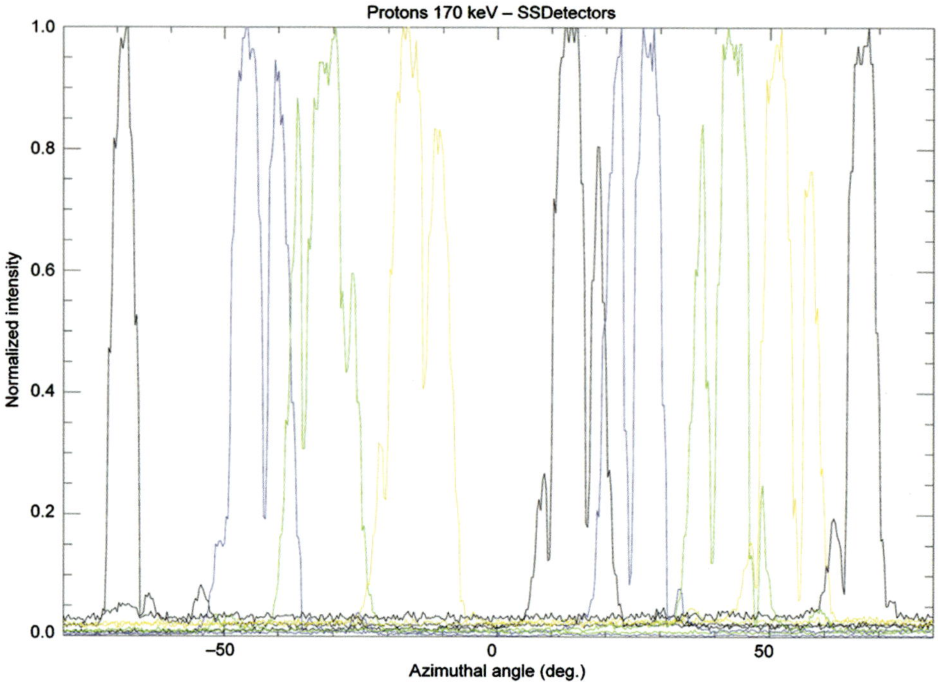

Fig. 21 Normalized responses of all SSDs to the same angular scan using 170 keV protons. All detectors showed response from the beam, except sector 1, 5, and 11 where the electrons SSDs are located. The color code is (cf. identification of detectors and positions in Fig. 11): I0—*black*, I1—*blue*, I2—*green*, I3—*yellow*, I4—*black*, I5—*blue*, I6—*green*, I7—*yellow*, I8—*black*; note the coverage void between I0 and I1 (E0), as well as I3 and I4 (E1) and the collimator blades absorption features in every detector. A ∼1° beam of protons was used due to lower noise; stimulation of the anodes (shown previously) used oxygen beams to provide larger pulses for the measurements

Fig. 22 The TOF spread of an 80 keV proton beam is shown in this figure. The fitted TOF FWHM is ∼2 ns, better than the expected 2.5 ns TOF resolution

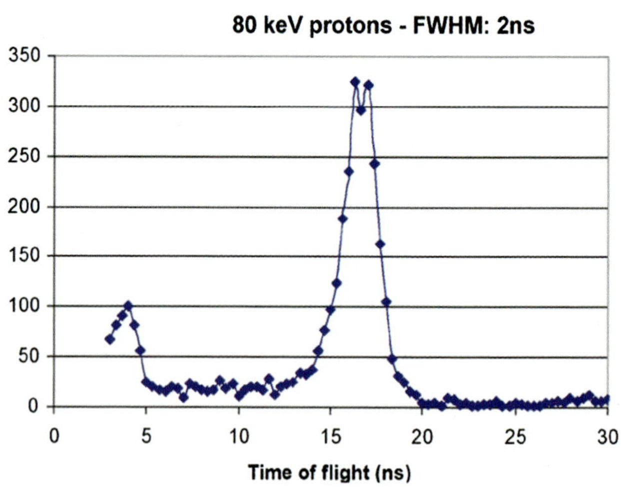

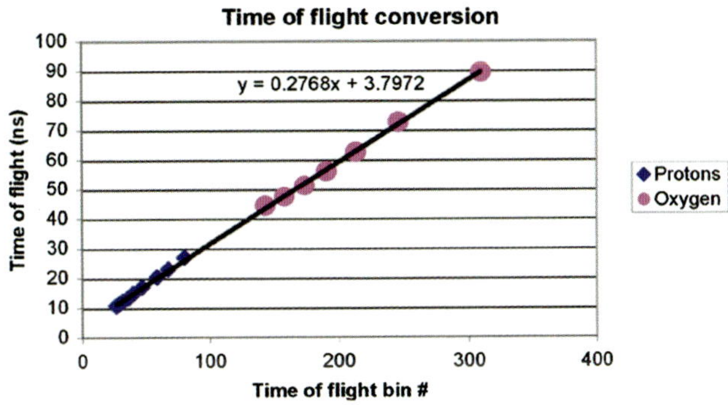

Fig. 23 The conversion coefficients relating the measured bin number and the actual TOF in nanoseconds. Conversion coefficients are calculated from the sensor simulation, taking into account pre-acceleration as well as losses through start foil

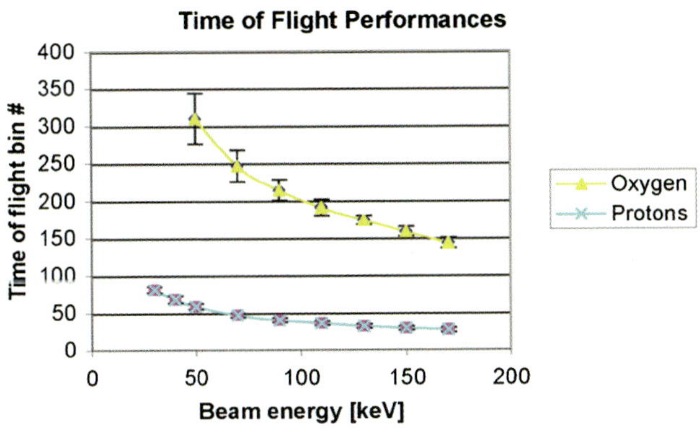

Fig. 24 TOF performance of all beam runs that were performed during the PEPSSI flight unit ground calibration. The lowest energy TOF spectrum that was observed for proton (oxygen) is 30 keV (50 keV)

4.3.3.3 Energy Resolution Figure 25 shows the energy spectrum of a 130 keV proton beam measured by PEPSSI. The upper panel shows the entire spectra from 0 to 200 keV, and the lower panel is a close-up of the spectrum. The measured energy peak is at 120 keV and the fitted FWHM is 8.9 keV in excellent agreement with the simulation. The conversion coefficients for relating energy bin number into actual energy in keV are shown in Fig. 26 and Fig. 27 summarizes the energy performance of PEPSSI during beam calibration; the lowest energy that has a resolvable energy spectrum for protons (oxygen ions) is 40 keV (150 keV).

4.3.3.4 Time-of-Flight versus Energy Events that have both TOF and energy measurement are called triple events. Depending on the ion mass, each species will have a distinct track in a TOF vs. E plot. Figure 28 shows a 100-keV proton and 170-keV oxygen beam and the expected location of the tracks through simulation. Both energies fall exactly where we expected them to be. Multiple energies and species were calibrated, and their combined results are shown in Fig. 29. Typical TOF (Energy) FWHMs are 2, 3.5, and 4.7 ns (10, 15, and 25 keV) for H, He, and O respectively.

A degraded radioactive source provides a broad energy spectrum that is difficult to reproduce using a mono-energetic accelerator beam. A Mylar-degraded ^{241}Am radioactive source was employed to extend the ~5 MeV alpha line into a broad energy spectrum all the way down to ~100 keV. Figure 30 shows the color spectrogram of the source spectrum as measured by PEPSSI during ground calibration. The y-axis shows the TOF in nanoseconds, while the x-axis shows the measured energy in keV. The spectrum falls exactly on top of the red curve, which is the expected TOF by E curve for helium calculated from the simulation.

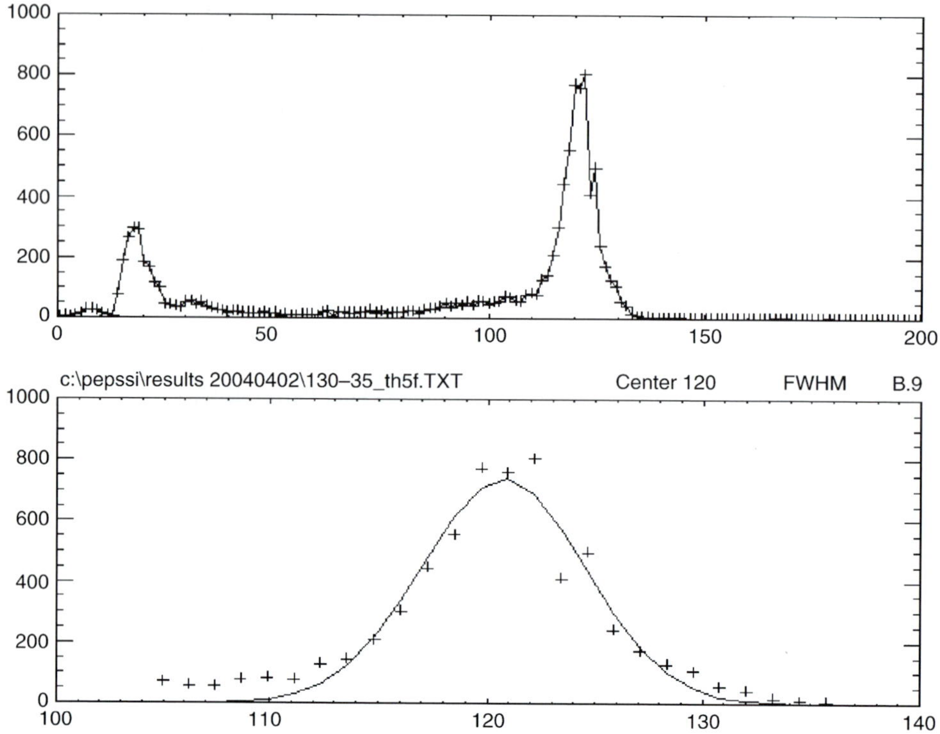

Fig. 25 The figure shows the energy spectrum of a 130 keV-proton beam measured by PEPSSI. The *upper panel* shows the energy spectra from 0 to 200 keV, and the *lower panel* is a close-up of the spectrum (100 to 140 keV). The measured energy peak is at 120 keV and the fitted FWHM is 8.9 keV in excellent agreement with the simulation developed and used for all preflight work

Fig. 26 Conversion coefficients for relating energy bin number into actual energy in keV. The coefficients are slightly detector chain dependent, however, the variation is $\sim \pm 2\%$

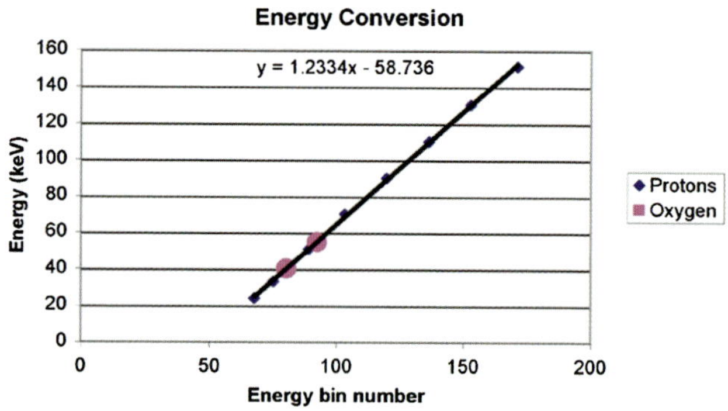

Additional testing with degraded alpha sources was performed during the instrument thermal vacuum tests. The goal was to characterize PEPSSI performance over a wide range of temperatures, confirming stability over the operational temperature range.

4.3.3.5 Pulse Height Analysis Due to the limited time available for calibration on the ground, the dependency of pulse height on the particle mass was not calibrated. Character-

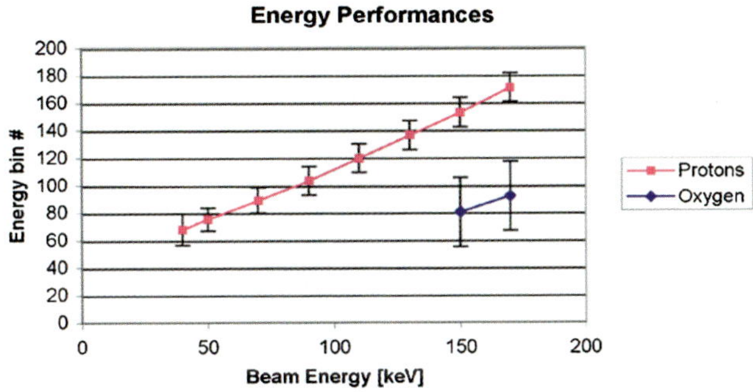

Fig. 27 Energy performance of all beam runs that were performed during the PEPSSI flight unit ground calibration. The lowest energy spectrum that was observed for protons (oxygen ions) is 40 keV (150 keV)

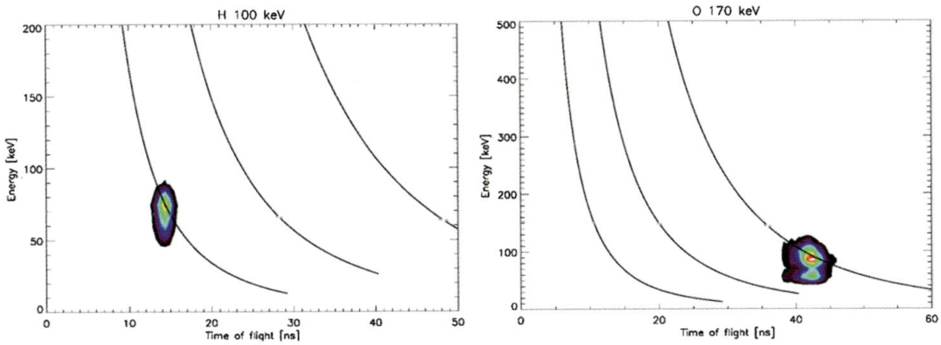

Fig. 28 Proton-beam and oxygen-beam results are shown in these TOF versus energy plots. The simulated proton, helium, and oxygen tracks are shown as a *thick line* in the figures. The measured TOFs and energies of both beams fall very close to the expected values (the simulation used a thinner stop foil than actually implemented in these calibration runs; the expected lack of sensitivity to foil thickness at these beam energies is confirmed by the good agreement)

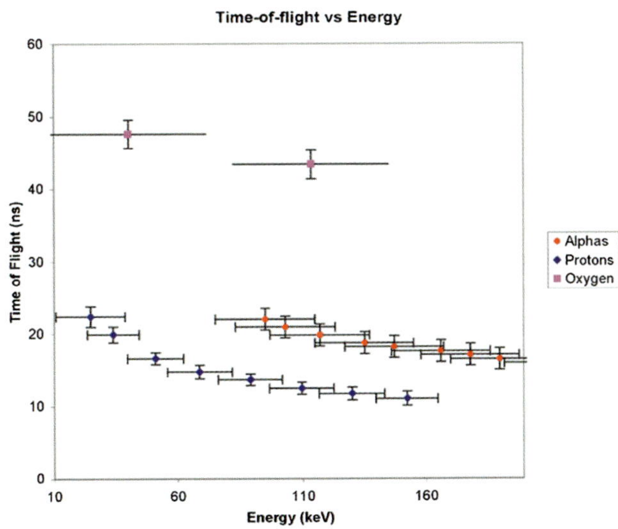

Fig. 29 Combined results from all the beam runs that were performed to characterize PEPSSI performance

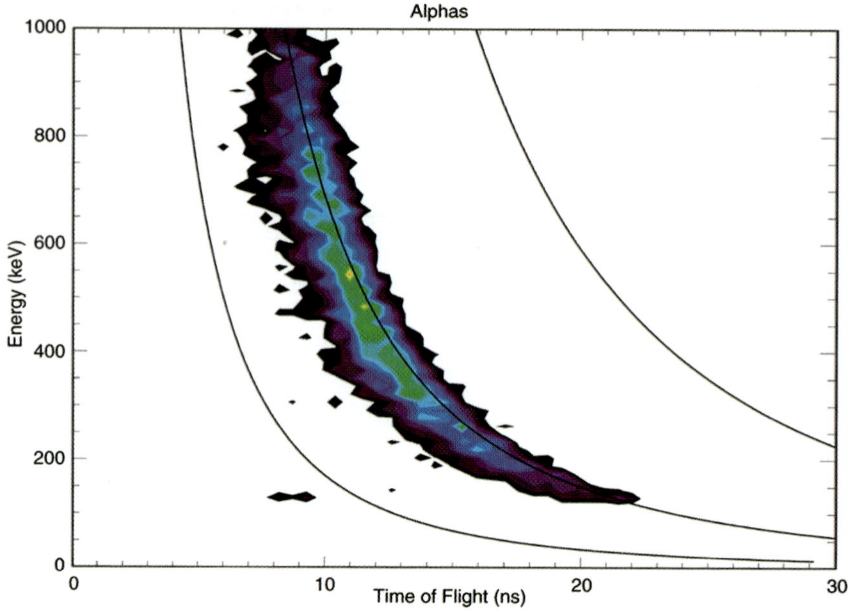

Fig. 30 Alpha-particle energy versus time-of-flight plot from the stimulation of the PEPSSI flight unit by a degraded ^{241}Am source. This produces a flat energy spectrum in the range of ~100 keV to ~5 MeV. The *central line* on the plot is from the simulated instrument response to He nuclei

ization of the heavy ion discriminators is being done during the commissioning phase in flight.

4.3.3.6 Efficiencies Figures 31 and 32 show the counting efficiency for both proton and oxygen beams as a function of the voltage across the MCP plates. The expected efficiency for 170 keV protons is 30–40% for the singles rate, and the measured stop foil efficiency is in agreement with that value. However, the start foil efficiency is clearly lower than expected, and it dominates the overall measured efficiency. The expected single efficiency for oxygen is in the 80–90% range. Here the efficiency is dominated by the start foil and does not reach a plateau even for relatively high values of the MCP voltage.

4.3.4 Summary of PEPSSI Flight Model Calibrations

Tables 3 and 4 summarize the results of the PEPSSI flight model ground calibration. Ion-beam calibrations of the flight PEPSSI instrument demonstrate it met most of the flight requirements. Most notably, the measured energy resolutions are in agreement with the expected simulated performance. The measured 2.5 to 3.5 ns TOF (FWHM) resolution is better than anticipated (probably through better than expected electronics performance).

The major outstanding issues following calibration of the PEPSSI flight model include: (1) lower than expected ion counting efficiencies, and (2) higher energy threshold of heavy ions. Major progress has been made in addressing both issues with the engineering model.

4.4 PEPSSI Engineering Model Calibrations

After the ground calibration and completion of all environmental and acoustic tests, PEPSSI was mounted on the spacecraft in December 2004. As noted previously, PEPSSI uses thin

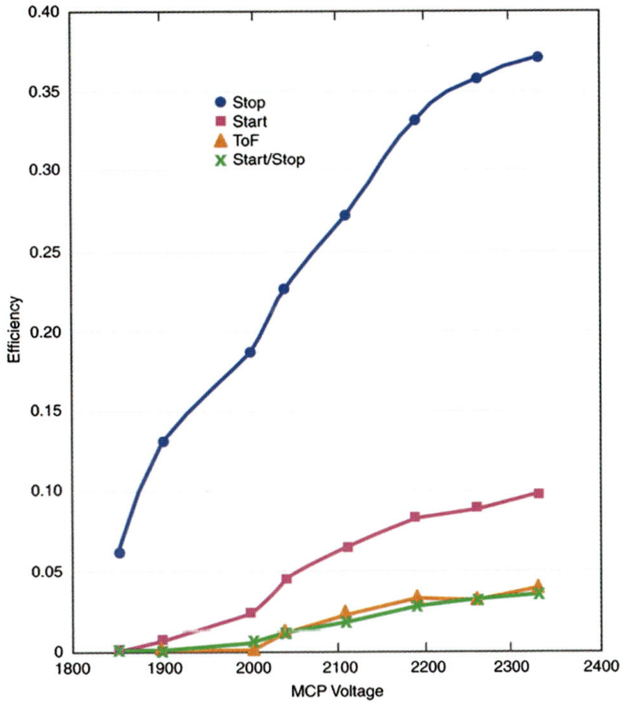

Fig. 31 The 170 keV proton efficiency plot as a function of the MCP bias voltage (across the plate). Expected efficiency for protons is 30–40%. Start foil efficiency is clearly lower than expected, and dominates the overall efficiency. These measurements are for the flight foils

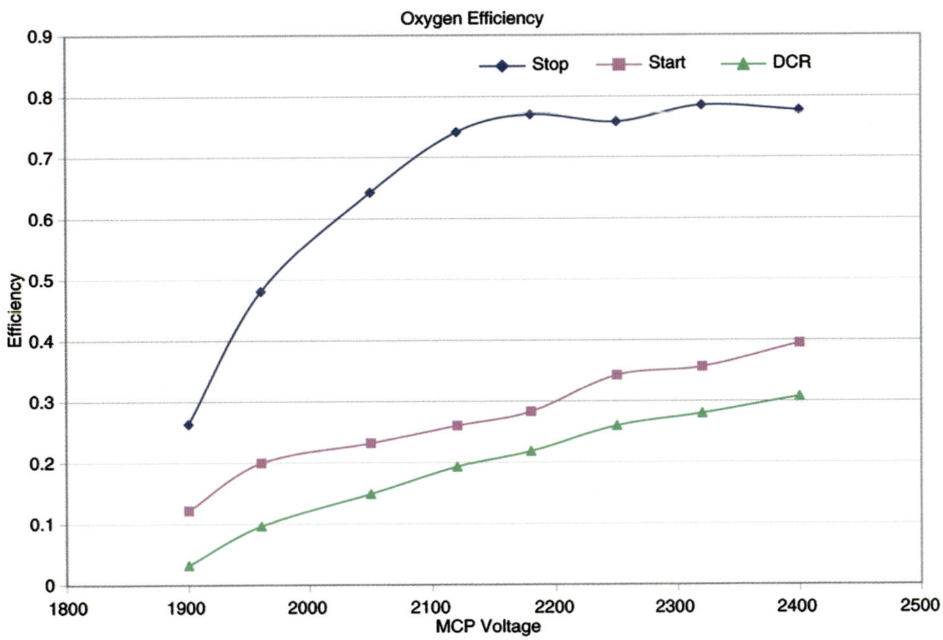

Fig. 32 The 170 keV oxygen efficiency as a function of the MCP bias voltage (across the plate). Expected efficiency for oxygen is 80–90%. Here the efficiency is dominated by the start foil efficiency, and does not reach a plateau even for relatively high values of the MCP voltage. These measurements are for the flight foils

foils to generate timing signals to gate the time needed for an ion to transverse a known path. The flight unit has a thicker back foil than the original design. The change was due to the fact that during the initial ground calibration on the PEPSSI engineering unit, we observed a lower than expected efficiency on the instrument TOF system. Subsequently, the problem was traced to both front and back foils being used on PEPSSI. Replacement of both front and back foils before instrument delivery to the spacecraft was not possible; hence, a stopgap measured was taken. A flight-qualified MESSENGER Energetic Particle Spectrometer (EPS) back foil (Andrews et al. 2007) was located and installed in the PEPSSI flight unit as the new back foil. It is this combination of PEPSSI original front foil and EPS back foil that has gone through ground calibration and is in the PEPSSI flight unit.

4.4.1 Original PEPSSI Foils

The original PEPSSI design uses aluminum-coated polyimide as the material for both the front and back foils. Both foils are made with 300-Å polyimide as support structure and then coated with 50 Å of aluminum on both sides to make it conductive. The total foil thickness is equivalent to 7 μg/cm^2. With both aluminum front and back foils, the expected TOF efficiency of a 170 keV proton beam was expected to be ~10%. However, less than 1% was measured on the engineering unit.

The discrepancy between the measured and expected efficiency is hypothesized to be produced by an oxidation layer on the thin aluminum coating on the foil. When aluminum is exposed to air, an oxide layer immediately forms. This oxide layer can be up to tens of Å in thickness, the same order of thickness as the aluminum coating. With such a relative thick oxide layer, the foil becomes an insulator, which affects both (1) the secondary electron production at the surface, and (2) the electrostatic optics within PEPSSI. These effects will be more pronounced for backward scattered (those secondary electron that travel opposite from the ion beam) electrons. We believe this is the reason why the measured stop efficiency is much lower than the expected value.

4.4.2 PEPSSI Start and EPS Stop Foil

The decision was made to replace the PEPSSI flight unit back foil with a spare MESSENGER EPS back foil, which is constructed with 500 Å of polyamide and 100 Å of palladium, while retaining the aluminum-coated front foil. This combination of foils has been calibrated fully in the flight unit that was then mounted on the New Horizons spacecraft in December 2004. The measured TOF efficiency of this combination for 170 keV protons is ~4%, which is dominated by the lower-than-expected start-foil efficiency since we still retain the original PEPSSI aluminum start foil.

4.4.3 Engineering Model Plans

Calibration and characterization of the PEPSSI instrument is an ongoing activity that has two major thrusts: (1) continued characterization and testing of the flight-like engineering model, and (2) extensive in-flight calibrations.

The engineering model is essentially complete with a full complement of SSDs. Preliminary testing has already begun with one SSD pair to verify the carbon foil performance. After the full complement of SSDs are installed on the unit, it will then be extensively exercised in both the ball jar vacuum chamber with radiation sources and with accelerator beams to help determine the right set of parameters for optimum efficiency, TOF and energy measurement performance (MCP voltage, CFD bias, CFD discriminators, Anode discriminators, etc). In addition, to characterize better the relative efficiencies of different mass species, work will investigate possible cross-talk in the valid-event identification circuitry that may occur when multiple energies and species are measured at the same time.

4.5 In-flight Calibration

Energetic particles in interplanetary space will be used to test the mass classification algorithm and verify the content of the various look-up tables. Because the interplanetary environment is variable, it is highly desirable to have the PEPSSI instrument operating substantially longer than that planned for the checkout period. Interplanetary energetic particles tend to vary on the time scale of a solar rotation: $\sim$27 days. It is highly desirable to operate the PEPSSI sensor for at least 27 days during each yearly checkout prior to the Pluto flyby in order to optimize the Pluto encounter measurements. The Jupiter flyby provided a key opportunity for checking out multiple instrument features as noted below.

4.5.1 Flight Performance

Following the launch of New Horizons on 19 Jan 2006, PEPSSI commissioning began with initial checkout on 20 Feb, a month later. Activities of spacecraft instruments are keyed to a Science Activity Plan (SAP) number. A total of 23 SAPs were executed in 2006. These included tests, the one-time door opening (3 May 2006), determination of thresholds, detector configuration, efficiency tests, and mapping of the orientation of the sensor head by monitoring when the Sun was in the sensor field of view. These tests are listed in Table 28.

PEPSSI was powered on again at 00:00 UTC 6 Jan 2007 and run through 19:20 UTC on 8 Jan; the instrument was powered on again at 00:09 UTC 9 Jan and run for seven and a half hours. On Jan 10 at 08:00 UTC PEPSSI was again turned on and run through 14 Jan 16:30 UTC. Each exercise was used to change instrument configuration for checkouts as Jupiter was approached. From 16 Jan 01:00 UTC through 20 Jan 23:00, 22 Jan 07:00 through 24 Jan 03:30, 24 Jan 06:00 though 31 Jan 05:00, and 31 Jan 07:00 though 10 Feb 12:30 further testing was conducted, including the uploading of new table parameters for the Jupiter flyby.

4.5.2 Jupiter Flyby

Coincident with the Jupiter flyby various setting changes were made to check the instrument response. Early on, PEPSSI high voltage was turned down to protect the microchannel plates from direct sunlight. The flyby provided significant science at Jupiter, especially on the Jovian magnetotail (McNutt et al. 2007), while also providing additional data on instrument performance and sensitivity. The scope of both efforts is shown in Fig. 33 that shows both energy spectra and tests that were conducted between days 40 and 140 (9 Feb–20 May 2007); scientific analysis of these data are provided in the referenced article. The instrument was finally fully powered down for the first New Horizons hibernation period on 21 Jun (day 172) at 19:40 UTC. Analysis of both science and instrument calibration data from this period is ongoing.

5 Operations and Science

5.1 Instrument Operations

After power-up and software boot, PEPSSI operates in either one of two operational modes—Standby and Science. Power requirements for each mode differ and depend upon

Table 28 PEPSSI commissioning activities

SAP No.	Stol/ Seq	Date Performed	Title	Description	Test Successful?
014	Stol	20-Feb-06	Initial check-out/functional	Run same test as pre launch ground CPT to verify the instrument is fully operational and has survived launch	Yes
004	Stol	22-Feb-06	Table load	Load new science classification tables	Yes
005	Stol	22-Feb-06	Heater test	Test flight cruise heater	Yes
006	Stol	1-Mar-06	SSD checkout	Power on bias voltage and monitor SSDs for 24 hours to characterize background with doors closed	Yes
012	Stol	2-Mar-06	Power off	Standard power off procedure, used for all commissioning activities.	Yes
016	Stol	27-Apr-06	Table load	Load new science classification tables	Yes
001	Stol	2-May-06	Power on for door opening	Power on and establish SSD background counts to observe difference when door is opened.	Yes
007	Stol	3-May-06	Open door	Open door, and verify with SSD counts, S/C primary current circuit (fire twice), S/C inertial data, and SDC sensor	Yes
008	Stol	16-May-06	Initial HV ramp	Ramp up HV to 2000 V, and manually verify current and voltage after each step	Test aborted due to Sun in FOV, only ramped to Sun safe levels. This was expected before running the test
008c	Stol	6-Jun-06	Load HV macros	New macros step HV in controlled sequence so a S/C reset can not occur between sending a HV command and re-enabling HV safing	Yes
008a	Stol	6-Jun-06	Initial HV ramp	Repeat 008 with new macro and S/C pointing so Sun is not in FOV	Yes
017	Stol	21-Jun-06	2000 V threshold scan	Determine optimal thresholds for most channels at 2000 V, also stayed on for 24 hours	Yes

Table 28 (*Continued*)

SAP No.	Stol/ Seq	Date Performed	Title	Description	Test Successful?
016	Stol	13-Jul-06	Table load	Load new science classification tables	Yes
018	Stol	14-Jul-06	Sun scan test	Scan when the Sun is in FOV by monitoring count rates and current when S/C is in spin mode	Yes
019	Stol	14-Jul-06	MCP plate scrub	Power MCPs to low voltage with Sun in FOV for 2 weeks to scrub plates	Yes
018	Stol	27-Jul-06	Sun scan test	Scan when the Sun is in FOV by monitoring count rates and current when S/C is in spin mode	Yes
015	Stol	11-Aug-06	Load macro 11	Load macro 11 to fully initialize PEPSSI at 2000 V, use data from 017 to set all thresholds	Yes
018	Stol	11-Aug-06	Sun scan test	Scan when the Sun is in FOV by monitoring count rates and current when S/C is in spin mode	Yes
018	Seq	26-Aug-06	Sun scan test	Scan when the Sun is in FOV by monitoring count rates and current when S/C is in spin mode	Yes
020	Seq	8-Sep-06	2150 V Threshold Scan	Found optimal thresholds for all channels at 2150 V	Yes
022	Seq	9-Sep-06	Enhanced Threshold Scan	Additional scans at 2000 V to determine thresholds for some detectors that were not determined during 017	Yes
021	Stol	25-Sep-06	Load and test macro 12	Load macro 12, verify it loaded correctly, then run and stay on for 24 hours	Yes
027		23-Nov-06	Efficiency test		Yes

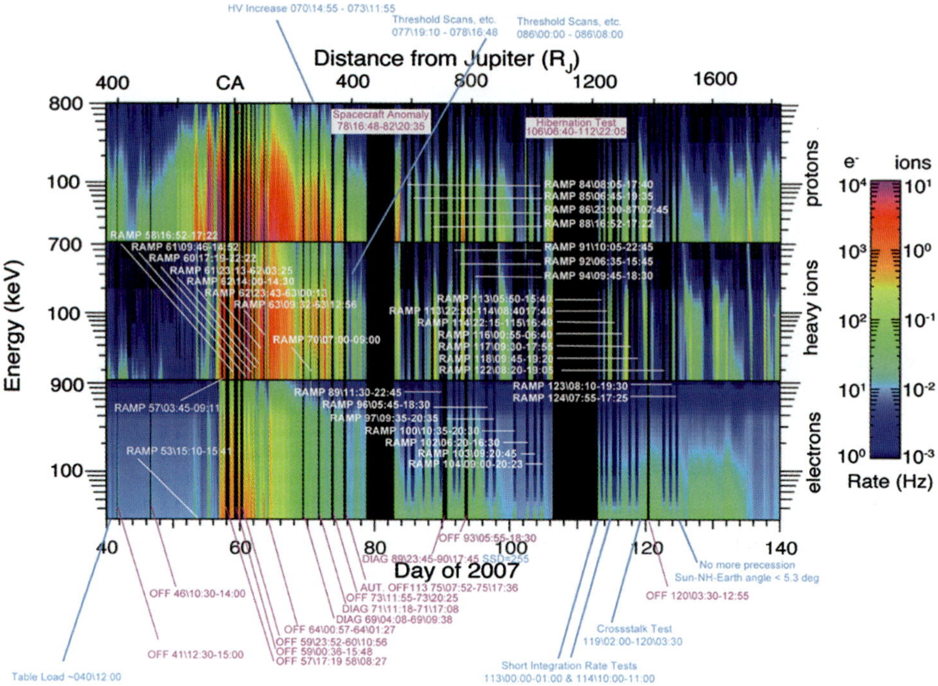

Fig. 33 Energy spectrograms from the electron, heavy ion, and proton channels during the Jupiter encounter showing variations from before the last inbound magnetopause crossing until after the first outbound crossing. Labels indicate instrument activity and show the scope of this effort, both for diagnostics and for instrument health (the high voltage on the MCPs was ramped down when sunlight could enter the instrument aperture. This is not an issue at Pluto)

the state of the HVPS as well as the rate at which data is collected. The not to exceed power allocation for PEPSSI is 2.55 watts.

The modes and the current best estimate (CBE) of power for each mode are:

- Boot/Standby: Power on, HVPS disabled (no science data).
  - Housekeeping status and instrument checkout—1.84 watts
- Science: Power on, HVPS enabled, science data sent to the C&DH.
  - Nominal Event Rate, measured (Pluto flyby)—2.43 watts
  - High Event Rate (measured at Jupiter flyby)—-2.46 watts

5.1.1 Cover Release In Space

The NH spacecraft was launched into heliocentric orbit with the PEPSSI instrument aperture covers closed. To prevent thermal damage to entry foils in the sensor module, the covers were not to be opened until the distance from the Sun to the NH spacecraft was least 2 AU. With no specific housekeeping flag to indicate the instrument covers are open the bias supply of the HVPS was activated and science data enabled. Open covers are evidenced by detection of background particle energy events. This operation showed that the covers were indeed open, but it also led to the discovery of the misalignment of the instrument on the spacecraft deck.

5.1.2 Idle State (HVPS Disabled)

At instrument power-up, the HVPS is always in the disabled state. The instrument idles indefinitely, waiting for command input from the spacecraft. In this state housekeeping data is output at a preset rate, and output science data packets contain only filler data. Prior to science data collection, the HVPS must be enabled and brought on-line.

5.1.3 HVPS Activation

To activate the HVPS, a preconfigured sequence of commands must be executed to enable HVPS operation, adjust the power supply clock rates, ramp up the output voltages, and set HVPS alarm limits. Once the HVPS is operating and functioning at the appropriate voltage levels, flowed science data is valid.

Following launch into space, prior to first activation of the HVPS, at least 14 days must be allowed to elapse to allow sufficient time for outgassing. In practice, the HVPS was not turned on until after the doors were open, much more than 14 days after launch.

Following HVPS activation, a number of HVPS parameters may be adjusted by command. These include commands to adjust the levels of the SSD bias voltage and the high voltage (HV) outputs. The HV voltages are adjusted during the cruise phase to compensate for known aging of the MCP electron gain characteristics. By adjusting to a slightly higher voltage, constant MCP gain (and hence TOF system characteristics) may be maintained. Similarly, the SSD bias voltage may be periodically adjusted to compensate for SSD aging characteristics.

HV and SSD bias supply clock frequencies may also be adjusted by command. As the HVPS components age, HVPS efficiency will be monitored. Slight adjustments to the HVPS clock frequencies may be made to retain power efficiency.

5.1.4 Science Mode Operation

In Science mode, particle TOF-plus-Energy, electron energy, and TOF-only measurements are made and output on a regular basis. Specific science mode data products are defined above. The rates at which science data products are collected are programmable (by adjusting the data integration time interval) and the range of data rates possible is outlined in prior sections that specify PEPSSI telemetry. Regardless of science data rates, the data collection intervals are synchronized to the MET 1PPS input from the spacecraft.

During PEPSSI science mode operation, the PEPSSI instrument requires few if any commands. However, there are a number of instrument operating adjustments, in addition to the data integration interval, that can be made by command. These adjustments are described below. These adjustments, if required, are typically made during early in-space operations. During cruise mode operations, and prior to any major instrument data collection campaign, instrument operation is monitored to determine if any further adjustments are required.

5.1.4.1 Energy Leading Edge Discriminator (LED) Threshold PEPSSI implements a Leading Edge Discriminator (LED) for each of the 12 SSD energy channels. Only particle energy events that exceed the LED threshold are counted and included in the telemetry science data output. The threshold for each LED is adjustable by instrument command over a range up to ~250 keV and to a quantization of ~1 keV.

The purpose of the LED is to provide a means to detect particles of minimum energy while minimizing false counts due to system noise. During ground calibration, the LED threshold of each channel is adjusted so that the false count rate is low, much less than one event per second, while maintaining the capability to detect particles at or below the energy thresholds required.

During initial in-space PEPSSI operations, the energy LED thresholds may be tweaked slightly to achieve optimum performance. After the initial tweaking, the threshold of any particular LED will be adjusted only if increased noise is detected in that channel.

5.1.4.2 Stop Anode Light and Heavy Discriminator Thresholds PEPSSI applies two discriminators to the TOF stop-anode signal. The discriminator thresholds are adjustable over a range of signal input levels by instrument command.

For TOF-only measurements, the magnitude of the respective anode signal is a coarse indicator of particle mass. The threshold levels are adjusted so that, for TOF-only measurements, the particles may be differentiated on the basis of light or heavy mass. The threshold level of one discriminator is set to a low value, low enough to detect H^+ ions, the other is set to a higher threshold, to detect CNO and higher mass particles. One is referred to as the light discriminator, the other the heavy discriminator.

Due to time constraints, the light and heavy discriminator thresholds were not calibrated and adjusted prior to launch. The thresholds will be adjusted for optimum performance as part of ongoing flight operations. It is anticipated that the final light and heavy thresholds will rarely, if ever, be changed; thresholds will be readjusted only in response to detected changes in TOF system noise or front-end sensitivity.

5.1.4.3 Constant Fraction Discriminator (CFD) Thresholds The composite TOF-start and TOF-stop signals are each passed through a CFD to generate digital level signals. Digital level TOF-start and TOF-stop pulses are required for input to the TOF chip. The CFD thresholds are adjustable over a range of values by instrument command; only TOF anode signals that exceed the CFD thresholds will result in a TOF measurement.

The CFD thresholds determine the lower signal threshold for TOF measurements. It is set during ground calibration to as low a threshold possible consistent with acceptable false TOF detection rates due to TOF system noise. During ground calibration, the threshold of each CFD was adjusted so that the false TOF event rate is low, much less than one event per second, while maintaining the capability to detect particles at or below the energy thresholds set previously. During initial in-space operations and checkout of PEPSSI, the CFD thresholds were tweaked to achieve optimum performance. After the initial tweaking, the CFD thresholds will be adjusted only if increased noise is detected in the TOF system.

5.1.4.4 Start Anode Discriminator Thresholds The PEPSSI design includes leading edge discriminators for each of the six start anode signals. The discriminator thresholds are adjustable over a range of values by instrument command. The discriminator outputs are used to determine the direction (i.e., sector) of the respective TOF measurement.

Only TOF anode signals that exceed the discriminator threshold will provide a signal output indicative of particle direction. During ground calibration, the threshold of each discriminator is adjusted to be consistent with CFD thresholds. During in-space operations, adjustments may be required if the CFD thresholds are changed.

5.1.4.5 Multiple Hit Check Enable/Disable As heritage from the MESSENGER EPS design, PEPSSI includes a multiple-energy-event detect/reject feature. This feature, enabled by command, causes the instrument to reject (not count) multiple energy events that occur closely in time. Multiple events occurring in close proximity may cause degradation in the accuracy of energy measurements, and the intent is to provide a means to prevent these types of events from being counted. If this feature is disabled, and multiple energy events occur closely in time, the multiple events are counted as a single event, and, if an event crosstalks to an adjacent detector, the wrong detector could be read out, misclassifying the energy.

5.1.5 Power Down Operation

PEPSSI is designed so that power to the instrument may be switched off while the HVPS is operating. No damage to the instrument can occur. However, for normal operations, the instrument HVPS is ramped down in voltage and turned off in a controlled sequence prior to switching off power to the instrument.

5.2 Data and Data Archiving

The New Horizons Science Operations Center (SOC) is part of the ground system that processes data returned from the New Horizons planetary spacecraft. Data down-linked from the spacecraft in raw packetized form are retrieved by the SOC from the Mission Operations Center (MOC) along with navigation and related ancillary data. The SOC generates the higher-level data products used by the instrument teams and science teams. In addition, the SOC performs archiving of data to the Planetary Data System (PDS). The science data processing component of the SOC is called the SOC pipeline.

The SOC pipeline is divided into three main parts: the Level 1 pipeline segment, the Pre-Level 2 pipeline segment, and the Level 2/3 segment. Pipeline processing is carried out sequentially. Results of the Level 1 pipeline are provided as inputs to the instrument Pre-Level 2 pipeline segment. The Level 2/3 segment combines the pre-Level 2 temporary data files into easily used (Level 2) and science/calibrated (Level 3) formats. The instrument pipeline generates Level 2 and Level 3 results that the SOC forwards to the PDS archiving process. All levels of science data files (Level 1, Pre-Level 2, Level 2, and Level 3) are stored in FITS format.

5.2.1 Level 1

The Level 1 data product is a FITS format data file and all data is contained in FITS extension Header Data Units (HDUs). Each HDU contains a PEPSSI science data telemetry block. The non-status HDUs are stored as 1-D FITS images containing the binary data of a de-commutated science data telemetry block. Status extensions have been "unrolled" into a normal 2-D FITS binary table with columns for each uncalibrated status quantity and a separate row for each measurement time. Level 1 files are not present in the PDS archive. Data from different telemetry classifications are placed in different files (Table 29).

5.2.2 Level 2

The Level 2/3 pipeline is run on the SOC processing station to transform Level 1 decommutated data into Level 2 and Level 3 calibrated science data. The instrument pipeline creates PDS standard, Level 2 and Level 3 provisional products in FITS format.

The Level 2 files "unroll" the data in the Level 1 files into flat, single datum per cell, tables that are easily human readable using standard FITS tools like fv (http://fv.gsfc.nasa.gov).

In the Level 2 files, all of the data for a single UTC day of observation are present in a single file. Each type of PEPSSI data: High Priority Rate Data (N1), Low Priority Rate Data (N2), Status Data, and the various forms of PHA data, in both normal and diagnostic modes is placed in a different FITS binary table with associated header keywords to create a separate FITS extension or header data unit (HDU).

Table 29 Data file layout

File Type	HDU Type
N1	Primary HDU: High Priority Telemetry Block
	Extension 1: PHA Telemetry Block
	Extension 2: Status Telemetry Block
N2	Primary HDU: Medium Priority Telemetry Block
	Extension 1: PHA Telemetry Block
	Extension 2: Status Telemetry Block
N3	Primary HDU: Low Priority PHA Telemetry Block

5.2.2.1 Summary of the Level 2 Header Data Units (HDUs) In PEPSSI Level 2 files, each HDU represents a different type of data; there is only one HDU of each type (EXTNAME). An HDU will only be present if there is data of that type taken during the time period covered by that file. Each file contains exactly one UTC observing day worth of data. The Primary HDU contains no data; it contains only informational header keywords identifying mission information, observational start time, and information about the file creation (date, software version, etc.).

The available HDUs are:

D_N1, D_N1_STATUS, D_N2, D_N2_STATUS, N1, N1_STATUS, N2, N2_STATUS, PHA_DIAG, PHA_ELECTRON, PHA_LOW_ION, PHA_HIGH_ION

The different extensions are described below.

When a charged particle enters the PEPSSI detector, we measure its energy and/or TOF. Because there is insufficient bandwidth to telemeter all of our events, we use a round-robin priority scheme to decide which PHA events to discard and which to telemeter. All events are counted in various histograms in the Rate data.

5.2.2.2 PHA HDUs The four PHA HDUs: PHA_DIAG, PHA_ELECTRON, PHA_LOW_ION, and PHA_HIGH_ION contain columns for the end time of the accumulating interval, the uncalibrated Energy and Time of Flight values and the detectors involved in measuring a given event (e.g. which solid state energy detector fired and which start anode fired).

Each row represents a separate charged particle event.

5.2.2.3 Rate HDUs The N1 and N2 (and D_N1 and D_N2) extensions contain several types of "Rate" data. The Rate data is accumulated in histograms that are then dumped at set intervals. For N1 data, usually the histograms are accumulated for 600 seconds. For N2 data, the accumulation time is usually 60 seconds except for the first hour of the day when it is 15 seconds. This can be changed, but during normal observing it is almost always true:

- B Rates: The number of high-energy ion events in the various "Rate Boxes". For example: a Rate labeled B01S03 represents Protons which deposited between 60 and 94 analog to digital units (ADUs) of energy in the solid state detector.
- C Rates: The contents of various hardware counters
- HK Rates: Various housekeeping quantities such as power levels and discriminator thresholds
- J Rates: Software counters that represent overall quantities like total number of Electron Events.
- L Rates: The number of low energy (TOF-only) ion events sorted into Boxes.
- R Rates: The number of electron events.

All of the rate column contents are specified in detail in the comment field of their respective TFORM keywords in the FITS Header.

The N1 and D_N1 data are identical in format; the D_N1 data are merely taken when the instrument is in diagnostic mode. The definitions of some of the Rate Boxes are different in diagnostic mode and normal mode. N2 (and D_N2) are identical to their N1 counterparts except that they are typically sampled much more frequently (every 15 or 60 seconds) and only a subset of the L and C rates are present.

5.2.2.4 Status HDUs All the STATUS HDUs contain the same quantities for their respective coverage periods and are the same values as in the Level 1 files calibrated to physically meaningful values.

The Level 2 data files represent all of the scientifically useable data from the PEPSSI instrument and are meant to be the starting point for science analysis of that data.

5.2.3 Level 3

The Level 3 data is a calibrated, scientifically useful subset of the Level 2 data. As with the Level 2 data, each Level 3 file covers all of a single day of observation. There are three basic types of data in the L3 files: Quick-Look, flux-calibrated Rate Data, and calibrated PHA data. No Diagnostic mode data is present in the L3 files.

The Level 3 files are meant to be, as much as possible, self-documenting. All calibration constants, calibration formulas, and physical units are present in the FITS header in an easily readable format.

The image in the primary array of the L3 file is a rate-weighted 2-D histogram of the PHA data for that day binned in calibrated deposited energy. The priority scheme distorts ion abundances, so we correct for that by using a "rate-weighted" rather than a single count histogram.

The next five HDUs: SPEC_Protons, SPEC_Helium, SPEC_Heavies, SPEC_Electrons, SPEC_LowIon, contain quick-look spectrograms of their respective species. These spectrograms present counts/second N2 data, averaged over 60 second intervals and summed over all incidence directions.

The FLUX HDU contains calibrated fluxes, uncertainties, and raw counts/sec rates for all of the High Energy Ion and Electron N2 Rate data. There is also an accumulation time column (**DT**) and three timing columns. Separate calibrations are given for different ion species for some of the rate boxes if the composition in that rate box is complicated (e.g. both oxygen and sulfur in a single box).

The three PHA extensions: PHA_ELECTRON, PHA_LOW_ION, and PHA_HIGH_ION contain the PHA event data telemetered in the N2 data. Each row represents a single PHA event. Cross-talk events are excluded. Quantities of limited usefulness (such as Heavy Ion Discriminator triggers) are excluded. Calibrated Deposited Energy and/or Time of Flight values are given. The linear calibration constants and formulas are in the FITS headers. A **Speed** column is calculated from the Time of Flight. The Rate Box classification for each event is given in the **Rate_Box** column.

The PHA_HIGH_ION HDU contains additional columns: **H_Incident_Energy**, **He_Incident_Energy, O_Incident_Energy**, and **S_Incident_Energy** columns contain the calculated Incident energy assuming that the event is of that (H, He, O, or S) species. The Rate_Normalized_Weight column has removed Priority Group artifacts from the PHA data. This column is usually used in making histograms of the High Energy Ion PHA data.

5.2.4 MIDL

PEPSSI data can be viewed using a software suite known as the Mission Independent Data Layer (MIDL). A full complement of analysis tools is available: line plots of individual particle rate channels, energy spectra and spectrograms, and histograms (one-dimensional and two-dimensional) of the PHA event data. MIDL also incorporates and makes available spacecraft ephemeris and pointing information vital to interpreting the *in situ* PEPSSI data. Each of the plotting tools also serves as a data access mechanism, because the data values used to make any plot can be extracted and saved separately. All discovery, downloading, and local caching of science and ancillary data are handled automatically, providing a seamless view of the entire PEPSSI dataset.

MIDL reads the Level 2 (or Pre-Level 2) data in FITS format as produced by the SOC, and applies the appropriate calibrations, allowing the user to select from a variety of units in which to view the data—everything from raw counts to fully calibrated intensity. Detailed calibration options are available for members of the instrument team, with reasonable defaults provided for non-expert users.

Created prior to PEPSSI, MIDL uses a modular approach in which plotting tools are isolated from dataset-specific read routines. New datasets can take advantage of the existing plotting tools by simply providing the appropriate set of read routines. Most plotting tools used by PEPSSI existed in MIDL prior to their use with PEPSSI. The exception is the PHA histogram tools, which were specifically developed for PEPSSI (but could also be used by other missions with PHA data). The original version of MIDL was funded as a research effort by NASA's Applied Information Science Research (AISR) program but has now been used as a primary analysis suite for three energetic particle instruments (others are Geotail/EPIC and Cassini/MIMI).

Written in Java, MIDL can be launched (via Java Web Start) as a client program on any modern operating system from any web browser.

6 Conclusion

The PEPSSI instrument on the New Horizons spacecraft will measure the outermost reach of Pluto's escaping atmosphere at the time of the flyby in 2015. Recent observations (Elliot et al. 2007) suggest that the atmospheric escape rate may not be significantly diminished from its value near Pluto's perihelion in 1988. Observations at comets and at Jupiter with PEPSSI indicate that the interaction region and pickup ions from Pluto's atmosphere should be easily discernable and diagnostic of the magnitude and extent of the interaction of Pluto with the solar wind.

The PEPSSI instrument is another step in an evolving suite of combined energy/time-of-flight instruments for probing the properties of multispecies energetic particles across the solar system, and, potentially, beyond. The origins of the these capabilities in a compact, low-mass, low-power design trace back to early planning for the inclusion of particle instruments on a Pluto flyby mission. The development and inclusion of PEPSSI on this mission to Pluto and the Kuiper Belt objects beyond can also be seen as a significant achievement of NASA's PIDDP efforts. With the successful measurements made near Jupiter and down the Jovian magnetotail, the instrument is fully checked out and calibrated for use during the encounter with Pluto in 2015.

With a trajectory near the same ecliptic longitude of Voyager 2, New Horizons has the capability to provide a platform for monitoring the solar wind flow and energetic particle propagation toward the heliopause during the regular annual checkouts and comparing

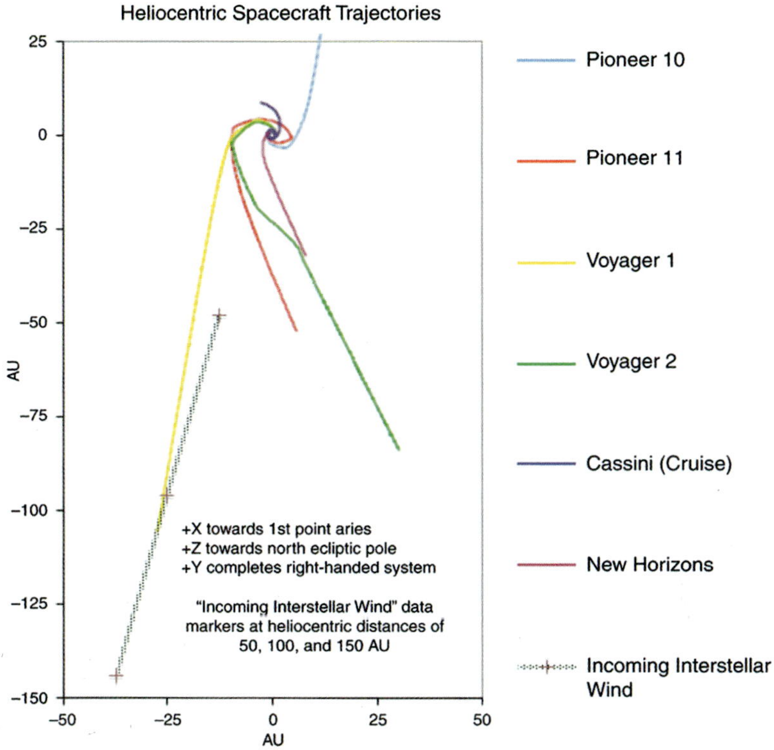

Fig. 34 Trajectories of all solar-system escaping spacecraft as projected into the plane of the ecliptic along with that for Cassini during its cruise to Saturn (for scale). Voyager 2 and New Horizons are escaping the solar system at roughly the same longitude and close to the flow direction of the very local interstellar medium

conditions with those at Voyager 2 (Fig. 34). In addition to probing the puzzle of Pluto's escaping atmosphere and its interaction with the solar wind in 2015, as well as potentially looking near a Kuiper Belt Object for cometary activity, PEPSSI will thus continue our research into how the solar system interacts with its environment of the very local interstellar medium.

Acknowledgements We would like to thank the support of the engineering support and technical services and support staffs at APL that made the design, fabrications, and the delivery of the PEPSSI instrument to the New Horizons spacecraft possible. We thank the two referees for their detailed, helpful comments. This work has been supported as part of the New Horizons effort under NASA Contract NAS5-97271.

References

S. Agostinelli et al., GEANT4—a simulation toolkit. NIM(A) **506**, 250–303 (2003)

G.B. Andrews, R.E. Gold et al., *Compact Particle Detector for Space Measurements: Prototype Performance* (SPIE, San Diego, 1998)

G.B. Andrews, T.H. Zurbuchen et al., The energetic particle and plasma spectrometer instrument on the MESSENGER spacecraft. Space Sci. Rev. **131**, 523–556 (2007)

F. Bagenal, T.E. Cravens et al., Pluto's interaction with the solar wind, in *Pluto and Charon*, ed. by S.A. Stern, D.J. Tholen (University of Arizona Press, Tucson, 1997), pp. 523–555

F. Bagenal, R.L. McNutt, Jr., Pluto's interaction with the solar wind. Geophys. Res. Lett. **16**, 1229–1232 (1989)

J.P. Biersack, L. Haggmark, Stopping and range of ions in matter. NIM **174**, 257 (1980)

N. Brosch, The 1985 stellar occultation by Pluto. Mon. Not. R. Astron. Soc. **276**, 571–578 (1995)

A.J. Coates, A.D. Johnstone et al., Pick-up water group ions at Comet Grigg-Skellerup. Geophys. Res. Lett. **20**, 483–486 (1993a)

A.J. Coates, A.D. Johnstone et al., Velocity space diffusion and non-gyrotropy of pickup water group ions at comet Grigg Skjellerup. J. Geophys. Res. **98**, 20985–20994 (1993b)

S.J. De Amicis, Instrumentation developed by the Johns Hopkins University Applied Physics Laboratory for Non-APL Spacecraft, Laurel, MD, JHU/APL, 1988

P.A. Delamere, F. Bagenal, Pluto's kinetic interaction with the solar wind. Geophys. Res. Lett. **31**, L04807 (2004). doi:10.1029/2003GL018122

M. Dryer, A.W. Rizzi et al., Interaction of the solar wind with the outer planets. Astrophys. Space Sci. **22**, 329–351 (1973)

J.L. Elliot, E.W. Dunham et al., Pluto's atmosphere. Icarus **77**, 148–170 (1989)

J.L. Elliot, A. Ates et al., The recent expansion of Pluto's atmosphere. Nature **424**, 165–168 (2003)

J.L. Elliot, M.J. Person et al., Changes in Pluto's atmosphere: 1988–2006. Astrophys. J. **134**, 1–13 (2007)

U. Fink, B.A. Smith et al., Detection of a CH4 atmosphere on Pluto. Icarus **44**, 62–71 (1980)

G.H. Fountain, D.Y. Kusnierkiewicz et al., The New Horizons spacecraft. Space Sci. Rev. (2008), this issue. doi:10.1007/s11214-008-9374-8

H.J. Frischkorn et al., Total yield and escape depth of electrons from heavy ion solid interactions. IEEE Trans. Nucl. Sci. NS-30 (1983)

A.A. Galeev, Encounters with comets: discoveries and puzzles in cometary plasma physics. Astron. Astrophys. **187**, 12–20 (1987)

A.A. Galeev, T.E. Cravens et al., Solar wind stagnation near comets. Astrophys. J. **289**, 807–819 (1985)

G. Gloeckler, F.M. Ipavich et al., The charge-energy-mass spectrometer for 0.3–300 keV/e ions on the AMPTE CCE. IEEE Trans. Geosci. Remote Sens. **GE-23**, 234–240 (1985)

G. Gloeckler, D. Hovestadt et al., Cometary pick-up ions observed near Giacobini-Zinner. Geophys. Res. Lett. **13**(3), 251–254 (1986)

R.E. Gold, S.C. Solomon et al., The MESSENGER mission to Mercury: scientific payload. Planet. Space Sci. **49**, 1467–1479 (2001)

E.M. Harnett, R.M. Winglee et al., Three-dimensional multifluid simulations of Pluto's magnetosphere: A comparison to 3D hybrid simulations. Geophys. Res. Lett. **32**, L19104 (2005). doi:10.1029/2005GL023178

D.E. Huddleston, A.J. Coates et al., Mass loading and velocity diffusion models for heavy pickup ions at comet Grigg-Skjellerup. J. Geophys. Res. **98**, 20995–21002 (1993)

K. Kecskemety, T.E. Cravens, Pick-up ions at Pluto. Geophys. Res. Lett. **20**, 543 (1993)

V.A. Krasnopolsky, Hydrodynamic flow of N2 from Pluto. J. Geophys. Res. **104**(E3), 5955–5962 (1999)

S.M. Krimigis, D.G. Mitchell et al., Magnetospheric imaging instrument (MIMI) on the Cassini mission to Saturn/Titan. Space Sci. Rev. **114**, 233–329 (2004)

L.J. Lanzerotti, R.E. Gold et al., Heliosphere instrument for spectra, composition, and anisotropy at low energies. Astron. Astrophys. Suppl. Ser. **92**, 349–363 (1992)

J.L. Lunine, D. Cruikshank et al., Pluto Express. Report of the Science Definition Team: 65, 1995

D.J. McComas, F. Allegrini et al., The solar wind around Pluto (SWAP) instrument aboard New Horizons. Space Sci. Rev. (2008), this issue. doi:10.1007/s11214-007-9205-3

R.W. McEntire, E.P. Keath et al., The medium-energy particle analyzer (MEPA) on the AMPTE CCE spacecraft. IEEE Trans. Geosci. Remote Sens. **GE-23**, 230–233 (1985)

S. McKenna-Lawlor, E. Kirsch et al., Energetic ions in the environment of comet Halley. Nature **321**, 347–349 (1986)

R.L. McNutt Jr., *Physics of Space Plasmas. SPI Conf. Proceedings* (Scientific Publishers, Cambridge, 1982)

R.L. McNutt Jr., Models of Pluto's upper atmosphere. Geophys. Res. Lett. **16**, 1225–1228 (1989)

R.L. McNutt, Jr., D.G. Mitchell et al., A compact particle detector. SPIE **2804**, 217–226 (1996)

R.L. McNutt, Jr., S.C. Solomon et al., The MESSENGER mission to Mercury: Development history and early mission status. Adv. Space Res. **38**, 564–571 (2006)

R.L. McNutt, Jr., D.K. Haggerty et al., Energetic particles in the Jovian magnetotail. Science **318**(5848), 220–222 (2007)

D.A. Mendis, E.J. Smith et al., Comet-solar wind interaction: Dynamical length scales and models. Geophys. Res. Lett. **13**, 239 (1986)

D. Morrison, D. Cruikshank et al., Nature **300**, 425 (1982)

U. Motschmann, K.-H. Glassmeier, Nongyrotropic distribution of pickup ions at comet P/Grigg-Skjellerup: A possible source of wave activity. J. Geophys. Res. **98**, 20977–20983 (1993)

M. Neugebauer, Spacecraft observations of the interaction of active comets with the solar wind. Rev. Geophys. **28**, 231–252 (1990)

M. Neugebauer, A.F. Cheng et al., Space physics objectives for the Pluto Fast Flyby mission. Report to NASA Space Physics Division, 1993

I.G. Richardson, S.W.H. Cowley et al., Three dimensional ion bulk flows at comet P/Giacobini-Zinner. Geophys. Res. Lett. **13**(4), 415–418 (1986)

T.R. Sanderson, K.-P. Wentzel et al., The interaction of heavy ions from comet P/Giacobini-Zinner with the solar wind. Geophys. Res. Lett. **13**(4), 411–414 (1986)

R. Sicardy, T. Widemann et al., Large changes in Pluto's atmosphere as revealed by recent stellar occultations. Nature **424**, 168–170 (2003)

A.J. Somogyi, K.I. Gringauz et al., First observations of energetic particles near comet Halley. Nature **321**, 285–288 (1986)

S.A. Stern, The New Horizons Pluto Kuiper Belt mission: An overview with historical context. Space Sci. Rev. (2008), this issue. doi:10.1007/s11214-007-9295-y

S.A. Stern, D.C. Slater et al., ALICE: The ultraviolet imaging spectrograph aboard the New Horizons Pluto-Kuiper Belt Mission. Space Sci. Rev. (2008), this issue. doi:10.1007/s11214-008-9407-3

F. Tian, O.B. Toon, Hydrodynamic escape of nitrogen from Pluto. Geophys. Res. Lett. (2005). doi:10.1029/2005GL023510

L.M. Trafton, D.M. Hunten et al., Escape processes at Pluto and Charon, in *Pluto and Charon*, ed. by S.A. Stern, D.J. Tholen (University of Arizona Press, Tucson, 1997), pp. 475–521

L. Tyler, I.R. Linscott et al., The New Horizons radio science experiment. Space Sci. Rev. (2008), this issue. doi:10.1007/s11214-007-9302-3

H.A. Weaver, W.C. Gibson et al., Overview of the New Horizons science payload. Space Sci. Rev. (2008), this issue. doi:10.1007/s11214-008-9376-6

D.J. Williams, R.W. McEntire et al., Energetic particles at Venus: Galileo results. Science **253**, 1525–1528 (1991)

D.J. Williams, R.W. McEntire et al., The Galileo energetic particles detector. Space Sci. Rev. **60**, 385–412 (1992)

D.J. Williams et al., GEOTAIL energetic particles and ion composition experiment. J. Geomagn. Geoelectr. **46**, 39–57 (1994)

The Student Dust Counter on the New Horizons Mission

M. Horányi · V. Hoxie · D. James · A. Poppe · C. Bryant · B. Grogan · B. Lamprecht · J. Mack · F. Bagenal · S. Batiste · N. Bunch · T. Chanthawanich · F. Christensen · M. Colgan · T. Dunn · G. Drake · A. Fernandez · T. Finley · G. Holland · A. Jenkins · C. Krauss · E. Krauss · O. Krauss · M. Lankton · C. Mitchell · M. Neeland · T. Reese · K. Rash · G. Tate · C. Vaudrin · J. Westfall

Originally published in the journal Space Science Reviews, Volume 140, Nos 1–4, 387–402.
DOI: 10.1007/s11214-007-9250-y

Abstract The Student Dust Counter (SDC) experiment of the New Horizons Mission is an impact dust detector to map the spatial and size distribution of dust along the trajectory of the spacecraft across the solar system. The sensors are thin, permanently polarized polyvinylidene fluoride (PVDF) plastic films that generate an electrical signal when dust particles penetrate their surface. SDC is capable of detecting particles with masses $m > 10^{-12}$ g, and it has a total sensitive surface area of about 0.1 m^2, pointing most of the time close to the ram direction of the spacecraft. SDC is part of the Education and Public Outreach (EPO) effort of this mission. The instrument was designed, built, tested, integrated, and now is operated by students.

Keywords Interplanetary dust · Instrumentation · Education and public outreach

1 Introduction

The Student Dust Counter (SDC) is an impact dust detector onboard the New Horizons Mission. It is designed to map the spatial and size distribution of interplanetary dust particles in order to verify the existence of the predicted structures in our dust disk, the Zodiacal cloud. Five spacecraft have carried dust detectors beyond the asteroid belt: Pioneers 10 and 11 (Humes 1980), Ulysses, Galileo (Grün et al. 1993), and Cassini (Srama et al. 2004). SDC will provide the first dust measurements beyond 18 AU, where the Pioneer sensors stopped working. After the Pluto–Charon flyby, SDC will continue to measure dust on into the Kuiper Belt. These observations will advance our understanding of the origin and evolution of our own solar system, and allow for comparative studies of planet formation in dust disks around other stars.

M. Horányi (✉) · V. Hoxie · D. James · A. Poppe · C. Bryant · B. Grogan · B. Lamprecht · J. Mack · F. Bagenal · S. Batiste · N. Bunch · T. Chanthawanich · F. Christensen · M. Colgan · T. Dunn · G. Drake · A. Fernandez · T. Finley · G. Holland · A. Jenkins · C. Krauss · E. Krauss · O. Krauss · M. Lankton · C. Mitchell · M. Neeland · T. Reese · K. Rash · G. Tate · C. Vaudrin · J. Westfall
Laboratory for Atmospheric and Space Physics, University of Colorado, Boulder, CO 80309-0392, USA
e-mail: mihaly.horanyi@colorado.edu

SDC is part of the Education and Public Outreach (EPO) effort of the New Horizons mission and is the first science instrument on a planetary mission to be designed, built, tested and operated by students. The SDC project has an unusual history. A similar professional dust instrument was part of a competing proposal to New Horizons in a parallel Phase A study. After the selection of New Horizons, motivated by the potential scientific contribution of a dust instrument, the idea emerged to redirect some of the funds from traditional EPO activities so that a group of students could try their hands at building space hardware. The advanced state of the rest of the New Horizons payload and the risk of involving unexperienced students made this request difficult. With the strong support of the mission PI, the NASA EPO board agreed to try the "SDC experiment."

To minimize the risk SDC might pose to the mission, all quality assurance inspections and the final flight assembly was done by NASA-certified personnel, and student activities were supervised by professionals. However, the student team, consisting of up to 20 engineering and physics undergraduate and graduate students, was responsible for the work done in all phases of this project, including presentations at all NASA milestone reviews. SDC was built and tested to the same NASA engineering standards as every other flight instrument. Due to the long duration of the New Horizons mission, generations of future students will continue to be involved, handing over their skills to the groups that follow them.

After six months of successful operations in space the SDC instrument was renamed, and the dedication reads:

> *New Horizons, the first mission to Pluto and the Kuiper Belt, is proud to announce that the student instrument (SDC) aboard our spacecraft is hereby named "The Venetia Burney Student Dust Counter" in honor of Ms. Venetia Burney Phair, who at age of eleven nominated the name Pluto for our solar system's ninth planet. May "Venetia" inspire a new generation of students to explore our solar system, to make discoveries which challenge the imagination, and to pursue learning all through their lives.*

2 Science Background

The interplanetary dust particle (IDP) population reflects a complex balance between sources, transport, and sinks which are all functions of the dust particle sizes and their positions in the solar system. IDPs are created primarily in asteroid-asteroid collisions, from the disintegration of comets near the Sun, from collisions between Kuiper Belt objects (KBOs), and the continuous meteoroid bombardment of the moons, rings, asteroids, and KBOs. Data from near-Earth and deep space missions suggest five distinct IDP populations (Divine 1993). Although this model makes predictions for IDP populations throughout the solar system, it has not been validated by direct measurements beyond 18 AU.

Beginning at about 3 AU from the Sun, the Pioneer 10 and 11 observations of particles larger than about 10 μm in radius showed a constant spatial density (number of particles per unit volume) out to 18 AU (Humes 1980). At this distance, the argon and nitrogen gas mixture in the Pioneer pressurized detector cells froze and no further measurements could be made.

IDPs born on the outskirts of the solar system slowly lose orbital energy due to Poynting-Robertson drag and migrate toward the Sun. A 10 μm radius grain born at 50 AU reaches the inner solar system in about 6.5 My. However, the continuous migration can significantly slow down or come to a halt due to mean motion resonances with the planets (Liou and Zook 1999; Moro-Martín and Malhotra 2002, 2003). Grains can also get ejected from the solar system via close encounters with the planets (Liou et al. 1996). Orbital integrations find that Neptune prolongs the lifetime of the inwardly migrating dust grains outside its orbit, imprinting its resonance structure on the spatial density of dust in the Kuiper Belt (Liou et al. 1999).

Jupiter, on the other hand, acts as a gate-keeper and keeps the inner part of our solar system relatively dust free by ejecting up to 80% of the particles that would cross its orbit. The nature of the inner planets effect on dust distribution is largely hidden due to dust production from active comets.

SDC measurements made as New Horizons crosses the solar system will help to understand the generation, the transport, and the loss processes of IDPs. The measurements of the density variations outside the orbit of Jupiter will allow for estimating the rate of dust production in the Kuiper Belt, the collisional history of the region, and the mass distribution of the primordial KBO population (Stern 1996).

3 Instrument Description

The SDC instrument (Fig. 1) consists of a set of polyvinylidene fluoride (PVDF) film impact sensors, carried on a detector support panel, which is mounted on the exterior of the New Horizons spacecraft. It is outside the spacecraft multi-layer insulating (MLI) blanket, facing the ram direction. Signals from the sensors are collected through an intra-harness that runs from the detector assembly into the spacecraft interior to the instrument electronics box mounted opposite the detector panel.

The measurement requirements of the SDC instrument were established by estimating the particle mass detection limit needed to resolve the expected features of the dust distribution. The characteristic width of the predicted resonance structures is on the order of 1 AU, hence a spatial resolution of $\approx$0.1 AU assures adequate sampling. With the characteristic spacecraft speed, during cruise, of $\approx$13 km/s, this is equivalent to an integration period of one week. Based on the current models (Divine 1993) outside of dust structures, a detector, with a lower mass detection limit of about 10^{-12} g and 0.1 m^2 surface area, is predicted to have on the order of 1 IDP hit per week. SDC was designed to resolve the mass of IDPs in the range of $10^{-12} < m < 10^{-9}$ g within factors of <2, covering an approximate size range

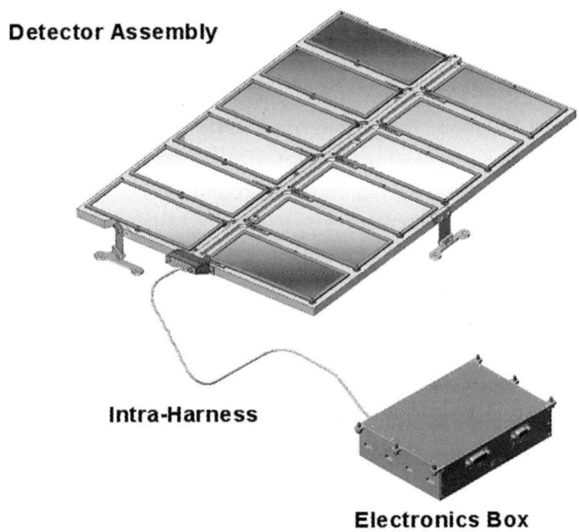

Fig. 1 The Student Dust Counter provides $\sim$0.1 m^2 of sensitive area. It was designed to measure the mass of IDPs in the range of 10^{-12} to 10^{-9} g. It continues registering the impacts of bigger dust particles without the ability to determine their mass. It weighs 1.6 kg and consumes 5.1 watts of average power

of 1–10 μm in particle radius. Bigger grains are also recorded without the ability of resolving their mass.

3.1 Sensor Design

SDCs dust impact detection is based on the use of permanently polarized PVDF films. An impacting particle causes a depolarization charge when it penetrates the film. PVDF sensors require no bias voltage, they are simple, inexpensive, reliable, electrically and thermally stable, mechanically rugged, radiation resistant, and do not respond to energetic ions or electrons. PVDF dust detectors have been extensively tested and calibrated in laboratory experiments (Simpson et al. 1989a, 1989b; Simpson and Tuzzolino 1985; Tuzzolino 1983, 1992, 1996) and have an excellent track record in space experiments. PVDF sensors were flown on the VEGA 1 and 2, Stardust, Cassini, and ARGOS missions (Tuzzolino et al. 2001, 2003; Srama et al. 2004).

3.1.1 PVDF Signal Generation

The magnitude of the PVDF depolarization charge depends on particle momentum as well as whether the particle penetrates into or through the film (Simpson and Tuzzolino 1985). SDC uses 28 μm thick PVDF films which have been shown to stop particles up to 10^{-10} g and speeds up to 20 km/s (Simpson and Tuzzolino 1985). For particles stopped in the film, the number of electrons generated is given by (Simpson and Tuzzolino 1985)

$$N_e = 3.8 \times 10^{17} m[\mathrm{g}]^{1.3} v[\mathrm{km/s}]^{3.0}, \tag{1}$$

where m is particle mass, in grams, and v is the impacting velocity in km/s. The speed of long-resident dust grains following circular Kepler orbits in the outer solar system is much less than the velocity of the spacecraft, hence we take v in (1) to be the spacecraft velocity.

For charge-sensing amplifiers, the noise floor is proportional to the detector capacitance (Spieler 2005) which is a function of the detector area, the material permittivity and the thickness ($C = \epsilon_0 \epsilon_r A/d$). The relative permittivity of PVDF, like many polymers, changes dramatically with temperature varying from $\epsilon_r = 11$ at 25°C to 2.5 at −120°C. For a given thickness of PVDF, and a given operating temperature, the required lower limit of detectable impact-induced charge determines the maximum allowable area of a single sensor. From (1) the SDC mass threshold of 10^{-12} g (sub-micron radius) requires a signal threshold of 6×10^6 e. To ensure a false detection rate of less than one per month, a signal-to-noise ratio (SNR) greater than five is required for an RMS noise level of $\approx 1 \times 10^5$ e. Given the characteristics of the charge-sensitive amplifier design used (see Sect. 3.3.1) the maximum allowable capacitance per sensor was determined to be ∼30 nF for an equivalent area of ≈80 cm^2.

To achieve a total sensitive area of 0.1 m^2 SDC is composed of 12 sensor patches, each 14.2 cm × 6.5 cm, mounted onto the top surface of a support panel. In addition, there are two reference sensor patches mounted on the back side of the detector support panel, protected from any dust impacts. These reference sensors, identical to the top surface sensors, are used to monitor the various background noise levels, from mechanical vibrations or cosmic ray hits of our electronics, for example.

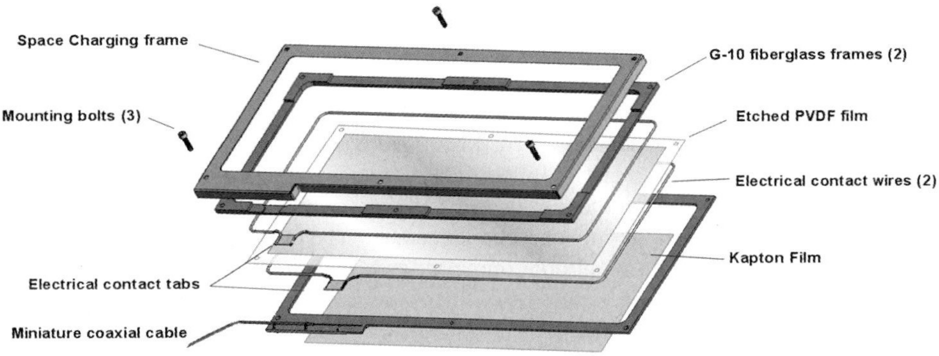

Fig. 2 Exploded view of a single SDC PVDF detector. The detector is mechanically bonded with structural epoxy. Electrical connection from the PVDF metalization to the signal wires is through silver-filled epoxy. The signal wires are soldered to miniature coaxial cables leading to the system electronics. Three mounting screws tie the space charging frame and the detector to the underlying honeycomb support panel

3.1.2 Sensor Construction

The SDC detectors (Fig. 2) are based on similar instruments from the University of Chicago. The PVDF film used by SDC is manufactured with 1,000 Å of Al/Ni electrode material on the top and bottom surfaces. The film is bonded between a pair of G-10 fiberglass frames and electrical contact to the PVDF is made through signal wires running around the interior of the frame apertures and bonded to the electrode material with conductive silver filled epoxy. The signal wires are routed to connection tabs where a miniature coaxial cable is attached.

3.2 Mechanical Design

The detector assembly of SDC is built on a ~1 cm thick aluminum composite honeycomb panel which provides mounting and support for the dust sensors onto the exterior spacecraft deck. The panel is attached to the spacecraft with a three-point compliant mount formed by titanium flexures. The flexures are sized to accommodate the thermal expansion mismatch between the spacecraft deck and detector assembly. Sensor wiring on the detector assembly is routed in a harness channel designed to withstand dust impacts.

3.2.1 Thermal Design

The thermal design of the SDC detector assembly was driven largely by the requirement that the detector panel be able to maintain passive thermal control while in view of the sun during early mission maneuvers. The PVDF sensor film itself has poor thermo-optical properties, and is prone to overheating when exposed to direct sunlight. PVDF will undergo permanent depolarization at temperatures over 85°C (Simpson and Tuzzolino 1985). To maintain the PVDF temperature below a design target of 65°C, a high emissivity polyimide tape is applied to the backside of the PVDF film to radiatively couple it to the honeycomb support panel below. The panel facesheet spreads out the heat from beneath the detectors. The top surface is covered with silverized Teflon tape that reflects 90% of the incident solar energy, preventing the panel itself from overheating. The detector support panel area was sized analytically to provide adequate radiating area around the 12 PVDF sensor patches.

3.3 Electrical Design

The SDC electronics are carried on two multi-layer printed wiring assemblies (PWA) housed in the SDC electronics box. Signals from the detectors come through the Intra-harness onto the analog PWA where they are amplified, conditioned, and converted to digital data. The digital data are collected to registers in the field programmable gate array (FPGA) on the digital PWA (Fig. 3) and from there to the microprocessor which adds time-stamps to the data and stores them in long-term, non-volatile memory. The digital PWA also contains the power supply, system health monitoring circuitry and interface electronics for spacecraft communications.

3.3.1 Analog Signal Chain

Given the large capacitance of the PVDF detectors, care was taken to simultaneously minimize the noise floor and to maximize the signal-to-noise ratio. Analysis showed that SNR is maximized when detector capacitance is equal to the input capacitance of the charge sensitive amplifier's (CSA) JFET (junction field effect transistor) (Radeka 1974; Spieler 2005). Earlier work was followed to address the problems of matching JFETs to very large capacitance detectors (Bertolaccini et al. 1988), and a process emerged that produces high capacitance JFETs with good transconductance.

A by-product of the specific JFET design is an increased gate volume sensitive to energetic particle charge deposition. A CRME96 radiation analysis using the gate dimensions and a critical deposited charge of 1×10^6 e gives 2.3×10^{-7} event s^{-1} during solar max and roughly doubles during solar min. In the nominal case the upset rate is well below the one per month science requirement. Using an ordinary solar flare model with mean composition, the upset rate rises to 1.1×10^{-5} event s^{-1}, suggesting that higher than expected event rates should be checked against possible space weather effects.

The CSA output is routed to the two-stage CR-RC shaper circuit (Fig. 4). Using the given CSA design values, theoretical noise limits can be calculated (Spieler 2005), allowing for the selection of an optimal shaper peaking time $\tau = 10$ μs. Measured SNR values from prototype circuits agree with the predicted values and show a SNR >10 for 10^6 electron input signals.

To accommodate signals over the ≈ 4 orders of magnitude required, the first stage of the shaper includes a logarithmic compression. To keep the circuit size small, the temperature dependence is uncompensated and must be corrected in post-processing.

Following the shaper section each analog signal chain has a peak-hold circuit that retains the shaper pulse maximum for sampling by the analog to digital converters (ADC). The 14 peak-hold outputs are divided into two groups of seven with each group routed to its own analog multiplexer (MUX) and 16-bit ADC. A side effect of the signal chain electronics is that low level signals produce a digital output number (DN) of $\sim$65,000 with increasing magnitude analog signals producing decreasing DN values.

The response curve of a representative analog chain through the ADC is shown in Fig. 5. The gain shift in the upper size range (lower right of the plot) is from the temperature sensitivity of the diodes in the logarithmic compression term of the first-stage shaper. This temperature dependency was characterized during pre-launch calibration (see Sect. 3.5.2) and the temperatures driving this shift are monitored during flight by temperature sensors on the shaper circuits. Additionally the thermal environment of the SDC electronics box is expected to be stable over long time scales.

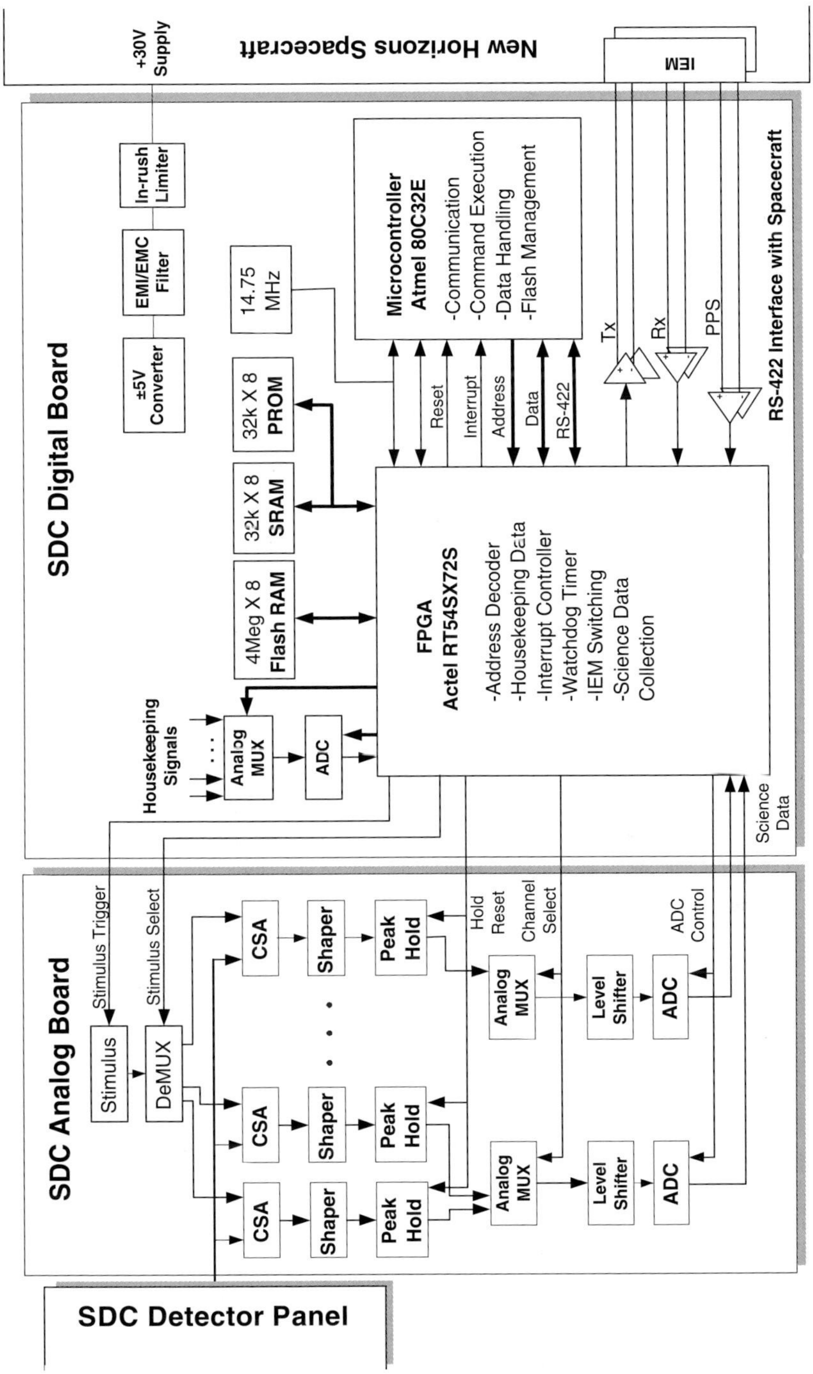

Fig. 3 SDC electronics block diagram. Signals from each PVDF detector are routed to individual analog electronics chains. Under FPGA control the science and housekeeping data are collected and passed to the system microprocessor which manages long-term storage of data into flash memory. The FPGA also acts as a switch between the redundant spacecraft communications lines and the microprocessor

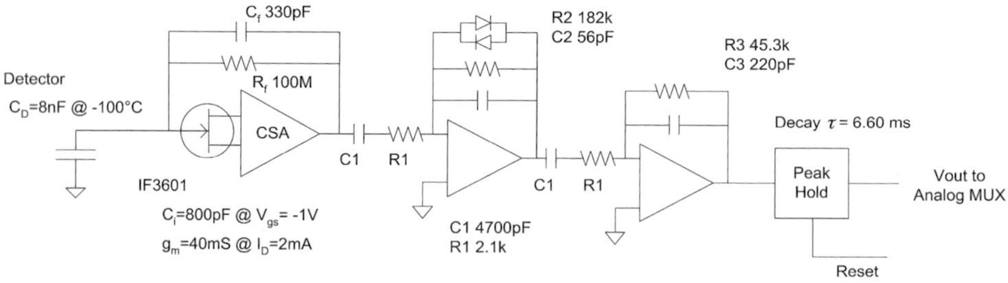

Fig. 4 Each of the 14 analog signal chains contains a JFET input charge sensitive amplifiers followed by a two-stage CR-RC shaper circuit. The first stage shaper includes a logarithmic gain term supporting a 2×10^5 to 5×10^{10} electron input range. The shaper output is captured in a peak-hold stage long enough to be sampled by a scanning analog to digital converter circuit

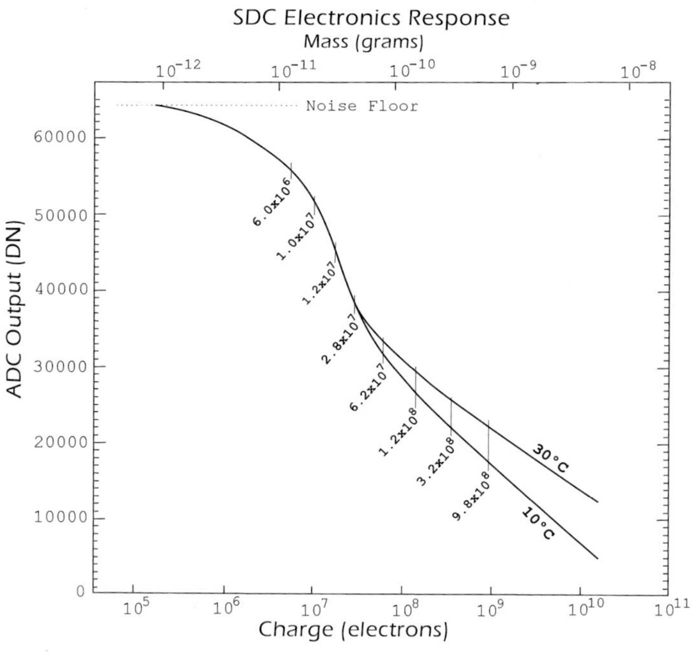

Fig. 5 A typical SDC charge to DN response curve. Temperature-dependent changes in gain are characterized during calibration. In-flight temperature monitoring and charge injection (the eight onboard stimulus levels are indicated as *vertical dashed lines*) support post-processing correction for gain curve shifts

3.3.2 Onboard Stimulus

To track changes in the gain of the signal chains during flight the SDC electronics includes stimulus circuitry that can inject selectable charge quantities into any of the signal chains at the JFET gates. The responses to these stimuli are collected in the same manner as science data and reported along with a measurement of the stimulus voltage. The nominal injection levels are shown on Fig. 5.

3.3.3 Digital Control Electronics

The two analog MUX and ADC sets are controlled in parallel by the FPGA on the digital PWA. The ADC conversion values from each channel are stored to registers and compared against individually programmable threshold values in the FPGA. Upon a valid event detection the FPGA executes one additional scan of every channel, to assure the shaper pulse has time to peak, then it stops scanning, leaving each channel's peak value in the FPGA registers. An interrupt is sent to the microprocessor to indicate that the science data is ready. The FPGA provides a control register which can be set to prevent any given channel from generating interrupts.

The system has a radiation tolerant 8032 microcontroller, 32 kbyte anti-fuse PROM for program storage, 32 kbyte rad-hard SRAM, and 4 Mbyte non-volatile flash RAM. The data retention time of the flash memory, rated at 10 years, is extended by providing circuitry to power it on only when the system is accessing it. The SDC flight software resides in ~24 kbyte of the 32 kbyte PROM. The system can be configured to load code from flash memory into SRAM and run it from there, supporting in-flight updates if necessary.

3.4 Software Design

For most of the long journey to Pluto, the New Horizons payload will be in "hibernation," with the exception of SDC, which was designed for stand-alone operations. Much of the SDC flight software is related to managing itself during these long periods of up to 500 days. In addition to data collection and storage, the software continuously executes a set of programmable "autonomy" rules that allow the instrument to adjust for many off-nominal conditions.

When SDC is turned on it immediately starts taking science data. Ground commands are required only to perform calibrations or reconfigure settings. The ground can also turn channels off or on, erase the flash memory, and request telemetry packets. SDC provides several data packet types including science data, housekeeping information, calibration results, memory dumps, and message logs. Communications, as defined by the spacecraft, is limited to one command sent per second, and one telemetry packet received per second. A telemetry packet can contain up to 1,024 data bytes.

3.4.1 Data Management

Whenever a science interrupt occurs the software compares each channel's value from the FPGA against that channel's current threshold value. For any channel exceeding its threshold, the channel number, the measured value, and the threshold value are stored to flash memory along with the mission elapsed time (MET).

In addition to science data, flash memory is used for storing system housekeeping data and maintaining system-critical variables, all of which are stored in triplicate. A simple majority vote is used to correct any corrupted values. A health check is performed on the flash memory every day to ensure it can be written to and erased. Flight software maintains a table of ill-performing flash memory blocks.

3.4.2 Autonomy Rules

The 4 Mbyte of flash memory is much larger than the total expected SDC data volume. However, it is possible for elevated noise levels to be mis-identified as science and potentially fill up the flash memory, leaving no room for real science. A high rate of noise events can also cause the software to hang while trying to process all the interrupts it receives. To mitigate the risk of filling up flash memory during the long periods of unattended operation, the software has two sets of autonomy rules to control the data rate from the channels. Set A monitors the number of interrupts received from each channel every second. Whenever a channel generates three interrupts in a second (the maximum reasonable interrupt rate) that channel is blocked at the FPGA from generating further interrupts for some programmable period, typically one hour. These rules accommodate transient noise periods, like spacecraft maneuvers.

Set B monitors the number of interrupts received from each channel in a day. If a channel reaches the daily limit, by default 20, that channel's threshold is changed to a previously selected, less sensitive level. There is a table of three decreasingly sensitive threshold values for each channel, the values of which can be changed by ground command. The threshold of each channel can be increased twice, and if it remains in violation of Set B, then the channel is blocked from generating interrupts for a longer period, typically 30 days. After this time the channel is turned on at its original threshold and allowed to go through the table one more time. Upon reaching the end of the threshold table a second time the channel is permanently blocked from generating interrupts until enabled by ground command. The intent of this rule set is to autonomously adjust the instrument's sensitivity to permanent changes in the noise environment, such as degradation of the electronics. The time-out period of rule A and all of the rule B default values are adjustable by ground command.

3.4.3 In-Flight Calibration Functions

The flight software manages the setup and data collection of the Stimulus Test (Sect. 3.3.2). In addition it can perform a Noise Floor calibration. This consists of temporarily suspending the autonomy rule set B and changing the thresholds of all the channels to a series of levels sensitive enough that electronics noise is detectable. By counting the number of science interrupts obtained at each threshold over specified times, the statistical distribution of the noise on each channel can be calculated. The flight software allows for up to five levels and test times. The test thresholds are configurable from the ground, but the durations are permanently set. At the completion of a test set, autonomy rule B is re-enabled and the channel thresholds are reset to their initial values.

3.5 Instrument Calibration

3.5.1 PVDF Sensor Characterization

In the summer of 2003 and 2006 a set of SDC/PVDF sensors were taken to the dust accelerator facility at Max Planck Institute for Nuclear Physics in Heidelberg, Germany, to verify that they show similar response to hypervelocity dust impacts (Fig. 6) as has been published for previous instruments using PVDF (Simpson and Tuzzolino 1985).

More than 80 iron dust particle impacts where collected from 4 separate prototype sensors (Fig. 7). At the lower particle sizes accurate measurements of CSA response was difficult. With these points excluded, the measured slope is close to that of Simpson and Tuzzolino (1) but with an offset. This offset could be due to differences in the thickness of the metallic layers, their exact composition, and/or differences in the manufacturing of the PVDF, including its level of polarization. The fit to our calibration data (using the same notation as in (1)) is

$$N_e = 5.63 \times 10^{17} m[\mathrm{g}]^{1.3} v[\mathrm{km/s}]^{3.0}. \qquad (2)$$

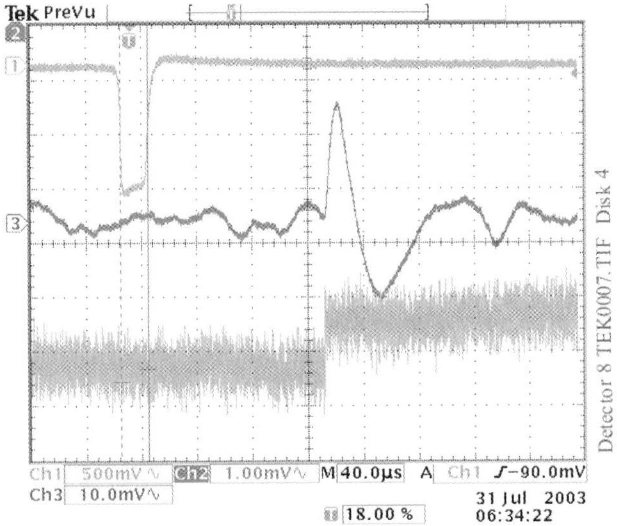

Fig. 6 Example of output pulse from a PVDF sensor to a 3.1×10^{-11} g iron particle impacting at 4.52 km/sec. The *top trace* is from the accelerator charge induction tube (100 mm, 6.1 V/pC), the *middle trace* is the output of the shaper (1.47×10^{-5} mV/e) and the *bottom trace* is from the CSA (7.3×10^{-7} mV/e). Note: gains are not flight values

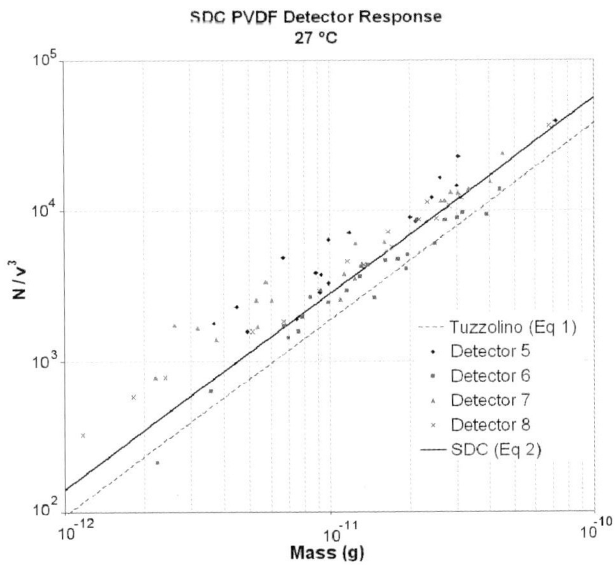

Fig. 7 Dust impacts on SDC prototype PVDF sensors. Plot of ratio N/v^3 vs projectile mass. N (electrons) is the measured detector signal and v (km/s) is projectile velocity reported by the dust accelerator

3.5.2 Electronics Calibration

The gain curve shown in Fig. 5 illustrates that the system response to sensor-produced charges is highly non-linear. A calibration function is needed to identify: (a) the compression stage gain variation with electronics box temperature, and (b) the possible change in CSA gain due to temperature-driven changes in PVDF capacitance. The calibration equation then becomes $Q(DN, T_b, C_{det}(T_d))$, where Q is the measured charge, DN is the reported value, T_b is the temperature of the electronics box, and $C_{det}(T_d)$ is the capacitance of the PVDF detector as a function of detector temperature, T_d.

PVDF detectors are very sensitive to acoustics and mechanical vibrations, making them impractical to use during calibration. From characterization of PVDF sensor capacitance over temperature, fixed capacitor values were selected to represent the detector capacitance, C_{det}, at 25, −30, and −100°C. A range of reference charge pulses were generated with a standard calibrated charge injection circuit, and the electronics box temperature was collected from SDC telemetry points, measured at the shaper circuitry (the same measurements that are reported in flight). The measured electronics temperatures, T_b, were 50, 40, 34, and −7°C.

For each electronics box temperature and C_{det} value, 21 different amplitude charges were injected 100 times into each of the 14 channels. For each charge pulse injected the resulting DN value and accompanying T_b were collected from the SDC telemetry. The pre-launch baseline calibration function for each channel, $Q_{ch}(DN, T_b)$, was produced by fitting that channel's data with a 10th order polynomial. C_{det} was dropped as a parameter to the calibration function after analysis showed that CSA gain changes due to variations in detector capacitance were insignificant.

4 Data Analysis

When SDC data are received on Earth, they are sent to the New Horizons mission operations center at the Johns Hopkins University's Applied Physics Laboratory. The SDC data, as well as several ancillary spacecraft data packets, are also sent to SDCs home institution, LASP, via the Ground System Equipment Operating System (GSEOS). Simultaneously, it is also received at the Tombaugh Science Operations Center (TSOC) of the Southwest Research Institute (SwRI) in Boulder.

The data are sent through "levels" of processing software that progressively decode more of the packets and send the decoded data to the database.

- Level 0 software extracts the data packets out of the GSEOS files.
- Level 1 software separates the data by packet type, such as Science, Housekeeping, Calibration, flash and SRAM memory dumps, and decommutates the type specific data into the Level 1 database.
- Level 2 software generates charge values from Level 1 DN values using $Q_{ch}(DN, T_b)$ from Sect. 3.5.2.
- Level 3 software uses (2) to convert charge to equivalent mass.
- Level 4 software removes noise using the reference detectors and other coincident events to produce validated IDP impact rates as function of dust mass.

The data products from Levels 0 through 2 are produced both at TSOC, as well as at LASP, using identical source controlled software. Data products from levels 3 and 4 are produced only at LASP.

5 Initial Results

5.1 Initial Turn-on and Checkout

On March 2, 2006, SDC was powered on and commanded to report back a full set of housekeeping values. With receipt of fully nominal values, a series of commands were sent instructing the instrument to initiate a set of Stimulus Tests (ST) (Sect. 3.3.2) and Noise Floor

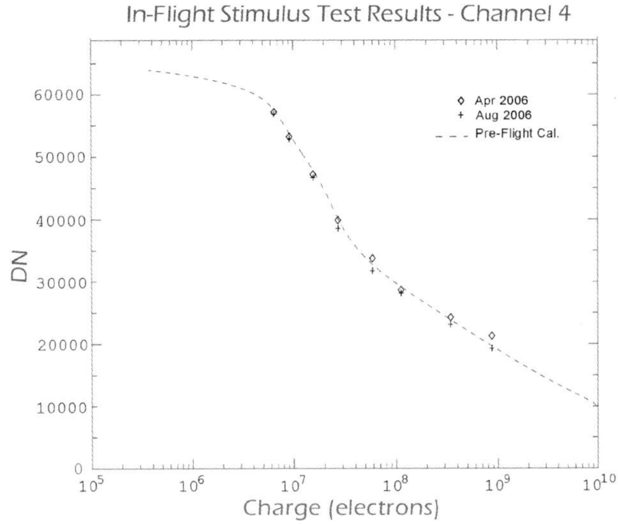

Fig. 8 Stimulus tests of channel 4 performed in April and August, 2006, and the pre-flight calibration curve

calibrations (NFC) (Sect. 3.4.3). All tests ran successfully and the results showed that SDC had survived the rigors of launch. Both ST and NFC were repeated several times over the next few months to better characterize the post-launch behavior of the instrument.

5.2 In Flight Calibration

5.2.1 Stimulus Calibration

ST results were compared to pre-launch baseline tests. Figure 8 shows the change in stimulus responses for one of the channels. As of the time of the test, with the exception of channel 11, all of the stimulus test results were holding steady within the variation expected from ground testing. Some time between instrument delivery to the spacecraft and launch, the detector on channel 11 began exhibiting symptoms of degraded electrical contacts to the PVDF. As a result, channel 11 is expected to have a reduced sensitivity.

5.2.2 Noise Floor Calibration

With the sensitivity of the PVDF detectors to acoustics and mechanical vibrations, the in-flight NFC was the first good measure of SDC systematic noise. For each channel the average noise rate (noise events collected divided by collection period) versus threshold is calculated. The logarithm of the rates is used to fit a second-degree polynomial to characterize the noise profile of that channel. The threshold value where the fit crosses the one hit per month rate is the lowest of the three thresholds for that channel to support autonomy rule B (Sect. 3.4.2). Figure 9 displays Noise Floor calibration data for a single channel and illustrates the threshold correction calculation.

A corrected set of thresholds were uploaded in late June 2006 and subsequently used for science collection through the fall. Two or three more rounds of calibrations are scheduled before entering cruise phase, during which the instrument must operate autonomously for up to 500 days.

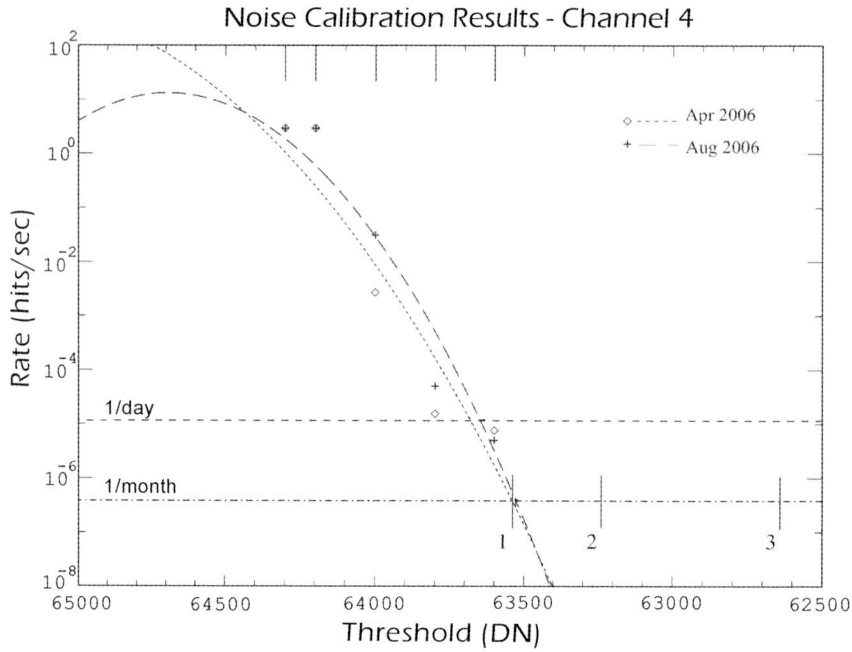

Fig. 9 SDC channel 4 noise collected in April and August of 2006 using the Noise Floor calibration routine (Sect. 3.4.3). The *curves* are fits to the collected rates at five separate thresholds (marked on the *upper horizontal axis*). The new science collection thresholds (*lower vertical lines* labeled *1*, *2*, and *3*) are also indicated

5.3 Dust Measurements

SDC was designed to take advantage of the quiet state of the spacecraft during cruise phase. Various active spacecraft operations cause mechanical shocks that are picked up by the PVDF sensors and registered by SDC as science events. This is particularly true during three-axis pointing and active spin mode when the spacecraft frequently fires short bursts of the attitude thrusters. Level four data reduction (Sect. 4) is used to filter out any hits that appear within a second of any thruster firing, thereby allowing science recovery between firings. However, because the thruster induced events are often frequent enough to violate the autonomy rule B (Sect. 3.4.2), during spacecraft maneuvers many SDC detector channels switch off for prolonged periods. During the first six months of flight, there are several periods, some weeks or months long, where SDC was completely off.

In spite of the initial difficulties of collecting science data, SDC does routinely take science during the short quiet periods available. As an example, Fig. 10 shows the number of events recorded on each channel during segmented periods between July 14 and August 16, 2006, totaling 2.5 weeks. Events that were coincident between channels were removed, and channel 11 was excluded in this analysis. The average number of hits on the front sensors was 19.7 ± 5.4, and on the reference sensors it was 18.0 ± 4.2. The difference of these are taken to be real dust hits of 1.73 ± 6.9 during this period, approximately consistent with our expectations of about 1 hit per week. The huge error bar is a consequence of the ongoing activities with other instruments on the spacecraft. Following the Jupiter encounter in February 2007, all other instruments will hibernate between scheduled yearly checkouts. SDC will stay on and start making measurements without interference.

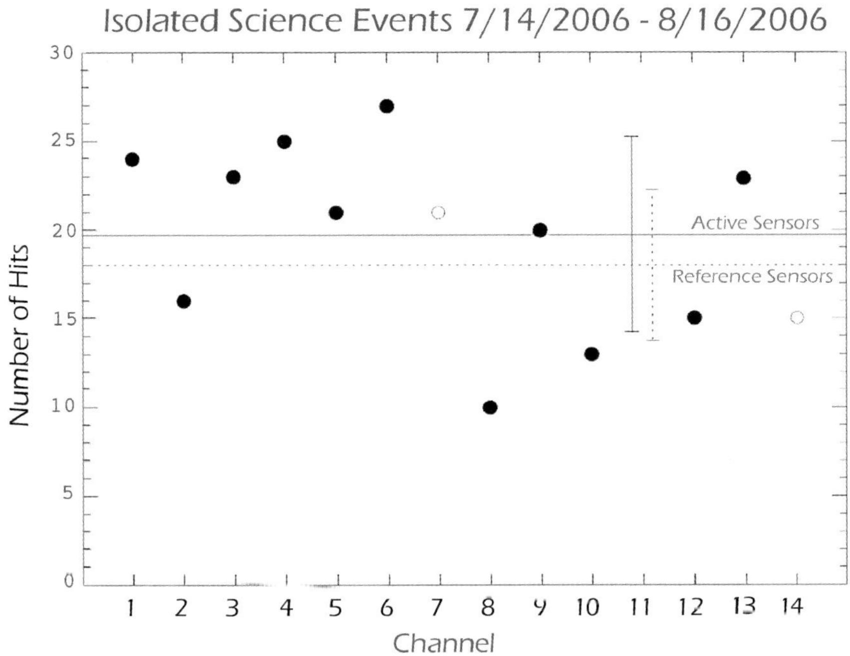

Fig. 10 The number of single events registered on each of the SDC channels (with the exception of channel 11). These were recorded between July 14, 2006, and August 15, 2006, during a total operating period of about 2.5 weeks. The average number of hits and the $\pm 1\sigma$ error bars are also indicated for both the front (*continuous line*) and the reference (*dashed line*) sensors

6 Conclusions

As part of the education and outreach effort of this mission, SDC provided a unique, hands-on opportunity for a group of undergraduate and graduate students to learn about building space hardware. Frequent "all-hands" meetings during the development phases of SDC ensured that students and professionals working on all the subsystems (mechanical, thermal, analog and digital electronics, software, and operations) continuously discussed the requirements, as well as the schedule and the budget. SDC was delivered on time and within budget. A large fraction of the initial SDC students are pursuing careers in space research. During the next decade a smaller rotating group of students will operate the instrument and analyze our data. The SDC project was the first of its kind to be supported by EPO funds. Based on its history to date, it proved to be an excellent investment as part of the EPO effort of New Horizons.

In-flight tests, calibrations, and the initial science data show that SDC is performing according to expectations. It will continue to make measurements to improve our understanding of the spatial distribution of dust in the solar system. SDC will be the first dedicated instrument to measure dust impacts outside 18 AU, and it is expected to continue operating for the entire duration of the mission, deep inside the Kuiper Belt.

Acknowledgements We are grateful to A. Stern for inviting SDC on the New Horizons Mission, and to NASA for accepting the risk of a student-built flight hardware as part of EPO. We thank APL for accommodating SDC as a late addition. We acknowledge significant initial help from A. Tuzzolino at the University of Chicago. We thank E. Grün and R. Srama for their contributions and use of the Dust Accelerator Facility at the Max Planck Institute for Nuclear Physics in Heidelberg, Germany. At LASP, R. Arnold, J. Johnson, P. Sicken, and J. Tracy assembled the flight version of SDC, and G. Lawrence guided the initial design of our electronics. M.H. was supported in part by the Alexander von Humboldt Foundation while on a sabbatical leave at the Max-Planck-Institute for Extraterrestrial Physics in Garching, Germany.

References

M. Bertolaccini et al., Nucl. Instrum. Methods **A264**, 399–406 (1988)
N. Divine, J. Geophys. Res. **98**, 17029–17051 (1993)
E. Grün et al., Nature **362**, 428–430 (1993)
D.H. Humes, J. Geophys. Res. **85**, 5841–5852 (1980)
J.C. Liou, H.A. Zook, S.F. Dermott, Icarus **124**, 429–440 (1996)
J.C. Liou, H.A. Zook, A.A. Jackson, Icarus **141**, 13–28 (1999)
J.C. Liou, H.A. Zook, Astron. J. **118**, 580–590 (1999)
A. Moro-Martín, R. Malhotra, Astron. J. **124**, 2305–2321 (2002)
A. Moro-Martín, R. Malhotra, Astron. J. **124**, 2255–2265 (2003)
V. Radeka, IEEE Trans. Nucl. Sci. **NS-21**, 51–64 (1974)
J.A. Simpson, D. Rabinowitz, A.J. Tuzzolino, Nucl. Instrum. Methods **A279**, 611–624 (1989a)
J.A. Simpson, D. Rabinowitz, A.J. Tuzzolino, Nucl. Instrum. Methods **A279**, 625–639 (1989b)
J.A. Simpson, A.J. Tuzzolino, Nucl. Instrum. Methods **A236**, 187–202 (1985)
H. Spieler, *Semiconductor Detector Systems* (Oxford University Press, New York, 2005), ISBN-0-19-852784-5
R. Srama et al., Space Sci. Rev. **114**, 465–518 (2004)
A. Stern, Astron. Astrophys. **310**, 999–1010 (1996)
A.J. Tuzzolino, Nucl. Instrum. Methods **212**, 505–516 (1983)
A.J. Tuzzolino, Nucl. Instrum. Methods **A316**, 223–237 (1992)
A.J. Tuzzolino, Adv. Space Res. **17**, 123–131 (1996)
A.J. Tuzzolino, R.B. McKibben, J.A. Simpson, S. BenZvi, H.D. Voss, H. Gursky, Planet. Space Sci. **49**, 689–703 (2001)
A.J. Tuzzolino et al., J. Geophys. Res. **108** (2003). doi:10.1029/2003JE002086